THE POLITICAL ECONOMY OF CLEAN ENERGY TRANSITIONS

UNU World Institute for Development Economics Research (UNU-WIDER) was established by the United Nations University as its first research and training centre and started work in Helsinki, Finland, in 1985. The mandate of the institute is to undertake applied research and policy analysis on structural changes affecting developing and transitional economies, to provide a forum for the advocacy of policies leading to robust, equitable, and environmentally sustainable growth, and to promote capacity strengthening and training in the field of economic and social policy-making. Its work is carried out by staff researchers and visiting scholars in Helsinki and via networks of collaborating scholars and institutions around the world.

United Nations University World Institute for Development Economics Research (UNU-WIDER)
Katajanokanlaituri 6B, 00160 Helsinki, Finland
www.wider.unu.edu

The Political Economy of Clean Energy Transitions

Edited by
DOUGLAS ARENT, CHANNING ARNDT,
MACKAY MILLER, FINN TARP,
AND OWEN ZINAMAN

A study prepared by the United Nations University World Institute for Development Economics Research (UNU-WIDER)

OXFORD
UNIVERSITY PRESS

OXFORD
UNIVERSITY PRESS

Great Clarendon Street, Oxford, OX2 6DP,
United Kingdom

Oxford University Press is a department of the University of Oxford.
It furthers the University's objective of excellence in research, scholarship,
and education by publishing worldwide. Oxford is a registered trade mark of
Oxford University Press in the UK and in certain other countries

First Edition published in 2017
Impression: 2

Published in the United States of America by Oxford University Press
198 Madison Avenue, New York, NY 10016, United States of America

British Library Cataloguing in Publication Data
Data available

Library of Congress Control Number: 2016953859

ISBN 978–0–19–880224–2

Printed in Great Britain by
CPI Group (UK) Ltd, Croydon, CR0 4YY

Foreword

Sustainable energy transitions involve the shift of resources between competing industrial sectors and political constituencies. Stakeholders in this process have varying degrees of political and economic power, and understanding how political economic factors influence clean energy transitions is crucial to effective policy formulation and facilitating transitions to sustainable energy systems. In partnership with the Joint Institute for Strategic Energy Analysis (JISEA), UNU-WIDER gathered together a substantial group of experts from around the world—from both developed and developing countries—to launch a multidisciplinary research project seeking to contribute to our enhanced understanding of these factors. The project sought to facilitate an energy transition that will generate very large environmental and economic benefits, particularly over the long run. The beneficiaries of clean energy transitions are highly diffuse and include future generations not yet born.

This book is the distilled essence of the cross-cutting academic project. I express my sincere and professional appreciation to the large group of expert authors for their dedication to the project, and to my fellow editors in helping bring together the book for readers to enjoy and absorb along with the findings and policy implications.

Finn Tarp
Helsinki, January 2017

Acknowledgements

The research project—The Political Economy of Clean Energy Transitions—was carried out in partnership between the United Nations University World Institute for Development Economics Research (UNU-WIDER) and the Joint Institute for Strategic Energy Analysis (JISEA),[1] which is housed within the National Renewable Energy Laboratory (NREL). We, the editors, are grateful to the staffs of the two institutions who have helped to make this collaboration smooth and productive.

We also thank the international team of authors—each of whom are experts in their own fields—for their dedication to the project, and their patience when reworking and revising the various versions of the individual studies which now make up the polished chapters of the book. This multi-authored book would not have been possible without their expert field knowledge and extremely valuable inputs.

Adam Swallow, Economics and Finance Commissioning Editor at Oxford University Press, provided invaluable guidance, and we are grateful for the anonymous referee reports that helped sharpen our focus. Particular thanks also go to Lorraine Telfer-Taivainen, UNU-WIDER's Senior Editorial and Publishing Assistant, for all her careful, critically needed, and sustained publication support, including the many contacts with OUP.

UNU-WIDER gratefully acknowledges the support and financial contributions to its research programme by the governments of Denmark, Finland, Sweden, and the United Kingdom.

Douglas Arent, Channing Arndt, Mackay Miller,
Finn Tarp, and Owen Zinaman

[1] The Joint Institute for Strategic Energy Analysis is operated by the Alliance for Sustainable Energy, LLC, on behalf of the US Department of Energy's National Renewable Energy Laboratory, the University of Colorado-Boulder, the Colorado School of Mines, the Colorado State University, the Massachusetts Institute of Technology, and Stanford University. The views and opinions of authors expressed herein do not necessarily state or reflect those of the United States government or any agency thereof.

Contents

VI. SECTOR REFORM

VII. SOCIAL INCLUSION

VIII. REGIONAL DYNAMICS

List of Figures

List of Tables

List of Abbreviations

21CPP	21st Century Power Partnership
ADB	Asian Development Bank
ADEME	Agence de maîtrise de l'énergie
AEC	ASEAN Economic Community
AFTA	ASEAN Free Trade Area
AGDI	Agencia Gaucha de Desenvolvimento e Promoção de Investimento (Brazil)
AGRESTE	Service de la statistique, de l'évaluation et de la prospective agricole (France)
AGW	anthropogenic global warming
ANBERD	Analytical Business Enterprise Research and Development
ANC	African National Congress
ANT	actor network theory
AOA	Agreement on Agriculture
APEC	Asia-Pacific Economic Cooperation
ARRA	American Recovery and Reinvestment Act 2009
ASEAN	Association of Southeast Asian Nations
ASTAE	Asia Sustainable and Alternative Energy Program
BAT	best available technology
BAU	business-as-usual
BCAs	border carbon adjustments
BDEW	Bundesverband der Energie- und Wasserwirtschaft
BECCS	bioenergy with carbon capture and storage
BEE	Black Economic Empowerment (S Africa)
BEE	Bundesverband Erneuerbare Energien
BEPE	Environmental Protection and Energy
BEST	Biomass Energy Strategy (Rwanda)
BJP	Bharatiya Janata Party (India)
BMU	Federal Ministry for the Environment, Nature Conservation and Nuclear Safety (Germany)
BMWI	Federal Ministry of Economics and Technology (Germany)
BNDES	Banco Nacional de Desenvolvimento Econômico e Social (Brazil)
BOF	Budget Office of the Federation (Nigeria)

BP	British Petroleum
BPD	barrels per day
BPE	Bureau of Public Enterprises (Nigeria)
BRICS	Brazil, Russia, India, China, and South Africa
BUSA	Business Unity South Africa
BWE	Bundesverband Windenergie
CAFTA	China–ASEAN Free Trade Agreement
CANACERO	Cámara Nacional del Acero (Mexico)
CAPEX	capital expenses
CAPS	Centre d'Analyse, de Prévision et de Stratégie
CAT	Climate Action Tracker
CBN	Central Bank of Nigeria
CCGT	combined cycle gas turbine;
CCS	carbon capture and storage
CDM	Clean Development Mechanism
CEEE	Companhia Estadual de Energia Elétrica (Brazil)
CEER	Council of European Energy Regulators
CEMAGREF	Centre d'Étude du Machinisme Agricole et du Génie Rural des Eaux et Forêts
CENACE	Centro Nacional de Control de Energía (Mexico)
CERI	Centre for Educational Research and Innovation
CERs	Certificates of Emissions Reductions
CESPEDES	Comisión de Estudios del Sector Privado para el Desarrollo Sustentable
CEST	Centro de Estudos Sociedade e Tecnologia (Brazil)
CfD	Contracts for Difference
CFE	Comisión Federal de Electricidad
CHESF	Hydroelectric Company of San Francisco
CHP	combined heat and power
CME	coordinated market economy
CNI	Confederação Nacional da Indústria (Brazil)
CO_2	carbon dioxide
CoP	Conference of Parties to UNFCCC
CoP1	First Conference of Parties to UNFCCC (1995)
CoP15	Fifteenth Conference of Parties to UNFCCC (2009)
CoP16	Sixteenth Conference of Parties to UNFCCC (2010)
CoP20	Twentieth Conference of Parties to UNFCCC (2014)
CoP21	Twenty-first Conference of Parties to UNFCCC (2015)

COSS	cost-of-service study
CPC	Communist Party of China
CPUT	Cape Peninsula University of Technology (S Africa)
CRE	Energy Regulatory Commission (Mexico)
CSIR	Council for Scientific and Industrial Research (S Africa)
CSP	concentrated solar power
CUB	Citizens' Utility Board
CUREJ	College Undergraduate Research Electronic Journal
DA	Democratic Alliance (S Africa)
DANIDA	Danish International Development Agency
DECC	Department of Energy & Climate Change (UK)
DEFG	Distributed Energy Financial Group
DER	distributed energy resources
DG	distributed generation
DGNREEC	Direktorat Jenderal Energi Baru Terbarukan dan Konservasi Energi
DISCOs	distribution companies
DME	Department of Minerals and Energy (S Africa)
DMEA	Department of Mineral and Energy Affairs (S Africa)
DMR	Department of Mineral Resources (S Africa)
DoE	Department of Energy (S Africa)
DPE	Department for Public Enterprises (S Africa)
DPI	Database of Political Institutions
DRS	deposit and refund schemes
DSB	Dispute Settlement Body
DSM	demand-side management
DSO	distribution service operator
DST	Department of Science and Technology (S Africa)
E3	Energy and Environmental Economics
E3G	Third Generation Environmentalism
EC	European Commission
ECN	Energy Commission of Nigeria
EDPRS	Economic Development and Poverty Reduction Strategy (Rwanda)
EDSO	European Distribution System Operators
EEAG	Environmental and Energy State Aid Guidelines
EEG	Energy Economics Group
EFF	Economic Freedom Fighters (S Africa)
EGSS	Environmental Goods and Services Sector
EGUs	electrical generating units

EIA	Energy Information Administration
EIAs	environmental impact assessments
EPA	Environmental Protection Agency
EPS	environmental policy stringency
EPSA	Electrical Power Supply Association
EPSRA	Electric Power Sector Reform Act (Nigeria)
ERC	Energy Research Center (S Africa)
EREC	European Renewable Energy Council
EROI	energy returned on investment
ESCS	Energy Security Cabinet Subcommittee (S Africa)
ETS	Emissions Trading System
EU	European Union
EU-ETS	European Union Emissions Trading System
EUAs	European Union Allowances
EUEI	European Union Energy Initiative
EVs	electric vehicles
EWEA	European Wind Energy Association
FAO	Food and Agriculture Organization
FCBA	Forêt Cellulose Bois-construction Ameublement
FDI	foreign direct investment
FEMSA	Fomento Económico Mexicano
FERC	Federal Energy Regulatory Commission
FFS	fossil-fuel subsidy
FFFSR	Friends of Fossil-Fuel Subsidy Reform
FFVs	flex-fuel vehicles
FIT	feed-in tariff
FS-UNEP	Frankfurt School-United Nations Environment Programme
FYPs	five-year plans
GATT	General Agreement on Tariffs and Trade
GCF	Green Climate Fund
GDP	gross domestic product
GEA	Global Energy Assessment
GGKP	Green Growth Knowledge Platform
GHG	greenhouse gas
GIZ	German Aid Organization
GMS	Greater Mekong Sub-Region
GNI	gross national income

GOI	Government of Indonesia
GoI	Government of India
GoR	Government of Rwanda
GSI	Global Subsidies Initiative
GTAI	Germany Trade & Invest
GTZ	Deutsche Gesellschaft für Technische Zusammenarbeit
GW	gigawatts
GWEC	Global Wind Energy Council
HHI	Herfindahl-Hirschman Index
ICEM	International Centre for Environmental Management
ICSID	International Centre for Settlement of Investment Disputes
ICT	information and communication technology
IDASA	Institute for Democracy in South Africa
IEA	International Energy Agency
IEEE	Institute of Electrical and Electronics Engineers
IFAD	International Fund for Agricultural Development
IFN	Inventaire Forestier National
IGN	Institut géographique national
IGS	Instituto Global para la Sosteniblidad
IMF	International Monetary Fund
INC	Indian National Congress
INDC	Intended Nationally Determined Contributions
INRA	Institut National de la Recherche Agronomique (France)
IPCC	Intergovernmental Panel on Climate Change
IPP	independent power producer
IPR	integrated resource plan
ICRG	International Country Risk Guide
IQR	interquartile
IRENA	International Renewable Energy Agency
IRP	Integrated Resource Plan
ISMO	Independent System Markets Operator
ISO	International Organization for Standardization
ITT	Ishpingo-Timbococha-Tiputini
IWES	Institute for Wind Energy and Energy System Technology
JIM	joint implementation mechanism
JISEA	Institute for Strategic Energy Analysis
JRC	Joint Research Centre of the European Commission

LAERFTE	Law for the Use of Renewable Energy and Financing of Energy Transition (Mexico)
LASE	Law for Sustainable Use of Energy (Mexico)
LCET	low-carbon energy technologies
LGCC	General Climate Change Law (Mexico)
LDCs	least developed countries
LIE	Law of Electric Industry (Mexico)
LMB	Lower Mekong Basin
LME	liberal market economy
LMP	localized marginal price
LNG	liquefied natural gas
LPG	liquefied petroleum gas
LSE	London School of Economics
LSEs	load-serving entities
LTE	Law on Energy Transition
MCTI	Ministry of Science, Technology and Innovation (Brazil)
MEC	minerals-energy complex
MEMR	Ministry of Energy and Mineral Resources (Indonesia)
MHI	Manitoba Hydro International
MINECOFIN	Ministry of Finance and Economic Planning (Rwanda)
MININFRA	Ministry of Infrastructure (Rwanda)
MINIRENA	Ministry of Environment and Natural Resources (Rwanda)
MIT	Massachusetts Institute of Technology
MMA	Brazilian Ministry of Environment (Brazil)
MME	Ministry of Mines and Energy (Brazil)
MoU	Memorandi of Understanding
MR	Mekong region
MRB	Mekong River Basin
MRC	Mekong River Commission
MRGP	maximum refinery gate price
MW	megawatt
MWh	megawatt hour
MYTO	multi-year tariff order
NAEC	Nigeria Atomic Energy Commission
NAFTA	North American Free Trade Area
NAMA	nationally appropriate mitigation actions
NAPTIN	National Power Training Institute of Nigeria

NBET	Nigerian Bulk Electricity Trading
NCP	National Council on Privatization (Nigeria)
NDA	National Democratic Alliance
NDBP	National Domestic Biogas Programme (Rwanda)
NDRC	National Development and Reform Commission
NEA	National Energy Administration
NEC	National Energy Commission
NECSA	Nuclear Energy Cooperation South Africa
NELMCO	Nigerian Electricity Liability Management Company
NEPA	National Electric Power Authority (Nigeria)
NERC	Nigerian Electricity Regulatory Commission
NERSA	National Energy Regulator South Africa
NFFO	non-fossil fuel obligation
NGO	non-governmental organizations
NIASA	Nuclear Atomic Industry Association
NIMBY	not in my back yard
NIPP	National Integrated Power Project (Nigeria)
NISP	National Improved Stove Programme (China)
NISR	National Institute of Statistics of Rwanda
NLGACCERCER	National Leading Group on Climate Change Energy Conservation and Emissions Reduction
NNEECC	Nuclear Energy Committee (S Africa)
NOx	nitrogen oxide
NPC	National Planning Commission
NPCC-RS	National Policy on Climate Change and Response Strategy
NPL	Northwest Power Ltd
NPV	net present value
NREL	National Renewable Energy Laboratory
NUM	National Unions of Mineworkers (S Africa)
NYMEX	New York Mercantile Exchange
OCGT	open-cycle gas turbine
OECD	Organisation for Economic Co-operation and Development
OFGEM	Office of Gas and Electricity Markets
OPEX	operational expenses
PAIA	Promotion of Access to Information Act
PBMR	pebble-bed modular reactor
PCT	Patent Cooperation Treaty

PDD	Project Design Documents
PECC	National Strategy on Climate Change (Mexico)
PEMEX	Petróleos Mexicanos
PHCN	Power Holding Company of Nigeria
PLB	Planbureau voor de Leefomgeving
PMP	predominant method of production
PPA	power purchase agreement
PPMs	processes and production methods
PRA	Participatory Rural Appraisal
PROINFRA	Programa de Incentivo a Fontes Alternativas de Energia Elétrica (Brazil)
PSCW	Public Service Commission of Wisconsin
PUC	Public Utilities Commission
PV	photovoltaic
PWR	pressurized water reactor
R&D	research and development
RACER	relevant, accepted, credible, easy to monitor, and robust
RAP	Regulatory Assistance Project
RD&D	research, development, and demonstration
RE	renewable energy
RE IPPPP	Renewable Energy Independent Power Producers' Programme
REDD+	Reducing Emissions from Deforestation and Forest Degradation
REFIT	renewable energy feed-in tariff
REG	Rwanda Energy Group
REN21	Renewables 2015 Global Status Report
RES	renewable energy sources
RESA	Renewable Energy Sources Act (Germany)
RETs	renewable energy technologies
RFS2	Renewable Fuel Standard Program
RO	renewables obligation
ROI	return on investment
RPS	renewable portfolio standard
RSA	Republic of South Africa)
SABC	South African Broadcasting Corporation
SCADA	supervisory control and data acquisition systems
SCC	social cost of carbon
SCM	Subsidies and Countervailing Measures
SCOT	social construction of technology

SEC	state-owned electricity company
SEMARNAT	Secretaría del Medio Ambiente y Rescursos Naturales
SENER	Secretariat of Energy (Mexico)
SERC	State Electricity Regulatory Commission (China)
SHP	solar heat and power
SIP	state implementation plan
SNV	Netherlands Development Organization
SO_2	sulphur dioxide
SOE	state-owned enterprise
SON	Standards Organisation of Nigeria
Sox	sulfur oxide
SPRU	Science Policy Research Unit
SSA	sub-Saharan Africa
SST	social shaping of technology
TCN	Transmission Company of Nigeria
tCO_2	tonne of carbon dioxide
TEPs	tradable emission permits
THC	thermohaline circulation
TOR	terms of reference
TSO	transmission system operator
TW	terawatt
TWh	terawatt hour
UFBA	Federal University of Bahia
UHV	ultra-high voltage
UK	United Kingdom
UMB	Upper Mekong Basin
UN	United Nations
UNCSD	United Nations Conference on Sustainable Development
UNDP	United Nations Development Programme
UNEP	United Nations Environment Programme
UNESCO	United Nations Educational, Scientific and Cultural Organization
UNFCCC	United Nations Framework Convention on Climate Change
UNU-MERIT	United Nations University–Maastricht Economic and Social Research Institute on Innovation and Technology
UNU-WIDER	United Nations University World Institute for Development Economics Research
US	United States
USAID	United States Agency for International Development

USDOE	United States Department of Energy
USGCRP	US Global Change Research Program
V2G	vehicle-to-grid project
VC	venture capital
VIUs	vertically integrated utilities
VoC	Varieties of Capitalism
WDI	World Development Indicators
WEC	Wisconsin Energy Corporation
WEPCO	Wisconsin Electrical Power Corporation
WGI	World Governance Indicators
WNN	World Nuclear News
WTI	West Texas Intermediate
WTO	World Trade Organization
WTP	willing to pay
WWF	World Wildlife Fund
YGCs	Yasuni Guarantee Certificates
YNP	Yasuni National Park

Notes on Contributors

Shahrouz Abolhosseini has a PhD in Technology Management (2014) from Seoul National University. He works as petroleum products pricing manager at National Iranian Oil Company. Also, he teaches as an invited lecturer at the College of Engineering, University of Tehran. His research interests include energy economics and policy, and renewable and non-renewable energy sources.

David Onyinyechi Agu is a researcher at the Department of Economics, University of Nigeria, Nsukka. He is also Research Fellow of Development Strategy Centre, Enugu. His research covers the areas of fiscal and monetary policies, international development finance, financial institutions, and trade and regional integration. He has published widely in these areas.

Celio Andrade is Associate Professor at Federal University of Bahia (UFBA), Brazil. He coordinates the Global Environmental Governance and Carbon Market Research Group at UFBA. His research revolves around low-carbon governance, especially with regard to integrative modes of analysis. He has published widely in the fields of global environmental governance, low-carbon economy, the global carbon market, in particular in the area of co-benefits of Clean Development Mechanism projects in the developing countries.

Douglas Arent is the Executive Director of the Joint Institute for Strategic Energy Analysis at the National Renewable Energy Laboratory. His research interests are centred on energy and sustainability, where he has been active for more than 30 years. He has published extensively on topics of clean energy, renewable energy, power systems, natural gas, and the intersection of science and public policy. He has a PhD from Princeton University, an MBA from Regis University, and a BSc from Harvey Mudd College in California. In addition to his JISEA/NREL responsibilities, Arent is Senior Visiting Fellow at the Center for Strategic and International Studies, serves on the American Academy of Arts and Sciences Steering Committee on Social Science and the Alternative Energy Future, is a member of the National Research Council Committee to Advise to US Global Change Research Program (USGCRP), and is a Member of the Keystone Energy Board.

Channing Arndt is Senior Research Fellow at the United Nations University World Institute for Development Economics Research (UNU-WIDER). He has substantial research management experience including leadership of interdisciplinary teams. His programme of research has focused on poverty alleviation and growth, agricultural development, market integration, gender and discrimination, the implications of the HIV/AIDS pandemic, technological change, trade policy, aid effectiveness, infrastructure investment, energy and biofuels, climate variability, and the economic implications of climate change.

Henok Birhanu Asmelash is Research Fellow at the Max Planck Institute Luxembourg for Procedural Law and a PhD Candidate in international law and economics at Bocconi University, Milan. His current research focuses on issues at the intersection of trade, energy, and the environment with a particular emphasis on the regulation of energy subsidies in the WTO.

Ross Astoria teaches public law, environmental policy, and political theory at the University of Wisconsin at Parkside. His research focuses on law and greenhouse gas mitigation policy. He has recently published on carbon taxes and international trade law.

Lucy Baker is Research Fellow in the Science Policy Research Unit (SPRU) at the University of Sussex. Her research is on the political economy of energy transitions. She has published extensively on this topic with regard to electricity, finance, and policy in South Africa.

Radhika Bhuyan is Senior Researcher at the Mapungubwe Institute for Strategic Reflection in South Africa. She is currently based in Johannesburg, and is completing doctoral research on firm innovation in alternative energy systems, from UNU–MERIT, the Netherlands. She has researched and worked on various issues and topics around renewable energy, including feed-in policies and sociot-echnological factors behind solar PV, wind, and hydrogen-fuel cells innovation, and understanding market competition and firm strategies in industry and global markets. She has authored and co-authored a number of articles, book chapters, and media articles on innovation, green economy, sustainable energy and technologies, and climate change.

Santiago J. Bucaram is Assistant Professor at the Department of Economics, and also Director of the Economic Studies Institute, Universidad San Francisco de Quito, Ecuador. His research focus on marine resource management, assessments of vulnerability to climate change and agricultural economics. He has published in the fields of fishery economics, in particular in the area of fishing decision behaviour and rights-based management.

Julien Bueb is an environmental economist. He currently works at the CAPS (Centre d'Analyse, de Prévision et de Stratégie) as a researcher and political advisor on economical, natural resources, and environmental matters. He is Associate Professor at University Paris 1,Panthéon-Sorbonne, and University of Franche-Comté.

Aron Buzogány is a researcher at the Otto Suhr Institute for Political Science, FreieUniversität Berlin. His research focuses on comparative analysis of political systems in both Western and Eastern Europe. He has published widely on the impact of Europeanization on environmental policy-making, on state–society relations as well as on executive–legislative relations.

Enrica de Cian is Senior Researcher at Centro Euro-Mediterraneo sui CambiamentiClimatici, and Fondazione Eni Enrico Mattei. Her research focuses on induced innovation and technological change, integrated assessment modelling, and climate change impacts and adaptation. She has published in the fields of climate change economics, integrated assessment modelling, energy economics, and environmental and resource economics.

Stefan Ćetković is Research Associate at the Environmental Policy Research Centre, FreieUniversität Berlin. His research examines the policy and politics of sustainability transitions in a comparative perspective. He has published on the topics of policy capacity for sustainable agriculture, and the political economy of climate and energy governance.

Alice Le Clézio is Associate Researcher, Policy Analyst, and PhD candidate in political science at CERI, Sciences Po Paris. She previously has served as a policy analyst (RAND Corp., UNESCO, MIT) and her areas of expertise are public policy, sustainable development, and migration, with a special focus on the environment–development–security nexus.

Shouro Dasgupta is Junior Researcher at Fondazione Eni Enrico Mattei and Centro Euro-Mediterraneo sui CambiamentiClimatici. His research interests include empirical investigation of the impacts of climate change on human health, impact of institutions and governance on environmental performance and policy, and political economy of energy innovation. He has published on climate change impacts, socioeconomic determinants of climatological disasters, impact of global financial crisis, and economic performance of developing countries.

Michael R. Davidson is a PhD candidate in engineering systems at the Massachusetts Institute of Technology, where he researches the long-term viability of wind energy in China from the perspectives of grid operations, electricity market design, and institutional change. His work builds on a range of methods including optimization, simulation techniques, and qualitative analysis. He has studied and worked on Chinese energy and environmental issues since 2008.

Jiska de Groot is Research Officer in the Energy, Poverty and Development Group at the Energy Research Centre in the University of Cape Town. Her research focuses on social science aspects of the deployment of renewable energy technologies, marine renewable energy, clean energy access, and poverty alleviation in developing countries.

Martin de Jong is Professor at the Delft University of Technology on the topic of China's urban and infrastructure developments. He is also Professor in Public Policy at Fudan University in China. He has published extensively in the fields of decision-making, cross-national institutional comparison, transport planning, urban planning, environmental planning, and low-carbon city developments in China.

Ariel Dinar is Professor of Environmental Economics and Policy at the School of Public Policy, University of California, Riverside. His work addresses various aspects of economic and strategic behaviour of management of natural resources and the environment. He has authored and co-authored nearly 150-peer-reviewed papers, and 24 books and textbooks. He founded two technical journals for which he serves as an Editor-in-Chief.

Cherrelle Eid is a PhD candidate at Delft University of Technology within the Sustainable Energy Technologies and Strategies PhD programme, a competitive Erasmus Mundus Joint Doctorate sponsored by the European Commission. Her research focuses on effective Smart Grid policy design. She holds a Bachelor's in Systems Engineering and Policy Analysis, and a Master's in Economics and Management of Network Industries.

Chika Ezeanya is Senior Lecturer in the College of Business and Economics at the University of Rwanda, Kigali. Her research focuses on understanding the role of home-grown, indigenous, and grassroots-based approaches to advancement in sub-Saharan Africa across sectors. She has worked as a consultant for the World Bank and the International Development Research Center, Canada.

Mario Andrés Fernández is a researcher at the Governance and Policy Team of Landcare Research, New Zealand. His research concentrates on mathematical modelling, economic impacts of climate change on agriculture, assessments of vulnerability to climate change and policy analysis. He is also active in the fields of non-market valuation, ecosystem services, and public health.

Erik Gawel is Professor of Economics, particularly institutional environmental economics, and Director of the Institute for Infrastructure and Resources Management at Leipzig University. He is Head of the Department of Economics of the Helmholtz Centre for Environmental Research in Leipzig. His main areas of research comprise environmental and energy economics, public finance and new institutional economics, particularly law and economics and public choice analyses.

Alexandra Goritz is a researcher at the think tank Third Generation Environmentalism (E3G). Her research interests are particularly in sustainable development, climate policy, and the political economy of energy transitions in emerging economies. She holds an MSc in Environment and Development from the London School of Economics and Political Science, and a BA in Politics and Public Administration from the University of Konstanz.

Diego Grijalva is Assistant Professor at the Department of Economics, Universidad San Francisco de Quito. His research fields are development economics and political economy. He has worked on topics ranging from the relationship between inequality and growth to the mechanisms of institutional change, including the role of beliefs and persuasion. He also worked at the Research Department of the Central Bank of Ecuador.

Rudi Hakvoort is Associate Professor in the faculty of Technology Policy and Management at Delft University of Technology. Furthermore he is a consultant for governments and energy utilities. His area of research involves smart grid technologies, innovation and strategy in the energy sector, market design, regulation, and system integration of renewables.

Lilian Richieri Hanania is an Attorney (admitted in Brazil and France), holds a PhD in International Law from University Paris 1, Panthéon-Sorbonne, a researcher at the Centro de Estudos Sociedade e Tecnologia at University of São Paulol, and an associate researcher at the IREDIES, University Paris 1, and CUREJ, University of Rouen, France. Her areas of expertise are international economic law, international cultural law, and sustainable development.

Almas Heshmati is Professor of Economics at Jönköping University and Sogang University. He held similar positions at the Korea University, Seoul National University, and University of Kurdistan Hawler. He was Research Fellow at UNU-WIDER during 2001–4. From 1998 to 2001, he was Associate Professor of Economics at the Stockholm School of Economics. He has a PhD from the University of Gothenburg (1994). His research interests include applied microeconomics, efficiency, productivity, and growth with application to manufacturing and services, inequality, poverty, and inclusive growth in developing countries.

Karoliina Isoaho is a PhD candidate in the Department of Social Research at the University of Helsinki. Her research interests include clean energy transition theories, renewable energy, and climate policy, and socio-cognitive aspects of framing in environmental governance. She holds an MSc in Environment and Development from the London School of Economics and Political Science, and a BA in Modern Languages and Contemporary European Studies from the University of Southampton.

Jesse Jenkins is a PhD candidate in engineering systems at MIT's Institute for Data Systems and Society, and a researcher with the MIT Energy Initiative. He applies an interdisciplinary approach to study energy economics, operations, regulation, and policy, with a focus on two overarching trends transforming the electricity sector: the transition to zero carbon power systems, and the proliferation of distributed energy resources.

Fredrich Kahrl is Director at Energy and Environmental Economics (E3), a San Francisco-based energy consulting firm. He advises energy developers, utilities, operators, and regulators on critical engineering and economic issues in the electricity and gas sectors. He has worked extensively on renewable energy market and regulatory design issues in western USA and China. He has worked on electricity policy issues in China for more than a decade.

Valerie J. Karplus is Professor in the Global Economics and Management group at the MIT Sloan School of Management, where she studies the dynamics of transition in global energy and resource-intensive systems, with a focus on emerging markets and the role of institutions and policy. Her work has focused on these themes as they relate to agriculture, transportation, energy, and entrepreneurship.

Abel Kennedy is a lecturer at the College of Business and Economics in the University of Rwanda, Kigali. His research interests are international trade and East Africa economic relations incorporating quantitative analysis.

Éric Nazindigouba Kere is Research Economist in the Complex of the Chief Economist of African Development Bank. His research focuses on energy and development economics. He has published mainly in the fields of bioenergy, sustainable management of natural resources (deforestation, protected areas, and biodiversity), climate change and development economics (food security, water, and sanitation).

Wikus Kruger is a researcher with the Management Programme for Infrastructure Reform and Regulation, based at the Graduate School of Business, University of Cape Town. His research focuses on energy access for poor households, as well as procurement models and policies for renewable energy independent power production in Africa.

Paul Lehmann is Research Fellow at the Department of Economics at the Helmholtz-Centre for Environmental Research, where he leads a social science energy research group. He is also affiliated with the Institute for New Economic Thinking at the University of Oxford. His research interests include the applied economic analysis of policy instruments and their combinations, particularly in the fields of climate, energy, and water policy.

Gilles Lepesant is Senior Researcher at the Centre National de la Recherche Scientifique in the research unit Géographie-Cités, Paris. He has published several books and papers on issues related to territorial development and to energy in Europe with an emphasis on the implementation of the related EU policies.

Sabrina McCormick PhD is a sociologist and filmmaker who investigates how to motivate climate mitigation and adaptation. She is Associate Professor in the Environmental and Occupational Health Department in the Milken Institute School of Public Health at George Washington University, and Senior Fellow at the Wharton Risk and Decision Center.

Mackay Miller, Professional Scholar, Joint Institute for Strategic Energy Analysis, National Renewable Energy Laboratory. He is Lead Analyst in the Technology & Strategy division of National Grid, where he scouts emerging technologies for application in utility planning and operations. Prior to National Grid, he worked for six years at the National Renewable Energy Laboratory, including an assignment as Senior Analyst at the US Department of Energy. He has written widely in areas of renewable energy integration, smart-grid deployment, and policy and regulatory dimensions of power system transformation. He holds an MBA from the University of Colorado-Boulder, and a BA in International Relations from Brown University.

Evelyn Nwamaka Ogbeide-Osaretin is Lecturer at the Department of Economics and Business Studies, Samuel Adegboyega University, Ogwa, Edo State. She is also Research Fellow of Development Strategy Centre, Enugu. Her research is in the areas of poverty, inequality, fiscal policy, and monetary policies. She has published widely in the area of poverty and inequality in Nigeria.

Eric Kehinde Ogunleye is Special Adviser (Policy and Planning) to the Nigerian Honourable Minister of Budget and National Planning at the Federal Ministry of Budget and National Planning. Until recently, he served as consultant Country Macroeconomist at the Nigeria Country Office of the African Development Bank. He has published widely in the fields of macroeconomics and development economics, especially in the area of structural and economic transformation in developing countries particularly Nigeria.

Kim Hang Pham Do is Senior Lecturer of Economics at School of Economics and Finance, Massey University, Palmerston North. Her current work addresses various aspects of multilateral set-ups with externalities and the role of issue linkage in managing conflicts of natural resources management and the environment. She has published widely in the fields of applied game theory, environmental, and development economics.

Jose A. Puppim de Oliveira is faculty member at the Getulio Vargas Foundation Federal University of Rio de Janeiro, and Fudan University, Shanghai. He is also a visiting researcher at the United Nations University International Institute for Global Health, Kuala Lumpur. His research and policy interests concentrate broadly in the area of the political economy of sustainable development. He is particularly interested in patterns of governance, institution-building, and policy implementation at different levels, looking at how global and national institutions are interlinked to local governance and action.

Britta Rennkamp is a researcher at the Energy Research Centre and Fellow at the African Climate and Development Initiative, University of Cape Town. She holds a PhD in Science and Technology Policy from the University of Twente, the Netherlands, and a Master's Degree in Latin American Studies from University of Cologne. Her research focuses on policy analysis of climate technology, climate change mitigation, and poverty eradication in developing countries. Previous work has analysed renewable energy and nuclear programmes, carbon taxation, and green industrial and innovation policies. She has published various papers and book chapters on the overall question of integrating policies on emissions reductions, energy supply, and development in Africa and Latin America.

Miranda Schreurs is Professor of Comparative Politics and Director of the Environmental Policy Research Centre at the Freie Universität, Berlin. Her research is focused on the comparative analysis of international political systems, environmental and energy policy, and the role of civil society, governments, and economic actors. She has expertise on Germany, the European Union, the USA, and East Asia—particularly Japan, China, and South Korea.

Nicolai Schulz is a PhD candidate in the Department of International Development at the London School of Economics (LSE) and Political Science. He holds an MSc in Development Studies from the LSE, and a BA in Politics and Public Administration from the University of Konstanz. His research focuses on the political economy of development and particularly on understanding the variation in governments' willingness and capacity to implement developmental policies.

Jonas Sonnenschein is a researcher at the International Institute for Industrial Environmental Economics, Lund University. His research is the evaluation of policy instruments targeting low-carbon energy technologies. His research aims at an enhanced understanding of how the design of policy interventions can be improved by taking into consideration empirical findings about the actual behaviour of firms and households.

Benjamin K. Sovacool is Professor of Energy Policy at the Science Policy Research Unit, School of Business, Management, and Economics, University of Sussex, UK. There he serves as Director of the Sussex Energy Group and Director of the Center on Innovation and Energy Demand, which involves the University of Oxford and the University of Manchester. He is also Director of the Center for Energy Technologies and Professor of Business and Social Sciences in the Department of Business Development and Technology at Aarhus University in Denmark. His research focuses on renewable energy and energy efficiency, the politics of large-scale energy infrastructure, designing public policy to improve energy security and access to electricity, and building adaptive capacity to the consequences of climate change.

Sebastian Strunz is a research fellow at the Department of Economics at the Helmholtz-Centre for Environmental Research in Leipzig. His research deals with the economics of sustainability transitions, in particular as regards the design of environmental policy. He has published in a range of ecological and institutional economics outlets with a focus on both conceptual analyses and applications to Germany's energy transition.

Isabel Studer is Director-General for International Economic Cooperation at the Mexican Agency of International Cooperation for Development. With a PhD from Johns Hopkins University-SAIS, she was Founding Director of the Global Institute for Sustainability at Tec de Monterrey.A Fulbright and Ford Foundation scholar, she has published largely on climate change, trade, and the environment, green growth, corporate sustainability, and regional integration.

Louise Tait is Research Officer in the Energy, Poverty, and Development group at the Energy Research Centre, University of Cape Town. Her research focuses on the nexus of energy and development in developing countries; energy access for poor households; a development-first approach to climate change mitigation and sustainable urban energy transitions.

Finn Tarp is Director of UNU-WIDER, and Coordinator of the Development Economics Research Group, University of Copenhagen. He is a leading international expert on issues of development strategy and foreign aid, with a sustained interest in poverty, income distribution, and growth. He has published widely in international academic journals alongside various books. He is a member of the World Bank Chief Economist's Council of Eminent Persons and is a resource person of the African Economic Research Consortium.

José María Valenzuela collaborates with the Energy Research Program at El Colegio de Mexico. His practice and academic research focuses on the institutional and political constraints for the development of clean energy transition, and therein power markets restructuring. He performed multiple responsibilities within Mexico's energy ministry, and has had extensive engagement with international organization and NGOs.

Elena Verdolini is Senior Researcher at Centro Euro-Mediterraneo sui Cambiamenti Climatici, and Fondazione Eni Enrico Mattei. Her research focuses on the analysis of innovation, adoption and transfer of energy-efficient and climate-friendly technologies, with a particular focus on the role played by research and development and public policies. She has published in the fields of international economics, energy, environmental and resource economics, and risk analysis.

Dewi Yuliani is an independent researcher and public policy analyst, focusing on sustainable energy. Her Master's thesis and doctoral dissertation from Institute Technology of Bandung dealt with various aspects of renewable energy development in Indonesia. Previously she worked at a regional government of West Java Province, where she was involved in hands-on field experience, as well as in public policy formulation for renewable energy development.

Owen Zinaman is the Deputy Lead for the 21st Century Power Partnership (21CPP), a multilateral initiative of the Clean Energy Ministerial; he also serves as a Power Sector Analyst at the US National Renewable Energy Laboratory, and has published widely on power system transformation issues, including the grid integration of bulk and distributed renewable energy resources, power system flexibility, and policy and regulatory issues across the clean energy spectrum.

Part I

The Political Economy of Clean Energy Transitions

1

Introduction and Synthesis

Douglas Arent, Channing Arndt, Mackay Miller, Finn Tarp, and Owen Zinaman

1.1 MOTIVATION

Climate change is frequently referred to as one of the defining challenges of the twenty-first century. We concur. In broad terms, the climate challenge is relatively straightforward. Global average temperatures are rising as a consequence of anthropogenic emissions of greenhouse gases. In the absence of deliberate and global action to first substantially reduce and then eliminate (or even turn net negative) greenhouse gas (GHG) emissions, global temperature rise within this century is very likely to surpass two degrees Celsius (IPCC 2014), which is the (somewhat arbitrary) threshold set by the international community as a tolerable level of warming.[1] Continuation of current levels of emissions or (worse) continued growth in emissions throughout the twenty-first century could result in warming far above the two-degree threshold with very bad implications for the environment of the planet and for human societies, particularly poor people.

These observations constitute the core arguments for serious efforts to reduce emissions, called mitigation policy, at the global level. A principal element to mitigation policy relates to energy use. Specifically, energy use must transition from technologies that emit substantial volumes of GHGs to technologies with limited or zero emissions. A 'clean energy transition' refers broadly to a substitution of technologies and associated fuel inputs across the full set of energy subsectors and consumers of energy, both as intermediates and final goods. This is the 'clean energy transition' referred to in the title of this book.

[1] A more ambitious 1.5-degree target has been set forth in recent negotiations arguing that it is 'a significantly safer defense line against the worst impacts of a changing climate' (UNFCCC 2015).

While the broad contours of the climate challenge, of which the mitigation challenge is a subset, are well understood, the specificities of almost all aspects of the climate challenge are deeply complex. Enormous efforts have been dedicated to the science of global change (IPCC 2014, 2013). While much remains to be learned, climate science provides solid foundations to the core arguments for serious efforts to reduce emissions. The technical challenge of inventing low emissions energy technology has been absorbing the attention of some of the world's top scientists and engineers for decades and has become increasingly commercial over the past decade. Further, a new wave of promising technologies is forming.

But, in the end, a solid foundation for action derived from climate science combined with an array of promising technologies for reducing emissions are not likely to be enough to catalyse a clean energy transition. A key phrase in the very first paragraph of this introductory chapter is 'deliberate and global action'. A clean energy transition is highly unlikely to occur on its own. Policies must be put in place that will foment a clean energy transition, and these policies must be effective globally (as opposed to just shifting emissions from one region to another). The challenge, perhaps the largest of them all, is implementing policies and programmes that actually achieve the necessary global emissions reductions. Here, political economy considerations take a leading role. These perspectives motivate our focus on the political economy of clean energy transitions.

1.2 A NEW ERA

For the purposes of avoiding the potentially execrable outcomes associated with climate change referred to in Section 1.1, a long series of global agreements and meetings have taken place under the auspices of the United Nations. In the jargon that sprouts from such efforts, the first conference of the parties (CoP1) to the United Nations Framework Convention on Climate Change (UNFCCC) took place in 1995 in Berlin. Through the twentieth meeting (CoP20) in 2014 in Lima, relatively little was accomplished in terms of actually changing the trajectory of the global emissions of GHGs that drive climate change. CoP21 in Paris in late 2015 represents a potential breakthrough that ushers in a new era for climate mitigation.

The successful conclusion of CoP21 reflects three features of the current landscape that differ significantly from previous major attempts to set the planet on a more desirable GHG emissions trajectory. The most recent major attempt, prior to CoP21, occurred at CoP15 in 2009 in Copenhagen. First, CoP21 represented the culmination of a shift in the negotiation framework. At CoP15, the negotiations retained more of a 'top-down' approach wherein, essentially, a global emissions trajectory was determined and negotiators sought to parse country-level responsibilities for achieving this path.[2] In contrast, CoP21 in Paris employed a 'bottom-up' offer system, wherein individual countries propose what they perceive to be achievable and fair emissions trajectories for their particular circumstances. These offers are formally called Intended Nationally Determined Contributions (INDCs). In this new negotiating framework, the resulting projected global emissions trajectory is the sum of individual country INDCs.

Second, the rapid pace of technological advances in renewable energy technologies and systems, even if one considers just the past six years, is in the process of influencing the political economy of clean energy transitions (USDOE 2015). Historically, governments aiming to take deliberate action to correct the colossal market failure of GHG emissions have suffered from a 'chicken and egg' problem. Specifically, many technologies that offered long-run potential to support a clean energy transition were also small-scale, immature, and relatively high-cost. As a result, they were largely unattractive to private investors. While these factors provide a solid economic rationale for government support, the politics of supporting small-scale, immature, and relatively high-cost technologies are nonetheless difficult. Difficult politics inevitably constrains the ambition of policies that are crucial for technology development.

In sum, a circle exists wherein politics drives policy, policy drives technology, and the state of technology circles back to influence politics. Today, from the perspective of advancing clean energy technologies, this circle shows evidence of becoming virtuous as opposed to vicious. Since 2008, the year before CoP15 notably failed to produce a move towards effective global mitigation, the global solar module price index has fallen by a factor of nearly four, a rate of technical advance vastly more rapid than nearly all predictions (Feldman et al. 2014). Declines in the cost of wind power—while not as dramatic—have been rapid by any common standard (Moné et al. 2015). These advances both spur private investment and generally ease the politics of supporting clean energy transitions. Investments in energy production have reflected these shifts. In 2014, for the first time in history, the amount of new renewable generation capacity surpassed that of new fossil fuel-based systems on a global basis (Sawin et al. 2015). This trend continued in 2015 with new renewable capacity outstripping fossil fuels again (Frankfurt School-UNEP Centre/BNEF 2016).

Third, the developing world confronts climate change issues with a far deeper and more sophisticated knowledge base than in 2009 (Arndt and Tarp 2015). In Copenhagen at CoP15, the critical role that developing countries must play in any effective global mitigation regime had become clear

[2] CoP15 also arguably seeded the approach taken in Paris at CoP21 through the discussions of nationally appropriate mitigation actions (NAMA).

simply as a matter of arithmetic. Yet, the complex implications of climate change impacts, adaptation policies, and mitigation policies had really only begun to penetrate the major decision-making apparatuses of developing countries. For instance, the World Bank's *Economics of Adaptation to Climate Change* study, which was meant to serve as a critical input to developing countries for CoP15 in 2009, was only published in 2010, *after* the Copenhagen CoP meeting had ended. In our experience at the time around CoP15, work on climate change issues, particularly when one spoke to personnel from the critical central finance and planning units in developing countries, frequently amounted to delivering primers on climate change and energy transition policy basics. The process of internalizing the information and assessing appropriate policy responses had only just begun.

It would be an overstatement to say today that climate change information has been fully internalized and appropriate policies assessed in developing countries. Nevertheless, the process of doing so is much more advanced than it was in 2009. In country after country, the central decision-making units have engaged. This is critical. The profound economic transformations inherent in a clean energy transition will need to be fully integrated into economic decision-making. The contributions from developing countries in this book are evidence of this increasingly sophisticated and nuanced view of the climate challenge. The more than 160 INDCs on the UNFCCC website are perhaps the most salient evidence.

India and China are cases in point. In 2009, it is fair to say that India's negotiation strategy aimed to position climate change as a developed country problem. In contrast, India's INDC offers serious attempts to reduce the carbon intensity of its GDP. China has gone further, offering to peak emissions by 2030 with declines thereafter. Taken as a whole, the INDCs presented at CoP21 represent *a decisive break from past emissions trends.* Recent analysis of the INDCs by the International Energy Agency (2015) indicates that nearly every country will have a strong focus on emissions mitigation, driving clean energy to more than 50 per cent of world energy by 2040. The scope and ambition of these offers stem from long and often difficult processes of internalization and policy option assessment that has taken place within both developed and developing countries.

These three shifts now combine to place country decision-making and country policies at centre stage. Like it or not, there is no current prospect for a unified global policy, such as a global carbon tax or cap-and-trade scheme, to which all nations agree to adhere. Rather, nearly all countries on the globe will set about to achieve their contributions in their own ways, and their means for achieving these ends will vary enormously. For example, the United States, a leading advocate in international fora for reliance on markets, looks set to pursue a domestic policy of regulatory edict. China, the paragon of the developmental state, announced intention for a nationwide cap-and-trade system in September 2015. Overall, the range of policies pursued, and hence the degree of policy experimentation, looks virtually certain to be very large.[3]

Before proceeding, the technology drivers mentioned in this section merit a closer examination.

1.3 TECHNOLOGY DRIVERS

> In controversies about technology and society, there is no idea more provocative than the notion that technical things have political qualities (Langdon Winner 1980).

The rate of technological advancement in the renewable energy space has been notably rapid. Established institutions, once isolated from rapid change, are now presented with a dynamic landscape of pathways for simultaneously achieving decarbonization goals and sustainable development objectives. With affordable low-carbon energy readily available or imminent in most contexts, institutional innovation is arising—out of necessity—across public policy, finance, business models, markets, planning, and other dimensions to promote deployment. These innovations—and the technical and political qualities they possess—are interacting with a range of incumbent actors and interests, and influencing the political economy of the clean energy transitions. Thus, a brief assessment of technology drivers is worthwhile.

The growing cost-competitiveness and advanced capabilities of renewable energy technologies, predominantly wind and solar, is a key pillar of clean energy innovation and technological advancement. We observe, in many contexts, the price of a newly constructed wind farm or solar plant is now at or below the cost of competing fossil fuel alternatives, even without considering the fuel price variability or environmental or health impacts (Stark et al. 2015). With their geographically diverse and variable nature, these resources are reshaping, in particular, how power systems are planned, operated, governed, and even conceptualized (Miller et al. 2015). Furthermore, the modularity of solar panels enables a viable alternative to the traditional provider–customer relationship, quite literally empowering consumers through technology, regulation, and business model innovation to create their own energy.

The qualities of clean energy technologies also have implications for energy security in both developed and developing country contexts. Renewable technologies offer the prospect of reducing dependence on fuel imports. Energy

[3] Backsliding in policies to achieve a clean energy transition is also a clear possibility in numerous countries.

trade between countries may or may not decline, however. There are portfolio effect gains from renewable energy generation over broad areas driven by the simple observation that it is likely to be windy and/or sunny somewhere (Keane et al. 2011). In addition, hydropower resources are often concentrated in a few locations. Both of these factors point to increased regional energy trade as a potential corollary to increased dependence on renewable energy sources. As a result, energy security under a renewable energy future may take on a much more regional hue (see Part VIII).

At the same time, the inherent dispersion of wind and solar resources, combined with new technologies and business models, present increasingly attractive pathways to expanding energy access from the bottom up, potentially leapfrogging the need for some of the cumbersome and difficult-to-finance infrastructure investments associated with traditional power systems. As will be discussed in Section 1.4, this dispersed nature of renewable energy may be particularly relevant for rural zones and smaller concentrations of demand located a distance from functional grids. Advances in data systems, communication technologies, and energy storage costs are accelerating decentralization and heterogeneity of the energy sector (Zinaman et al. 2015).

While technology is a fundamental driver, it has become increasingly clear that the availability of technology is not in itself sufficient to accelerate a clean energy transition; innovative and nationally-customized deployment strategies—hinging on public policy and regulation, market reforms, private sector engagement, and strong analytical tools and data—remain important factors.

More often than not, regulation and governance lag behind technology innovation, compelling forms of institutional innovation in order to play catch up. Ongoing innovations in energy systems often require either adaptations of established regulatory constructs to accommodate new technologies (a form of incremental change) or broad-based reform of the regulatory constructs themselves (perhaps via more reconstructive or evolutionary approaches) (see, for example, Zinaman et al. 2015). Across all contexts, addressing the techno-institutional complex perpetuating carbon-intensive systems—termed by some as 'carbon lock-in' (see, for example, Unruh 2000)—is a common theme.

Technology is highly likely to remain one of the key driving factors influencing climate commitments and energy-related development goals, both in terms of goal-setting and implementation. What is technically possible and economically attractive today vis-à-vis decarbonization and sustainable development is much greater than it was during (for example) the Kyoto Protocol era. Continued rapid rates of technical advance are expected. In order to seize the opportunities offered by this technical advance, equally innovative approaches to regulation and policy are likely to be required. This highlights the inherent political economic factors to be considered, as various pathways are weighed and implementation efforts are mounted.

1.4 CHALLENGES IN DEVELOPING VERSUS DEVELOPED ECONOMIES

The political economy of energy transitions is of interest across both the developed and developing worlds, and a defining feature of this book is a review of experiences from a diversity of contexts. As emphasized, the mitigation challenge cannot be addressed by developed countries alone. The volume of current emissions from developing countries combined with their rapid growth trajectories highlight the importance of developing countries in any effective global mitigation regime.

Developing countries simultaneously confront enormous development challenges. Eliminating absolute poverty is also a defining challenge of the twenty-first century, as set forth in the Sustainable Development Goals. Developing countries are highly unlikely to shelve their developmental aspirations in favour of mitigation objectives. Thus, the political economy of clean energy transitions in the developing world present some of the thorniest and most important challenges.

With respect to the developed world, their historical emissions, relatively comfortable material circumstances, institutional capabilities, and technical knowhow lead to the expectation that they will lead the energy transition. This means reducing absolute emissions in the near term and achieving very deep cuts by mid-century. This change must be undertaken by energy systems characterized by weak or even negative energy demand growth as well as deeply entrenched actors and interests.

In sum, the challenges facing both developing and developed countries are not to be taken lightly. While developed countries are expected to lead—for example, with respect to government commitments to research, development, demonstration, and deployment activities for new technologies—the critical role of regulatory frameworks, policies, and institutions have already been emphasized. These require localized solutions in both developing and developed country contexts. The dividing line between these two broad country groups is neither clear nor fast in other respects as well. Citizens of developed countries expect economic progress through time along with environmental stewardship, and developing countries certainly have their share of entrenched interests.

Nevertheless, the broadly defined challenges facing developed and developing economies do differ in important ways. In particular, driven by population/labour force growth, technological catch-up, a relatively high marginal product of capital and substantial growth aspirations, developing countries' economies can be expected to grow more rapidly than developed economies. Accordingly, the demand for new energy supply is likely to be much greater in the developing than developed world.

There are multiple edges to this challenge. On the one hand, the INDCs set forth by developing countries point to a reorientation away from the well-trodden path of employing massive fossil energy to fuel development. This charting of a new path, or new paths, is almost surely less straightforward than following prior recipes. As institutional and human capabilities in developing countries are characteristically weak relative to developed countries, the need to chart new paths and confront new challenges provokes legitimate concern.

On the other hand, fossil-based systems have a series of, by now, well-known shortcomings. First, developing countries frequently encounter difficulties implementing fossil-fuel-based systems, particularly for electricity generation. These difficulties arise from numerous factors. The bottom line is that unreliable power supply has long been a hallmark of many developing country cities and is frequently pointed to as a substantial brake on economic development (see Foster 2008). While intermittency in output is a characteristic of many renewable generators, that variability reduces substantially at a system level; and meeting or improving upon the reliability levels currently attained in many developing country contexts is often a fairly low bar of accomplishment. The relatively modular nature and short investment lead times of wind and solar power generation systems also favour developing countries where demand growth tends to be much more variable and much less predictable than in developed country contexts.

Second, fossil-fuel-based systems are poorly suited to rural areas. This is particularly true of electricity generation. Around 1.2 billion people (about 17 per cent of the world's population) lack access to electricity, and the vast majority of these people live in rural areas of developing countries (IEA 2015). Rural inhabitants in zones that lack access to electricity are frequently absolutely poor. In short, existing fossil-based power systems serve the least well off of the world's population very badly. Various renewable technologies have been shown to scale effectively in these areas. Biopower systems currently serve dozens of villages in South Asia (Bhattacharyya 2014), and next generation bioenergy systems also hold out additional promise for rural zones. With the rapid advances in solar and battery technology, distributed solar systems provide a potentially unprecedented opportunity to extend electricity access to some of the world's poorest citizens.

Third, localized pollution impacts of fossil-fuel-based systems can be intense. Poor air quality gives rise to serious health concerns. New Delhi and Beijing are just the most recent examples of places where low air quality seriously impacts wellbeing. Clean energy systems have the potential to diminish or even effectively remove these real costs.

Fourth, fossil-based systems both fuel and disrupt development. Experience in countries with fossil fuel endowments indicate that they are not an unalloyed boon for their economies in general and the welfare of their citizenry in particular. The vagaries of fossil fuel prices, and concomitant macroeconomic instability, combined with the tendency for revenues derived from sale of fossil resources to concentrate in a few hands have not been helpful for development patterns in many countries leading some authors to proclaim a 'resource curse' (Frankel 2010). For most fossil fuel importers, variations in fossil fuel prices have large impacts, often with implications for political stability (e.g., Arndt et al. 2012).

Finally, developing countries may possess inherent advantages in terms of clean energy endowments. Many developing countries are relatively well endowed with sun, wind, and unexploited hydropower potential. In a world dominated by clean energy systems, many developing countries may possess an inherent comparative advantage in energy-intensive activities.[4]

For these reasons, a clean energy transition is not necessarily an impediment to the growth aspirations of the developing world. And, there are a series of solid rationales for developed countries to assist developing countries in realizing a clean energy transition. Not least, a failure on the part of developing countries to transition to cleaner energy sources implies a failure to stabilize the global climate, with negative implications for everyone.

Developed countries are also responsible for a disproportionate share of the stock of greenhouse gases in the atmosphere. This would be highly problematic if the lack of space for even greater stocks of atmospheric GHGs imposed a tight trade-off between the development aspirations of the citizens of developing countries over the next few decades and a permanent alteration of the global climate. The fact that the developed world has effectively claimed squatters' rights on the global atmospheric commons becomes a lot less problematic if new paths to fuelling development are opened as the fossil fuel pathway is foreclosed.

The practical and ethical arguments for assisting developing countries in taking these new pathways are strong. At the same time, it is not a question of simply willing a clean energy system into place whatever the cost. As emphasized, the changes inherent in a clean energy transition are profound, involving the full economic system with implications for competiveness and economic growth. Improperly done, those costs could easily be very high and would likely sap the will for undertaking that very transition.

Hence, economic efficiency and reasonable equity are key. Efficient and relatively low-cost transitions to a stable global climate are widely viewed as

[4] Of course, whether developing countries are capable of actually capitalizing on these advantages (if they indeed exist) is another question. This is an important area for future research.

imminently possible.[5] The cost estimations in the Fifth Assessment Report of the IPCC indicate approximately a year or two of global growth by around mid-century. In other words, global GDP per capita with mitigation would reach the same level in 2055 as it would have attained in about 2053 without mitigation. These calculations typically ignore the benefits of mitigation in terms of climate change impacts avoided as well as health benefits from reduced pollution. Also, there are real possibilities to enhance the equity of the energy transition through, for example, more rapid rural electrification and better urban air quality in developing countries.

1.5 THIS BOOK

This book takes as a starting point that a new era of reducing emissions at scale has begun. The proximate challenges of this new era are codified in the emissions reductions offers (INDCs) from 165 countries available on the UNFCCC website. To date, the scale of emissions reductions efforts has been nowhere near adequate to the task. But, this does not mean that nothing has been tried. Considerable experience has been gained, and many features of the political economy of clean energy transitions have been revealed. It makes sense to profit from this experience in order to help meet the challenge of greatly scaling up emissions reductions efforts.

As countries and regions grapple with the complex task of reducing emissions in accordance with their INDCs while providing better lives for their citizenry, the demand for sharing of experience and lessons learned looks set to increase dramatically. This applies both to successes and to failures. Advancing this process of knowledge-sharing, to the benefit of all, but especially the most vulnerable of present and future generations, is the *raison d'être* of this book.

This book presents 27 cases, reviewing country experience, regional experience (e.g., the European Union), and international experience/cross-cutting issues, with a focus on the political economy aspects of the clean energy transition.

The book's parts are organized by major political economy subject matter areas germane to characterizing clean energy transitions. While many of the individual chapter topics are cross-cutting in nature, we, the editors, believe this organizing framework to be a useful construct. A short introduction to each part highlights the issues and the main points drawn from the constituent chapters.

[5] Llavador et al. (2015) disagree. They find that global mitigation objectives can only be met through reductions in the rate of growth of GDP.

There are eight major parts.

In Part I, Chapter 1 introduces readers to the layout of the book. Chapter 2 examines the history and politics of energy transitions and draws lessons for today.

Part II features policies designed to advance clean energy and combat climate change from a global or general perspective.

Part III features chapters that explore how institutions and governance influence the processes of energy innovation, deployment, and policy formation.

Part IV features chapters that raise key political questions about the role of actors, interests, and institutions in the energy sector: who has the power to change, who sets the terms of transition, and for whom?

Part V features chapters that explore relationships and tensions between emerging clean energy sectors and incumbent stakeholders.

Part VI features chapters that discuss the drivers, obstacles, and implications of energy sector reforms which shift the balance of public and private participation in clean energy transitions.

Part VII features chapters that explore the role of clean energy, as an enabler of economic growth and development, and social inclusion.

Part VIII features chapters that explore how clean energy transitions challenge traditional national boundaries and differentially impact regions within national boundaries.

1.6 LOOKING FORWARD

A clean energy transition is not easy. This is amply illustrated in the case studies. Even if the technical path is clear and fully illuminated, a clean energy transition will involve the shift of resources between competing economic sectors and political constituencies alongside changes in institutional and policy frameworks. Stakeholders in this process have varying degrees of political and economic power. Regardless of the society or the political system, understanding how political economy factors influence clean energy transitions is crucial to effective policy formulation and facilitating transitions to sustainable energy systems.

Despite the challenges, this introductory chapter has adopted a purposefully optimistic tone. This seems appropriate. CoP21 does represent a substantial break from the past. Technological change in clean energy sectors has been very rapid. Institutional and policy changes are evident in many countries. And, resource allocations are shifting as evidenced by the large investments in clean energy systems that are occurring worldwide. In effect, the set of INDCs derived from CoP21 pledge an essentially global transition towards clean energy systems. Put differently, global mitigation efforts have begun in earnest.

While the first steps have been taken, much more effort is required. Over the next few years, countries need to follow through on their INDCs. Looking further ahead, it is well known that the sum of the commitments in the INDCs does not result in an energy system that is sufficiently environmentally benign as to be compatible with a stable global climate. Even more ambitious commitments/transformations will be necessary in future.

While a freewheeling 'bottom-up' approach appears to have been well suited to getting started, it is likely that limitations to the highly dispersed approach adopted in Paris at CoP21 will become apparent. For example, the solicitation of INDCs is not an approach that is particularly well suited to addressing the vexing and inter-related issues of international trade, carbon trade, and footloose industries/carbon leakage. Thoughts on future stages of the clean energy transition are discussed in Chapter 29. Chapter 29 also provides forward perspectives on the research agenda.

REFERENCES

Arndt, C., M. A. Hussain, E. S. Jones, V. Nhate, F. Tarp, and J. Thurlow (2012). 'Explaining the Evolution of Poverty: The Case of Mozambique'. *American Journal of Agricultural Economics*, 94(4): 854–72.

Arndt, C. and F. Tarp (2015). 'Climate Change Impacts and Adaptations: Lessons Learned from the Greater Zambeze River Valley and Beyond'. *Climatic Change*, 130(1): 1–8.

Bhattacharyya, S. C. (2014). 'Viability of Off-Grid Electricity Supply Using Rice Husk: A Case Study from South Asia'. *Biomass and Bioenergy*, 68: 44–54.

Feldman, D., G. Barbose, R. Margolis, T. James, S. Weaver, N. Darghouth, and R. Wiser (2014). 'Photovoltaic System Pricing Trends: Historical, Recent, and Near-Term Projections'. 2014 Edition. Presentation by SunShot, US Department of Energy. NREL/PR-6A20-62558. National Renewable Energy Laboratory, Golden, CO.

Foster, V. (2008). 'Overhauling the Engine of Growth: Infrastructure in Africa'. African Infrastructure Country Diagnostic. The World Bank, Washington, DC.

Frankel, J. A. (2010). 'The Natural Resource Curse: A Survey' (No. w15836). National Bureau of Economic Research.

Frankfurt School-UNEP (United Nations Environment Programme) Centre/BNEF (Bloomberg New Energy Finance) (2016). 'Global Trends in Renewable Energy Investment'. Frankfurt School of Finance and Management, Frankfurt.

International Energy Agency (2015). 'World Energy Outlook Special Report 2015: Energy and Climate Change'. International Energy Agency, Paris.

IPCC (Intergovernmental Panel on Climate Change) (2013). *Climate Change 2013: The Physical Science Basis. Contribution of Working Group I to the Fifth Assessment*

Report of the Intergovernmental Panel on Climate Change. Cambridge and New York: Cambridge University Press.

IPCC (2014). 'Climate Change 2014: Synthesis Report. Contribution of Working Groups I, II and III to the Fifth Assessment Report of the Intergovernmental Panel on Climate Change'. IPCC, Geneva.

Keane, A., M. Milligan, C. J. Dent, B. Hasche, C. D'Annunzio, K. Dragoon, and M. O'Malley (2011). 'Capacity Value of Wind Power.' *IEEE Transactions on Power Systems*, 26(2): 564–72.

Llavador, H., J. E. Roemer, and J. Silvestre (2015). *Sustainability for a Warming Planet*. Cambridge, MA: Harvard University Press.

Miller, M., E. Martinot, S. Cox, B. Speer, O. Zinaman, S. Booth, R. Zissler, J. Cochran, S. K. Soonee, P. Audinet, L. Munuer, and D. Arent (2015). 'Status Report on Power System Transformation', NREL/TP-6A20-63366. National Renewable Energy Laboratory, Golden, CO.

Moné, C., A. Smith, B. Maples, and M. Hand (2015). '2013 Cost of Wind Energy'. National Renewable Energy Laboratory: Golden CO.

Sawin, J. L., F. Sverrisson, W. Rickerson, C. Lins, L. E. Williamson, R. Adib, and A. Valero (2015). 'Renewables 2015 Global Status Report: Annual Reporting on Renewables: Ten Years of Excellence'. REN Secretariat, Paris.

Stark, C., J. Pless, J. Logan, E. Zhou, and D. J. Arent (2015). 'Renewable Electricity: Insights for the Coming Decade'. Joint Institute for Strategic Energy Analysis, Golden, CO. Available at: <http://www.nrel.gov/docs/fy15osti/63604.pdf> (accessed 12 October 2016).

UNFCCC (United Nations Framework Convention on Climate Change) (2015). 'Historic Paris Agreement on Climate Change: 195 Nations Set Path to Keep Temperature Rise Well Below 2 Degrees Celsius'. Press release on 12 December. <http://newsroom.unfccc.int/unfccc-newsroom/finale-cop21/> (accessed 12 October 2016).

Unruh, G. (2000). 'Understanding Carbon Lock-In'. *Energy Policy*, 28(12): 817–30.

USDOE (United States Department of Energy) (2015). 'Revolution ... Now: The Future Arrives for Five Clean Energy'. United States Department of Energy, Washington, DC.

Winner, L. (1980). 'Do Artifacts Have Politics?' *Daedalus*, 109(1): 121–36.

Zinaman, O., M. Miller, A. Adil, D. J. Arent, J. Cochran, R. Vora, S. Aggarwal, M. Bipath, C. Linvill, A. David, R. Kauffman, M. Futch, E. Villanueva Arcos, J. M. Valenzuela, E. Martinot, M. Bazilian, and R. K. Pillai (2015). 'Power Systems of the Future', NREL/TP-6A20-62611. National Renewable Energy Laboratory, Golden, CO.

2

The History and Politics of Energy Transitions

Comparing Contested Views and Finding Common Ground

Benjamin K. Sovacool

2.1 INTRODUCTION

Transitioning away from our current global energy system is of paramount importance (Riahi et al. 2012: 1203–306). As Grubler (2012: 8) has compellingly written, 'the need for the "next"' energy transition is widely apparent as current energy systems are simply unsustainable on all accounts of social, economic, and environmental criteria'. Miller et al. (2013: 136) add that, 'The future of energy systems is one of the central policy challenges facing industrial countries'. Unfortunately, however, neither private markets nor government agencies seem likely to spur a transition on their own (Fri and Savitz 2014). Moreover, shifts to newer, cleaner energy systems such as sources of renewable electricity (Painuly 2001; Sovacool 2009) and electric vehicles (Sovacool and Hirsh 2009; Nielsen et al. 2015) often require significant changes not only in technology, but also in political regulations, tariffs, and pricing regimes, and the behaviour of users and adopters.

Thus, the speed at which a transition can take place—its timing, or temporal dynamics—is a critical element of consideration (Sovacool 2016). According to the International Energy Agency (2012: 3), for example, if 'action to reduce CO_2 emissions is not taken before 2017, all the allowable CO_2 emissions would be locked-in by energy infrastructure existing at that time'. In other words, if a transition does not occur quickly, or soon, it may be too late. Giddens (2009) went so far as to call this the 'climate paradox', the fact that by the time humanity may come to realize fully how much they need to shift to low-carbon forms of energy, they will have already passed the point of no return.

The notion of 'energy transitions' sits at the heart of this polemical discussion. O'Connor (2010) once defined an energy transition as 'a particularly significant set of changes to the patterns of energy use in a society, potentially affecting resources, carriers, converters, and services'. In other words, to those subscribing to this definition, an energy transition refers to the time that elapses between the introduction of a new primary energy source, or prime mover, and its rise to claiming a substantial share of the overall market. According to one view, energy transitions take an incredibly long time to occur. As the geographer Vaclav Smil (2010a: 141–2) writes, 'all energy transitions have one thing in common: They are prolonged affairs that take decades to accomplish, and the greater the scale of prevailing uses and conversions, the longer the substitutions will take.' Fast transitions, when they occur at all, are anomalies, limited to countries with very small populations or unique contextual circumstances that can hardly be replicated elsewhere.

Another view argues the opposite. Broadening the discussion beyond simply national sources of energy supply and substantial shifts of their composition, it suggests that there have been many transitions—at varying scales, involving different things including fuels, services, and end-use devices—that have occurred quite quickly, that is, between a few years and a decade or so, or within a single generation. At smaller scales, the adoption of cookstoves, air conditioners, and flex-fuel vehicles (FFVs) are excellent examples. At the state or national scale, almost complete transitions to oil in Kuwait, natural gas in the Netherlands, and nuclear power in France took only a decade, roughly, to occur. Indeed, the second part of this chapter presents ten case studies of energy transitions that, in aggregate, affected almost 1 billion people and needed only 1–16 years to unfold. Clearly, this antithetical view proposes that some energy transitions can occur much more quickly than commonly believed.

Which side is right? Similar to other controversies in the energy studies literature (Sovacool et al. 2016), this chapter holds that both are. After presenting evidence in support of both theses, it elucidates a common ground consisting of four arguments. First, sometimes the 'speed' or 'scale' at which an energy transition occurs has less to do with what actually happened and more to do with what or when one counts. Second, what may seem a sweeping transition can actually be a bundle of more discrete minor conversions or substitutions. Third, energy transitions are complex, and irreducible to a single cause, factor, or blueprint. Fourth, most energy transitions have been, and will likely continue to be, path dependent rather than revolutionary, cumulative rather than fully substitutive.

2.2 ONE SIDE: ENERGY TRANSITIONS ARE LONG, PROTRACTED AFFAIRS

This view holds that energy transitions—defined by some as the time that elapses between the introduction of a new fuel or technology (sometimes called a 'prime mover') and its rise to 25 per cent of national market share—takes a significant amount of time (Smil 2010a). The Global Energy Assessment (GEA) (2012: 788), a major international, interdisciplinary effort to understand energy systems, notes that 'transformations in energy systems' are 'long-term change processes' on the scale of decades or even centuries. This view holds that, as two Stanford University scientists write, 'it appears that there is no quick fix; energy system transitions are intrinsically slow' (Myhrvold and Caldeira 2012: 1). Support for this side comes from (1) the historical record, (2) the validity of looking at the 'big picture', and (3) the literature on 'lock-in' and 'path dependency'.

2.2.1 History Shows Major Transitions Taking Decades to Centuries

In the USA, crude oil took half a century from its exploratory stages in the 1860s to capturing 10 per cent of the market in the 1910s, then 30 years more to reach 25 per cent. Natural gas took 70 years to rise from 1 per cent to 20 per cent. Coal needed 103 years to account for only 5 per cent of total energy consumed in the USA and an additional 26 years to reach 25 per cent (Smil 2012). Nuclear electricity took 38 years to reach a 20 per cent share, which occurred in 1995.

Globally, we see even longer time frames involved with energy transitions. Coal surpassed the 25 per cent mark in 1871, more than 500 years after the first commercial coal mines were developed in England. Crude oil surpassed the same mark in 1953; about nine decades after Edwin Drake drilled the first commercial well in Titusville, Pennsylvania, in 1859. Hydroelectricity, natural gas, nuclear power, and 'other' sources such as wind turbines and solar panels *still* have yet to surpass the 25 per cent threshold—as Figure 2.1 depicts—with only nuclear reaching the meagre 5 per cent mark.

Assessing prime movers rather than fuels, Smil (2010b) adds that steam engines were designed in the 1770s, but didn't take off until the 1800s, and the gasoline-powered internal combustion engine, first deployed by Benz, Maybach, and Daimler in the middle of the 1880s, reached widespread acceptance in the USA only in the 1920s, and even later for Europe and Japan. As Smil (2012: 3) deduces from these examples:

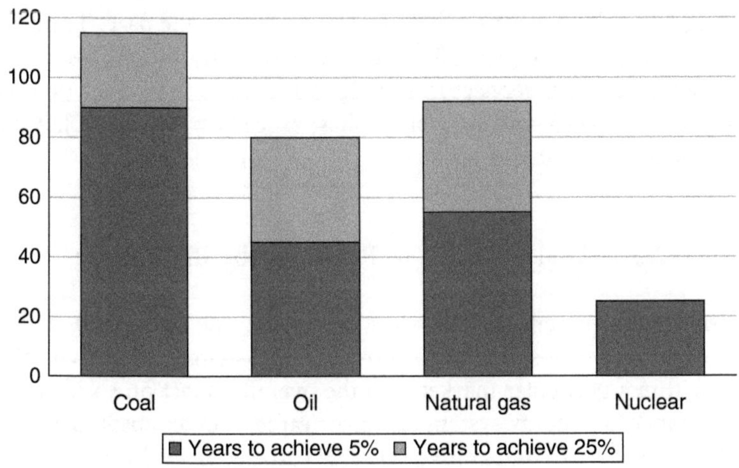

Figure 2.1. Major transitional shifts in global energy supply, 1750–2015.
Source: Author's illustration based on Smil (2012).

> Energy transitions have been, and will continue to be, inherently prolonged affairs, particularly so in large nations whose high levels of per capita energy use and whose massive and expensive infrastructures make it impossible to greatly accelerate their progress even if we were to resort to some highly effective interventions.

This is why he calls energy systems 'a slow-maturing resource' and jokes that 'energy sources, they grow up so...slowly' (Smil 2012: 2–3). As he remarks, 'it is impossible to displace [the world's fossil fuel-based energy] super-system in a decade or two—or five, for that matter. Replacing it with an equally extensive and reliable alternative based on renewable energy flows is a task that will require decades of expensive commitment. It is the work of generations of engineers' (Smil 2012: 3).

The notion that energy transitions are inherently lengthy events finds further support from energy analysts Peter Lund and Roger Fouquet. Lund (2006) found that market penetration of new energy systems or technologies can take as long as 70 years. Short 'take-over times' of less than 25 years are limited to a few end-use technologies such as water heaters or refrigerators, and are not common for major infrastructural systems like those involving electricity or transport. A second study of Lund's (2010a: 650) exploring 'how fast new energy technologies could be introduced on a large scale' estimated that the earliest wind could produce more than 25 per cent of world electricity, and solar 15 per cent, would be 2050—40 years from the date of his study. As Lund (2006: 3318) noted, 'the inertia of energy systems against changes is large, among others because of the long investment cycles of energy infrastructures or production plants' and the 'rate of adoption of these new [renewable energy] technologies would not exceed that of oil or nuclear in the past' (Lund 2010b: 3580). Analogously, Fouquet (2010) studied various transitions between both energy fuels and energy services from 1500 to 1920, and found that, on average, each single transition has an innovation phase exceeding 100 years followed by a diffusion phase approaching 50 years.

2.2.2 Analysts Need to Focus on the Big Picture

Furthermore, proponents of this view argue that one must look at the 'big picture', that is, the absolute change in energy systems, rather than discrete growth within a particular market, and the overall impact on society.

For instance, an energy system can grow rapidly, in an absolute sense, but still fail to grow in a comparative sense. Hydroelectricity in the USA was a low-cost source of energy in the 1950s and 1960s, where it grew in capacity *threefold* from 1949 to 1964. However, during this time, because other sources of energy (and demand for electricity) grew faster, hydropower's overall national share dropped from 32 per cent to 16 per cent (O'Connor 2010). Similarly, from 2000 to 2010, global annual investment in solar photovoltaic (PV) power increased by a factor of 16, investment in wind grew fourfold, investment in solar heating threefold. This sounds impressive—yet the overall contribution of solar (heating and PV) and wind to total final energy consumption grew from less than one-tenth of 1 per cent to slightly less than 1 per cent over the same period (Sovacool 2016), hardly a drop in the bucket.

Furthermore, part of a big picture approach means realizing that energy transitions do not always produce desirable results. The massive energy transitions that occurred in Japan from 1918 to 1945, North Korea in the 1990s, and Cuba in the 1990s saw societies grapple with sudden shifts in the availability of energy. Japan lost upwards of 70 per cent of its oil imports due to the US trade embargo of 1941, North Korea dropped 90 per cent of their oil imports from the Soviet Union in 1991, and Cuba saw a decline of energy imports from the Soviet Union of 71 per cent between 1989 and 1993. In each case, national planners responded to energy scarcity with military force (Japan) or by preserving the privileges of the elite at the expense of ordinary people (North Korea and Cuba) (Friedrichs 2013).

2.2.3 'Path Dependency' and 'Lock-In' Make Future Transitions Difficult

A final thread of this thesis is that desirable energy transitions are so difficult to achieve because of the momentum, path dependency, or obduracy of the existing system exerts on actors. In the case of national energy systems, such large sums of labour, capital, and effort are 'sunk' into them that they create their own 'inertia' (Knox-Hayes 2012; Steinhilber et al. 2013). On top of that, institutional legacies protect the status quo, and political regulations, tax codes, and even banks and educational institutions come to support a particular energy pathway, along with associated coalitions (Goldthau and Sovacool 2012). The result is that energy transitions, breaking out of these embedded systems, require a 'long-term transformation' that is 'a messy, conflictual, and highly disjointed process' (Meadowcroft 2009). Collectively, these technological and behavioural forces 'lock' us into a carbon-dependent energy system that highly resists change (Unruh 2000). In the case of prime movers, we see similar resistance. As Smil (2010a: 140) writes, 'There is often inertial reliance on a machine that may be less efficient (steam engine, gasoline-fueled engine) than a newer machine but whose marketing and servicing are well established and whose performance quirks and weaknesses are known. The concern is that rapid adoption of a superior converter may bring unexpected problems and setbacks.'

In order to counteract this inertia, scholars looking at energy transitions have argued that truly 'transformative change' must be the result of alterations at *every* level of the system, simultaneously, that is, one must alter technologies, political and legal regulations, economies of scale and price signals, and social attitudes and values together, making transition a gruelling process. Or to use parlance from sociotechnical systems theory, it is rare that innovation niches become regimes and rarer still for those regimes to influence the broader, global landscape (Geels and Schot 2007; Schot and Geels 2008). This parallels what feminist scholar, Eve Kosofsky Sedgwick (1993) termed, the 'Christmas Effect' to describe the way that institutions, technology, and behaviour can coalesce around a common goal. During the holidays, the institutions of Western society come together and speak 'with one voice' for the Christmas holiday. Christian churches build nativity scenes and hold a greater number of masses; state and federal governments establish school and national holidays; and the media 'rev up the Christmas frenzy' and 'bark out the Christmas countdown' (Sedgwick 1993). Such sociotechnical inertia favouring the Christmas holiday exerts profound and lasting influence over our behaviour, and the argument runs that a similar alignment of values and incentives occurs with energy. This could be why, in their forecasts about the future, the US EIA (2013) still predicted in 2013 that in 2040, three-quarters of energy in the USA would come from oil, coal, and natural gas. The International Energy Agency (2012: 51) similarly projected that in 2035, under their 'Current Policies' scenario, 80 per cent of total primary energy supply worldwide would come from 'traditional' fossil fuels.

2.3 THE OTHER SIDE: ENERGY TRANSITIONS CAN HAPPEN QUICKLY

Contrary to those emphasizing the longevity or difficulty of energy transitions, an alternate view is that under certain conditions (or, if one chooses to count different things), energy transformations can occur rather speedily. Arguments in support of rapid transitions hold that (1) we have seen numerous fast transitions in terms of energy end-use, (2) plentiful examples of national-scale transitions litter the historical record, and (3) we can sufficiently learn from these trends so that favourable future energy transitions can be expedited. This section of the chapter explores no less than ten 'quick' energy transitions, five of them focused on end-use devices such as lighting and air conditioning, and five of them focused on national systems such as oil in Kuwait and nuclear power in France. Table 2.1 provides an overview of these cases, which collectively involved almost a billion people.

2.3.1 History Shows Speedy Transitions in Energy End-Use Devices

At least five transitions in end-use devices, or prime movers, have occurred with remarkable rapidity: lighting in Sweden, cookstoves in China, liquefied

Table 2.1. Overview of rapid energy transitions

Country	Technology/fuel	Period of transition	Number of years (from 1 to 25 per cent market share)	Approximate size (population affected in millions of people)
Sweden	Energy-efficient ballasts	1991–2000	7	2.3
China	Improved cookstoves	1983–1998	8	592
Indonesia	Liquefied petroleum gas (LPG) stoves	2007–2010	3	216
Brazil	Flex-fuel vehicles (FFVs)	2004–2009	1	2
USA	Air conditioning	1947–1970	16	52.8
Kuwait	Crude oil	1946–1955	2	0.28
Netherlands	Natural gas	1959–1971	10	11.5
France	Nuclear electricity	1974–1982	11	72.8
Denmark	Combined heat and power (CHP)	1976–1981	3	5.1
Canada (Ontario)*	Coal	2003–2014	11	13

Note: * The Ontario case study is the inverse, showing how quickly a province went from 25 per cent coal generation to zero.

Source: Author's compilation.

petroleum gas (LPG) stoves in Indonesia, ethanol vehicles in Brazil, and air conditioning in the USA.

Sweden was able to phase in an almost complete shift to energy-efficient lighting in commercial buildings in about nine years (Lund 2007). Swedish Energy Authorities arranged for the procurement of high-frequency electronic ballasts for lights in office buildings, commercial enterprises, schools, and hospitals, devices which saved 30–70 per cent compared to ordinary ballasts, in 1991. They used a multipronged approach of standardization and quality assurance, direct procurement, stakeholder involvement, and demonstrations to disseminate those ballasts. They began by collaborating with experts to develop a list of lighting quality factors for commercial buildings, and then asked for competitive tenders from manufacturers that met these standards. Then, the government directly purchased almost 30,000 units in a pilot phase, and worked with real estate management companies (for new buildings) and owners of public, commercial, and industrial buildings (for retrofits) to ensure that they were installed (Ottossen and Stillesjo 1996). After the pilot phase, they promoted distribution through government subsidies, sponsored demonstrations of the technology among the commercial sector, and involved consumer groups in discounted bulk purchases. Due to these concerted efforts, self-supporting volume effects were reached as early as 1996, catalysing very rapid market penetration, which jumped from about 10 per cent that year to almost 70 per cent by 2000. In essence, this meant that between 1991 and 2000, 2.3 million Swedish workers experienced changes in their office lighting.

The Chinese Ministry of Agriculture managed an even more impressive National Improved Stove Programme (NISP), managed by the Bureau of Environmental Protection and Energy (BEPE), from 1983 to 1998 (Smith et al. 1993; Brown and Sovacool 2011a: 292–301). The BEPE adopted a 'self-building, self-managing, self-using' policy focused on having rural people themselves invent, distribute, and care for energy-efficient cookstoves, and it set up pilot programmes in hundreds of rural provinces. From the start of the programme until 1998, the NISP was responsible for the installation of 185 million improved cookstoves and facilitated the penetration of improved stoves from less than 1 per cent of the Chinese market in 1982 to more than 80 per cent by 1998—reaching half a billion people, as Table 2.2 shows. The cookstoves being installed in China in 1994, during the height of the programme, were equivalent to 90 per cent of all improved stoves installed globally. As a consequence, Chinese energy use per capita declined in rural areas at an annual rate of 5.6 per cent from 1983 to 1990.

Indonesia also ran a large household programme focusing on the conversion from kerosene stoves to LPG stoves to improve air quality. Under leadership from their vice president, Jusuf Kalla, the Indonesian 'LPG Mega-project' offered households the right to receive a free 'initial package' consisting of a 3 kilogram LPG cylinder, a first free gas-fill, one burner stove, a hose,

Table 2.2. Households adopting improved stoves under the Chinese National Improved Stove Programme (NISP) and affiliated provincial programmes

	NISP households (million)	Households under provincial programmes (million)	Total households/ year (million)	Total people/ year (million)
1983	2.6	4.0	6.6	21.1
1984	11.0	9.7	20.7	66.2
1985	8.4	9.5	17.9	57.3
1986	9.9	8.5	18.4	58.9
1987	8.9	9.1	18.0	57.6
1988	10.0	7.5	17.5	56.0
1989	4.5	5.0	9.5	30.4
1990	3.6	7.8	11.4	36.5
1991–1998	7.8	57.2	65.0	208.0
Total	66.7	118.3	185.0	592.0

Source: Author's compilation based on Brown and Sovacool (2011a).

and a regulator. The government, in tandem, lowered kerosene subsidies (increasing its price) and constructed new refrigerated LPG terminals to act as national distribution hubs. Amazingly, in just three years from 2007 to 2009, the number of LPG stoves nationwide jumped from a mere 3 million to 43.3 million, meaning they served almost two-thirds of Indonesia's 65 million households (or about 216 million people). Six entire provinces, including that of Jakarta, the capital, were declared 'closed and dry'—meaning that the programme reached all of its targets, and that all kerosene subsidies were withdrawn (Budya and Arofat 2011).

Brazil has perhaps the fastest energy transition on record, though (to be fair) it depends on what one counts. Brazil created its Proálcool programme in November 1975 to increase ethanol production and substitute ethanol for petroleum in conventional vehicles, and in 1981, six years later, 90 per cent of all new vehicles sold in Brazil could run on ethanol—an impressive feat. However, a more recent transition, connected in part to the Proálcool programme, is even more noteworthy. The Brazilian government started incentivizing FFVs in 2003 through reduced tax rates and fuel taxes. These Brazilian FFVs were capable of running on any blend of ethanol from zero to 100 per cent, giving drivers the option of switching between various blends of gasoline and ethanol depending on price and convenience. The first year FFVs entered the market in 2004, they accounted for 17 per cent of new car sales but they rapidly jumped to 90 per cent in 2009—meaning 2 million FFVs were purchased in total over the first five years of the programme (Brown and Sovacool 2011b).

Air conditioning in the USA is a final example. In 1947, mass-produced, low-cost window air conditioners became possible, enabling many people to enjoy air conditioning without the need to buy a new home or completely renovate their heating system (National Academy of Engineering 2013). That year, only 43,000 units were sold, but by 1953, the number had jumped to 1 million, as air conditioners became endorsed by builders eager to mass produce affordable, yet desirable, modern homes and electric utilities that wanted to increase electricity consumption throughout the growing suburbs (Rosen 2011). Consequently, more than 12 per cent of people (occupying 6.5 million housing units) reported to the US Census in 1960 that they owned an air conditioner, rising to 25 per cent in 1963, and 35.8 per cent in 1970, representing 24.2 million homes and more than 50 million people (US Census Bureau 1960, 1970). Since then, the presence of air conditioning in single-family homes jumped from 49 per cent in 1973 to 87 per cent in 2009 (US EIA 2011). In hot and humid places such as southern Florida, its use grew from 5 per cent in 1950 to 95 per cent in 1990. American motorists also use 7–10 billion gallons of gas annually to air condition their cars. In aggregate, the USA on an annual basis now consumes more electricity for air conditioning than the entire continent of Africa consumes for all electricity uses (Cox 2012). Or, in other terms, the USA currently utilizes more energy (about 185 billion kWh) for air-conditioning than all other countries' air conditioning usage combined (Sivak 2013).

2.3.2 Fast Transitions in National Energy Supply Have Occurred

Proponents of this alternative view can also point to five other transitions that have occurred at the national level: to crude oil in Kuwait, natural gas in the Netherlands, nuclear electricity in France, combined heat and power (CHP) in Denmark, and coal retirements in Ontario, Canada.

Two concurrent modifications, in electricity and transport, catalysed an almost complete shift in Kuwait's national energy profile in about nine years. Oil use catapulted from constituting a negligible amount of total national energy supply in 1946 to 25 per cent in 1947, and more than 90 per cent in 1950 (Kuwait Ministry of Planning 1988). In 1938, when Kuwait was still a small, impoverished British protectorate, geologists discovered the Burgan oil field, which proved to be the world's second largest accumulation of oil following Saudi Arabia's Ghawar oil field. Commercial exploitation began in earnest (after a suspension of operations due to the Second World War) in 1946, increasing from 5.9 million barrels that year to 16.2 million barrels in 1947, and 398.5 million barrels in 1955, in tandem with the development of other oil fields (Al-Marafie 1989). Within five years, 1945–49, the Kuwaiti oil industry was transformed from one dependent on five gallon barrels being distributed manually to customers, carried on camels, donkeys, or wooden push carts to one characterized by huge volumes and scale economies that were dependent on motorized trucks and tankers, pipelines, and filling stations. Simultaneously, Kuwait began using oil for electricity generation. The Kuwait Oil Company obtained and commissioned its first 500 kW generator in 1951 and in 1952, built a 2.25 MW steam power station at Al-Shewaikh, essentially tripling national electricity capacity in three years. Demand for such electricity grew considerably, doubling again by 1960, and then increasing (in per capita terms) from 1473 kWh to 9255 kWh in 1985 (Al-Marafie 1988). Thereafter, a rapid expansion of distillation units, refineries, petrol stations, and the establishment of the Kuwait National Petroleum Company in 1960, the same year Kuwait helped form the Organization of Petroleum Exporting Countries, saw oil's rise continue so that in 1965, Kuwait became the world's fourth largest producer of oil (behind the USA, the Soviet Union, and Venezuela, and ahead of Saudi Arabia). As even energy transition sceptic Smil (2010b: 55) concedes, 'In energy terms Kuwait thus moved from a pre-modern society dependent on imports of wood, charcoal, and kerosene to an oil superpower in a single generation'.

The Netherlands—thanks in large part to the discovery of a giant Groningen natural gas field in 1959—started a rapid transition away from oil and coal to natural gas (Smil 2010b). That year, coal supplied about 55 per cent of Dutch primary energy supply followed by crude oil at 43 per cent and natural gas at less than 2 per cent. In December 1965, however, one year after gas deliveries began from Groningen, natural gas supplied 5 per cent of the Netherland's primary energy, rising quickly to 50 per cent by 1971, an ascent visually depicted in Figure 2.2. To facilitate the transition, the government decided in December 1965 to abandon all coal mining in the Limburg province within a decade, doing away with some 75,000 mining-related jobs impacting more than 200,000 people. What made the transition successful was

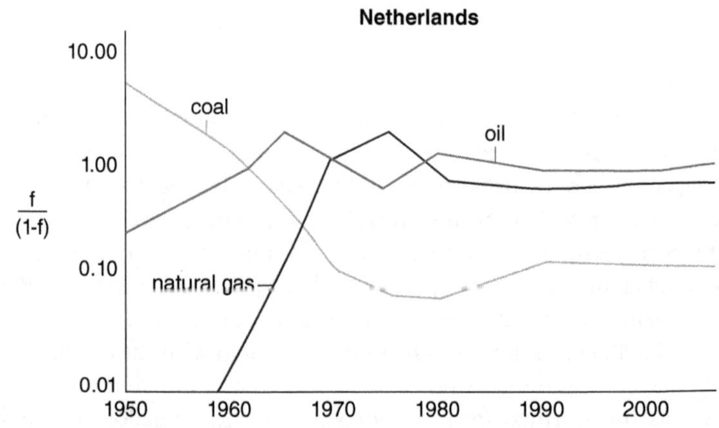

Figure 2.2. Coal, natural gas, and oil supply in the Netherlands, 1950–2010.
Source: Author's illustration based on Smil (2010a).

that the government strategically implemented countermeasures such as subsidies for new industries, the relocation of government industries from the capital to regions of the country hardest hit by the mine closures, retraining programmes for miners, and offering shares in Groningen to Staatsmijnen (the state mining company). After its peak output in the mid-1970s, extraction of gas at Groningen was purposely scaled back to maximize the lifetime of the field, though natural gas continued to play a prominent role in the nation's energy mix. In 2010, for instance, natural gas still provided 45 per cent of total primary energy supply, larger than any other source (EC 2010).

The French transition to nuclear power was also swift. Following the oil crisis in 1974, Prime Minister Pierre Messmer announced a large nuclear power programme intended to generate all of France's electricity from nuclear reactors to displace the Republic's heavy dependence on imported oil. As the maxim went at the time, 'No coal, no oil, no gas, no choice!' The 'Messmer Plan' proposed the construction of 80 nuclear power plants by 1985 and 170 plants by 2000. Work commenced on three plants—Tricastin, Gravelines, and Dampierre—immediately following the announcement of the plan and France ended up constructing 56 reactors in the period 1974–89. As a result, nuclear power grew from 4 per cent of national electricity supply in 1970 to 10 per cent in 1978 and almost 40 per cent by 1982 (Araujo 2013). As Grubler (2010: 5186) has noted, 'the reasons for this success lay in a unique institutional setting allowing centralized decision-making, regulatory stability, dedicated efforts for standardized reactor designs and a powerful nationalized utility, EDF, whose substantial in-house engineering resources enabled it to act as principal and agent of reactor construction simultaneously'.

Though Denmark is perhaps more famous for a transition to wind energy, a far more accelerated transition occurred in the 1970s and 1980s. This transition, also partially in the electricity sector, was away from oil-fired electricity to other fossil fuels and CHP plants. From 1955 to 1974, almost all heating in Denmark was provided by fuel oil, which meant the oil crisis had particularly painful impacts on the country's economy (Sovacool 2013). The Danish Energy Policy of 1976 therefore articulated the short-term goal of reducing oil dependence, and it stated the importance of building a 'diversified supply system' and meeting two-thirds of total heat consumption with 'collective heat supply' by 2002. Moreover, it sought to reduce oil dependence to 20 per cent, an ambitious goal that involved the conversion of 800,000 individual oil boilers from natural gas and coal. In a mere five years, 1976–81, Danish electricity production changed from 90 per cent oil-based to 95 per cent natural gas- and coal-based. Stipulations in favour of CHP were further strengthened by the 1979 Heat Supply Act, whose purpose was to 'promote the best national economic use of energy for heated buildings and supplying them with hot water and to reduce the country's dependence on mineral oil' (Sovacool 2013: 833). As a result, CHP production increased from trivial amounts in 1970 to supply 61 per cent of national electricity and 77 per cent of the country's district heating in 2010.

A final example is intriguing because rather than transitioning towards something it involves transitioning away. In 2003, the government of Ontario in Canada committed to retiring all coal-fired electricity generation by 2007, something they accomplished a few years behind schedule. Ontario's oldest coal plant, the 1140 MW Lakeview facility, was closed in April 2005 followed by sequential closures of Thunder Bay (306 MW), Atikokan (211 MW), Lambton (1972 MW), and Nanticoke (3945 MW) from 2007 to 2014. Coal generation thus declined from 25 per cent of provincial supply in 2003 to 15 per cent in 2008, 3 per cent in 2011, and zero in 2014, as Figure 2.3 illustrates. The primary justification for the closure, apart from its obvious climate change benefits, was public health. A government study estimated that shifting away from coal would reduce some 333,660 related illnesses and more than 700 deaths related to coal pollution to fewer than 6 deaths and only 2,460 illnesses. Put into monetary terms, the 'coal switch' was estimated to save US$4.4 billion per year in health, environmental, and financial damages along with US$95 million in displaced operating and maintenance costs (Office of the Premier 2013). To achieve this transition, Ontario invested more than US$21 billion in cleaner sources of energy including wind, hydroelectricity, solar, and nuclear power, as well as US$11 billion in transmission and distribution upgrades (Toronto Ministry of Energy 2013). Ontario is on track to see renewable sources of electricity grow to 46 per cent of supply by 2025, and typical residential customers are expected to save US$520 on their bills, and large industrial customers to save US$3 million each on their bills, from 2013 to 2017.

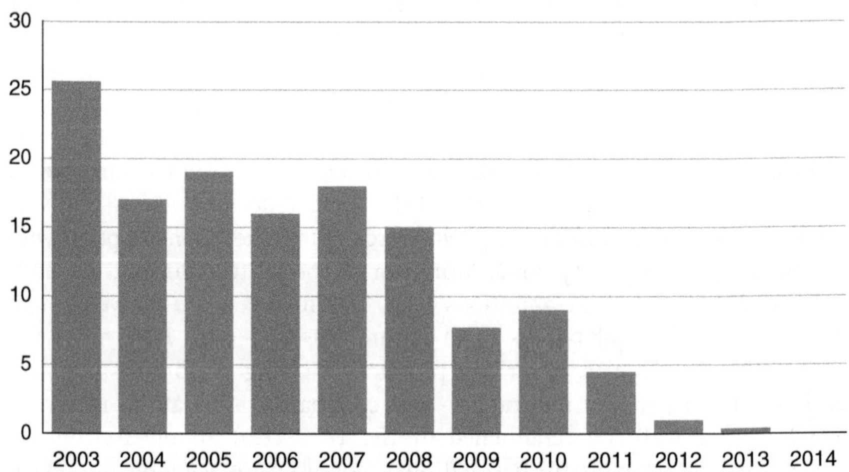

Figure 2.3. Generation from coal in Ontario from 2003 to 2014 (in %).
Source: Author's illustration based on the Office of the Premier (2013).

2.3.3 Future Energy Transitions Can Be Expedited

The final argument in favour of expeditious transition is that although some may have taken a great deal of time, we have learned a sufficient amount from them so that contemporary, or future, energy transitions can be expedited. As Araujo (2013: 12) has argued:

> Countries can, in fact, alter their energy balance in a significant way—stressing low carbon energy sources—in much less time than many decision-makers might imagine. Critical substitution shifts within [Brazil, France, Denmark, and Iceland] were accomplished often in less than 15 years. Moreover, these transitions were effectuated even amidst circumstances at times involving highly complex energy technologies.

Former Vice President Al Gore encapsulated this type of thinking when he argued, in 2008, that 'today I challenge our nation to commit to producing 100 percent of our electricity from renewable energy and truly clean carbon-free sources within 10 years' (Revkin 2008: 2). Gore went on to say that a complete change in energy production was 'achievable, affordable and transformative' within the course of one decade.

Therefore, it is possible that innovations in both technology and policy design can accelerate technological change, and achieve an energy transition, in ways not possible even just a few decades ago. Put another way, technological learning and innovation can result in new technologies and systems with the potential for exponential growth. Previous transitions such as that from wood to coal or coal to oil occurred without the accumulation of knowledge we have currently about the sociology, politics, and economics of energy transitions, that is, without the complex historical analyses conducted by the likes of those such as Smil, Lund, and Fouquet arguing for caution. Because we now possess this knowledge, we can apply it going forward to minimize the unnecessary lag or delay of a future energy transition. Even Fouquet and Pearson (2012: 2) write that 'past energy transitions may not be the best analogies for a future low carbon energy transition'.

2.4 CONCLUSIONS: ENERGY TRANSITIONS ARE PATH DEPENDENT AND CUMULATIVE

How can these two almost incommensurable views be reconciled? This final part of the chapter offers four synthetic conclusions.

First, sometimes the 'speed' or 'scale' at which an energy transition occurs has less to do with what actually happened and more to do with what or when one counts. The American transition to oil, according to Smil, took about 80 years to reach a 25 per cent share, yet during the most accelerated phase of that transition—from 1900 to 1925—oil grew from 2.4 per cent to 24 per cent, justifying those who would call it 'quick' (Pratt 1981: 9–34). For air conditioning, whether one takes the time of first conception (Nikola Tesla developed electric motors that made possible the invention of oscillating fans in 1885), first invention (Willis Carrier invented the first modern system in 1902), or first successful commercial application (when Henry Galson developed an affordable mass produced system in 1947) greatly alters the perceived rate of market penetration (Oremus 2013). Brazil's transition to FFVs, arguably, took one year (from the start of the national programme to large-scale diffusion), more than twenty years (from the first invention of a FFV in 1980), almost thirty years (from the start of their national ethanol programme), or more than eight decades (from the first invention of a Brazilian engine capable of using ethanol in the 1920s).

In the case of national transitions, we see similar ambiguity. Kuwait's transition to oil can be said to have begun in 1934, with the first concession given to the Kuwait Oil Company; or in 1937, when the first exploratory wells were drilled in the Burgan field; or in 1946, when commercial production began (the starting point taken here); or even in 1949, when the first refinery was established. Similarly the French nuclear power programme could have defensibly begun in 1942 with the first chain reaction under the Manhattan Project; or in 1945, with the formation of the Commissariat à l'Énergie Atomique; or in 1948, when their first research reactor was commissioned; or in 1974, with the launch of the Messmer Plan (or with the 1938 experiments by Lise Meitner and Otto Hahn, or even earlier!). Deciding what one counts includes within it normative assumptions about what an energy transition is; the problem is that most analysts do not make these assumptions transparent.

Second, what may seem a sweeping transition can actually be a bundle of more discrete conversions. As O'Connor (2010: 34) concludes, 'Big transitions are the sum of many small ones. Looking at overall energy consumption will miss the small-scale changes that are the foundation of the transitions.' This is because sometimes choices underlying technology adoption and fuel choice occur in technology-fuel bundles—shifts to facilitate use of higher quality or more advanced forms of the same or substitute technology. For instance, the big ascent of oil discussed at the start of this chapter can also be interpreted as a series of less grand changes involving:

- The switch from animal power to internal combustion engines for private vehicles, and the non-adoption of electric vehicles.
- The conversion of steam engines on ships and locomotives to diesel for marine vessels and trains.
- The shift from candles and kerosene for lighting to oil-based lamps.

- The adaptation of coal boilers to oil boilers for the generation of electric power.
- The exchange of wooden fireplaces and coal stoves to oil and gas furnaces in homes.

Similarly, the air conditioning transition in the USA was actually the result of concurrent innovations in air circulation, heat exchangers, heat pumps, halocarbon refrigerants, customization and mass production, and marketing (National Academy of Engineering 2013). It is often these 'minor transitions' that, when they occur in a concerted manner, create the 'major transitions' that are so easily identifiable.

Third, energy transitions are complex, and irreducible to a single cause, factor, or blueprint. They can be influenced by endogenous factors within a country, such as aggressive planning from stakeholders in France (Messmer Plan), Denmark (regulations for district heating), Indonesia (a high-level initiative backed by the prime minister's office), or Ontario (city and municipal planners), accelerated by political will and stakeholder involvement, or exogenous factors outside of a country, such as military conflict (think of the world wars spawning the French nuclear programme), a major energy accident (Chernobyl, Fukushima), or some global crisis (the oil shocks of the 1970s, the collapse of communism in the early 1990s). Other transitions, such as the adoption of air conditioning or ascendance of oil in Kuwait, were almost entirely market-driven.

The implication is that, apart from transitions directly steered by strong policy programmes in choreographing the diffusion of technologies, most energy transitions have no magic formula. The UK, for instance, had the same access to natural gas that the Netherlands did, yet it was unable to cultivate the same type of changeover (Smil 2010b). The experience of tiny, affluent countries such as Denmark and Kuwait may be relevant for countries in a similar class (such as Belgium, Brunei, and Qatar), but less so for India or Nigeria. Moreover, the sociocultural or political conditions behind transitions in Brazil and China, at the time military dictatorships and communist regimes (respectively), are incompatible with the governance norms espoused in modern democracies across Europe and North America. Furthermore, history seems to suggest that past transitions—including many of the case studies presented here—are based on either discoveries of new, significant, and affordable forms of energy or technology, or scarcity conditions created by oil embargos or rapid political revolutions (Pratt 1981; Friedrichs 2013). Both abundance and scarcity have played, and will continue to play, an important role in future transition trajectories.

Fourth, and lastly, is that given these attributes of specificity, accretion, and complexity, most energy transitions have been, and will likely continue to be, path dependent rather than revolutionary, cumulative rather than fully substitutive. Older sources of energy—such as muscle power, animate power, wood power, and steam power—still remain in use throughout the world today; they have not entirely been replaced by fossil, nuclear, and modern renewable energy. One analyst at Massachusetts Institute of Technology (MIT) recently commented that, 'we'll use renewable energy more as technology makes it cheaper, but we're likely to keep using more of the other sources of energy, too' (Bullis 2013: 4). The motorized automobile behind (in part) the transition to oil in Kuwait and FFVs in Brazil is actually an amalgamation of earlier inventions fused together: the internal combustion engine, the wheel, the casting of steel, electric lights, tyres, the assembly line, and so on. The CHP, geothermal, biomass, wind, and solar technology behind the transitions in Denmark, and Ontario have benefited from advances in the fossil-fuel chain including combined cycle turbines, batteries, and compressed air energy storage. Thus, transitions often appear not as an exponential line on a graph, but as a punctuated equilibrium, which dips and rises. To borrow from Strunz (2014), such dips and rises could even be interpreted as forms of highly nonlinear pathways, such as switching from one equilibrium to another.

Perhaps future energy transitions, because they can draw on synergistic advances in multiple domains at once—cutting across materials, computing, combustion, gasification, nanotechnology, biological and genetic engineering, three-dimensional (3D) printing, and the Industrial Internet—can truly be accelerated in ways that past transitions have (generally) not been, even if it is the twin (and contradictory) factors of abundance and scarcity driving them. In addition, past interpretations of the pace of energy transitions, once reconciled, should affect how we view the prospects for conceiving and enabling a large-scale, low-carbon energy transition on a global scale.

REFERENCES

Al-Marafie, A. M. R. (1988). 'Assessment and Prospects for Energy Resources in Kuwait'. *Energy*, 13(8): 647–56.

Al-Marafie, A. M. R. (1989). 'Assessment of Oil and Natural Gas Resources in Kuwait'. *Energy*, 14(1): 1–7.

Araujo, K. M. (2013). 'Energy at the Frontier: Low-Carbon Energy System Transitions and Innovation in Four Prime Mover Countries'. PhD dissertation. Cambridge, MA: MIT.

Brown, M. A. and B. K. Sovacool (2011a). 'China's National Improved Stove Program, 1983–1998'. In M. A. Brown and B. K. Sovacool (eds), *Climate Change and Global Energy Security: Technology and Policy Options*. Cambridge, MA: MIT Press, pp. 292–301.

Brown, M. A. and B. K. Sovacool (2011b). 'Brazil's Proalcohol Program and Promotion of Flex-Fuel Vehicles'. In M. A. Brown and B. K. Sovacool (eds), *Climate Change and Global Energy Security: Technology and Policy Options*. Cambridge, MA: MIT Press, pp. 260–74.

Budya, H. and M. Y. Arofat (2011). 'Providing Cleaner Energy Access in Indonesia through the Megaproject of Kerosene Conversion to LPG'. *Energy Policy*, 39: 7575–86.

Bullis, K. (2013). 'How Energy Consumption Has Changed since 1776'. *MIT Technology Review*. Available at: <http://www.technologyreview.com/view/516786/how-energy-consumption-has-changed-since-1776/> (accessed 11 March 2016).

Cox, S. (2012). 'Climate Risks Heat Up as World Switches on to Air Conditioning'. *The Guardian*, 10 July.

European Commission (EC) (2010). *Netherlands: Energy Mix Fact Sheet*. Brussels: European Commission.

Fouquet, R. (2010). 'The Slow Search for Solutions: Lessons from Historical Energy Transitions by Sector and Service'. *Energy Policy*, 38(11): 6586–96.

Fouquet, R. and P. J. G. Pearson (2012). 'Past and Prospective Energy Transitions: Insights from History'. *Energy Policy*, 50: 1–7.

Fri, R. W. and M. L. Savitz (2014). 'Rethinking Energy Innovation and Social Science'. *Energy Research & Social Science*, 1: 183–7.

Friedrichs, J. (2013). *The Future Is Not What It Used to Be: Climate Change and Energy Scarcity*. Cambridge, MA: MIT Press.

Geels, F. W. and J. W. Schot (2007). 'Typology of Sociotechnical Transition Pathways'. *Research Policy*, 36: 399–417.

Giddens, A. (2009). *The Politics of Climate Change*. New York: Polity.

Global Energy Assessment (GEA) (2012). *Global Energy Assessment—Toward a Sustainable Future*. Cambridge and New York: Cambridge University Press.

Goldthau, A. and B. K. Sovacool (2012). 'The Uniqueness of the Energy Security, Justice, and Governance Problem'. *Energy Policy*, 41: 232–40.

Grubler, A. (2010). 'The Costs of the French Nuclear Scale-up: A Case of Negative Learning by Doing'. *Energy Policy*, 38: 5174–88.

Grubler, A. (2012). 'Energy Transitions Research Insights and Cautionary Tales'. *Energy Policy*, 50: 8–18.

International Energy Agency (2012). *World Energy Outlook*. Paris: OECD.

Knox-Hayes, J. (2012). 'Negotiating Climate Legislation: Policy Path Dependence and Coalition Stabilization'. *Regulation & Governance*, 6(4): 545–67.

Kuwait Ministry of Planning (1988). *Statistical Review, 1950–1970*. Kuwait City: Kuwait Ministry of Planning, Statistics and Census Sector.

Lund, P. (2006). 'Market Penetration Rates of New Energy Technologies'. *Energy Policy*, 34: 3317–26.

Lund, P. (2007). 'Effectiveness of Policy Measures in Transforming the Energy System'. *Energy Policy*, 35: 627–39.

Lund, P. (2010a). 'Exploring Past Energy Changes and Their Implications for the Pace of Penetration of New Energy Technologies'. *Energy*, 35: 647–56.

Lund, P. (2010b). 'Fast Market Penetration of Energy Technologies in Retrospect with Application to Clean Energy Futures'. *Applied Energy*, 87: 3575–83.

Meadowcroft, J. (2009). 'What about the Politics? Sustainable Development, Transition Management, and Long-Term Energy Transitions'. *Policy Sciences*, 42(4): 323–40.

Miller, C. A., A. Iles, and C. F. Jones (2013). 'The Social Dimensions of Energy Transitions'. *Science as Culture*, 22(2): 135–48.

Myhrvold, N. P. and K. Caldeira (2012). 'Greenhouse Gases, Climate Change and the Transition from Coal to Low-Carbon Electricity'. *Environ. Res. Lett.*, 7 014019.

National Academy of Engineering (2013). 'Air Conditioning and Refrigeration Timeline'. Washington, DC: National Academies.

Nielsen, J. R., H. Hovmøller, P.-L. Blyth, and B. Sovacool (2015). 'Of "White Crows" and "Cash Savers": A Qualitative Study of Travel Behavior and Perceptions of Ridesharing in Denmark'. *Transportation Research*, Part A, 78: 113–23.

O'Connor, P. A. (2010). 'Energy Transitions'. The Pardee Papers, No. 12. Boston: Boston University, The Frederick S. Pardee Center for the Study of the Longer-Range Future. Available at: <https://www.bu.edu/pardee/files/2010/11/12-PP-Nov2010.pdf> (accessed 11 March 2016).

Office of the Premier (2013). 'Ontario – First Place in North America to End Coal-Fired Power'. Available at: <http://news.ontario.ca/opo/en/2013/11/ontario—first-place-in-north-america-to-end-coal-fired-power.html> (accessed 11 March 2016).

Oremus, W. (2013). 'A History of Air Conditioning'. *Slate Magazine*, 15 July.

Ottossen, A. and S. Stillesjo (1996). 'Procurement and Demonstration of Lighting Technologies for the Efficient Use of Electricity'. Stockholm: Royal Institute of Technology and NUTEK.

Painuly, J. P. (2001). 'Barriers to Renewable Energy Penetration: A Framework for Analysis'. *Renewable Energy*, 24: 73–89.

Pratt, J. A. (1981). 'The Ascent of Oil: The Transition from Coal to Oil in Early Twentieth-Century America'. In L. J. Perelman, A. W. Giebelhaus, and M. D. Yokel (eds), *Energy Transitions: Long-Term Perspectives*. Boulder, CO: AAAS.

Revkin, A. (2008). 'The Annotated Gore Energy Speech'. *New York Times*, 17 July.

Riahi, K., F. Dentener, D. Gielen, A. Grubler, Z. Klimont, V. Krey, and D. L. McCollum (2012). 'Energy Pathways for Sustainable Development'. In T. B. Johansson, N. Nakicenovic, A. Patwardan, and L. Gomez-Echeveri (eds), *Global Energy Assessment: Toward a More Sustainable Future*. Cambridge and New York: Cambridge University Press.

Rosen, R. A. (2011). 'Keepin' it Cool: How the Air Conditioner Made Modern America'. *The Atlantix*, 14 July.

Schot, J. W. and F. W. Geels (2008). 'Strategic Niche Management and Sustainable Innovation Journeys: Theory, Findings, Research Agenda, and Policy'. *Technology Analysis and Strategic Management*, 20(5): 537–54.

Sedgwick, E. K. (1993). *Tendencies*. New York: Duke University Press.

Sivak, M. (2013). 'Will AC put a Chill on the Global Energy Supply?' *American Scientist* (September/October).

Smil, V. (2010a). *Energy Myths and Realities: Bringing Science to the Energy Policy Debate*. Washington, DC: Rowman and Littlefield.

Smil, V. (2010b). *Energy Transitions: History, Requirements, Prospects*. Santa Barbara, CA: Praegar.

Smil, V. (2012). 'A Skeptic Looks at Alternative Energy'. *IEEE Spectrum*. Available at: <http://spectrum.ieee.org/energy/renewables/a-skeptic-looks-at-alternative-energy> (accessed 11 March 2016).

Smith, K. R., G. Shuhua, H. Kun, and Q. Daxiong (1993). 'One Hundred Million Improved Cookstoves in China: How Was It Done?' *World Development*, 21(6): 941–61.

Sovacool, B. K. (2009). 'Rejecting Renewables: The Socio-Technical Impediments to Renewable Electricity in the United States'. *Energy Policy*, 37(11): 4500–13.

Sovacool, B. K. (2013). 'Energy Policymaking in Denmark: Implications for Global Energy Security and Sustainability'. *Energy Policy*, 61: 829–41.

Sovacool, B. K. (2016). 'How Long Will It Take? Conceptualizing the Temporal Dynamics of Energy Transitions'. *Energy Research & Social Science*, 13: 202–15.

Sovacool, B. K., M. A. Brown, and S. V. Valentine (2016). *Fact and Fiction in Global Energy Policy: Fifteen Contentious Questions*. Baltimore, MD: Johns Hopkins University Press.

Sovacool, B. K., and R. F. Hirsh (2009). 'Beyond Batteries: An Examination of the Benefits and Barriers to Plug-In Hybrid Electric Vehicles (PHEVs) and a Vehicle-to-Grid (V2G) Transition'. *Energy Policy*, 1095–103.

Steinhilber, S., P. Wells, and S. Thankappan (2013). 'Socio-Technical Inertia: Understanding the Barriers to Electric Vehicles'. *Energy Policy*, 60: 5319.

Strunz, S. (2014). 'The German Energy Transition as a Regime Shift'. *Ecological Economics*, 100: 150–8.

Toronto Ministry of Energy (2013). 'Achieving Balance: Ontario's Long-Term Energy Plan'. Available at: <http://www.energy.gov.on.ca/en/ltep/> (accessed 11 March 2016).

Unruh, G. C. (2000). 'Understanding Carbon Lock-In'. *Energy Policy*, 28: 817–30.

US Census Bureau (1960). *US Census of Population and Housing 1960*, Vol. 1. Washington, DC: United States Census Bureau.

US Census Bureau (1970). *US Census of Housing 1970*, Vol. 1. Washington, DC: United States Census Bureau.

US Energy Information Administration (EIA) (2011). 'Air Conditioning in Nearly 100 Million U.S. Homes'. Available at: <http://www.eia.gov/consumption/residential/reports/2009/air-conditioning.cfm> (accessed 11 March 2016).

US Energy Information Administration (EIA) (2013). *Annual Energy Outlook*. Washington, DC: Department of Energy.

Part II

Climate Policy

This part of the book offers unique insights into the complex terrain of climate and clean energy policies. While many prior works have addressed the efficacy of carbon-pricing via either tax or cap-and-trade mechanisms and evaluated sectoral, national, and cross-border implications, this new work adds insights on three particular aspects. Recognizing that the transitions toward clean energy futures are diverse, and a multitude of policy approaches have been implemented in the past decades and that we will likely continue to experience a broad range of approaches, including many that incorporate carbon pricing, and some that may not, this part enhances our understanding of the dynamics behind climate-related policy choices and trade-offs.

The chapters in this part capture the essence of this topic through thoughtful illustration of political constraints, international/cross-border issues, and the complexities of renewable energy policy within the broader climate policy portfolio. In particular, Jenkins and Karplus evaluate climate policy in the face of binding political constraints. They employ a stylized model of the energy sector and offer intuition about the welfare-maximizing combination of CO_2 price, subsidy for clean energy production, and lump-sum transfers to energy consumers or producers under each constraint. Secondly, Bueb, Hanani, and Le Clézio offer new insights related to cross-border issues and outline, especially in how border carbon adjustments (BCAs) may be used as an engine of economic and energy transition, for developed and developing countries equally. Finally, Gawel, Strunz, and Lehmann evaluate the challenges of realizing critical steps in implementing a multifaceted approach to climate policy and how clean energy policies have been considered as part of a policy portfolio.

3

Carbon Pricing under Political Constraints

Insights for Accelerating Clean Energy Transitions

Jesse D. Jenkins and Valerie J. Karplus

3.1 INTRODUCTION

For decades, the economically efficient prescription for the severe consequences of global climate change has been clear: establish a price on emissions of carbon dioxide (CO_2) and other greenhouse gases (GHGs) that internalizes the far-reaching external costs of climate change in market transactions (e.g., Nordhaus 1992; Stavins 1997; Stern 2007). In sharp contrast to this prescription, a diverse patchwork of climate policy measures has proliferated, and where CO_2 pricing policies do exist, the prices established typically fall far short of the levels necessary to fully internalize the estimated marginal social cost of climate damages.

The failure of governments to establish a pricing (or equivalent market-based) approach to climate change mitigation—or to adequately price carbon when they succeed in doing so—can be largely attributed to a variety of persistent political economy challenges. In particular, climate change mitigation is a global collective action challenge (Olson 1984), demanding coordination among many disparate stakeholders (e.g., nations, emitting industries, individual consumers). Meanwhile, the benefits of climate mitigation are uncertain, unevenly distributed, and accrue primarily to future generations (IPCC 2014), while the costs of climate mitigation are born immediately, with acute distributional impacts for particular constituencies (Burtraw et al. 2002; Bovenberg, Goulder, and Gurney 2005; Rausch and Karplus 2014). Climate mitigation thus has all the hallmarks of an intergenerational principal agent problem (Eisenhardt 1989), with private costs of mitigation out of proportion to the private benefits for many actors. Furthermore, climate policy must be established through political processes, which invoke classic challenges in public choice (Arrow 1970; Black 1987; Buchanan and Tullock 1999; Downs 1957) and are vulnerable to capture by vested interests (Stigler 1971). Voters frequently express limited tolerance for measures that have salient impacts on their private welfare (such as tax or energy price increases) (Kotchen, Boyle, and Leiserowitz 2013; Johnson and Nemet 2010). Industrial sectors with high concentrations of assets that would lose considerable value under carbon pricing policies (e.g., fossil energy extraction, fossil electricity production, fuel refining, concrete production, and energy-intensive manufacturing) have also mounted vociferous and often effective opposition to climate policies. As a result of these public choice dynamics, policy-makers tend to support policies that minimize salient impacts on businesses and households, minimize burdens on strategically important sectors, and/or redistribute rents in a manner that secures a politically-durable coalition. In practice, policy-makers have thus preferred command-and-control regulations that are narrowly targeted (and thus allow for regulatory capture while reducing scope for opposition) and subsidies (which allow for transfers of rents while spreading policy costs broadly and indirectly across the tax base), rather than uniformly pricing CO_2 (Gawel, Strunz, and Lehmann 2014; Karplus 2011).

These persistent political economy constraints motivate a search for climate policies that are politically feasible, environmentally effective, and economically efficient (Jenkins 2014). As in many other domains of economic regulation, second best (Lipsey and Lancaster 1956) climate policy mechanisms abound. By paying close attention to the distributional impacts of different climate policy instruments and their interaction with potentially binding political constraints, economists, political scientists, and policy-makers can help design climate policy responses that are both palatable enough to be implemented today and economically superior to politically feasible alternatives.

In light of these challenges, this chapter aims to develop general insights about the design of climate policy in the face of binding political constraints. We employ a stylized partial-equilibrium model of the energy sector to explore the welfare implications of combining a CO_2 price with the strategic application of revenues to compensate for and/or relieve several potential political constraints on carbon pricing policies. Specifically, we implement constraints of varying severity on: 1) the maximum feasible CO_2 price itself; 2) the maximum tolerable increase in final energy prices; 3) a maximum tolerable decline in energy consumer surplus; and 4) a maximum decline in fossil energy producer surplus. Under each political constraint, we identify the CO_2 price, subsidy for clean energy production, and lump-sum transfers to energy consumers or fossil energy producers that maximizes total welfare.

This chapter begins by contrasting the range of carbon pricing policies implemented across the world with estimates of the full social cost of carbon (Section 3.2). We then introduce our model formulation and stylized representations of four political constraints that could explain the relatively low carbon prices that have been achieved to date in real world policy-making contexts (Section 3.3). We then present numerical results demonstrating that improvements in total welfare and carbon abatement can be achieved by the strategic application of carbon pricing revenues under each of the four political constraints considered (Section 3.4). Finally, we discuss the implications of these findings for climate policy and ongoing research (Section 3.5).

3.2 CARBON PRICING IN THEORY AND PRACTICE

Economists generally conceptualize climate change as a conventional environmental externality caused by emissions of GHGs, which are globally-acting stock pollutants. As such, the traditional economic prescription involves establishing a Pigouvian fee (Pigou 1932) on GHG emissions that corrects for the unpriced externality, either via an emissions tax (Metcalf and Weisbach 2009) or a market-based emissions cap and permit trading mechanism (Coase 1960; Stavins 2008). While there are conceptual and practical differences between CO_2 taxes and emissions trading programmes (Aldy et al. 2010; Weitzman 1974), here we will refer to both instruments collectively as 'carbon pricing policies'. If these instruments successfully establish a carbon price that internalizes the full climate-change-related external damages associated with emissions of CO_2 and other GHGs, the private costs of GHG emitting activities will reflect their marginal social costs, theoretically restoring a level of emissions that is Pareto optimal.

Marginal damage estimates for climate change are expressed in terms of the social cost of CO_2 emissions, or the 'social cost of carbon' (SCC). There is great uncertainty surrounding the true estimate of the SCC, both because damages from climate change under a given level of warming are uncertain and because calculating such figures involves normative judgements such as the appropriate inter-generational discount rate. As shown in Figure 3.1, a review of the literature (Tol 2011) suggests a price on the order of \$75 per tonne CO_2 (tCO_2) in constant 2015 US dollars is necessary in order to internalize the full SCC. The US Environmental Protection Agency also estimates the SCC under different discount rates, which federal agencies apply to estimate the climate benefits of regulations. Average estimates assuming a 3 per cent discount rate increase over the period 2015–50 from \$41 to \$80 per tCO_2 (EPA 2015).

While a variety of jurisdictions have implemented some form of carbon pricing instrument, real-world examples of CO_2 prices that fall squarely within the central range of SCC estimates are few and far between (Figure 3.1). Sweden (\$130 per tCO_2), Switzerland (\$62), Finland (\$47–62, depending on the fuel), and Norway (\$53) are all at the very high end of the spectrum. Each of these nations is relatively wealthy and has abundant supplies of low-carbon

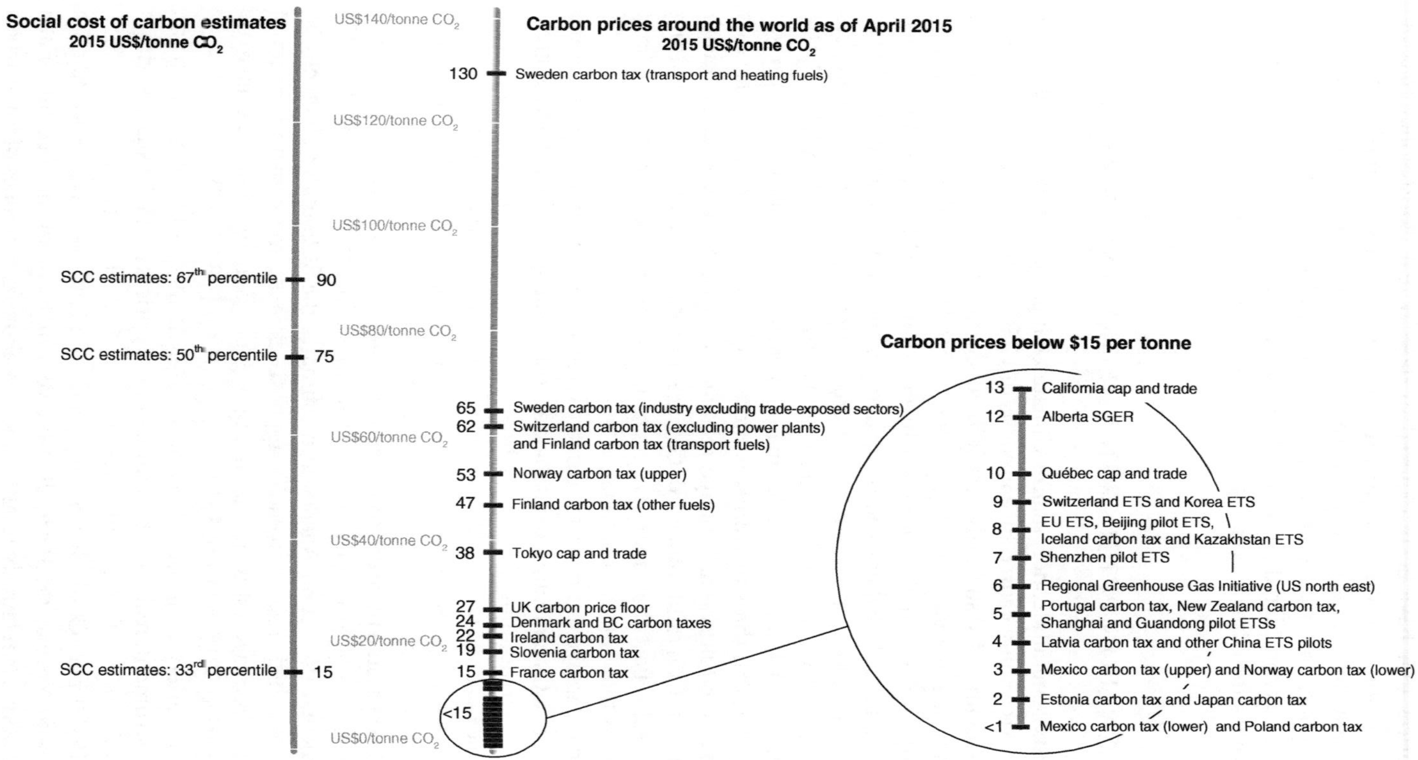

Figure 3.1. CO_2 prices in markets around the world, compared to the social cost of carbon.

Note: Values adjusted to 2015 US dollars; by authors using the US Bureau of Labor Statistics inflation index.

Sources: Authors' illustration. Social cost of carbon estimates from Tol (2011); CO_2 prices from Kossoy et al. (2015).

electricity. Yet even these nations frequently adjust carbon pricing policies in light of political constraints. Sweden, for example, appears to have the highest carbon price in the world. Yet the carbon tax was implemented as part of a series of reforms in 1991 that simultaneously reduced existing energy taxes by 50 per cent. The total effect was to lower overall tax rates on fossil energy consumption (Johansson 2000). Furthermore, Sweden exempts trade-exposed, energy-intensive industries such as pulp-and-paper and mining from the carbon tax entirely, while other industrial emitters pay only half the tax rate. Power plants and district heating are also exempt from the tax and instead fall under the European Union's Emissions Trading System (EU-ETS), which imposes a price of just \$8 per tCO_2 (World Bank 2014). Switzerland similarly allows industrial emitters to opt out of the carbon tax if they participate in the country's own ETS, in which CO_2 permits trade for just \$9 per tonne. Meanwhile, most countries and regions that have implemented CO_2 prices to date have established prices below \$15 per tonne (Kossoy et al. 2015), including the most significant carbon pricing policies established by the world's largest emitters: the EU-ETS, China's ETS pilots, Japan's carbon tax, and two regional programmes in the United States, the US north east's Regional Greenhouse Gas Initiative and California's cap-and-trade programme.

A central premise of this chapter is that political economy constraints explain why the majority of carbon pricing policies around the world today fall well below the central range of estimates of the full social cost of carbon. Any effort to transform the energy system will create economic and political winners and losers, and introducing a CO_2 price is no exception. Climate policy design and instrument choice must therefore contend not only with efficiency concerns, but also with distributional impacts and the resulting implications for political feasibility and durability. Attention to how clever policy design can manage the distributional impacts and costs associated with a clean energy transition while maximizing the efficiency of policy measures is an important (and elusive) challenge.

3.3 MODEL AND SCENARIO IMPLEMENTATION

In this section, we present a stylized model of the energy sector to simulate CO_2 pricing and policy strategies under political economy constraints. The model is based on a single aggregate energy demand function and two energy supply sub-sectors: a CO_2-emitting fossil energy sector and a zero-emissions clean energy sector (e.g., renewable and nuclear energy). For analytical tractability, we assume constant linear slopes for both supply and demand curves.

We further assume the two energy supply sub-sectors are perfectly competitive and are perfect substitutes.

We parameterize the model to roughly approximate the current US energy sector, with 100 Quadrillion British thermal units (Quads) of energy supplied, 80 per cent of which is initially supplied by the fossil energy sub-sector and 20 per cent by the clean energy sub-sector. The initial energy price is $10 billion per Quad (or $10 per million British thermal units), yielding an aggregate annual energy expenditure of $1 trillion. The fossil energy sector emits 5,276 million metric tonnes of CO_2, equivalent to 2013 US energy-related emissions (EIA 2014).

Policy decisions include the level of CO_2 price established, a subsidy per unit of energy supplied by the clean energy sub-sector, and lump-sum transfers to fossil energy producers or energy consumers to compensate for the private welfare impacts of policy decisions. The model is solved to maximize aggregate social welfare over a single time period and is subject to market clearing constraints and one of four stylized representations of commonly-encountered political economy constraints: a direct constraint on the CO_2 price; a constraint on the increase in final energy prices; a constraint on the decrease in net energy consumer surplus; or a constraint on the decrease in net fossil producer surplus. The remainder of this section describes the mathematical formulation of the core model (Section 3.3.1) and the political constraint scenarios explored (Section 3.3.2).

3.3.1 Model Formulation

Energy demand and consumer surplus—The aggregation of household, commercial, and industrial demand for energy is represented as a single aggregate inverse demand function representing the marginal benefit of consumption:

$$MB(q) = d^{-1}(p) = \alpha_d + \beta_d q, \tag{1}$$

where $q = q_f + q_c$ or the sum of both fossil (q_f) and clean (q_c) energy consumed and p is the market clearing price of energy. The marginal benefit of consumption is declining in the quantity consumed ($\beta_d < 0$) and β_d is parameterized based on a plausible initial point estimate of the elasticity of demand. The intercept, α_d, is then set to yield 100 Quads of total consumption in the no-policy case at an initial price of $10 billion per Quad.

Consumer surplus is then expressed as the cumulative benefit of consumption less expenditures on energy and net of the welfare value of any lump-sum transfers (r_d):

$$CS(q, r_d) = \int_0^q MB(q)dq - pq + \varphi_d r_d = \alpha_d q + \frac{1}{2}\beta_d q^2 - pq + \varphi_d r_d. \tag{2}$$

The parameter φ_d captures the 'efficiency' at which sums are transferred to consumers. If this value is set to 1.0, each unit of revenues transferred to consumers translates directly to one unit of increase in consumer surplus. Alternatively, if $\varphi_d < 1.0$, consumers do not value transfers equivalently to the benefits of consumption, requiring greater lump-sum transfers to offset initial private surplus losses. This parameter can therefore be used to capture loss aversion (Kahneman and Tversky 1984) on the part of consumers.

Fossil energy supply and fossil producer surplus—Fossil energy supplies are represented via a linear marginal cost curve with final cost sensitive to the imposition of a CO_2 price (τ):

$$MC_f(q_f, \tau) = \alpha_f + \tau\rho_f + \beta_f q_f, \tag{3}$$

where ρ_f is the CO_2 emissions rate of fossil energy supply. Marginal costs are increasing with the quantity produced ($\beta_f > 0$) and, as with consumer demand, β_f is parameterized based on an initial point estimate of the elasticity of supply. α_f is then set to yield 80 Quads of total fossil energy production in the no-policy case at an initial price of $10 billion per Quad.

Fossil producer surplus is expressed as the sum of revenues less cumulative production costs and tax payments and net of any lump-sum transfers (r_f):

$$\begin{aligned} PS_f(q_f, \tau, r_f) &= pq_f - \int_0^{q_f} MC(q_f, \tau)dq_f + \varphi_f r_f \\ &= pq_f - \alpha_f q_f - \frac{1}{2}\beta_f q_f^2 - \tau\rho_f q_f + \varphi_f r_f. \end{aligned} \tag{4}$$

As with lump-sum transfers to consumers, φ_f represents the 'efficiency' at which lump-sum transfers to producers offset producer surplus losses due to climate policy decisions.

Clean energy supply and clean producer surplus—Clean energy supply is likewise represented as a linear marginal cost curve with final costs adjusted by a per-unit production subsidy (σ) applied to all clean energy production:

$$MC_f(q_f, \sigma) = \alpha_c - \sigma + \beta_c q_c. \tag{5}$$

Marginal costs are increasing with the quantity produced ($\beta_c > 0$), and β_c is again parameterized based on an initial elasticity of supply with α_c then set to yield 20 Quads of total clean energy production in the no-policy case at an initial price of $10 billion per Quad.

Clean energy producer surplus is the sum of revenues and subsidy payments less cumulative production costs:

$$PS_c(q_c, \sigma) = pq_c - \int_0^{q_c} MC(q_c, \sigma)dq_c = pq_c - \alpha_c q_c - \frac{1}{2}\beta_c q_c^2 + \sigma q_c. \tag{6}$$

Note that this formulation applies subsidies to both inframarginal and marginal clean energy production. A more targeted policy measure could reduce the required revenues by applying to marginal production only, reducing the required revenues (and the total transfer to clean energy producers).

Aggregate supply function—The aggregate supply curve corresponding to the marginal cost of supplying an additional unit of energy is the horizontal sum of fossil and clean energy marginal cost functions:

$$MC_t(q,\tau,\sigma) = \left(\frac{\alpha_c - \sigma}{\beta_c} + \frac{\alpha_f + \tau\rho_f}{\beta_f}\right)\left(\frac{\beta_f\beta_c}{\beta_f + \beta_c}\right) + \left(\frac{\beta_f\beta_c}{\beta_f + \beta_c}\right)q. \tag{7}$$

Government revenues and climate damages—Net government revenues produced by the CO_2 tax after transfers to consumers and fossil producers or used to fund clean energy subsidies contribute to overall welfare as follows:

$$R(r_f, r_d, \sigma, \tau) = \varphi_g(\tau\rho_f q_f - \sigma q_c - r_f - r_d). \tag{8}$$

In this case, $\varphi_g > 1.0$ indicates that government revenues offset other distortionary taxes elsewhere and therefore deliver a 'double dividend' (Goulder 1998), increasing their net impact on social welfare. Alternatively, if net revenues are assumed to be utilized inefficiently, this value can be set such that $\varphi_g < 1.0$.

Climate-related damages associated with CO_2 emissions are a simple function of the quantity of fossil energy supplied:

$$E(q_f) = \eta\rho_f q_f, \tag{9}$$

where η is the full social cost of carbon.

Objective function and constraints—The objective function (10) maximizes total social welfare given as the sum of consumer and producer surplus and the welfare value of government revenues less climate-related damages from CO_2 emissions. The model is subject to equilibrium market clearing constraints (11–12).

$$\begin{aligned} Max\, W(\cdot) = {} & CS(q, r_d) + PS_f(q_f, \tau, r_f) + PS_c(q_c, \sigma) \\ & + R(r_f, r_d, \sigma, \tau) - E(q_f) \end{aligned} \tag{10}$$

$$s.t.\, p = MB(q) = MC_t(q) = MC_f(q_f, t) = MC_c(q_c, s) \tag{11}$$

$$q = q_f + q_c. \tag{12}$$

3.3.2 Political Economy Constraint Scenarios and Analytical Solutions

Direct CO_2 price constraint—The first political economy constraint considered is a direct constraint on the level of the CO_2 price of the form:

$$\tau \leq \bar{\tau}, \tag{13}$$

where $\bar{\tau}$ is the maximum politically feasible carbon price level and where $\bar{\tau}<\eta$ (the full SCC).

In this case, social welfare (10) is maximized when the CO_2 price approaches the SCC as closely as possible (i.e., $\tau^* = \bar{\tau}$). However, due to the political economy constraint, the carbon price remains below the full SCC (i.e., $\tau^*<\eta$). Therefore, un-internalized climate-related damages remain, which can be reduced further by using revenues to subsidize clean energy adoption and reduce fossil energy consumption. However, the imposition of a subsidy creates several distortions in the market, including a distortion in total consumption, a distortion in fossil energy production, and a distortion in clean energy production. The optimal clean energy subsidy under this constraint is thus the value that equalizes the marginal increase in deadweight loss due to distortions introduced by the subsidy and the marginal decrease in unpriced external damage from CO_2 emissions due to the reduction in fossil fuel consumption driven by the subsidy. See Jenkins and Karplus (2016) for a full derivation of the optimal subsidy level and analysis of comparative statics for this case.

Energy price constraint—The second political economy constraint we consider is a constraint on the change in the equilibrium energy price after policy decisions. This constraint takes the form:

$$p(\tau, \sigma) \leq p^0(1 + \overline{\Delta p}), \tag{14}$$

where $p(\tau, \sigma)$ is the equilibrium energy price as a function of the CO_2 price and clean energy subsidy policy decisions, p^0 is the equilibrium energy price absent policy intervention (i.e., $p(\tau = 0, \sigma = 0)$), and $\overline{\Delta p}$ is the maximum per cent change in energy price permitted by political economy considerations.

Under such a constraint, a CO_2 pricing instrument alone would be suboptimal. The CO_2 price would be allowed to rise only until it exhausts the political tolerance for energy price increases, internalizing a limited portion of the climate-related externality. In this case, however, welfare could be further improved by combining the carbon price with a clean energy subsidy, which by reducing final energy prices *ceteris paribus*, allows for a larger CO_2 price to be established than would otherwise be possible. At the same time, as in the direct CO_2 price constraint case, the subsidy itself leads to substitution of clean energy for fossil energy, further reducing deadweight loss associated with any remaining unpriced climate externality. The welfare-maximizing CO_2 price and clean energy subsidy level under this constraint is thus the combination that internalizes a greater share of the climate externality and induces further reductions in unpriced damages while balancing these benefits against deadweight loss due to market distortions induced by the clean energy subsidy. Again, see Jenkins and Karplus (2016) for a full derivation of the optimal subsidy level and analysis of comparative statics for this case.

Consumer surplus constraint—Limits on the decrease in energy consumer surplus due to climate policy form an additional political economy constraint, captured in our model as follows:

$$CS(\tau, \sigma, r_d) \geq CS^0(1-\overline{\Delta CS}), \tag{15}$$

where $CS(\tau, \sigma, r_d)$ is final consumer surplus as a function of the carbon price and clean energy subsidy decisions and net of any lump-sum transfers, CS^0 is the consumer surplus absent policy intervention, and $\overline{\Delta CS}$ is the maximum per cent change in producer surplus allowed by political economy considerations.

Assuming efficient transfers, the first-best solution is within reach under a constraint of this form. The welfare-maximizing strategy under this constraint is to establish a CO_2 price equal to the full SCC ($\tau^* = \eta$) while offsetting the impact on consumer surplus via lump-sum transfers (r_c). While a clean energy subsidy can also reduce the final impact on consumer surplus by reducing the final energy price paid by consumers, this strategy is less efficient than a lump-sum transfer, as the subsidy introduces several distortions into the market.

In the case that either $\varphi_c < 1.0$ or $\varphi_g > 1.0$, this strategy incurs additional efficiency losses, which must be balanced against the reduction in climate-related deadweight loss that results from relaxing the indirect constraint on carbon prices. If $\varphi_c < 1.0$, representing loss aversion on the part of energy consumers, the most efficient strategy to mitigate the impact on consumer surplus will include a non-zero clean energy subsidy, as the subsidy also mitigates consumer surplus loss by reducing final energy prices. Indeed, the welfare-maximizing strategy when $\varphi_c < 1.0$ would equalize the marginal deadweight loss associated with distortions due to the clean energy subsidy with the deadweight loss associated with the inefficiency of compensatory transfers to consumers. Cases where $\varphi_c < 1.0$ could therefore also be considered a hybrid of the energy price and consumer surplus constraints.

Fossil producer surplus constraint—The final political economy constraint we consider is a constraint on the decline in fossil energy producer surplus induced by climate policy decisions:

$$PS_f(\tau, \sigma, r_f) \geq PS_f^0(1-\overline{\Delta PS_f}), \tag{16}$$

where $PS_f(\tau, \sigma, r_f)$ is final fossil producer surplus as a function of carbon tax and clean energy subsidy decisions and net of any lump-sum transfers, PS_f^0 is the producer surplus absent policy intervention, and $\overline{\Delta PS_f}$ is the maximum per cent change in producer surplus allowed by political economy considerations.

As with the consumer surplus constraint, assuming transfers are frictionless, the welfare-maximizing strategy is to impose a CO_2 price equal to the full SCC ($\tau^* = \eta$) while compensating fossil energy producers as required to satisfy the political economy constraint via lump-sum transfers (r_f). As a clean energy subsidy would only further reduce fossil producer surplus and introduce market distortions, $\sigma^* = 0$ under this constraint.

Again, if either $\varphi_f < 1.0$ or $\varphi_g > 1.0$, transfers to producers incur additional welfare losses. In this case, the optimal transfer would equalize the marginal reduction in climate-related deadweight loss achieved by offsetting producer surplus impacts and relaxing the indirect constraint on the carbon price on the one hand, and the marginal deadweight loss associated with the inefficiency of compensatory payments and the impact of distortionary taxes elsewhere in the economy on the other.

3.4 RESULTS

In this section, we present results for a numerical simulation using the model presented in Section 3.3. To demonstrate the mechanisms by which strategic allocation of carbon pricing revenues achieves superior performance, we compare two cases for each of the four political constraint scenarios defined in Section 3.3.2. a case in which a CO_2 price is introduced and all revenues collected are retained by the state, and a case in which some portion of the revenues from the CO_2 charge are used to achieve either additional CO_2 reductions by subsidizing clean energy or to offset the burden on producers or consumers through government transfers. Figures 3.2 and 3.3 demonstrate the improvement in total welfare and CO_2 emissions reductions, respectively, under each of the four forms of political economy constraints considered herein.

In all cases, we assume the full SCC is $75 per tCO_2 (as per the median estimate from Figure 3.1), initial elasticities of demand and supply of −0.8 and 0.8 respectively, and that φ_g, φ_f and φ_c equal 1.0 (i.e., all transfers are frictionless). See Jenkins and Karplus (2016) for analysis of the sensitivity of outcomes to alternative values for the price elasticities of supply and demand.

3.4.1 Direct Constraint on the CO_2 Price

In a world where the politically-feasible CO_2 price remains below the full SCC, using revenues to subsidize clean energy results in additional welfare gain and

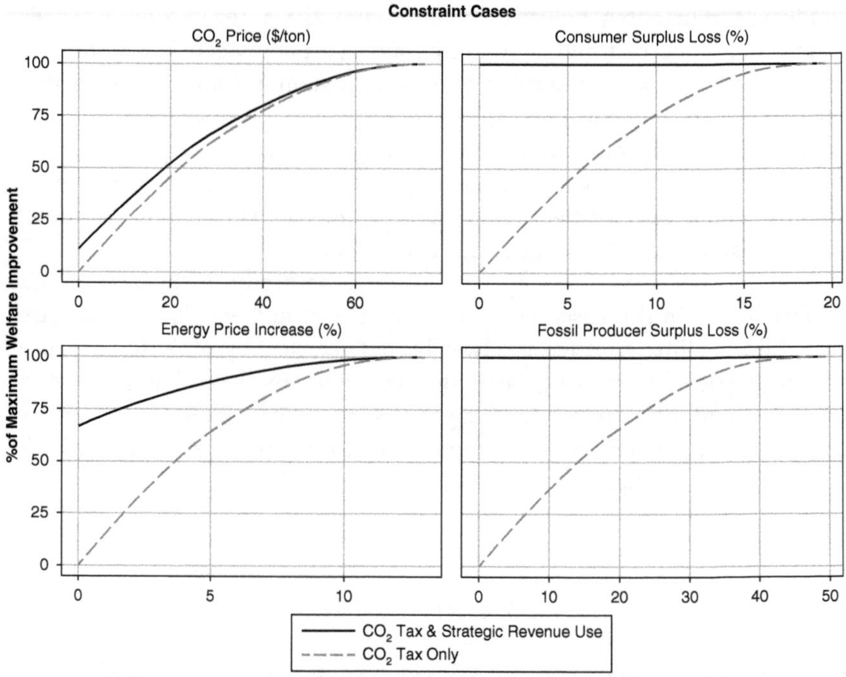

Figure 3.2. Total welfare gain under four political constraint scenarios.
Source: Authors' analysis and illustration.

CO_2 emissions reduction, relative to the constrained no-subsidy case, as shown in Figures 3.2 and 3.3. The largest welfare gains from the subsidy occur when the CO_2 price constraint binds at low levels. In the absence of any carbon price at all, the welfare-maximizing clean energy subsidy achieves 11 per cent of the maximum reduction in CO_2 and improvement in welfare achievable under the first-best carbon pricing level, given these parameters. If the allowable carbon tax is constrained at very low levels, funding the optimal clean energy subsidies may require additional revenues from elsewhere in government budgets, and the policy as a whole will be revenue consuming (see Figure 3.5). When the CO_2 price rises, the welfare and emissions performance improvements from the clean energy subsidy decline. This is because the optimal subsidy level decreases as the damages associated with emissions are steadily internalized by the carbon price. In all cases, a non-zero subsidy improves overall welfare unless the carbon price equals the full SCC. In addition, as revenues from the tax increase, the optimal policy becomes revenue generating (see Figure 3.5).

The direct constraint on CO_2 prices is in many ways the most challenging constraint to overcome via the strategic use of carbon pricing revenues.

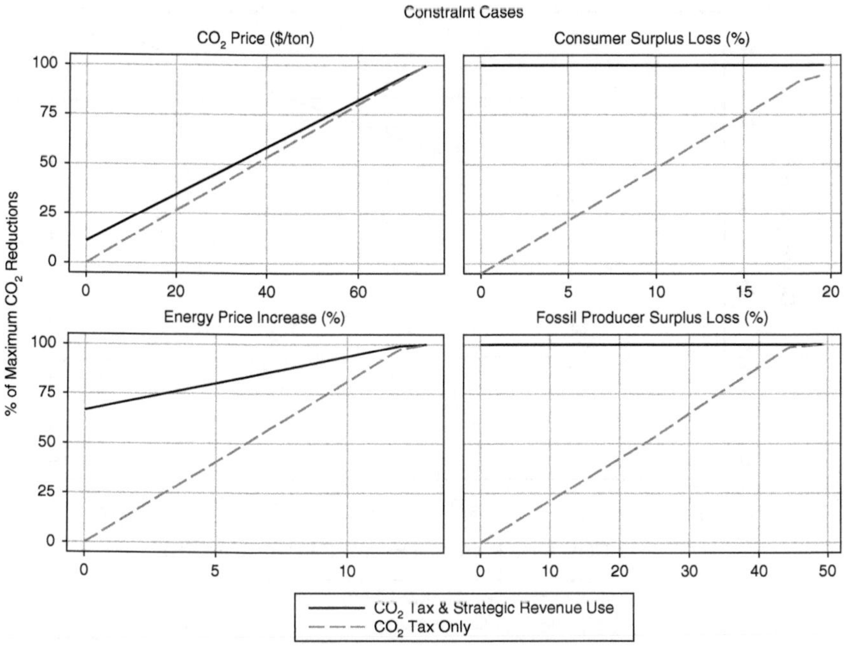

Figure 3.3. Total CO_2 emissions under four political constraint scenarios.
Source: Authors' analysis and illustration.

Subsidizing clean energy in this case does not *relax* the constraint itself, but merely compensates for the low carbon price by delivering additional abatement. However, this abatement comes at the cost of economic distortions introduced by the subsidy, delivering relatively modest improvements in overall welfare. By contrast, under the other constraints, use of revenues not only generates additional abatement but also directly relaxes the constraint itself, allowing for higher carbon prices to be achieved than would otherwise be possible.

3.4.2 Constraint on Final Energy Price Increases

Under a constraint on the allowable energy price increase, employing carbon pricing revenues to subsidize clean energy enables a significantly higher price of CO_2, as demonstrated in Figure 3.4. As clean energy subsidies reduce final energy prices, *ceteris paribus*, deploying revenues to subsidize clean energy alternatives effectively relaxes the constraint on the energy price increase. For example, using clean energy subsidies to offset the rising costs of energy enables a carbon price of \$35 per tCO_2 even when only a negligible increase

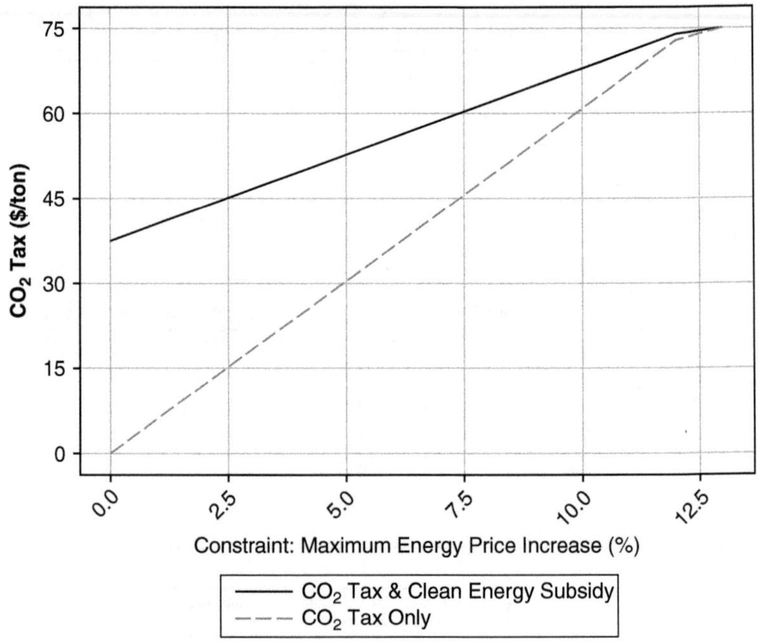

Figure 3.4. CO_2 price achieved under binding constraint on energy price increases, with and without employing revenues to subsidize clean energy.

Source: Authors' analysis and illustration.

in final energy prices is permitted. In addition, as in the carbon price constraint case, the clean energy subsidy drives additional abatement that would not be achieved via the carbon price alone, further improving overall welfare. These benefits again trade off against the deadweight loss due to distortions induced by the clean energy production subsidy.

In combination, the carbon price and clean energy subsidy deliver much greater CO_2 reductions than a carbon price alone, especially when the energy price increase is constrained at low levels (Figure 3.3). Given the parameters assumed here, nearly two-thirds of the optimal reduction in CO_2 emissions can be achieved without increasing final energy prices at all. Employing revenues to fund clean energy subsidies improves the environmental performance of the policy intervention until the full social cost of carbon is internalized. Overall welfare improves similarly when revenues are used to subsidized clean energy production, achieving two-thirds of the optimal welfare gain even when no increase in energy prices is permitted, rising to nearly 90 per cent when a 5 per cent increase in final energy prices is tolerated (Figure 3.2).

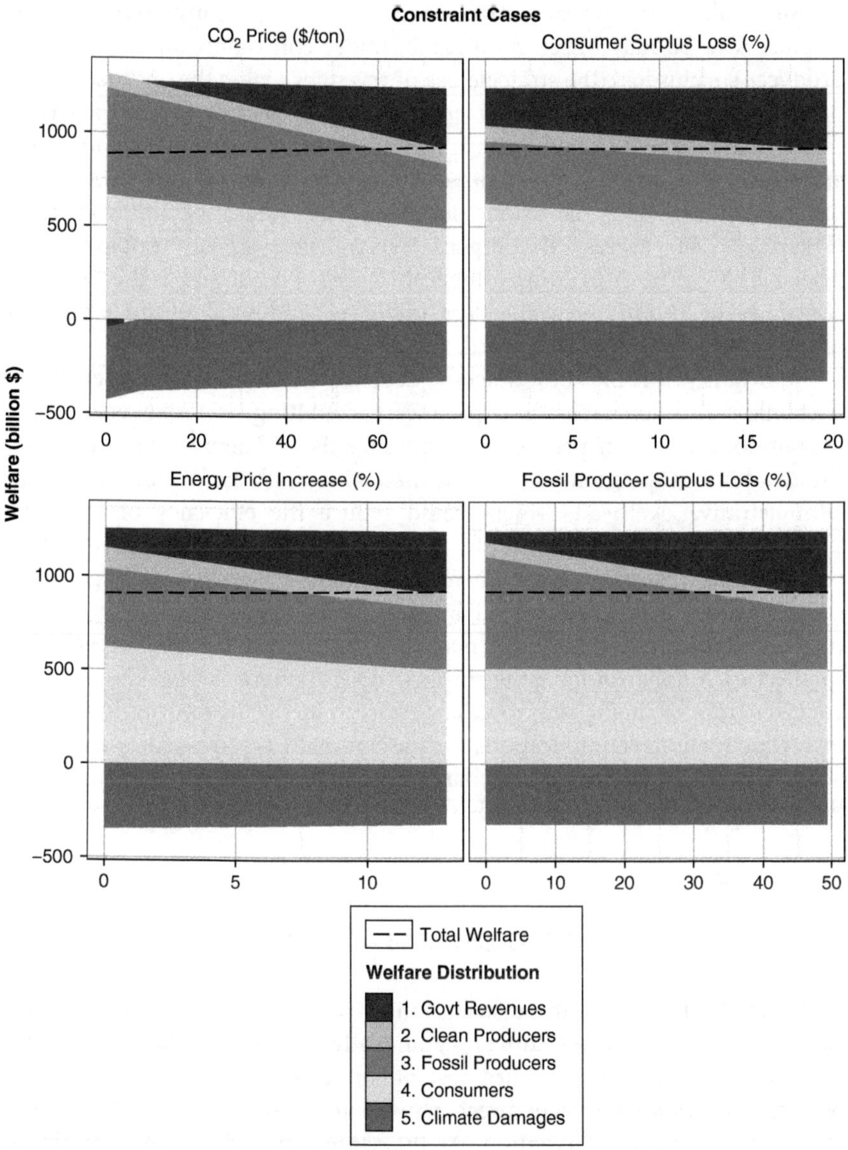

Figure 3.5. Disposition of welfare under four political constraint scenarios.
Source: Authors' analysis and illustration.

3.4.3 Constraints on Net Energy Consumer and Fossil Producer Surplus Loss

Unlike the prior cases, where political constraints continue to result in a second-best CO_2 pricing level, redistributing carbon revenues as lump-sum transfers allows the private surplus losses for energy consumers or fossil producers to be fully offset. As a result, under constraints on consumer or producer surplus loss, the strategic use of revenues makes the optimal carbon price immediately feasible, provided transfers are frictionless and available funds are sufficient. When compensatory transfers are utilized, the CO_2 externality can be fully internalized, maximizing welfare (Figure 3.2) and driving optimal CO_2 emissions levels for all values of the constraint (Figure 3.3). In contrast, if compensatory transfers are not employed, the available CO_2 price rises linearly under this form of constraint as the allowable consumer or producer surplus loss increases, and welfare and emissions outcomes are similarly constrained.

Importantly, this first-best outcome depends on lump-sum transfers being frictionless and consumers and producers exhibiting no loss aversion, two assumptions which in practice may be unrealistic. These results thus raise three important questions: 1) what is the real loss, if any, due to frictions or administrative overhead, which would reduce the efficiency of transfers; 2) what is the opportunity cost of using revenues for transfers rather than to reduce other distortive government taxes; and 3) what is the additional compensation, if any, demanded by loss-averse consumers and producers (i.e., do recipients of transfers demand more than a dollar of compensation to offset each dollar of foregone surplus)? Our framework provides a way to consider transfer inefficiency and loss aversion in calculations of deadweight loss, which will have implications for the optimal CO_2 price, CO_2 emissions abatement, and distribution of welfare impacts. We will leave a full analysis of these implications for future work.

3.4.4 Disposition of Welfare

As Figure 3.5 illustrates, the distribution of welfare under the four political economy constraint cases differs significantly. As one might expect, consumers and fossil producers are best off under the respective cases where political constraints motivate direct transfers to offset any surplus losses they incur due to policy intervention. At the same time, consumers are almost equally well off when revenues are used to subsidize clean energy in the face of a constraint on energy price increases. Here, clean energy subsidies drive incremental substitution of clean for dirty energy and keep final energy prices low, insulating energy consumers from welfare losses. Similarly, as total reductions in fossil energy use are modest under the case where the CO_2 price is directly constrained, fossil producers are nearly as well off in this case as they are under the direct constraint on fossil producer losses. Political constraints on the carbon price or energy price increases may therefore be interpreted as the indirect expression of concern about producer or consumer surplus losses, respectively, particularly in cases where consumers and producers exhibit significant loss aversion and thus view compensatory payments in an inferior light.

Clean energy market share and the growth of clean energy producer surplus is most significant under the energy price constraint (Figure 3.5). If the size and relative economic importance of clean energy production sectors positively affects the political durability of coalitions that support climate mitigation policy and increases tolerance for future increases in carbon prices, combining a carbon price with subsidies for clean energy producers may yield additional dynamic benefits. Similarly, incrementally higher deployment of clean energy in the near term could drive learning-by-doing, economies of scale, or induced research and innovation, decreasing the cost of clean energy supply in the future, although the magnitude of these benefits is uncertain. Over time, the result would be greater mitigation at a given cost, an important dynamic benefit to consider.

3.5 CONCLUSION AND IMPLICATIONS FOR POLICY AND RESEARCH

Global experience to date suggest that the distributional impacts of carbon pricing policies on energy producers and consumers make it difficult to legislate CO_2 price levels needed to fully internalize the climate change externality. This reality points to two important ongoing agendas for research: one aimed at improving on existing estimates of the social cost of carbon and evaluating the impacts of fully internalizing these damages through CO_2 pricing, and another that starts from the presently feasible set of alternatives, taking political constraints as binding in the near term, and evaluates options for improving welfare and expanding this feasible set over time. In the latter case, the goal is to identify policy designs that are not too distant from the efficient frontier and that alter the relative influence of actors in ways that support gradual convergence towards a socially optimal CO_2 price. Although methods for estimating the SCC are still hotly debated, as long as prevailing CO_2 prices remain below the lower end of the SCC range, as they do in many CO_2 pricing systems at present, focusing on political constraints is important to answering the critical question: how do we begin to address climate change-related externalities as efficiently and effectively as possible given today's political realities?

In this analysis, we investigated the impact of four different political economy constraints on carbon pricing, focusing on how the stringency of the constraints affect the welfare gain associated with alternative uses of CO_2

price revenues. We find that in all cases, using revenues to subsidize additional abatement or offset private surplus loss improves total welfare, relative to a constrained case where revenues are simply used for general government purposes. We show that compensating for a direct constraint on the CO_2 price delivers modest gains, because the benefits associated with additional abatement are offset by deadweight loss resulting from over-consumption induced by clean energy subsidies. In this respect, a constraint on the absolute level of the CO_2 price constitutes the most restrictive case. By contrast, greater welfare gains are possible under a constraint on energy price increases, as carbon pricing revenues can be used to subsidize clean energy and keep final energy prices low, allowing a higher carbon price to be achieved than would otherwise be possible. Indeed, when revenues are deployed to subsidize clean energy, a substantial CO_2 price is possible even if no increase in final energy prices is tolerated at all. Finally, using revenues to offset consumer and producer surplus loss supports a return to optimal CO_2 price levels and a first-best solution—with the important caveat that compensatory transfers must be frictionless and consumers and producers do not exhibit loss aversion.

While the analysis presented herein develops intuition about how constraints function individually and in an idealized context, reality is inevitably more complex. An important question for decision-makers and political scientists involves establishing which political economy constraints bind in the jurisdiction in question and through which mechanisms they operate. In practice, multiple political economy constraints may bind at the same time—for example, a high CO_2 price may be unavailable because covered parties are concerned about the resulting energy price increase, or the magnitude of the impact on consumer and producer surplus, or all of these. In the face of multiple political economy constraints, one potential solution would be to dip into government budgets to further subsidize CO_2 abatement or to offset reductions in consumer and producer surplus. However, this option requires careful consideration of the opportunity cost of channelling additional funds to relieve political economy constraints, as potential second-best solutions will compete with each other, and with other possible uses of public funds, for available government revenues. Ultimately, the political feasibility of this path is constrained by public decision-making on appropriate spending priorities, and the nature of the climate change problem is such that near-term public investments with more concrete benefits may be preferred.

Our analysis shows that it is possible to achieve the first-best CO_2 price if revenues can be used to offset consumer and producer surplus losses. In reality, however, none of the transfers discussed here are likely to be frictionless. It is important therefore to also understand the real and perceived value of these transfers to recipients and the general equilibrium implications of changes in government revenues. Transfers to support clean energy subsidies may also have associated frictions, which will magnify the relative inefficiency of the subsidy. On the other hand, more targeted subsidies which only apply to supra-marginal suppliers could reduce the overall revenues required to drive clean energy adoption and associated mitigation, an important consideration in cases where subsidy programmes entail additional efficiency losses (i.e., due to foregone opportunities to reduce other distortionary taxes). The nature and magnitude of these frictions and their efficiency implications will be specific to particular contexts, increasing the importance of understanding and quantifying their impact on interests and incentives.

The main objective of this exercise was to put an analytical framework around the question of how we can get started down a relatively efficient path to a lower carbon world. The answer will be different, depending on the unique political economy of the climate issue across nations and regions. We conclude by briefly illustrating the guidance this framework would offer policy-makers under different prevailing political constraints.

First, in jurisdictions without significant domestic fossil energy production sectors, political constraints are likely to centre on concerns about the impact of climate mitigation policies on household incomes and the economic competitiveness of domestic industries. In such cases, the prevailing constraint is likely to be the unwillingness of energy consumers to bear the burden of higher energy prices and associated surplus loss. Our results suggest that an effective policy strategy in the face of such constraints would be to establish a carbon price while employing revenues to make clean energy cheaper and mitigate the impact on final energy prices (i.e., via subsidies). Any remaining politically-salient losses to energy consumers could be offset with lump-sum transfers as needed (e.g., to trade-exposed industries or low-income households).

Second, in jurisdictions where influential fossil energy producers and industrial energy consumers are aligned in opposition to CO_2 pricing, neutralizing opposition from industrial energy consumers by subsidizing clean energy adoption and keeping energy input prices low could remove a major barrier to CO_2 pricing, while allowing the CO_2 price to rise to a meaningful level. Remaining resistance from the fossil energy industry could then be addressed through transfer payments—either taken from CO_2 price revenues or elsewhere in the government budget. This strategy may be most viable in jurisdictions with strong domestic fossil energy sectors and relatively large energy-intensive industrial sectors, such as steel, aluminum, concrete, or pulp and paper production.

Under either case, if political constraints relax over time, whether as an endogenous outcome of policy or a shift in stakeholder preferences, CO_2 prices could rise towards the full social cost of carbon, achieving further welfare gains. The dynamic impacts of near-term policy decisions on political constraints over time is thus an additional key consideration worthy of future research. For example, encouraging near-term deployment of clean energy to an extent that realizes benefits from scale economies, learning, and a growing clean energy constituency with a strong interest in its own continued survival and growth could have significant impacts on the political durability of climate policy over time.

Clean energy transitions will inevitably create winners and losers. The scenarios and analysis presented herein suggest illustrative paths by which the costs and distributional impacts of a clean energy transition could be smoothed over time, gradually nudging the possible in the direction of the optimal.

REFERENCES

Aldy, J. E., A. J. Krupnick, R. G. Newell, I. W. H. Parry, and W. A. Pizer (2010). 'Designing Climate Mitigation Policy'. *Journal of Economic Literature*, 48(4): 903–34.

Arrow, K. J. (1970). *Social Choice and Individual Values*. Second Edition. New Haven, CT: Yale University Press.

Black, D. (1987). *The Theory of Committees and Elections*. Dordrecht: Springer.

Bovenberg, A. L., L. H. Goulder, and D. J. Gurney (2005). 'Efficiency Costs of Meeting Industry Distributional Constraints under Environmental Permits and Taxes'. *RAND Journal of Economics*, 36(4): 951–71.

Buchanan, J. and G. Tullock (1999). *The Calculus of Consent*. Collected Edition. Indianapolis, IN: Liberty Fund, Inc.

Burtraw, D., K. Palmer, R. Bharvirkar, and A. Paul (2002). 'The Effect on Asset Values of the Allocation of Carbon Dioxide Emission Allowances'. *The Electricity Journal*, 15(5): 51–62.

Coase, R. H. (1960). 'The Problem of Social Cost'. *Journal of Law and Economics*, 3: 1–44.

Downs, A. (1957). *An Economic Theory of Democracy*. First Edition. New York: Harper and Row.

EIA (2014). 'U.S. Energy-Related Carbon Dioxide Emissions, 2013'. US Energy Information Administration.

Eisenhardt, K. M. (1989). 'Agency Theory: An Assessment and Review'. *The Academy of Management Review*, 14(1): 57–74.

EPA (Environmental Protection Agency) (2015). 'Technical Support Document: Technical Update of the Social Cost of Carbon for Regulatory Impact Analysis'. US Environmental Protection Agency.

Gawel, E., S. Strunz, and P. Lehmann (2014). 'A Public Choice View on the Climate and Energy Policy Mix in the EU—How Do the Emissions Trading Scheme and Support for Renewable Energies Interact?' *Energy Policy*, 64: 175–82.

Goulder, L. H. (1998). 'Environmental Policy Making in a Second-Best Setting'. *Journal of Applied Economics*, 1(2): 279.

IPCC (Intergovernmental Panel on Climate Change) (2014). 'Climate Change 2014: Impacts, Adaptation, and Vulnerability'. IPCC, Geneva.

Jenkins, J. D. (2014). 'Political Economy Constraints on Carbon Pricing Policies: What Are the Implications for Economic Efficiency, Environmental Efficacy, and Climate Policy Design?' *Energy Policy*, 69: 467–77.

Jenkins, J. D. and V. J. Karplus (2016). 'Carbon Pricing under Binding Political Constraints'. UNU-WIDER Working Paper No. 44/2016. Available at: <https://www.wider.unu.edu/publication/carbon-pricing-under-binding-political-constraints> (accessed 12 October 2016).

Johansson, B. (2000). 'Economic Instruments in Practice 1: Carbon Tax in Sweden'. OECD Workshop on Innovation and the Environment, Paris, 19 June.

Johnson, E. and G. F. Nemet (2010). 'Willingness to Pay for Climate Policy: A Review of Estimates'. University of Wisconsin-Madison, La Follette School Working Paper No. 2010-011.

Kahneman, D. and A. Tversky (1984). 'Choices, Values, and Frames'. *American Psychologist*, 39(4): 341–50.

Karplus, V. J. (2011). 'Climate and Energy Policy for U.S. Passenger Vehicles: A Technology-Rich Economic Modeling and Policy Analysis'. PhD Thesis, Massachusetts Institute of Technology.

Kossoy, A., G. Peszko, K. Oppermann, N. Prytz, A. Gilbert, N. Klein, L. Lam, and L. Wong (2015). 'Carbon Pricing Watch 2015'. World Bank Group.

Kotchen, M. J., K. J. Boyle, and A. A. Leiserowitz (2013). 'Willingness-to-Pay and Policy-Instrument Choice for Climate-Change Policy in the United States'. *Energy Policy*, 55: 617–25.

Lipsey, R. G. and K. Lancaster (1956). 'The General Theory of Second Best'. *The Review of Economic Studies*, 24(1): 11–32.

Metcalf, G. E. and D. Weisbach (2009). 'The Design of a Carbon Tax'. *Harvard Environmental Law Review*, 33(2): 499–556.

Nordhaus, W. D. (1992). 'An Optimal Transition Path for Controlling Greenhouse Gases'. *Science (New York, N.Y.)*, 258(5086): 1315–19.

Olson, M. (1984). *The Rise and Decline of Nations: Economic Growth, Stagflation, and Social Rigidities*. New Haven, CT, and London: Yale University Press.

Pigou, A. C. (1932). *The Economics of Welfare*. Fourth Edition. London: Macmillan.

Rausch, S. and V. J. Karplus (2014). 'Markets versus Regulation: The Efficiency and Distributional Impacts of U.S. Climate Policy Proposals'. *The Energy Journal*, 35(1): 199–228.

Stavins, R. N. (1997). 'Policy Instruments for Climate Change: How Can National Governments Address a Global Problem?' *The University of Chicago Legal Forum*, 6: 293–329.

Stavins, R. N. (2008). 'Addressing Climate Change with a Comprehensive US Cap-and-Trade System'. *Oxford Review of Economic Policy*, 24(2): 298–321.

Stern, N. (2007). *The Stern Review on the Economics of Climate Change*. Cambridge: Cambridge University Press.

Stigler, G. J. (1971). 'The Theory of Economic Regulation'. *The Bell Journal of Economics and Management Science*, 2(1): 3–21.

Tol, R. S. (2011). 'The Social Cost of Carbon'. *Annual Review of Resource Economics*, 3(1): 419–43.

Weitzman, M. L. (1974). 'Prices vs. Quantities'. *Review of Economic Studies*, 41(January): 477–91.

World Bank (2014). 'Putting a Price on Carbon with a Tax'. World Bank.

4

Border Adjustment Mechanisms

Elements for Economic, Legal, and Political Analysis

Julien Bueb, Lilian Richieri Hanania, and Alice Le Clézio

4.1 INTRODUCTION

Paris COP21 has been presented as both a diplomatic success and a victory for common sense, representing considerable progress in overcoming the deadlock the United Nations negotiation process had been in since Copenhagen COP15. Its ambition is indeed great and unprecedented. However, upon closer examination of the contents of the agreement, the COP21 epitomizes old, and perhaps even growing difficulties that negotiators meet when discussing an urgently needed global, strong, and binding compromise.

Proposed by many economists but absent from the Paris Agreement, is the definition of an international price of carbon emission, which appears in practice extremely hard to achieve and insufficient to counter biodiversity loss or achieve broad environmental goals. It is, nevertheless, very much a precondition for international sustainable production and consumption, since price is the foremost economic signal to help orient economic players. It stimulates investments and innovation in cleaner production processes and provides a new source of financing for public policies, be they environmental or developmental.

As of this writing, the most important regional greenhouse gas (GHG) pricing system is the European Union Emissions Trading Scheme (EU ETS). Spearhead of the European climate policy, this carbon market has since its inception been used as a model by other countries and regions. However, tradable emission markets are ridden with many structural flaws, which result in a final pricing of the carbon tonne that is much too low. Prices thus lose their incentive function and virtuous players—those economic agents who have already started to invest in clean technologies—are found to be at a disadvantage.

Among the many identifiable reasons accounting for international inertia in the field, it appears that the weakness of prices (i.e., taxes and carbon market prices) might be related to a reasonable fear of competitiveness loss, notably in heavy industries (Branger, Quirion, and Chevallier 2013). Indeed, one of the predominant political arguments for such low prices to internalize environmental externalities[1] is the (real or conjectured) loss of competitiveness for local businesses in a globalized economy. According to this argument, local enterprises are confronted with international competitors that would indeed be either exempt of environmental charges, or, if they exist, under less constraining regulations. Businesses respectful of environmental regulations are therefore forced to accept international competitiveness loss or offshore their production to countries with less constraining frameworks, thus annihilating efforts made in virtuous countries.

In order to avoid such perverse economic and environmental effects, several developed countries, such as France and the United States (US), have been devising adjustment systems to guarantee the efficiency of their environmental policies. These mechanisms are called 'border carbon adjustments' (BCAs). A BCA is a tariff measure that internalizes carbon emissions into the price of a given imported product. It is, therefore, a fiscal instrument, which is used in complement of carbon emissions reduction tools, such as a carbon market or a carbon tax, and targets internationally traded goods. Its primary objective is to level the playing field between domestic producers facing costly climate change measures and foreign producers facing very few.

France and Italy have repeatedly demonstrated their interest in such mechanisms at the European level: BCAs would be enforced in the most energy-intensive sectors (e.g., those that bear simultaneously the weight of environmental regulation and international competition), and would target selectively those countries that have not undertaken appropriate commitments, according to their capacity and responsibility in climate change, within the international legal framework on climate change (France 2009: article 2). Another example is the 2009 US Waxman–Markey Bill, which set both a BCA mechanism and a tradable emissions permit market.[2] Other developed countries, such as Japan and Australia, are studying diverse solutions, and will inevitably be influenced by their European and American counterparts. In Europe, however, strong opposition to such initiatives persists, most notably in the United Kingdom. The same obviously applies to developing countries, although some voices have been raised in favour of

[1] Externalities are external effects created by an economic agent's activity trading a free advantage, or a damage (climate degradation for instance), without monetary compensation.

[2] The tradable GHG emissions permits market could have been created in 2012 at the federal level in the US, but was rejected by the Senate (Schott and Fickling 2010).

BCAs (Mattoo and Subramanian 2013) amidst a growing recognition of the need for a satisfying global framework.[3]

In the scientific literature, BCAs are also controversial. The foremost argument against their enforcement contends that such unilateral measures are mere kludges. The universal enforcement of a carbon price would be more relevant and efficient (Markusen 1975). Houser et al. (2008), as well as Dröge et al. (2009), further argue that it could slow down international cooperation on climate change, trigger economic retaliation, and foster a return to protectionism, or stimulate emission reduction costs transfers through trade effects. However, considering the weakly binding and limited country-level mitigation provisions reached under the United Nations Framework Convention on Climate Change (UNFCCC), unilateral action is likely more useful than the status quo (OECD 2014). Moreover, their virtuous effect on competitiveness,[4] and even in increasing global efficiency (Böhringer, Balistreri, and Rutherford 2012), is well documented in the growing economic literature (OECD 2014).

Based on the discrepancy between the urgency of climate issues and the meagre results achieved in international negotiations, this chapter weighs the usefulness of BCAs as a complement to strong regional or domestic environmental regulation. This leads us to discuss the interplay of economic competitiveness and climate change, before the economic challenges posed by BCAs in order to reach fairness in its design and implementation (Section 4.2). Section 4.3 sheds light upon the legality of BCAs according to international trade law, while Section 4.4 provides an assessment of policy-related implications. It outlines, in particular, how BCAs could be used as an engine of a necessary economic transition, for developed and developing countries equally, according to the common but differentiated responsibilities principle.

4.2 BCA IMPLEMENTATION

Environmental policy instruments fall into two categories: they are either of a regulatory (e.g., quotas, standards, certifications) or economic (e.g., taxes, tradable emission permits—TEPs)[5] nature. In both cases, they intend to internalize environmental externalities. BCAs can be perceived as commercial

[3] In 2010, India declared its intention to bring BCA measures before the World Trade Organization (WTO) Dispute Settlement Body, but has since softened its stance, based on the idea that developed countries cannot simultaneously commit to a significant reduction of their emissions and bear the weight of a large reduction of their economic activity (OECD 2014).

[4] See, in particular, Demailly and Quirion (2008: 497–504). Other economists defend a contrary position, as Weber and Peters (2009: 432–40).

[5] TEPs can also be referred to as 'tradable emission quotas' or 'carbon markets'.

measures devised to level the playing field between domestic and foreign players whose position before environmental regulation is asymmetrical. For instance, if European producers, notably those from energy-intensive sectors, compensate rising costs linked to the implementation of the EU ETS on the final customer, they might lose domestic market shares to foreign competitors. Conversely, if they do not apply such compensation, their profits, and over time, their investments will shrink, which will reduce their ability to win. In the end, it may lead to a transfer of investments from European companies to foreign markets, as well as a joint loss of market shares on domestic and foreign markets, resulting in severe job loss and a rise of GHG emissions outside Europe—in other words, in carbon leakage, with its two main components:

- An 'operational leakage'; that is, the operational delocalization of a domestic business, induced by environmental constraints, to less regulated foreign markets.
- An 'investment leakage'; that is, the redirection of investment flows from environmentally regulated domestic markets to foreign countries where such regulatory framework is absent.

The implementation of a BCA requires a clear definition of its field of application, in terms of product scope, instruments, and countries, leading to the following questions.

4.2.1 Which Regulating Instrument(s)?

Most climate policies use a combination of instruments, but market signals have often been a major segment of such measures.[6] Indeed, when dealing with environmental issues and in debates regarding the implementation of a BCA, economic instruments have largely been favoured instead of regulatory instruments. They aim to encourage, through price signals, the voluntary modification of behaviours to reduce pollution.

In general, BCAs may consist in (i) a tax on imported products based on taxes applied on similar domestic products (tax adjustment on imports); (ii) tax credits on exported products (tax adjustment on exports); (iii) mandatory acquisition of emission permits in sectors where carbon leaks have been identified; or (iv) the allocation of free permits for those sectors subject to high competition. The definition of a BCA depends as well on pre-existing instruments in the regulated zone. A border carbon tax may supplement a carbon tax on polluting activities of a given country or zone, or complement a carbon market. Tradable permits cannot be added to a carbon tax because they

[6] For an analysis of environmental taxes and carbon markets, see Bueb (2014).

EXISTING SYSTEMS	BORDER ADJUSTMENT MECHANISM	IMPLEMENTATION	EFFECTS & LIMITS
CARBON TAX	TAX	**T1 CONSUMPTION** Tax on final product	T1 Final tax, similar to VAT, for goods sold on domestic territory. Its amount is proportionate to the carbon content of goods sold on domestic territory.
		T2 PRODUCTION Tax at each step of production process	T2 Paid at each step of the production process,the costs of this tax can be transferred to the consumer of the product, thus significantly encouraging 'green' processes through cumulative effects.
		T3 EXPORTS Tax credits for exporters	T3 Although carbon constraint exemptions can eliminate incentives to shift to environmental-friendly processes, it minimizes domestic disadvantages for domestic companies on international markets.These credits also encourage businesses with different production sites to export the most emission-intensive goods, while keeping the less-intensive for their domestic market.
TRADABLE EMISSION PERMIT	TAX	See T1, T2, T3	See T1, T2, T3 + added difficulty in coordinating both mechanisms
	TRADABLE EMISSION PERMIT	**TEP 1** TEP overall quantity is stabilized & importers included.	TEP 1 Increase in TEP demand sholud be followed by an increase in prices, and therefore, in marginal costs for domestic businesses. Price effects, nonetheless, depend on elasticities and market's structure. Such a measure will mostly benefit businesses holding quotas surpluses or those who have adopted environmental-friendly practices. In this scenario, where permits are auctioned, importers and domestic players bear an equal environmental burden. Free permit issuance could indeed be considered discriminating and assimilated to a subsidy.
		TEP 2 Creation of additional TEPs specifically for importers.	TEP 2 This option is less disadvantageous for domestic players as it does not penalize supply and demand. It can, however, constitute a discrimination betweeen domestic and foreign players if TEP allocation is even partially free.
		TEP 3 'Carbon credits', are accepted to enter a market. They can be obtained through CDM/JIM.	TEP 3 Holds difficulties related to CDM/JIM, problems mostly due to carbon credit attribution and opacity of the process.
		TEP 4 Restitution of emission quotas to exporters	TEP 4 See T3.

Figure 4.1. Tradable emission permits.

Note: CDM: Clean Development Mechanism; JIM: Joint Implementation Mechanism.

Source: Authors' illustration.

require an operating carbon market—they would take the form of purchased, allocated, or restituted permits instead (see Figure 4.1).

4.2.2 How to Evaluate the Carbon Content of Imported Products?

Whatever form a BCA might take, it bears on processes and not products—it is blind to the product itself. The global political sensitivity of BCA enforcement renders precise definition of sectors and products extremely critical. It is important to evaluate correctly the carbon content of products in order to implement fair BCA measures. Carbon content is highly variable according to countries' regulations and throughout production processes and across sectors:

- Considering country regulations, environmental constraints differ and the application of distinct BCAs according to each target country is difficult due to the amount of information needed to correctly evaluate and categorize environmental regulations.
- On the product side, two similar products from the same sector, produced by similar companies, can have starkly different emission rates. Within a single sector, similar goods can also be produced with very different methods and thus have very different carbon contents: the quantity of energy used depends on where it is produced and/or transported; the quality of energy used relates to the primary source of energy (gas or coal), even if the final energy used is the same.

These differences of carbon content would entail varied BCAs on similar products, which would be very difficult to implement.

Moreover, all importers should be subjected to emissions reduction obligations to prevent environmental dumping. Yet, prices (i.e., tax value or the purchase price of a permit bought by importers) need to be uniform (for instance, based on average values). In practice, such a uniform price may suppress individual countries' and/or individual companies' initiatives regarding emissions reduction, with virtuous players remaining comparatively unrewarded for their efforts. There would therefore be no incentive to invest in clean technology. A differentiated BCA system would reward efforts in emissions reduction, but in practice still appears unmanageable.

To lessen the problem, a first alternative could be to encourage importing companies to willingly pass an audit on the carbon content of their products. Another option would be demanding that companies give verifiable evidence of environmental regulation compliance, for instance through a certifying body. Such a disposition might, however, prove difficult to require from small importers considering the high administrative costs involved, in addition to possibly being considered discriminating, since not all companies would be subjected to this requirement.

Two other approaches, applied in conjunction or independently, are proposed in the literature: 'top-down' methods, using input–output analysis to estimate embodied energy, CO2 emissions, pollutants, and land appropriation of international trade activities, and 'bottom-up' approaches, based on the calculation of embodied carbon through the examination of production processes of specific products. The former seems too broad and, therefore, not easy to apply. The latter, more realistic, involves two methods: the 'best available technology' (BAT) and the 'predominant method of production' (PMP). According to the PMP method, the country of import would apply a BCA by assessing the carbon embedded in an import in relation to its own domestic production processes (Zhang 2010: 19). As for the BAT method, the country of import would apply a border adjustment by assessing the carbon embedded in an import compared to what the carbon emissions would be if the best available technology had been used (Ireland 2010).

The BAT is likely the best method. It serves as a reference to assess the carbon content of imported goods, making carbon content evaluation by an independent organism easier through the application of a single procedure used as an international standard. Furthermore, such a measure would elude the issue of cleaner foreign goods than those manufactured in the regulated zone.[7] Offering an impartial criterion, the BAT could possibly obtain international approval and allow for the definition of international standards of production. Such approach to evaluate the carbon content of products leads, however, to a comparative assessment of production methods and processes, which might admittedly also be difficult to put into practice. Bearing in mind that BAT references do not take indirect emissions into account (for instance, emissions due to transportation) and also because of the difficulties firms have in evaluating their energy mix, the relevance of the BAT method for audits may be challenged.

4.2.3 Which Countries Would Be Affected?

Due to the existing differences in climate change regulation between developed and developing or emerging economies, BCAs would likely primarily affect the so-called BRICS and other developing countries.

Least developed countries (LDCs) could nevertheless be exempted, in order to protect their fragile development. However, exceptions of this kind are probably not the best course of action because LDCs would thus run the risk of being altogether excluded from the decision-making process. Besides, risks of trade infringement through LDCs are real in globalized and fragmented production processes, where it is difficult to track products and component parts, and this would considerably weaken such a BCA mechanism.

4.2.4 Which Flows?

In order to address climate change effectively, most sectors should in theory be subjected to environmental policies. Whether a given imported product is liable in whole or in part to a BCA will, however, naturally depend on the instrument in use (tax or TEP). Carbon markets cannot deal with all

[7] An over-allocation of TEPs or a subsidization of these products could be complementarily considered in order to ensure positive discriminatory treatment to those products and avoid discouraging progressive investment in cleaner technologies.

emissions, in particular small firms' emissions, given the need to standardize TEP. Taxing by tonne of emitted carbon equivalent would therefore seem more appropriate, allowing for larger product coverage.

Attention should also be drawn to exports outside an environmentally regulated area. A BCA can only restore a fair level of competition within a regulated area. Protecting environmentally regulated businesses may, for instance, involve a total exemption on their exports. Such a measure, however, would most likely encourage big emitter plants to produce for export markets, while lesser emitter plants would concentrate on domestic regulated markets. An option to avoid such effect might be to base the adjustment level for exports on a uniform benchmark, such as a BAT.

BCA implementation encounters, therefore, numerous practical and economic difficulties, carbon cost evaluation being particularly challenging.[8] However, these challenges, far from hampering BCA implementation, can steer the design of such measure towards the most efficient and fairest emission reduction mechanism, on the way to clean energy transition. Due to its effects on international competition and the international market, a BCA would, of course, also need to be compatible with countries' obligations in respect to International Trade Law.

4.3 BCAS AND INTERNATIONAL TRADE LAW

A BCA adopted by a member of the WTO should comply with the obligations undertaken by such member within the organization. The following subsections briefly present insights on the ways the General Agreement on Tariffs and Trade (GATT)[9] allows contesting or legitimizing BCA mechanisms.[10]

4.3.1 Challenging a BCA under the GATT Principle of Non-Discrimination

A core obligation established among WTO members is the non-discrimination principle. This principle is notably reflected in the 'national treatment' and the

[8] To avoid these manageable challenges, France has devised the path to determine the level of a BCA using the global GHG emissions volume per capita or per unit of gross domestic product (GDP), rather than the amount of carbon contained in a product (OECD 2014).

[9] For clarification purposes, questions regarding climate change within the WTO go beyond the GATT framework, also reaching other WTO agreements. Moreover, a country wishing to implement a BCA in its territory will need to comply with obligations it has also undertaken under other international treaties, including bilateral and regional trade agreements.

[10] As recalled by the OECD, 'measures that attempt to correct for carbon leakage, such as BCAs, are likely to be challenged by WTO members with export-oriented economies and relatively lax carbon regulation' (OECD 2014: 15).

'most favoured nation treatment' obligations. With respect to trade in goods, which is governed by the GATT, such obligations may be summarized as follows:

- National treatment (article III of the GATT) prohibits discriminatory treatment through internal taxes or other internal charges (article III: 2) between national products and foreign products that are considered like products or directly competitive or substitutable products. It also prevents a WTO member from applying a discriminatory treatment between like national and foreign products through laws, regulations, and requirements (article III: 4) (GATT 1947: article III);
- The Most Favoured Nation treatment (article I of the GATT) requires that a preference accorded to a product originating in or being exported to a WTO member be extended to similar products originating in or destined for the territories of all other members (GATT 1947: article I).

Besides the non-discrimination principle, and in order to ensure market opening and transparency, tariff restrictions are preferred to quantitative restrictions (e.g., quotas, import or export licences). Indeed, according to article XI of the GATT and subject to the exclusions foreseen in that provision, WTO members are not allowed to adopt quantitative restrictions to trade and quantitative restrictions must be administered in a non-discriminatory manner (article III of the GATT). BCAs might be associated to a quantitative restriction if a member decided to impose, exclusively on importers, the obligation to buy quotas on the carbon market, for example. It is, however, likely that a member would rather in such a case establish a regulatory scheme applicable to its entire market, including domestic producers, which would be examinable under article III (OECD 2014: 19).

In case a BCA adopted by a WTO member is considered by another member as altering the terms of competition existing in the international market by granting benefits exclusively to domestic producers, to the disadvantage of foreign producers, such BCA mechanism might be contested before the WTO Dispute Settlement Body (DSB) for alleged violation of the national treatment obligation under the GATT. The WTO judge will firstly need to assess whether the products at issue are 'like products' or 'directly competitive or substitutable products' according to article III of the GATT. The 'likeness' of products is evaluated on a case-by-case basis in WTO law, according to criteria taken from the GATT 1947 case law: 'the product's end-uses in a given market; consumers' tastes and habits, which change from country to country; the product's properties, nature and quality' (WTO 1996a: 20). Carbon-intensity may be raised as a distinctive feature between a domestic and a foreign product if a member demonstrates, for instance, that consumers have a preference for products with a lower carbon print, even if the products are physically identical.[11]

It is also tenable that, as long as the BCA entails the same level of constraint on importers and domestic producers, a BCA would not be discriminatory under the GATT national treatment obligation and would comply with WTO law in this regard. For instance, the GATT allows WTO members to apply a tax adjustment at the border when it authorizes members to levy, at the importation of a product, a charge equivalent to an internal tax, which directly or indirectly affects national like products (article II: 2 (a) of the GATT). The same rationale applies to tax adjustments at exportation when these adjustments simply correspond to the exemption or refund of internal duties or taxes applied to like products destined for consumption in the country of origin or exportation. In practice, depending on the type of BCA, such equivalency of treatment between domestic and foreign products can, however, be difficult to demonstrate. A tax imposed on importers of goods while national producers of like products are subject to a carbon market (and not a tax) would probably be more difficult to justify or to prove equivalency of.

In the event a BCA adopted by a WTO member was contested by another member before the WTO judge based on the GATT national treatment obligation, the difficulties mentioned in Section 4.2 of this chapter would also undoubtedly complicate the parties' arguments and the assessment to be made by the WTO judge (e.g., WTO-UNEP 2009: 101–3). In fact, in order to evaluate the impact of a specific BCA on competitive foreign products, it would, for instance, be necessary to measure precisely the carbon content of a product and the foreign product it is compared to, as well as to justify the determination of sectors and imported products, which are targeted by that BCA. On the other hand, it is also defensible that those difficulties might become less significant with the progressive development and larger acceptance of norms and objective BAT references for the calculation of carbon prints.

Based on the most favoured nation treatment obligation, violation of the GATT might in addition be alleged with respect to differences in BCAs applied according to the (foreign) targeted country of origin or destination of a product. One of the exceptions to the most favoured nation treatment in the WTO is the special and differential treatment in favour of developing and least developed countries, enshrined in varied provisions of the WTO Agreements.[12] Its implementation is, nevertheless, dependent on certain conditions. In fact, according to the WTO special and differential treatment,

[11] This might be a difficult task in practice, though. See OECD (2014: 16), which recalls that 'some authors argue that market studies most often will show that consumers generally ignore the processes and production methods (PPMs) of products'.

[12] For an overview of the WTO special and differential treatment and recent discussions on this theme, see WTO (2015).

a member wishing to favour LDCs when defining its BCA mechanism would need to employ well-defined and reasonable criteria, which would furthermore need to be applied in a non-discriminatory manner to every developing country presenting the same circumstances.[13]

Finally, the choice of countries targeted by a BCA could also possibly be challenged under the principle of 'shared but differentiated responsibilities', recognized nowadays as a component of sustainable development (United Nations 1992: Principle 7). In fact, this principle has been consecrated in the UNFCCC and it is defendable that it has been integrated as well into WTO law through the concept of sustainable development, which is acknowledged as an objective of the organization (WTO 1994: Preamble, emphasis added). Such principle requires that different countries' capacities, social and economic situation, and historical contribution to climate change be taken into account. It would therefore not be unlikely for a WTO judge assessing the design and implementation of a BCA to consider those elements in their legal reasoning and admit that the BCA may apply dissimilar treatment to different foreign countries according to their specific circumstances and their historical contribution to climate change.[14]

4.3.2 Resorting to the GATT General Exceptions Regarding Health and Environment Protection

In the advent of a contested BCA brought before a WTO judge and considered discriminatory by the latter, and therefore in violation of the GATT, the member having adopted the BCA would likely attempt to justify it on the grounds of the general exceptions under article XX of the GATT. These general exceptions are part of the general balance of rights negotiated by WTO members (*EC—Hormones* (DS26 and DS48) and *US—Shrimp* (DS58)) (WTO 1998a, 1998b) and aim, among others, to allow members to adopt measures for health and environmental protection that might otherwise be considered contrary to GATT obligations. Article XX (b) and (g) of the GATT establish exceptions for measures '(b) necessary to protect human, animal or plant life or health' and '(g) relating to the conservation of exhaustible natural resources if such measures are made effective in conjunction with restrictions on domestic production or consumption'.

Article XX (b) requires the fulfilment of a necessity criterion, implying a proportionality control of a measure with respect to the objective that a

[13] See WTO (2004: §§ 160–7). For a brief chronological overview of the special and differential treatment at the WTO, see Richieri Hanania (2009: 137–40).

[14] According to Morosini (2010: 717–25), border adjustment measures in the Waxman-Markey Bill were inconsistent with WTO law, inter alia, because they did not respect the principle of shared but differentiated responsibilities.

member is pursuing, as well as the availability of alternative reasonable measures that might be less restrictive to trade but as effective. Such interpretation has been applied in the well-known cases *Korea—Beef* (DS161 and DS169), *EC—Asbestos* (DS135), *US—Gambling* (DS285), and *Brazil—Retreaded Tyres* (DS332), where the WTO judge explained that a measure may be considered necessary based on the importance of the interest or value it protects, its contribution to the materializing of the objective pursued (effectiveness of the measure), and its degree of restriction to trade (WTO 2000: §§ 160). Once these conditions are all met, article XX (b) may be considered applicable. When defending the non-applicability of article XX (b) to a BCA, a WTO member would thus need to demonstrate that other alternative, as effective and less trade restrictive options existed.

Regarding article XX (g), the expression 'relating to' has been interpreted as meaning 'primarily aimed at' (*US—Gasoline* (DS2)) (WTO 1996b: 15–19), which is larger than the notion of 'necessary'. If a BCA is demonstrated to be primarily aimed at environmental objectives (and not only competiveness issues), article XX (g) might be considered applicable. Moreover, in the case *US—Shrimp* (DS58), the WTO Appellate Body considered that the notion of 'exhaustible natural resources' should be read according to contemporary concerns regarding the protection of the environment, as defined in recent international treaties (WTO 1998b: § 129). Such decision allows for an evolutionary interpretation of general exceptions in the GATT and demonstrates the WTO judge's wish to take into account environmental objectives as they are progressively established in International Law under the concept of sustainable development. Environmental interests reflected in a BCA might therefore be positively considered by the WTO DSB based on contemporary environment-related concerns and recent climate change discussions.

However, the exceptions listed in article XX are to be interpreted together with its introductory statement, the *chapeau* of article XX: a measure must not be applied 'in a manner which would constitute a means of arbitrary or unjustifiable discrimination between countries where the same conditions prevail, or a disguised restriction on international trade' (GATT 1947: article XX). In fact, should the challenged WTO member succeed in demonstrating that its BCA mechanism conforms to one of the exceptions of article XX, the *chapeau* of that same article would still require the WTO judge to verify how the contested measure has been applied. In a few words, the WTO judge would need to examine, cumulatively, if:

- The defendant has attempted to cooperate with other members in a non-discriminatory manner in order to diminish the trade restrictive effects that the contested measure may have.
- The measure is sufficiently flexible to account for different conditions existing among countries, even if it is not required from a WTO member to expressly anticipate individual conditions of each member.

- The conception, the base principles, and the structure of the measure do not testify to protectionist purposes. The environmental objectives of a BCA will need to be particularly highlighted in order for a member to respond to this condition.

While the GATT principles exposed in this section may guide the design of a BCA, its consistency with WTO rules may only be considered on a case-by-case basis. In case of a dispute on the matter, the WTO judge will necessarily examine the economic features and obstacles identified in the preceding sections of this chapter for each type of BCA, in order to assess their conformity to International Trade Law. Since their effects on international competition and their complexity may be significant, it is likely that, while certain BCAs might be considered as responding to the conditions required under one of the paragraphs of article XX, their acceptability according to the *chapeau* of that same article might be problematic. Any attempt to elaborate a BCA that respects WTO Law should therefore take into account the three cumulative criteria listed here (pp. 71–2).

Based on the economic and legal aspects put forward in Sections 4.2 and 4.3, the following section discusses the acceptability of BCAs from a political sciences standpoint.

4.4 THE POLITICAL IMPLEMENTATION OF BCAS IN THE CONTEXT OF ENERGY TRANSITION

The objectives stated in the COP21 agreement—that is, capping at 1.5°C–2°C the global temperature increase, as well as the reference to carbon neutrality—seem vague since no specific emissions reduction targets have been agreed upon by the Parties to respond to the urgency in limiting global warming. The only measure which can be viewed as relatively constraining are the 'Intended Nationally Determined Contributions' (INDCs).

Consequently, the implementation of BCAs remains an effective tool for countries ready to implement, or having already committed to, a bold energy transition. BCAs can level the environmental commitment (INDC) among countries facing similar responsibilities in climate change. Moreover, a well-designed BCA should aim at answering the concerns of developed countries while promoting growth in developing countries, and should be complemented by sobriety policies.[15]

[15] Referring to the principle of primary energy consumption reduction by using less of an energy service, also called energy sparing, 'sustainable lifestyles and sustainable patterns of consumption and production' (COP21 Paris Agreement, Preamble), behavioural change, energy saving, energy soberness, energy sobriety, and so on. The authors have chosen to use 'sobriety' in

4.4.1 Energy Transition, BCAs, and Developing Economies

As BCAs are gaining more interest from both governments and academia, the growing body of scientific literature has mainly focused on carbon leakage and competitiveness issues. However, another obstacle to BCAs implementation involves their potential consequences on developing economies. Indeed, BCAs have also been interpreted as a commercial sanction towards developing countries, making international negotiations related to this measure more complicated. This question has become a recurring argument against BCAs, but requires a more careful examination of the stakes for developing countries. In an insightful analysis of climate change and trade policy, Mattoo and Subramanian (2013) evaluate the different forms of BCAs and their consequences from both a trade and an environmental perspective. The authors suggest the 'least undesirable' form of a BCA: '[a] border tax adjustment based on the carbon content in domestic production would broadly address the competitiveness concerns of producers in high-income countries while inflicting less damage on developing-country trade' (Mattoo and Subramanian 2013: 113). To ensure developing countries do not remain vulnerable to trade policy, the international climate change debate, rather than WTO negotiations alone, should also cover international agreements on such trade impact.

At the global level, matters are much more intricate when considering the energy transition and economic development nexus. Reframing the problem, and therefore looking for viable solutions without hampering development opportunities, have become both necessary and urgent. Mattoo and Subramanian propose to challenge what they dub the 'narrative' problem, that is the historical perspective on the industrialized countries' responsibilities in climate change and the past 'colonization of emissions space' (Mattoo and Subramanian 2013: 1).

A decisive climate change issue for developing countries relates to technology generation and transfer. Technology generation, diffusion, and management policies are a key component of climate change mitigation and adaptation at the global scale (IPCC 2014). Technological change, accelerated through international cooperation, could enable developing countries to meet emission reduction goals. This could be achieved by shifting from the sole financing of emission cuts to an investment in technology generation (Mattoo and Subramanian 2013: 18). However, reaching consensus on international technological cooperation might prove to be as difficult as achieving an agreement in international climate negotiations. Considering the urgency of climate change matters, the unilateral implementation of BCAs could represent, beyond a

the present study, the translation of the French term 'sobriété', because it best describes our current energy overconsumption and its 'hangover' results. Sobriety differs from efficiency (or efficient energy use), which refers to using less energy for a constant service.

palliative solution, a necessity. In fact, this type of measure could serve as an instrument to discourage environmental dumping and promote international involvement and significant mitigation commitments (beyond INDCs) in the framework of the Paris Agreement. Furthermore, resulting in an increase in prices in developed countries, BCAs limit consumption in the North and leave resources available for the South. As inequalities (of access to goods and of income) exacerbate environmental damage, in turn further increasing inequalities (UNDP 2011), BCAs could play an important role in promoting equity and development. A BCA would, therefore, be one step among others towards energy transition and serious efforts in addressing climate change.

4.4.2 BCAs, an Available Tool to Reshape the World

Climate change issues, and environmental concerns at large, also present a social dimension, which includes inequality and development issues related to burden sharing, ultimately translated into the principle of common but differentiated responsibilities towards a carbon-free world.

In order to address that objective, green growth is often advocated, not as a means but as an end, in different forms (sustainable and inclusive growth or green economy). It entails the 'greening' of economic activities according to a particular logic that starts with renewable energy and ends with energy efficiency facilitated by technology. Yet this reasoning overlooks four essential elements. First, economic activity is parameterized by our ability to produce energy: the more energy per capita, the higher the growth potential in developed countries (Giraud and Kahraman 2014). Second, the ability to substitute the primary carbon energy consumed for renewable energy depends on the availability of materials needed for windmills or photovoltaic panels, as well as on the pace of deployment of these systems (Roman 2014). Actually, the historical energy return on investments rates are decreasing, all energies considered, and more energy is needed to produce energy and to extract raw materials, since the most accessible deposits have been or are becoming depleted. Furthermore, if renewable energy systems are deployed too fast, the newly created energy will be used for the production of other units of carbon-free energy production, rather than substituting current polluting energies (Pearce 2009). Third, the focus on renewable energy and energy efficiency conceals the issues of energy embedded in products[16] and 'grey energy'.[17] Fourth, technological solutions may prove hazardous. The benefits in energy efficiency achieved through technological progress are real, but remain

[16] Embedded energy is the energy needed to produce a given good.

[17] Grey energy is the sum of embedded energy and all other forms of energy used throughout the lifecycle of a given good until its recycling.

insufficient, particularly given the environmental urgency. In addition, the rebound effect, process by which greater energy efficiency may lead to an increased energy use ('buy a more fuel-efficient car, drive more' rationale) is another issue for technological developments. Finally, since innovation cycles continue to be variable and unpredictable, waiting for a major innovation that could completely change energy use seems too precarious.

For these reasons, sobriety policies are as necessary as renewable energies and energy efficiency in order to achieve energy transition. Often disregarded because it entails a strong political will, this policy concept challenges certain aspects of economic growth. Sobriety policies aim indeed at classifying products according to utility and regulating their production and use. For instance, organic farming based on agroecological principles should be promoted, given the fact that such type of production of agricultural goods involves lower inputs and fewer cleanup operations. This is where sobriety collides with growth: any activity involving an over-consumption of energy or even environment cleanup operations contributes positively to the gross domestic product (GDP).

The implementation of sobriety policies, followed by energy efficiency policies, and the deployment of renewable energies,[18] reverses the traditional reasoning applied to climate policy and underlines the need for a strong, compelling regulation. Dealing with the complexity of the economical–environmental–social nexus, itself embedded in a fluctuating international context, national regulators must resort to various instruments to foster definite changes in behaviour. BCAs are an essential component of this array of instruments. They contribute to the development of more local, thus 'sober' activities.

In addition, limiting transportation has become a necessity, as the COP21 French ambassador underlined (Tubiana 1991). Sea transportation is still not subjected to emissions reduction and a recent emissions stabilization agreement for air transportation agreed to at the International Civil Aviation Organization contains no mention of objectives of 1.5°C–2°C, despite the need for greater regulation of those sectors. Transport reduction policies, perhaps supplemented by trade reduction policies, would encourage local production, circular economy,[19] economy of functionality,[20] and thus development, at any development

[18] Sobriety policies and energy efficiency would reduce primary energy consumption by 66 per cent. As a result, the need for renewable energies would be considerably reduced, paving the way to smoother raw materials sharing at the international level (Association NégaWatts 2011).

[19] The main goal of circular economy is to enable the production of goods and services while preventing excessive consumption and squandering of raw materials, and as to facilitate recycling.

[20] A functional economy is one that optimizes the use (or function) of goods and services and thus the management of existing wealth (goods, knowledge, and nature). The economic objective

stage. Again, BCAs may be seen as regulatory instruments available for the national regulator to foster behaviour change.

Moreover, the implementation of BCAs transcends the stakes of competitiveness and level playing field issues. Seeking to reach an equitable distribution of commitment, and considering the principle of common but differentiated responsibilities, it appears necessary to shift from national production-based emissions counts to a consumption-based approach. Considering the amount of GHG consumed rather than produced opens a fair route, between countries as well as within social categories, and encourages sobriety behaviours. Additionally, BCAs become incentive schemes for countries to adapt production to local needs and rethink their environmental policy. BCAs would thus become an essential component of social and environmental justice,[21] helping to depart from a business and macroeconomic logic.

In sum, BCA enforcement would be, in all likeliness, necessary, together with paradigm shifts, breeding strong incentives to ease developing countries' acceptance of this instrument. Also, funds collected via BCAs should be reallocated to objectives such as the production of more sustainable goods, innovation, technology, or skills transfer. The Green Climate Fund (GCF) could equally be made the recipient of all or parts of these funds.

4.5 CONCLUSION

The commitments agreed upon by the Parties during the COP21 should lay the foundations for any energy transition strategy but also for economic transition policies. However difficult to meet, these commitments entail a profound, and voluntary, transformation of our societies. The BCA mechanism can enable our (over)consumption societies to move towards a sober, low-carbon society, which uses energy and resources thoughtfully. BCAs turn out to be more than a simple restraint towards a given trade partner. It is, actually, a powerful means to share resources.

In fact, obstacles to climate negotiations are elsewhere. Developed countries must assume their historical responsibilities in climate change through an effective constraining commitment to limit emissions at the national level, which entails important internal transformations to achieve energy transition

of the functional economy is to create the highest possible use value for the longest possible time while consuming as few material resources and energy as possible.

[21] Poverty and environmental destruction are highly correlated (Laurent 2011; Keucheyan 2014).

effectively. They should respond to demands from developing and some emerging economies for financing, technology transfer, and sobriety policies. Such requirements could be accommodated through the implementation of a global carbon price. BCAs would be useful in balancing environmental efforts with growing equity. Difficulties in implementation, which appear to be more of an economic nature than legal or political, can be overcome. The French proposal to apply BCAs according to the global GHG emissions volume per capita or per unit of GDP, rather than the amount of carbon contained in a product, seems to be not only feasible, but also fair since calculation would be favourable to developing countries. The emission ratio by MWh created by the International Energy Agency, which tracks average GHG emissions per country for each megawatt produced, could also be employed.

BCA implementation requires, nevertheless, a strong and durable political commitment to environmental protection and its design calls for a careful assessment of the numerous constraints exposed in this chapter.

ACKNOWLEDGEMENTS

The thoughts and opinions expressed here are those of the individual contributors alone and do not necessarily reflect the views of their employers. The work was supported in part by the CEST (Centro de Estudos Sociedade e Tecnologia), University of São Paulo, Brazil.

REFERENCES

Association NégaWatts (2011). 'Manifeste NégaWatts, Domaine du possible'. Actes Sud/Colibris/Association NégaWatt, France.

Böhringer, C., E. J. Balistreri, and T. F. Rutherford (2012). 'The Role of Border Carbon Adjustment in Unilateral Climate Policy: Results from EMF 29'. *Energy Economics*, 34(2): 97–110.

Branger, F., P. Quirion, and J. Chevallier (2013). 'Carbon Leakage and Competitiveness of Cement and Steel Industries under the EU ETS: Much Ado about Nothing'. Cired Working Paper 53–2013, Cired, Paris.

Bueb, J. (2014). 'Marché de quotas d'émission négociables ou taxe carbone? Un enjeu en termes d'efficacité'. *IHEJ*, La Note de ConventionS, n°13. Available at: <http://convention-s.fr/notes/marche-de-quotas-demission-negociables-ou-taxe-carbone-un-enjeu-en-termes-defficacite/> (accessed 16 November 2016).

Demailly, D. and P. Quirion (2008). 'Concilier compétitivité industrielle et politique climatique: faut-il distribuer les quotas de CO2 en fonction de la production ou bien les ajuster aux frontières?' *La Revue Économique*, 59(3): 497–504.

Dröge, S., H. van Hasselt, T. Brewer, M. Grubb, R. Ismer, Y. Kameyama, M. Mehling, S. Monjon, K. Neuhoff, P. Quirion, K. Schumacher, L. Mohr, W. Suwala,

Y. Takamura, T. Voituriez, and X. Wang (2009). 'Tackling Leakage in a World of Unequal Carbon Prices'. Synthesis Report, Climate Strategies, University of Cambridge, Cambridge.

France (2009). 'LOI n° 2009-967 du 3 août 2009 de programmation relative à la mise en œuvre du Grenelle de l'environnement'. JORF no. 0179 du 5 août 2009 page 13031, texte no. 2. Available at: <https://www.legifrance.gouv.fr/eli/loi/2009/8/3/DEVX0811607L/jo/texte> (accessed 21 March 2016).

GATT (1947). 'The General Agreement on Tariffs and Trade'. Available at: <https://www.wto.org/english/docs_e/legal_e/gatt47_01_e.htm> (accessed 14 March 2016).

Giraud, G. and Z. Kahraman (2014). 'How Dependent Is Growth from Primary Energy?' CNRS Working Paper, CNRS, France.

Houser, T., R. Bradley, B. Childs, J. Werksman, and R. Heilmayr (2008). 'Leveling the Carbon Playing Field: International Competition and US Climate Policy Design'. Peterson Institute for International Economics and World Resources Institute, Washington, D.C. Available at: <http://pdf.wri.org/leveling_the_carbon_playing_field.pdf> (accessed 11 March 2016).

IPCC (Intergovernmental Panel on Climate Change) (2014). 'Climate Change 2014: Synthesis Report'. Contribution of Working Groups I, II, and III to the Fifth Assessment Report of the Intergovernmental Panel on Climate Change, IPCC, Geneva.

Ireland, R. (2010). 'Implications for Customs of Climate Change Mitigation and Adaptation Policy Options: A Preliminary Examination'. *World Customs Journal*, 4(2): 21–36.

Keucheyan, R. (2014). 'La nature est un champ de bataille'. Essai d'écologie politique. Zones, La Découverte, Paris.

Laurent, E. (2011). *Social écologie*. Paris: Flammarion.

Markusen, J. R. (1975). 'International Externalities and Optimal Tax Structures'. *Journal of International Economics*, 5(1): 15–29.

Mattoo, A. and A. Subramanian (2013). 'Greenprint: A New Approach to Cooperation on Climate Change'. Center for Global Development, Washington, D.C.

Morosini, F. (2010). 'Trade and Climate Change: Unveiling the Principle of Common but Differentiated Responsibilities from the WTO Agreements'. *George Washington International Law Review*, 42(4): 717–25.

OECD (Organization for Economic Co-operation and Development) (2014). 'Border Carbon Adjustment and International Trade: A Literature Review'. Joint Working Party on Trade and Environment, COM/TAD/ENV/JWPTE(2014)51/FINAL, OECD Trade and Agriculture Directorate and Environment Directorate, Paris.

Pearce, J. M. (2009). 'Optimizing Greenhouse Gas Mitigation Strategies to Suppress Energy Cannibalism'. Second Conference on Climate Change Technologies, Hamilton, Ontario, Canada.

Richieri Hanania, L. (2009). 'Diversité culturelle et droit international du commerce'. CERIC, La Documentation française, Paris.

Roman, P. (2014). 'Le découplage PIB-matières'. In T. Caminel, G. Giraud, P. Frémeaux, A. Lalucq, and P. Roman (eds.), *Produire Plus, Polluer Moins: l'Impossible Découplage?*. Paris: Les Petits Matins, pp. 45–62.

Schott J. J. and M. Fickling (2010). 'Revisiting the NAFTA Agenda on Climate Change'. Peter G. Peterson Institute for International Economics, PB 10-19. Available at: <http://www.piie.com/publications/pb/pb10-19.pdf> (accessed 22 March 2016).

Tubiana L. (1991). 'La mondialisation contre l'écologie'. *Le Monde diplomatique.* Available at: <http://www.monde-diplomatique.fr/1991/01/TUBIANA/43229> (accessed 11 March 2016).

UNDP (United Nations Development Programme) (2011). 'Human Development Report 2011. Sustainability and Equity: A Better Future for All'. Available at: <http://hdr.undp.org/sites/default/files/reports/271/hdr_2011_en_complete.pdf> (accessed 14 March 2016).

United Nations (1992). 'Rio Declaration on Environment and Development'. Report of the United Nations Conference on Environment and Development, Rio de Janeiro, 3–14 June 1992, Annex I. Available at: <http://www.un.org/documents/ga/conf151/aconf15126-1annex1.htm> (accessed 14 March 2016).

Weber, C. L. and G. P. Peters (2009). 'Climate Change Policy and International Trade: Policy Considerations in the US'. *Energy Policy*, 37(2): 432–40.

WTO (World Trade Organization) (1994). 'Marrakesh Agreement Establishing the World Trade Organization'. Available at: <https://www.wto.org/english/docs_e/legal_e/04-wto_e.htm> (accessed 14 March 2016).

WTO (1996a). 'Appellate Body Report, Japan—Taxes on alcoholic Beverages'. WT/DS8/AB/R; WT/DS10/AB/R; WT/DS11/AB/R. World Trade Organization, Geneva.

WTO (1996b). 'Appellate Body Report, United States—Standards for Reformulated and Conventional Gasoline'. WT/DS2/AB/R. World Trade Organization, Geneva.

WTO (1998a). 'Appellate Body Report, European Communities—Measures Concerning Meat and Meat Products (Hormones)'. WT/DS26/AB/R and WT/DS48/AB/R. World Trade Organization, Geneva.

WTO (1998b). 'Appellate Body Report, United States—Import Prohibition of Certain Shrimp and Shrimp Products'. WT/DS58/AB/R. World Trade Organization, Geneva.

WTO (2000). 'Appellate Body Report, Korea—Measures Affecting Imports of Fresh, Chilled and Frozen Beef'. WT/DS161/AB/R and WT/DS169/AB/R. World Trade Organization, Geneva.

WTO (2004). 'Appellate Body Report, European Communities—Conditions for the Granting of Tariff Preferences to Developing Countries'. WT/DS246/AB/R.World Trade Organization, Geneva.

WTO (2015). 'Special and Differential Treatment Provisions'. World Trade Organization, Geneva. Available at: <https://www.wto.org/english/tratop_e/devel_e/dev_special_differential_provisions_e.htm> (accessed 16 September 2015).

WTO-UNEP (2009). 'Trade and Climate Change'. Joint Report by the United Nations Environment Programme and the World Trade Organization. Available at: <https://www.wto.org/english/res_e/booksp_e/trade_climate_change_e.pdf> (accessed 16 September 2015).

Zhang, Z. X. (2010). 'Climate Change Meets trade in Promoting Green Growth: Potential Conflicts and Synergies'. Fondazione Eni Enrico Mattei Working Papers 408. FEEM, Venice.

5

Support Policies for Renewables: Instrument Choice and Instrument Change from a Public Choice Perspective

Erik Gawel, Sebastian Strunz, and Paul Lehmann

5.1 THE ENERGY TRANSITION AS A POLITICAL CHALLENGE

In order to limit the risk of climate change, mankind needs to initiate the transition away from fossil towards clean energies now—delaying mitigation would imply 'much more rapid scale-up of low carbon energy' over the period 2030–50 if any chance is to remain of restricting global warming to 2° Celsius, or even less (COP 2015: 2), compared to pre-industrial levels. From an economic point of view, consequently, postponing policy interventions that facilitate the transition is costly (Acemoglu et al. 2012; Kalkuhl, Edenhofer, and Lessmann 2012). However, timely and rigorous policy efforts to curb fossil energies are hampered by vested interests and institutional inertia (e.g., Helm 2010; Spash 2010).

Against this backdrop, Germany's energy transition is a particularly fertile research topic. Germany has set itself very ambitious transition goals that would, inter alia, completely restructure the electricity system around renewable energy sources (RES) by the middle of this century (Bundesregierung 2010). So far, the German transition pathway closely follows the envisaged trajectory as regards RES electricity expansion.[1] The RES share, measured

[1] Outside the electricity sector, the transformation does not unfold equally rapidly. To the contrary, progress in transforming transport and heating is rather slow (cf. Monopolkommission 2013): within the heating sector, at least some transformation efforts are present, while the transport sector displays strong inertia. The variety of both stakeholders and policy instruments across the different sectors is considerable. Hence, a detailed comparison between the different sectors lies beyond the scope of this chapter, which focuses on the electricity sector.

against gross electricity consumption, has almost quadrupled from 7 per cent in 2000 to 27.8 per cent in 2014 and will increase up to 80 per cent by 2050 (BDEW 2015: 15). Strong international attention is focused on the transition process—it has been referred to as the 'possibly globally catalytic "*Energiewende*"' (Stirling 2014: 87). There are three reasons a successful transition in Germany might serve as a leading international example. First, Germany represents a highly industrialized economy (industry's share at gross value added was 25.5 per cent in 2013 as compared to the European Union [EU] average of 19.1 per cent).[2] Second, the starting position for the transition has not been particularly good, with only 4 per cent of gross electricity consumption being renewable electricity during the 1990s. Third, similarly ambitious long-term goals are not common in many other nations so far.

When initiating the energy transition, which policies did German regulators choose and for what reasons? The theoretical framework that we employ to address this question is the Public Choice perspective: it holds that policy choices derive from the self-interest driven behaviour of political stakeholders (e.g., McCormick and Tollison 1981; Niskanen 1971; Olson 1971; Stigler 1971; Tullock 1967). The transition towards clean energy is no different in this respect: energy policy choices also involve 'rent management', that is, (re) distribution of resources among stakeholders (Schmitz, Johnson, and Altenburg 2013). In consequence, the internalizing of climate and other environmental externalities from conventional energies (e.g., nuclear risks, leaking gas pipelines, local air pollution) might not be feasible in a direct and timely way. The first objective of this chapter, therefore, consists in explaining why specific instrument choices initiating the energy transition have been made in Germany and why they have turned out to be successful so far.

Yet, increasing the share of RES is not sufficient to achieve the sustainable transformation of the electricity system. For instance, the reduction of energy consumption via efficiency measures and altered consumption patterns constitutes a complementary measure. Furthermore, the rise of RES brings along new challenges, such as controlling the costs of RES deployment, the need to adapt the system to fluctuating feed-in from wind and solar energy or the mitigation of environmental impacts from RES generation. In consequence, 'first-generation' RES policies should evolve into 'next-generation integrated power system policies' (Miller et al. 2013). From the regulator's perspective, however, this task is fraught with uncertainties related to, inter alia, RES generation costs, patterns of technological change, external environmental and system integration costs associated with RES deployment, or the response of RES investors to public policies (Purkus et al. 2015). These uncertainties

[2] See: <http://ec.europa.eu/eurostat/statistics-explained/index.php/File:Gross_value_added_at_basic_prices,_2003_and_2013_%28%25_share_of_total_gross_value_added%29_YB15-de.png>.

imply that RES policy decisions taken today may turn out to be erroneous in the future, as new information becomes available to regulators.

The second objective of this chapter, therefore, consists in shedding some light on the question of how the chosen support instrument can be adapted to the manifold uncertainties of the transition. While this sounds like an unequivocal call for political flexibility, potential trade-offs arising from reduced policy stability need to be acknowledged. Stability is a key ingredient of successful economic policy (Weingast 1993) since uncertainty may lead to inefficiently low private investment (Dixit 1992; Hepburn 2006). Moreover, the adaptation of instruments over time may be impaired by the fact that they inevitably create technological and institutional path dependencies. Accordingly, instruments have to keep up with changed circumstances in order to avoid carbon lock-in (Unruh 2000) merely being replaced with a new RES lock-in.

We are thus faced with a challenging and rather complex set of requirements for policy intervention to facilitate the transition towards clean energies: substantial transition policies need to be launched right away to ensure timely climate mitigation; the emerging policy framework should be flexible enough to enable adaptation to unforeseen challenges (for example, technological surprises) but it should also be stable enough not to curb the deployment of clean RES. Even for a 'benevolent dictator', this task would be far from trivial.

Summing up, we frame the overall challenge of 'how to facilitate the transition towards clean energies' as a two-stage Public Choice problem applied to the case of Germany's *Energiewende.* The first stage represents the *instrument choice* problem of a regulator addressing climate and environmental externalities of energy use in the electricity sector. We demonstrate that support policies for RES better conform to the Public Choice framework than alternative policy options. An overview of the introduction and development of Germany's support scheme for RES corroborates the theoretical predictions. The second stage concerns the problem of *instrument change* regarding necessary adaptations of the RES support scheme: in the long run, fundamental changes are inevitable to promote market and system integration of RES. One crucial question here is which possible solutions the Public Choice framework offers for the trade-off between political stability and flexibility (i.e., rigidity vs. uncertainty; see Acemoglu et al. 2008; Rodrik 1996): the design of transition policies, or 'green industrial policies', as they have been called, needs to take the self-interested motivations of political stakeholders into account—otherwise, policy recommendations may well turn out to be practically irrelevant (see Rodrik 2014; Schmitz, Johnson, and Altenburg 2013). Certainly, the Public Choice perspective is only one among a number of possible ways to frame the transition towards clean energies; yet it brings with it the particular merit of highlighting the politico-economic restrictions for any transition strategy.

The remainder of this chapter is organized as follows. In Section 5.2, the problem of instrument choice is analysed against the background of Germany's support scheme for RES. Section 5.3 addresses the problem of instrument change: how to adapt existing RES support schemes to enable market and system integration of RES. In Section 5.4, we discuss our findings from the German case and draw conclusions for the global challenge of clean energy transitions.

5.2 SUPPORT FOR RENEWABLES AS A SOLUTION TO CRUCIAL CLIMATE AND ENERGY POLICY CHALLENGES

5.2.1 The Problem of Instrument Choice

In principle, policy makers could choose a range of instruments to initiate the phase-out of fossil energies. As Table 5.1 illustrates, these instruments differ with respect to their rent management potential, redistributing resources in a targeted way by 'providing (and withdrawing) opportunities for above-average profits and investment' (Schmitz, Johnson, and Altenburg 2013: 5). For instance, a carbon tax might be considered the most direct way to internalize climate damage. Yet carbon taxes trigger strong resistance by emitting industries that generally prefer cap-and-trade schemes because they provide more possibilities to extract rents—(e.g., grandfathering of emission permits; see also Spash (2010)). In other words, getting the carbon prices right might prove to be impossible because of the bargaining power of emitting industries: carbon pricing puts abatement costs directly on those stakeholders represented by well-organized interest groups.

From a public choice perspective, then, regulators can be expected to choose those instruments that maximize stakeholder support by enabling the most targeted distribution of rents: politicians act as transfer brokers, redistributing welfare between different stakeholders so as to maximize their chances of electoral success and their personal income (McCormick and Tollison 1981). Eventually, the best organized interest groups extract rents by steering regulation in their favour (Olson 1971; Stigler 1971; Tullock 1967). This perspective implies strong restrictions on the feasibility of textbook climate and energy policy instruments: climate policy essentially constitutes a 'carbon pork barrel' that waits to be distributed among stakeholders (Helm 2010), all of whom strive to obtain the largest possible share. Once a policy instrument has been chosen, its implementation offers another possibility of rent-seeking: bureaucracies also follow their special interests, such as budget maximization, and therefore may use any discretionary power over regulatory details in their favour (Gawel 1995; Niskanen 1971). In sum, the *instrument choice* problem, as framed through the Public Choice lens, reads: how can regulators meet the demand for transition policies in a way that maximizes stakeholder support?

Against this backdrop, support policies for RES promise two main political benefits. First, by encouraging the switch away from fossil fuels without directly increasing the emitters' abatement burden, they face less political resistance from organized fossil interest groups. Instead, the burden of the transformation is to be borne by less organized groups (e.g., electricity consumers). Within the EU emissions trading scheme, RES support lowers the allowance price and therefore makes stricter emission caps even easier to negotiate (Gawel, Strunz, and Lehmann 2014). Second, RES subsidies constitute rents to be distributed; hence, they extend the 'rent management' (Schmitz, Johnson, and Altenburg 2013) potential of decision makers. In addition to these political benefits, support for RES may also increase the overall efficiency of energy provision insofar as such energy sources help to address other environmental externalities beyond climate change, for which direct policy instruments with burdening effects may not be politically feasible (Sijm et al. 2014).

Overall, there is a wide variety of ways to organize RES support. The annual Global Status Reports on RES development catalogue all pertinent instruments applied worldwide (see REN21 2015: 99ff.). As demonstrated in Table 5.1, the decision to support RES also needs to be accompanied by a series of more specific instrument design choices—for instance, regarding the level of support granted to RES or the financing mechanism. For all of these specific choices about RES support, the targeted distribution of rents provides a major design incentive. In the following, we will take a more specific look at the characteristics of Germany's RES support framework for RES as it has emerged over the last 25 years.

5.2.2 Germany's Support Policies for Renewables

Until 2012, RES support was exclusively based on a feed-in tariff, complemented by prioritized feed-in for RES-generated electricity. This implied a very high degree of investment security for RES producers in two respects. First, they received a fixed remuneration for every kWh produced over 20 years. Second, even when the electricity produced was 'superfluous' (such as in times of negative prices on the spot market, or if looming grid overload necessitates disconnection of a wind generator from the grid), the RES producers have been (and still are being) compensated. Efforts recently initiated to adapt this scheme will be addressed in Section 5.3.2 on *instrument change*, while this section focuses on the introductory phase of *instrument choice*.

Table 5.1. Transition instruments and their rent management potential: schematic overview

Categories of instrument design	Characteristics of transition instruments	Theoretical prediction: targeted redistribution of rents	Empirical observation in Germany
Internalizing instruments: Carbon tax, emissions cap-and-trade			
Price and/or quantity regulation	Price regulation (tax) vs. quantity regulation (emissions cap)	Quantity regulation more susceptible to regulatory capture than price regulation	No general CO_2 tax, only sector-specific (e.g., gasoline), emissions trading scheme on EU level
Differentiation and exemptions	Auctioning or grandfathering of emission permits, tax discounts	Cap-and-trade with grandfathering	Cap-and-trade with initial grandfathering and loopholes (e.g., Clean Development Mechanism credits)
Support instruments for clean energy: Tax credit, feed-in tariff, feed-in premium (fixed or flexible), quota scheme, tender scheme			
Support level	RES production costs as benchmark	Production cost plus mark-up to enable rent management	Production cost plus mark-up
Exposure to market risks	From low (feed-in tariff) to high (tender schemes) exposure	RES groups favour no exposure to market risk	1991–2012: feed-in tariff (no market risk); from 2012 on: phasing out of feed-in tariff (slow increase in market risk)
Differentiation within finance mechanism	Financing source (e.g., public budgets or levy on electricity prices) and differentiation within eligible group	Lobby groups aim at minimizing their constituents' contribution, politicians aim at concealing costs	RES levy on electricity prices; generous exemptions for energy-intensive industries
Command-and-control instruments: Emission standards, technical standards			
Performance and/or behaviour regulation	Mandate a certain standard (e.g., prohibition of incandescent light bulbs, emission limits for car engines)	Efficiency standards (performance) instead of consumption quotas (behaviour)	No consumption quotas; successful lobbying against strong emission standards for cars

Source: authors, based on a review of public choice literature (e.g., McCormick and Tollison 1981; Schmitz et al. 2013) and previous original work (e.g., Gawel et al. 2014).

At first glance, it might seem puzzling that RES support has been introduced at all (see Strunz, Gawel, and Lehmann (2016) for a more detailed account of the following argument). Industry interest groups are mostly better organized than environmental interests (Kirchgässner and Schneider 2003), and therefore they enjoy excellent access to policy makers. This asymmetry in political clout has also paid off in financial terms: conventional energies have benefitted from cumulative subsidies that amount to several times the cumulated RES subsidies (1970–2014: €327 billion for hard coal, €219 billion for nuclear power, €102 billion for RES; Küchler and Wronski 2015: 7). So why did or could conventional utilities not prevent the rise of a competing RES industry? When RES first received direct (non-R&D) subsidies under a support scheme in 1991, they were only minor niche technologies. Hence the 'big four', the oligopoly of conventional utilities that dominated the German electricity market, underestimated the long-term threat posed by RES. Legislators, in turn, by positively supporting RES (and thus bringing purchasing power to the benefitting sectors at the public expense) rather than heavily taxing fossil resources (and thus cutting rents for powerful pressure groups), could circumvent the opposition of well-organized conventional industry groups. Thus, the RES support represented a seemingly harmless, symbolic way of currying favour with the environmental movement that had been growing since the 1980s in Germany.

Yet the extension of RES support in 2000 crucially affected economic and subsequently political dynamics. The RES share, measured against gross electricity consumption, has risen from 7 per cent in 2000 to 27.8 per cent in 2014 (BDEW 2015: 15). In economic terms, the increased financial subsidies for RES contributed to relative price changes to the disadvantage of conventional energy. As a result, the conventional utilities' profits have plummeted in recent years, whereas the sums distributed by the RES scheme are continuously increasing. Over time, self-reinforcing dynamics emerged (see Strunz 2014): RES support created RES constituencies that benefit from and rely on the support scheme: 'green jobs' in the RES industry, small-scale RES producers such as PV (photovoltaic) homeowners, and so on. Consequently, the political leverage of the RES sector now rivals the influence of the conventional industry, which serves to perpetuate RES subsidies (see Section 5.3). But while the RES industry has developed into an important political player (Sühlsen and Hisschemöller 2014), the 'big four' are struggling to survive within a fundamentally changed business environment (Kungl 2015).

As regards the specific design of RES support policies, the public choice framework points to a very clear incentive structure—the more differentiated a support scheme is, the more the associated rents may be distributed in a targeted way. Hence, regulators have an incentive to devise a highly differentiated scheme that allows for active rent management. We maintain that Germany's feed-in tariff results from such rent management: in contrast to quota schemes and feed-in premiums, which do not guarantee a targeted allocation of a fixed and secure level of rents, feed-in tariffs enable targeted rent management, thereby maximizing stakeholder support (see Table 5.1). What is more, the feed-in tariff is further differentiated along three dimensions: technology, spatial scale, and financing of the scheme.

First, technology differentiation ensures that support not only reaches the cheapest RES at a given point in time, but all the technologies of the desired portfolio. In other words, green industrial policy may be actively differentiated into several strands. Indeed, Germany's feed-in tariff is highly differentiated with respect to technology. Differentiation occurs not only between technologies (PV, wind, biomass, geothermal energy) but also within technologies: the tariff distinguishes, for instance, between on- and offshore wind, or between small PV installations and large PV installations. Hence, the basic set-up allows for pinpoint distribution of rents to separate constituencies. A major benefit for RES industries from such a differentiated scheme is that it prevents competition between RES technologies. The structural impacts of this 'green industrial policy' may be sizeable: in 2011 there were 128,000 jobs within Germany's PV industry (BSW Solar, cited in Hoppmann, Huenteler, and Girod 2014: 1430).

Second, in Germany's federally organized system, regional differentiation may cater to spatially heterogeneous interests. Although no formal differentiation takes place (the scheme is implemented uniformly at the national level), there exist indirect ways to account for regional stakes by affecting consecutive reforms of the RES scheme (see Strunz, Gawel, and Lehmann 2016). In particular, the German *Länder* seize every opportunity to promote regional development through RES deployment without impacting on their own state budgets. Conversely, in cases where regional costs exceed regional benefits—this may materialize in the form of NIMBY (not in my back yard) protests—the *Länder* aim at altering or affecting national policies in favour of the constituencies concerned. Consider the example of Bavaria, which recently lobbied for amendments to the federal grid expansion plan in reaction to local NIMBY protests.

Third, regarding the financing mechanism, differentiated consumer surcharges may shield well-organized interests from fully contributing to the scheme—at the expense of less organized groups. This is precisely the situation with Germany's feed-in-tariff: it is financed via a levy on electricity prices characterized by generous exemptions for energy-intensive industries (cf. Gawel and Klassert 2013). Due to the energy-intensive industries' organizational advantage compared to dispersed consumers/voters, the former succeeded in steering regulation in their favour. In fact, high-volume users are all but exempt from paying the RES levy: while they represent only 4 per cent of all companies, their respective electricity share, which is burdened with a minor fraction of the full levy (as low as 1 per cent), accounts for 41 per cent of all electricity consumed within the industry sector (BDEW 2015). The official criterion for the 'special exemption' from the RES levy refers to 'high exposure to international competition'. In practice, however, the energy-intensive industries have fully captured the scheme in this respect by transferring the burden to small- and medium-sized companies and household consumers (Gawel and Klassert 2013).

The gist of this *instrument choice* overview: specific instrument design choices within Germany's *Energiewende* project follow the presumptions of Public Choice theory as laid out in Section 5.2.1. Continuous lobbying efforts by affected stakeholders (such as energy-intensive industries, the states) with heterogeneous interests and bargaining power account for the observed characteristics of a highly differentiated RES support scheme.[3]

5.3 THE LONG-RUN PERSPECTIVE: HOW TO INTEGRATE RENEWABLES IN ENERGY MARKETS?

5.3.1 The Problem of Instrument Change

The widespread deployment of RES means that they have grown out of their niche existence. This success yields new challenges. These, in turn, open up the problem of how to adequately adapt RES support over time—what North (1990) refers to as 'adaptive efficiency'. There are two main reasons, why the energy transition requires adaptations of the policy framework (see Table 5.2). First, multiple technological changes, and possibly surprises, have characterized

Table 5.2. Schematic overview of major transition challenges beyond RES deployment

Time horizon of transition challenges Categories	Short run	Long run
Economic	Cap expenses for RES deployment	Market integration: expose RES producers to market risks
Systemic	Avoid grid congestion	System integration: adapt to technological innovations; ensure security of supply with volatile RES
Political	Avoid lock-in, maintain capacity to adapt	Find the right balance between stability/ rigidity and flexibility/uncertainty

Source: Authors, based on a review of energy transition literature (e.g., Miller et al. 2013).

[3] This is not to say that that the public choice assumptions exclusively explain all aspects of empirically observed RES support schemes. The variety of specific instrument configurations in the EU demonstrates that country-specific factors, such as natural conditions or political culture, may also bear a substantial impact.

the energy sector and will continue to do so. This concerns not only the most imminent aspect of RES production costs but also prospective developments of storage options and demand-side management, as well as developments in the fossil energy sector (see the shale gas boom). So, beyond the question of how to adequately account for RES cost decreases in RES remunerations, the issue of how RES can be aligned with complementary technologies needs to be addressed. This leads to the second point, namely that RES support policies cannot stop at the single aim of increased deployment of RES (see Miller et al. 2013). Immature niche technologies might necessitate a narrow policy focus on capacity increases, but as RES mature, new priorities emerge: market and system integration. With increasing shares of RES, volatile feed-in from wind and photovoltaic power, as well as cumulating expenses for RES support are to be dealt with. In the long run, volatile RES must guarantee security of supply, which is not merely a technological issue but also one of incentives—at some point in time, RES producers will have to face price and quantity risks. Consequently, RES support needs to evolve beyond simple deployment of RES by fixed feed-in tariffs, considering all relevant systemic repercussions (both from a technological and a socio-economic point of view).

In other words, the danger arises that energy transition policies replace the 'carbon lock-in' (Kalkuhl, Edenhofer, and Lessmann 2012; Unruh 2000) with a RES lock-in. At the same time, however, political stability constitutes an important factor for ensuring successful economic policy in general (Weingast 1993) and technology policy in particular (Grubler et al. 2012). While avoiding lock-ins, political flexibility gives rise to political uncertainty and may lead to inefficiently low private investment (Dixit 1992; Hepburn 2006). Thus, a trade-off holds, similar to the discussion about discretionary vs. rule-based approaches in monetary policy (see Lohmann 1992). Optimally, the RES support scheme would be flexible enough to account for unexpected developments concerning technologies and market structures, but also stable enough so as not to undermine private investors' contributions to the transition process via including RES in their portfolios.

This 'adaptive efficiency' perspective needs to be linked with the Public Choice framework. Politicians are primarily motivated to win elections, so their incentive to conceive of a coherent mix of systemic and anticipatory policies (see Sovacool 2009) is low compared to the overwhelming incentive to secure instantaneous stakeholder support. Given that the status quo of RES support represents a balanced compromise resulting from careful political brokering of different interests, path dependency looms. Furthermore, politics generally responds to challenges in an ad hoc way because it is intently fixed on the media attention cycle. Consequently, challenges are not met according to the objective risk they carry but according to the actors' capacities to respond to the media-induced level of stress (see Prittwitz 1990). As regards the transition challenges, the advisable 'integrated power system policy' (Miller et al. 2013) remains elusive. In sum, one important prediction of the Public Choice framework is that some degree of path dependency of the main RES policy characteristics can be expected.

Yet, interestingly, it is, even in principle, unclear what the optimal policy solutions to the aforementioned challenges of market and system integration would be. For instance, can spatial issues such as system-friendly allocation of new wind parks be adequately addressed *within* the framework initially set up for launching RES or does this require a wholly new support scheme? In the following, we analyse the German RES scheme with respect to its history of adaptation. On this basis, we set out future regulatory options that take restrictions of political feasibility into account.

5.3.2 Adapting Germany's Support Policies for Renewables

Since the introduction of the current feed-in tariff for RES in 2000, it has been regularly updated and reformed. Overall, the observed reform process confirms the public choice framework's predictions. The chosen policy pathway is highly inert in that changes to fundamentally different policy regimes are not politically feasible. Although recent reforms have been advertised as important breakthroughs with respect to increasing the cost efficiency of the scheme (in a narrow, short-term sense), they do not expose RES producers to market risks in any significant way. Consider the 'market premium' approach to encourage direct marketing by RES producers, which was initially introduced in 2012 as an option (cf. Gawel and Purkus 2013). From 2012 to 2014, RES producers could choose between the fixed feed-in tariff and a sliding feed-in premium. The latter was calculated as the difference between the average value of the electricity (monthly spot market averages) and the tariff level. In addition, participating producers received a so-called 'management premium' to compensate for costs incurred through market participation. As a consequence, RES producers made significant windfall profits but overall market and system integration was hardly improved.

The latest policy overhaul of the support scheme occurred in 2014 (see Gawel (2014) and Gawel and Lehmann (2014) for a more detailed analysis). While the reform was boldly labelled as the Renewables Support Act 2.0 (Gabriel 2014), no far-reaching or even game-changing innovation has occurred. Among the innovations are target corridors for PV and wind deployment and a prototype PV tender to gather experience with tendering schemes. Moreover, the 'management premium' has been repealed and tariffs for new installations have been reduced. The association of RES producers has scorned the reform and warned that it might 'choke' the further deployment of RES (BEE 2013). This is clearly exaggerated. Most likely, the direct effect of these reforms will be minor as regards the overall trajectory of the transition. The prototype tender may, in the long run, constitute one possible avenue for deeper market integration of RES. It is important to note, in this respect, that the EU Commission is increasingly pushing for tenders as standard for RES support schemes (European Commission 2014).

Thus reforms of the RES support scheme proceed only incrementally. Drastic policy changes, such as the harsh retroactive cuts in Spanish RES support in 2011–13 in reaction to the financial crisis and the change to a conservative government, are not conceivable within Germany's political culture—unless there was some major shift in the economic environment. In particular, retroactive cuts to RES support are all but ruled out because the principle of *pacta sunt servanda* (agreements must be kept) enjoys inviolable status. This might also be seen from the fact that even a rather soft form of 'disguising' rather than rescinding old contractual RES remuneration obligations has met sharp public criticism: some politicians had put forward the idea of a partly debt-financed fund to cover all old support grants. However, due to the public outcry, this proposal to essentially transfer costs into the future has not been implemented (see Gawel and Lehmann 2014).

In conclusion, the experience of past reforms suggests a narrowly restricted future for any reform efforts in Germany. Generally, every policy change implies a redistribution of rents and therefore needs to be implemented against the protest of the losing stakeholders. At the same time, politicians clearly feel the need to display reactive capacity and self-assertion—even if via symbolic measures only (see Edelman 1964). Against this backdrop, how might the reiterative process of lobbying and policy making unfold? Two countervailing forces affect the long-run adaptation process. On the one hand, beneficiaries of the conventional energy system still attempt to slow down or even derail the transition towards RES. For instance, there have been successful efforts of political agenda-setting, as demonstrated by recent public discussions about an alleged 'cost-tsunami' due to the expenses for PV power (Frondel, Ritter, and Schmidt 2010). On the other hand, as described in Section 5.2.2, positive feedback, arising from the feed-in tariff scheme distributing financial benefits to a sizeable portion of the electorate, makes the RES support scheme politically resilient (Strunz 2014).

The uncertainty over the best way to fully integrate volatile RES in the electricity market (e.g., Winkler and Altmann 2012) while phasing out fossil and nuclear electricity provides new opportunities to extract rents. For instance, the alleged increased risk of black-outs due to volatile RES is brought forward as an argument for capacity markets—an instrument to the particular benefit of conventional power stations, whose introduction would, however, constitute an essentially irreversible act from a public choice perspective (Lehmann et al. 2015). A related example of vested interests' regulatory impact can be seen in recently failed efforts to introduce a 'climate levy' to be borne by

Germany's oldest lignite power stations. Since the EU emissions trading scheme sends only insufficient decarbonization signals (due to the huge oversupply of emission permits), Germany's lignite industry has enjoyed an upturn in the first half of the 2010s. In consequence, so as to ensure that Germany's greenhouse gas emission reduction target for 2020 will not be missed, a climate levy had been proposed. However, through a concerted lobbying action by industry associations and unions, the *polluter pays* principle was converted to *polluter profits* in that the planned levy was replaced by 'capacity payments' which will be awarded to owners of the oldest lignite plants in exchange for taking their plants off the grid while keeping them in some kind of strategic reserve—in other words, a classic 'golden handshake' (see Gawel and Strunz 2015).

So at every incremental transition step, rent-seeking by affected stakeholders will impact on the process. As long as none of the crucial stakeholders succeeds in full regulatory capture, this need not damage the process as a whole. Rather, one might conceive of the process as a selection between several possible scenarios: technologically, there might be both centralized and decentralized ways to engineer a fully RES-based system. Yet it is primarily a question of economic power and political influence within a highly dynamic environment that determines which of the scenarios materializes. Schmid, Knopf, and Pechan (2015) identify necessary conditions that would bring different actors within Germany's energy sector into a dominant position so as to steer the overall process in a more (de)centralized direction.

5.3.3 Systemic Challenges

The energy transition does not only require RES deployment but also the integration of newly built RES with other elements of the electricity system such as remaining conventional production capacities and storage options. Yet the most prominent related aspect concerns transmission grids. Rising RES shares put pressure on the existing grid, particularly as regards electricity generated in wind parks along the North Sea coast that needs to be transported to consumption centres in southern Germany. Bottlenecks loom and the transmission system operators increasingly need to 're-dispatch' power in order to prevent grid overload (Bundesnetzagentur 2014: 16). The precise relations between RES deployment and grid extension are, however, debated in both an academic context and the public domain. In fact, Nordensvärd and Urban (2015) claim that Germany's energy policy already suffers from a lock-in that diverts investment away from the pivotal issue of grid extension and retrofitting. Yet the argument that wind support is responsible for delays in grid extension seems much too narrow because it neglects that grid extension itself is a public choice issue.

Although from a systemic point of view grid extension and RES deployment are complementary, the distributional effects of upgrading and enlarging transmission grids are complex: 'expanding and interconnecting grids typically works against the interests of some producers, by eroding the ability of incumbents to extract rents. It also gives politicians less scope to interfere with their national and local energy systems' (*The Economist* 2015: 10). One case in point seems to be Bavaria's efforts to prevent new transmission lines crossing into this southernmost state. Instead, the Bavarian government aims to increase biomass and gas capacities—thereby obviously favouring the local distribution of rents and backtracking on previous agreements (Hecking 2015).

Yet while this citation suggests that new transmission lines only disadvantage the incumbents (insofar as the new lines serve to transport electricity from RES), they may also serve the incumbents: a heated discussion in Germany as to whether new connections to the south from eastern Germany, where a lot of electricity is generated from wind but also from lignite, attests to this point. Gerbaulet et al. (2013) reject the idea that the new lines promote the transition, as the Federal Network Agency holds, and argue that they would rather foster lignite production in the eastern states. Sure enough, local NIMBY activists who oppose these new lines jumped upon this argument. Whether an objective answer to this dispute exists, remains doubtful. Each stakeholder naturally uses those arguments that serve to back up his private interests.

These issues indicate a crucial insight: the market and system integration of RES cannot be treated as a purely technical problem for which an efficient policy solution needs to be found. On the contrary, stakeholder interests, systemic requirements, and policy options are intertwined. Only very broad formulations of that *instrument change,* which is needed, may find consensus. It is clear that the fixed tariff scheme, combined with prioritized feed-in, has been very effective in ramping up RES deployment *without regard to systemic consequences.* While this is exactly what niche support is all about, with RES becoming a major pillar of the electricity system, the systemic effects of RES deployment need to be addressed. Beyond this general statement, the issue becomes one of public choice rather than technology.

Given this assessment, what might realistic reform pathways look like? In order to illustrate, let us consider the spatial allocation of new wind farms. The existing feed-in tariff includes a small element of spatial regulation in that less windy locations receive remunerations that are a bit higher than their more windy peers: hence, there is a small incentive not only to build wind farms in the north but also in the more southern back country. Now a range of options might lend themselves to mitigating bottlenecks on the north–south axis: market splitting (different price zones in the north/south to incentivize capacity building in the high-price zone), grid extension, even higher differentials for windy/less windy locations, or exposure of RES producers to quantity risks.

The latter option could mean, for instance, that RES producers receive no compensation when wind farms have to be taken off the grid to ensure grid stability. This would incentivize RES producers to search for storage options and/or to locate new installations only in areas that are not prone to grid overload. Certainly, stakeholders would protest against a decrease in the expected value of their payoff. However, such a policy adjustment would be broadly in line with the overall requirement of our framework: it would not fundamentally alter but only incrementally change the existing support scheme. Furthermore, it would align with what is envisaged in the EU Commission's latest State Aid Guidelines (2014 C 200/01). In case of negative spot market prices occurring more often due to high RES feed-in (cf. Bundesnetzagentur 2014: 123), the stakeholders' argumentative leverage against such measures would also decline.

5.4 CONCLUSION

In this chapter we have framed the transition towards clean energies as a two-stage public choice problem of *instrument choice* and *instrument change*. Both stages of the transition involve active rent management to reconcile the interests of different stakeholders. The first stage—instrument choice—concerns the question how to incentivize substitution away from conventional fossil energies. From the regulator's perspective, support policies for RES bring a twofold political benefit as compared to taxing conventional energies or capping emissions: most importantly, RES support creates new rents for (new) power producers, rather than reaping existing rents of incumbent producers. Thus the transition faces less opposition from conventional energies' (or, more generally, emitting industries') lobbying protests and stakeholder support is maximized. Moreover, RES support reduces abatement costs within existing cap-and-trade schemes, which further improves the regulator's negotiation position against the conventional energies' lobbies (Gawel, Strunz, and Lehmann 2014). An analysis of the introduction, specific set-up, and development of Germany's RES support scheme corroborates the main premises of the public choice framework: the empirically observed, highly differentiated support scheme enables targeted rent distribution to a variety of stakeholders.

The second stage—instrument change—addresses the adaptation of the chosen policy framework in line with new transition challenges. One important issue is how to integrate high shares of volatile RES into an existing system without compromising security of supply. Adapting the RES support scheme might prove difficult due to path dependencies: the beneficiaries of RES support constitute political stakeholders who would lobby against any change of the support scheme leaving them worse off. Indeed, our investigation of

Germany's current efforts to foster the market integration of RES shows that policy adaptations have been rather symbolic so far. Moreover, RES support establishes a specific technological transformation pathway (i.e., the system is adapted to the needs of these technologies), which, in turn, disadvantages potential technological alternatives. In consequence, the *Energiewende* trajectory exhibits self-stabilizing tendencies. Despite the risks of path dependencies, one might argue that there is also a positive aspect to this inertia in that the transition pathway becomes more resilient (Strunz 2014).

Acknowledging that under normal circumstances instrument change proceeds incrementally does not imply predictability. In the case of exogenous shocks or major political shifts, windows of opportunity may open and facilitate rapid policy change—Germany's nuclear phase-out in reaction to the Fukushima disaster in 2011 is a case in point.[4] Overall, one might thus conclude that policy change unfolds erratically, rather than in a planned and purposeful way, and depends both on windows of opportunity and politicians' capacity to seize them (see Prittwitz 1990). What are we to infer from this for the transition process? The answer, crucially, depends on our reference point. Judging single policy decisions against the standards of static and adaptive efficiency would yield a bleak picture. Yet, should stakeholder rent-seeking be framed as an aberration or rather, more realistically, as an essential characteristic of pluralist democracies? There exists no silver-bullet against regulatory capture—even deliberative accounts of democracy, which strongly value public debates on an egalitarian footing, cannot get around addressing self-interest and power (see Mansbridge et al. 2010). Hence, the important criterion for judging transition policies simply reads: does the transition succeed? In the long run, the crucial issue is not whether the transition follows some hypothetical ideal path,[5] but whether the challenges are addressed in a way that enables the transition to proceed.

What policy advice can be drawn from this analysis for the specific task of adapting Germany's RES support scheme? Having already passed the first stage of niche development, the 'next generation' transition challenges become more prominent (see Section 5.3.1; also Miller et al. 2013). Germany needs to increasingly expose RES producers to market risks (even if only incrementally so). One next step within a pragmatic rent management strategy could make RES producers bear quantity risks: so far, RES producers are compensated even when installations have to be taken off the grid in order to avoid system overload. Exposing RES producers to the risk of remaining uncompensated in

[4] To be sure, the anti-nuclear movement has a long history in Germany and a first nuclear phase-out law had been implemented in 2000. However, Chancellor Merkel had previously watered down the old phase-out and performed a political U-turn in response to Fukushima.

[5] We have repeatedly emphasized that the surrounding uncertainty cannot be resolved ex-ante, so the 'optimal' transition path remains elusive.

times of negative electricity prices would reward system-friendly strategies (e.g., it offers an incentive to store electricity locally). By keeping the overall policy framework of technology-differentiated feed-in tariffs intact in general, such a policy might avoid strong resistance from concerned RES industry stakeholders. As long as the effects are not concentrated regionally, it would also be compatible with Germany's federal organization, which constrains every legislative proposition.

While the chapter has focused on Germany as an empirical case study, other EU member states face very similar challenges because of the EU's 'Roadmap 2050', its common long-run decarbonization pathway. What is more, feed-in tariff schemes in favour of RES have spread internationally, and so has the need for market and system integration of renewables, which are expanding worldwide. This does not only hold for high-income countries; in fact, 62 per cent of low-income countries have implemented some form of RES policy (as compared to 82 per cent of high-income countries; see REN21 2015: 91). Thus, the framework of instrument choice and instrument change might also be fruitfully applied to other contexts than that analysed here.

REFERENCES

Acemoglu, D., P. Aghion, L. Bursztyn, and D. Hemous (2012). 'The Environment and Directed Technical Change'. *American Economic Review*, 102(1): 131–66.

Acemoglu, D., S. Johnson, P. Querubín, and J. A. Robinson (2008). 'When Does Policy Reform Work? The Case of Central Bank Independence'. *Brookings Papers on Economic Activity*, 39: 351–429.

BDEW (Bundesverband der Energie- und Wasserwirtschaft e.V [German Association of Energy and Water Industries]) (2015). 'Erneuerbare Energien und das EEG: Zahlen, Fakten, Grafiken (2015)'. Available at: <https://www.bdew.de/Internet.nsf/id/DE_Erneuerbare-Energien> (accessed 12 October 2016).

BEE (Bundesverband Erneuerbare Energien [German Renewable Energy Federation]) (2013). 'Erneuerbare-Energien Branche warnt vor Abwürgen der Energiewende'. Pressemitteilung 04/13, 14 February, Berlin.

Bundesregierung (German Federal Government) (2010). *Energiekonzept für eine umweltschonende, zuverlässige und bezahlbare Energieversorgung*. Available at: <https://www.bundesregierung.de/ContentArchiv/DE/Archiv17/_Anlagen/2012/02/energiekonzept-final.pdf?__blob=publicationFile&v=5> (accessed 8 November 2016).

Bundesnetzagentur (German Federal Network Agency) (2014). 'Monitoringbericht 2014'. Available at: <http://www.bundesnetzagentur.de/SharedDocs/Downloads/DE/Allgemeines/Bundesnetzagentur/Publikationen/Berichte/2014/Monitoringbericht_2014_BF.pdf?__blob=publicationFile&v=4> (accessed December 2015).

COP (Conference of the Parties) (2015). 'Adoption of the Paris Agreement'. 12 December 2015. Available at: <http://unfccc.int/resource/docs/2015/cop21/eng/l09r01.pdf> (accessed 12 October 2016).

Dixit, A. K. (1992). 'Investment and Hysteresis'. *Journal of Economic Perspectives*, 6(1): 107–32.

Edelman, M. (1964). *The Symbolic Uses of Politics*. Urbana, IL: University of Illinois Press.

European Commission (2014). 'Communication from the Commission: Guidelines on State Aid for Environmental Protection and Energy 2014–2020, April 2014'. Available at: <http://ec.europa.eu/competition/sectors/energy/eeag_en.pdf> (accessed 17 December 2015).

Frondel, M., N. Ritter, and C. M. Schmidt (2010). 'Die Förderung der Photovoltaik: Ein Kosten-Tsunami'. *Energiewirtschaftliche Tagesfragen* 60(12): 36–44.

Gabriel, S. (2014). 'Eckpunkte für die Reform des EEG'. 21 January, Berlin. Available at: <http://www.bmwi.de/BMWi/Redaktion/PDF/E/eeg-reform-eckpunkte,property=pdf,bereich=bmwi2012,sprache=de,rwb=true.pdf> (accessed 12 October 2016).

Gawel, E. (1995). 'Bürokratietheorie und Umweltverwaltung. Ökonomische Einsichten in verwaltungsrechtliches Handeln im Umweltschutz'. *Journal of Environmental Research (Zeitschrift für angewandte Umweltforschung ZAU)*, 8(1): 79–89.

Gawel, E. (2014). 'Eckpunkte zur EEG-Reform: Der Energiewende nächster Akt'. *Wirtschaftsdienst*, 94(2): 82–3.

Gawel, E. and C. Klassert (2013). 'Probleme der besonderen Ausgleichsregelung im EEG'. *Zeitschrift für Umweltrecht*, 24(9): 467–80.

Gawel, E. and P. Lehmann (2014). 'Support for Renewable Energy in Germany after the 2014 Renewable Energy Sources Act'. *Wirtschaftsdienst*, 94(9): 652–8.

Gawel, E. and A. Purkus (2013). 'Promoting the Market and System Integration of Renewable Energies through Premium Schemes: A Case Study of the German Market Premium'. *Energy Policy*, 61: 599–609.

Gawel, E. and S. Strunz (2015). 'Klimaabgabe für Kohlekraftwerke: Ein richtiger Schritt zur Erreichung des Klimaziels?' *Ifo-Schnelldienst*, 68(14): 8–11.

Gawel, E., S. Strunz, and P. Lehmann (2014). 'A Public Choice View on the Climate and Energy Policy Mix in the EU: How Do Emissions Trading Scheme and Support for Renewable Energies Interact?' *Energy Policy* 64: 175–82.

Gerbaulet, C., F. Kunz, C. von Hirschhausen, and A. Zerrahn (2013). 'German Electricity Transmission Grid Remains Robust'. *DIW Wochenbericht* 20/21: 3–12. Available at: <http://www.diw.de/documents/publikationen/73/diw_01.c.421217.de/13-20.pdf> (accessed 12 October 2016).

Grubler, A., F. Aguayo, K. Gallagher, M. Hekkert, K. Jiang, L. Mytelka, L. Neij, G. F. Nemet, and C. Wilson (2012). 'Policies for the Energy Technology Innovation System (ETIS)'. In *Global Energy Assessment—Toward a Sustainable Future*. Cambridge: Cambridge University Press, and Laxenburg: International Institute for Applied Systems Analysis.

Hecking, C. (2015). 'Der Energiewendehals'. *Die Zeit*, 6 March. Available at: <http://www.zeit.de/2015/10/stromtrasse-bayern-horst-seehofer-energiewende-kernkraftwerk> (accessed 12 October 2016).

Helm, D. (2010). 'Government Failure, Rent-seeking, and Capture: The Design of Climate Change Policy'. *Oxford Review of Economic Policy*, 26(2): 182–96.

Hepburn, C. (2006). 'Regulation by Prices, Quantities, or Both: A Review of Instrument Choice'. *Oxford Review of Economic Policy*, 22(2): 226–47.

Hoppmann, J., J. Huenteler, and B. Girod (2014). 'Compulsive Policy-making: The Evolution of the German Feed-in Tariff System for Solar Photovoltaic Power'. *Research Policy*, 43(8): 1422–41.

Kalkuhl, M., O. Edenhofer, and K. Lessmann (2012). 'Learning or Lock-in: Optimal Technology Policies to Support Mitigation'. *Resource and Energy Economics*, 34(1): 1–23.

Kirchgässner, G. and F. Schneider (2003). 'On the Political Economy of Environmental Policy'. *Public Choice*, 115(3): 369–96.

Küchler, S. and R. Wronski (2015). 'Was Strom wirklich kostet', Forum Ökologisch-Soziale Marktwirtschaft, Berlin. Available at: <http://www.greenpeace-energy.de/uploads/media/Greenpeace_Energy_Was_Strom_wirklich_kostet_2015.pdf> (accessed 12 October 2016).

Kungl, G. (2015). 'Stewards or Sticklers for Change? Incumbent Energy Providers and the Politics of the German Energy Transition'. *Energy Research & Social Science*, 8: 13–23.

Lehmann, P., R. Brandt, E. Gawel, S. Heim, K. Korte, A. Löschel, P. Massier, M. Reeg, D. Schober, and S. Wassermann (2015). 'Capacity Payments to Secure Electricity Supply? On the Future of Germany's Power Market Design'. *Energy, Sustainability and Society*, 5:15. Available at: <http://link.springer.com/article/10.1186/s13705-015-0039-7> (accessed 16 November 2016).

Lohmann, S. (1992). 'The Optimal Degree of Commitment: Credibility versus Flexibility'. *American Economic Review*, 82(1): 273–86.

McCormick, R. E. and R. D. Tollison (1981). *Politicians, Legislation and the Economy: An Inquiry into the Interest-Group Theory of Government*. Boston, MA: Martinus-Nijhoff.

Mansbridge, J., J. Bohmann, S. Chambers, D. Estlund, A. Follesdal, A. Fung, C. Lafont, B. Manin, and J. L. Marti (2010). 'The Place of Self-interest and the Role of Power in Deliberative Democracy'. *Journal of Political Philosophy*, 18(1): 64–100.

Miller, M., L. Bird, J. Cochran, M. Milligan, M. Bazilian, E. Denny, J. Dillon, J. Bialek, M. O'Malley, and K. Neuhoff (2013). 'RES-E-Next: Next Generation of RES-E Policy Instruments'. Study commissioned by IEA-RETD, 4 July.

Monopolkommission (2013). 'Wettbewerb in Zeiten der Energiewende'. Sondergutachten 65, Bonn.

Niskanen, W. A. (1971). *Bureaucracy and Representative Government*. Chicago, IL: Aldine-Atherton.

Nordensvärd, J. and F. Urban (2015). 'The Stuttering Energy Transition in Germany: Wind Energy Policy and Feed-in Tariff Lock-in'. *Energy Policy*, 82: 156–65.

North, D. C. (1990). *Institutions, Institutional Change and Economic Performance*. Cambridge: Cambridge University Press.

Olson, M. (1971). *The Logic of Collective Action: Public Goods and the Theory of Groups*. Revised Edition. Cambridge, MA: Harvard University Press.

Prittwitz, V. V. (1990). *Das Katastrophenparadox. Elemente einer Theorie der Umweltpolitik*. Opladen: Leske und Budrich.

Purkus, A., M. Röder, E. Gawel, D. Thrän, and P. Thornley (2015). 'Handling Uncertainty in Bioenergy Policy Design: A Case Study Analysis of UK and German Bioelectricity Policy Instruments'. *Biomass and Bioenergy*, 79: 64–79.

Rodrik, D. (1996). 'Understanding Economic Policy Reform'. *Journal of Economic Literature*, 34(1): 9–41.

Rodrik, D. (2014). 'Green Industrial Policy'. *Oxford Review of Economic Policy*, 30(3): 469–91.

REN21 (2015). 'Renewables 2015 Global Status Report'. Paris: REN21 Secretariat. Available at: <http://www.ren21.net/gsr> (accessed 12 October 2016).

Schmid, E., B. Knopf, and A. Pechan (2015). 'Putting the German Energiewende into Practice: An Analysis of Actors and Future Requirements for Electricity Infrastructures'. *Energy Research and Social Science*, 11: 263–75.

Schmitz, H., O. Johnson, and T. Altenburg (2013). 'Rent Management—The Heart of Green Industrial Policy'. IDS Working Paper 418. Institute for Development Studies, University of Sussex, Brighton. Available at: <https://www.ids.ac.uk/files/dmfile/Wp418.pdf> (accessed 12 October 2016).

Sijm, J., P. Lehmann, U. Chewpreecha, E. Gawel, J.-F. Mercure, H. Pollitt, and S. Strunz (2014). 'EU Climate and Energy Policy beyond 2020: Are Additional Targets and Instruments for Renewables Economically Reasonable?' UFZ Discussion Paper 2014–4. Available at: <http://www.ufz.de/export/data/global/63246_DP_3_2014_Sijmetal.pdf> (accessed 12 October 2016).

Sovacool, B. K. (2009). 'The Importance of Comprehensiveness in Renewable Electricity and Energy Efficiency Policy'. *Energy Policy*, 37(4): 1529–41.

Spash, C. (2010). 'The Brave New World of Carbon Trading'. *New Political Economy*, 15(2): 169–95.

Stigler, G. J. (1971). 'Theories of Economic Regulation'. *Bell Journal of Economics*, 2(1): 3–21.

Stirling, A. (2014). 'Transforming Power: Social Science and the Politics of Energy Choices'. *Energy Research & Social Science*, 1: 83–95.

Strunz, S. (2014). 'The German Energy Transition as a Regime Shift'. *Ecological Economics*, 100: 150–8.

Strunz, S., E. Gawel, and P. Lehmann (2016). 'The Political Economy of Renewable Energy Policies in Germany and the EU'. *Utilities Policy*, 42: 33–41.

Sühlsen, K. and M. Hisschemöller (2014). 'Lobbying the "Energiewende". Assessing the Effectiveness of Strategies to Promote the Renewable Energy Business in Germany'. *Energy Policy*, 69: 316–25.

The Economist (2015). 'Puffs of Hope'. 1 August.

Tullock, G. (1967). 'The Welfare Costs of Tarriffs, Monopolies and Theft'. *Western Economic Journal*, 5(3): 224–32.

Unruh, G. C. (2000). 'Understanding Carbon Lock-in'. *Energy Policy*, 28(12): 817–30.

Weingast, B. R. (1993). 'Constitutions as Governance Structures: The Political Foundations of Secure Markets'. *Journal of Institutional and Theoretical Economics*, 149(1): 286–311.

Winkler, J. and M. Altmann (2012). 'Market Designs for a Completely Renewable Power Sector'. *Zeitschrift für Energiewirtschaft*, 36(2): 77–92.

Part III

Institutions and Governance

The introductory chapter to this book highlighted the importance of 'deliberate and global action' to foment a clean energy transition. This naturally engages institutions—both public and private—and places demands on governance systems. The chapters in this part of the book focus on 'deliberate action' with examples from both developed and developing economy settings.

Two chapters focus on the political economy of energy innovation, with the natural focus being in developed countries. Dasgupta, De Cian, and Verdolini focus on factors that affect incentives to devote resources to energy research and development. In a complementary chapter, Sonnesenschein considers how policies to support energy innovation should be evaluated. Remaining in a developed country setting, Ćetković, Buzogány, and Schreurs consider a specific technology: wind power. Their chapter highlights the importance of state–market coordination in both onshore wind, a relatively mature technology, and offshore wind, which is much more nascent.

The developing country applications are notably diverse. Yuliani examines feed-in tariff policies in Indonesia. Kere takes a cross-country approach to consider the factors affecting bioenergy production. Finally, Agu and Ogbeide-Osaretin focus on the fiscal implications of a clean energy transition for Nigeria with a focus on state governments. These chapters are a reminder that a clean energy transition will both bring in the new, as in feed-in tariffs and opportunities in bioenergy production, and require adjustments within existing frameworks, including such major items as the financing of state-level government in an oil-dependent economy.

6

Varieties of Clean Energy Transitions in Europe

Political-Economic Foundations of Onshore and Offshore Wind Development

Stefan Ćetković, Aron Buzogány, and Miranda Schreurs

6.1 INTRODUCTION

There is a broad consensus among scholars and practitioners that a wide range of new renewable energy technologies must be standardized and become cost-effective if energy systems are to be decarbonized and excessive climate change avoided (Verbong and Loorbach 2012). The question remains regarding the motives and capacities of different countries to advance new technologies 'to the shelf' and adopt the already advanced technologies 'from the shelf' (Sandén and Azar 2005). An extensive body of literature has emerged, particularly within the studies on sustainability transitions (Smith and Raven 2012) and national innovation systems (Suurs and Hekkert 2009), exploring how and why renewable energy innovations develop in a certain context and what factors determine their successful diffusion to other institutional settings. The major shortcoming of the existing literature is that national political-economic institutions and interests and processes which underpin them have often been neglected or discussed in a non-systemic manner. In addressing this gap, a recent trend in studying sustainable energy policies has been to identify common trends across countries and explain them in relation to the particular type of national political-economic 'logic'. The Varieties of Capitalism framework (VoC) (Hall and Soskice 2001) has been suggested as a promising approach for capturing and investigating the common types of national market economies and how they influence technology and policy choices in energy transitions (Ćetković and Buzogány 2016).

This chapter contributes to the literature which adopts the comparative capitalism approach to clean energy transitions in two respects. First, it develops further the theoretical rationale for applying the VoC framework to understand and compare national renewable energy transition pathways. It does so by bringing back the state in the analysis and enriching the VoC framework with the literature on national innovation systems, state–industry relations and the corporatism vs. pluralism debate. In addition, the chapter extends the so-far applied categorization based on the distinction between Liberal Market Economies and Coordinated Market Economies, by including a largely neglected type of what we termed 'simple Coordinated Market Economies'. Second, the chapter demonstrates the value of the proposed framework by exploring the fate of two related renewable energy technologies which are at the different stage of development: one mature (onshore wind) and one infant renewable energy technology (offshore wind). Both technologies are seen as central for achieving a clean energy transition. This should allow for discerning the relative importance of national vs. sector-specific features in influencing the diffusion of renewable energy technologies.

6.2 THEORETICAL FRAMEWORK

Our starting point for analysing and comparing how different national political-economic structures influence renewable energy transitions is comparative research that emphasizes the role of state–market relations as drivers of sectoral innovation and policy change. We draw here mainly on the 'varieties of capitalism' perspective that stresses the institutional differences among developed capitalist market economies (Hall and Soskice 2001). These differences concern the question of how the political institutional context helps or hinders firms to solve their cooperation problems. The focus of the approach is on firms and the institutional setting they are embedded into. According to Hall and Soskice, most developed market economies cluster into two distinct types of capitalism, 'Coordinated market economy' (CME) and 'Liberal market economy' (LME).

The distinctions between the two main VoCs enable the systematic identification of different characteristics and likely problems that occur in LMEs or CMEs when firms innovate and new economic sectors emerge. LMEs favour radical, path-breaking innovation processes and hold comparative advantages in innovation-intensive high-tech industries and services (Hall and Soskice 2001: 40–1). Innovation in CMEs is mainly taking place in traditional industry fields, such as machinery or chemical production. In contrast to LMEs, innovation in CME countries is rather small-scale, incremental, but also more continuous. It is often based on path-dependent cooperation between firms and the banking sector as well as science. In general, long-term perspectives dominate over concerns of immediate profitability, a feature that is typical of LMEs. The path-dependent innovation process in CMEs is to a large extent structurally predetermined: industrial relations in CMEs are more oriented towards employee participation, trade unions have a stronger say, and labour law gives less chances for hiring and firing. The innovation path is supported also by the system of vocational training, which emphasizes interactions between industry and research and is able to produce highly skilled workforce.

By focusing on firms and sector-level innovation, the VoC framework disregards the role of national institutions, which often play a central role in the development and diffusion of new technologies (see Mazzucato 2015). Emphasizing institutional stability rather than change, the VoC framework is criticized for being static and deterministic in character (Hancké 2009). To address this shortcoming, we complement the VoC framework with insights from the literature on national innovation systems on the one hand, and comparative political science literature differentiating state–industry–society relations on the other hand.

The literature on national innovation systems suggests that the linear model of innovation provides an inadequate picture of how technology innovations emerge and become widespread (Bergek et al. 2015). This particularly holds true for new renewable energy technologies. Although the provision of new knowledge through research and development (R&D) spending ('push mechanism') is critical for developing and improving renewable energy technologies (Ragwitz and Miola 2005), their market success is contingent on a number of additional factors. These factors are described as 'pull mechanism' and serve to facilitate market formation for new technologies. This includes stable and long-term market demand, provision of necessary skills, financing, and supportive legal conditions (Fagerberg 2015). One should also add the necessity of legitimization for new technologies (Jacobsson and Bergek 2004).

The second relevant literature, focusing on neo-corporatism, welfare states, and political systems emphasizes the political conditions for policy stability and strategic coordination (Esping-Andersen 1990; Lijphart 1999). There are important affinities between Lijphart's classification of political systems as majoritarian vs. consensus-oriented and the VoC framework's LME vs. CME typology (Schneider and Soskice 2009). Taking into account Colomer's observation that the number of parties increases policy stability (Colomer 2012), we can summarize that LMEs can be typically found in two party-systems with high levels of policy instability, whereas CMEs usually have multi-party systems and feature higher levels of policy continuity. Thus, institutional and party system factors can help explain the long-term and stable support for certain policy issues or the lack of it.

One possibility to further differentiate the political institutional context has been suggested by Schmidt (2012), who takes a polity-oriented perspective on

European political systems and argues that the effects of institutional arrangements can be conceptualized as being 'simple' and 'compound' polities. Whereas in simple polities the state structure is centralized and governing is concentrated in a single authority, compound polities feature multiple authorities. Drawing on these insights, we distinguish between simple and compound varieties of LME and CME (see Ćetković, Buzogány, and Schreurs 2016: 5).

We restrict our focus to West European states across three subtypes of capitalist market economies: Simple LMEs (the UK), Simple CMEs (Denmark, Sweden, Netherlands), and compound CMEs (Germany).[1] The case selection of countries is influenced by the fact that they all have significant onshore and offshore wind potential. We make two central assumptions. First, due to the mechanism of strategic coordination and consensus-based policy style, both simple and compound CMEs are more conducive to exploring economic opportunities through strategically advancing new renewable energy technologies than LMEs. Second, emerging renewable energy technologies require sufficiently plural environments in which they can grow and mobilize political support. This implies that compound CMEs are more open than simple CMEs to experimenting with and embracing new energy technologies.

6.3 EMPIRICAL ANALYSIS

6.3.1 Onshore Wind

6.3.1.1 Compound CME (Germany)

Wind power has played a central role in renewable energy development in Germany. Initially, the wind energy enthusiasts experimented with small-size turbines for local use while the national government directed support towards large-scale wind power (Bruns et al 2011: 265). Eventually, the government-funded R&D programmes encompassed a variety of wind turbine models, from small to large installations, supplied dominantly by German manufacturers (Jacobsson and Lauber 2006: 263). In 1989, the government began supporting market creation by subsidizing 100 MW of wind power which in 1991 extended to 250 MW (Lauber and Metz 2004: 201). The first Electricity Feed-in Law adopted in 1991 laid the ground for rapid market expansion of wind power by introducing fixed preferential tariffs (Feed-in Tariff—FiT) for renewable electricity producers. This was accompanied by supporting programmes for research, demonstration, and project implementation at the

[1] For a more extensive analysis which also includes the case of Norway, see Ćetković, Buzogány, and Schreurs (2016).

national, federal, and local levels (Lauber and Metz 2004). The Renewable Energy Law of 2000 further improved the overall framework guaranteeing fixed technology-specific tariffs for renewable energy producers for the period over 20 years. In 2010, the government adopted the Energy Concept, which stressed renewable energy as a cornerstone of the future energy mix and set the target of 80 per cent renewable energy in the electricity mix by 2050 (BMWI/BMU 2010). The Renewable Energy Law has been amended on several occasions, most recently in 2012 and 2014. Although the fundamentals of the support scheme have largely remained intact, the regulatory adjustments in 2014 entailed important changes designed to ensure better planning and more efficient integration of renewable electricity into the market (Bundestag 2014). The R&D spending on wind energy has been consistently high, both from the public and corporate sources (European Commission 2013: 38–9).

The growth of onshore wind installations has witnessed a continuous annual increase ever since the early 1990s (IWES n.d.). Despite regulatory changes, Germany added 3.5 GW of onshore wind power in 2015, the second best annual record after 2014 (BWE 2016). By 2016, Germany had as much as 41.6 GW of the installed onshore wind power (BWE 2016) (see Ćetković, Buzogány, and Schreurs 2016: 8). Not only has the German policy approach been effective in creating a vibrant domestic market for wind energy, but it has also significantly enhanced industrial competitiveness, innovations, job creation, and environmental benefits (Pegels and Lütkenhorst 2014; Boeckle et al. 2010). Based on the data from 2012, the wind industry in Germany employs 117,900 people, out of which 18,000 are in offshore wind (GTAI 2014: 7) (see Ćetković, Buzogány, and Schreurs 2016: 8).

All this signifies remarkable policy stability and overall policy success. It would be wrong to conclude, however, that renewable energy policy in Germany has not faced challenges, opposition, and setbacks. In fact, it was the political contestations and 'battle over institutions' (Jacobsson and Lauber 2006) that made the emergence of the wind energy sector possible. The pluralist and federal political landscape allowed the early wind energy advocates to experiment with wind energy technologies, attract the attention of policy-makers at different levels of governance (local, federal, national), and gradually mobilize government support. Once wind energy became more widely embraced and began producing economic gains for both citizens and the industry, its legitimacy became increasingly difficult to dispute. Jacobsson and Lauber (2006) illustrate how the alliance of wind energy supporters from different spheres of politics and society successfully persuaded parliament members to refrain from reducing wind energy subsidies. This was particularly critical in the formative phase of wind energy during the 1990s. As the wind energy industry matured, it was able to benefit from the already present mechanisms of strategic state–market interaction and industrial upgrading in Germany, typical for CMEs. This included long-term technology-specific government support, export promotion, vocational training, local and national financing instruments, and close science–industry collaboration (Ćetković and Buzogány 2016). Overall, it can be said that the pluralist political environment has launched wind energy on the political agenda whereas the strategic and locally embedded state–market interactions strengthened the legitimacy of wind energy and turned it into a successful industrial policy.

6.3.1.2 Simple LME (UK)

The UK government began supporting research on wind energy in the mid-1970s. Whereas the basic research and education on wind energy has been assessed as good, there has been limited collaboration and networking among scientific institutions as well as between science and industry (Simmie, Sternberg, and Carpenter 2014). The key obstacle to the growth of the wind energy sector and domestic wind power technologies has been the lack of a credible and long-term pull mechanism from the government that would facilitate market formation.

In 2000, the government announced plans for achieving 10 per cent of electricity from renewable energy by 2010 'as long as the cost to consumers is acceptable' (DTI 2003: 45). In 2009, the EU set the target for the UK of 15 per cent of energy from renewable sources by 2020. The development of both onshore and offshore wind is essential for achieving the national target (DECC 2011: 14). The government support for renewable energy evolved from technology-neutral support schemes, embodied in the Non-Fossil Fuel Obligation (NFFO) of 1989 and Renewables Obligation (RO) of 2002, to the more technology-specific RO model adopted in 2009 and FiT for small renewable energy plants introduced in 2010 (Simmie, Sternberg, and Carpenter 2014). In 2013, the government enacted the Electricity Market Reform, which foresees a gradual replacement of the RO model with the market-based support scheme called Contracts for Difference (CfD) by 2017. Although RO was planned to last until 2017, the government has announced the decision to close RO for onshore wind already in 2016. It has even been suggested that onshore wind will be removed from future auctions under the CfD scheme implying the end of subsidies for onshore wind projects (Howard and Drayson 2015: 8). Other regulatory changes have further reduced the profitability of renewable energy projects, such as the decision in 2015 to remove the exemption from the climate change levy for electricity from renewable sources (HM Revenue and Customs 2015). Long and cumbersome administrative procedures have posed further barriers to onshore wind development (IRENA 2012: 130).

The first commercial onshore wind farm in the UK was built in 1991. The deployment of onshore wind energy was slow during the 1990s. Real progress occurred in the second half of the 2000s due to a more supportive national policy context and the adoption of the EU Renewables Directive in 2009.

In 2010, R&D investments in wind energy reached the highest level (see Ćetković, Buzogány, and Schreurs 2016: 7). The largest share of R&D spending on wind energy was performed by the state rather than by corporate actors (European Commission 2013: 38–9). 2012 was the record year in terms of the annually installed onshore wind capacity with 1,937 MW of onshore wind power being connected to the grid (RenewableUK 2014: 19). The overall installed capacity in 2015 was 8.5 GW (RenewableUK 2016) (see Ćetković, Buzogány, and Schreurs 2016: 8). The UK domestic wind energy industry has struggled to develop (IRENA 2012: 130). In terms of job creation, the figures from 2015 estimate that the number of workers directly employed in the wind industry in the UK was 15,500 (RenewableUK 2015: 3). The number of registered patents has also been considerably lower than in Germany and Denmark (see Ćetković, Buzogány, and Schreurs 2016: 8).

The centralization of political power and insufficient state–market coordination have decisively constrained the market and industrial expansion of onshore wind in the UK. The unitary state structure and majoritarian political system failed to provide policy certainty and offered little space for wind energy stakeholders to win the hearts and minds of decision makers and the local population. Unsurprisingly, the rapid deployment of onshore and offshore wind in the UK between 2010 and 2015 occurred under one of the rare coalition governments composed of the Conservative Party and the Liberal Democrats. Given the lack of policy commitment and long-term strategy at the national level, onshore wind projects were mainly developed at the initiative of foreign investors which often faced resistance by local communities (Simmie, Sternberg, and Carpenter 2014). The weak legitimacy and local acceptance of wind energy technology by local communities were, in turn, used as an argument by political elites to justify the lack of support for onshore wind development (Hackley 2015). The devolution of political power and increasing authority of local and regional governments in energy planning have proven favourable for onshore wind. This is particularly the case for Scotland, which has been the most supportive of wind power as part of its economic and industrial policies (IRENA 2012: 128–9; RenewableUK, 2015: 4). Overall, the absence of credible market formation from the side of the UK government, combined with inadequate provision of skills, knowledge, and finance have constrained the emergence of domestic wind industrial capacities and local value creation.

6.3.1.3 Simple CMEs (Denmark, Sweden, Netherlands)

The policy strategy for promoting wind energy has been most comprehensive in Denmark. The oil crisis hit Denmark particularly hard given that the country was highly dependent on foreign oil and lacked reliable domestic energy resources. Somewhat similar to the German case, in Denmark two technological wind energy subsystems have developed in parallel: 1) small-scale wind turbines promoted by farmers, local communities, and wind energy enthusiasts, and 2) large-scale wind turbine demonstration and production supported by the state (Kamp 2008). However, it was the small-scale wind energy that accelerated the legitimacy and technological development of wind energy in Denmark. Private owners of wind turbines mobilized and established in 1978 the Danish Windmill Owners' Association (*Danmarks Vindmølleforening*). In the same year, the Windmill Manufacturers Association was created. The members of the two associations collaborated closely in gradually improving the reliability and effectiveness of wind turbines (Garud and Karnøe 2003). In 1979, the government introduced investment subsidies to individuals and cooperatives for installing wind turbines, which enhanced the domestic wind power market (Buen 2006: 3890). In 1981, the long-term goal of installing 60,000 small wind turbines by 2000 was adopted (Buen 2006: 3890). In 1993, Denmark introduced a fixed FiT for renewable energy power producers (IRENA 2012: 56). The onshore wind power installations peaked in 2000 and 2002 followed by several years of stagnation due to the government's decision to liberalize the electricity market and replace FiT with more market-based instruments (IRENA 2012: 57). In 2009, the new support scheme for renewable energies was adopted and in 2011 the government outlined a strategy to become independent from fossil fuels by 2050 with the interim goal of 30 per cent of energy use supplied by renewable sources by 2020. Wind energy is central for achieving these goals (IRENA 2012: 58–9). At the end of 2015, Denmark had 3.8 GW of the total installed onshore wind power capacity (EWEA 2016a, 2016b). The wind industry (both onshore and offshore) employs nearly 29,000 people and contributes to more than 5 per cent of the country's exports (Denmark 2015) (see Ćetković, Buzogány, and Schreurs 2016: 8).

The oil crisis and the referendum decision to phase out nuclear energy in 1980 led to increasing public support for R&D in wind energy in Sweden. However, market creation instruments and long-term targets were lacking. This prevented the uptake of the domestic wind energy market (Söderholm et al. 2007: 368). The R&D support was rather narrow focusing only on large-scale wind power facilities (Jacobsson and Bergek 2004: 221). It was only during the mid-1990s and early 2000s that the installed wind power capacity started to grow due to the introduced technology-neutral green certificate scheme (Söderholm et al. 2007: 369). The size of the domestic wind energy market was still considerably smaller than in Germany and Denmark (Söderholm et al. 2007: 370) and lacked domestic suppliers of wind energy components (Jacobsson and Bergek 2004: 223). The initial impact of the support scheme was limited but wind power deployment rates have seen a steady increase in recent years. Several factors contributed to this: the EU renewable energy directive, the expansion of the green electricity scheme to Norway, cost reduction of wind energy and more active state role (see Giest 2015). By the end of 2015, Sweden had 6 GW of the total installed wind power (EWEA 2016b), of which 0.2 GW was offshore (EWEA 2016a) (see Ćetković, Buzogány, and Schreurs 2016: 8). Sweden has thus been able to develop its domestic wind market recently but with the limited involvement of the domestic wind industry and consequently small economic benefits in terms of innovations, job creation, and exports.

The promotion of wind energy in the Netherlands during the 1970s and 1980s resembled the policy approach taken in Germany and Denmark, characterized by broad R&D support, inclusion of local suppliers, and investment subsidies (Jacobsson and Bergek 2004: 222). The political commitment decreased in the following period and the problems of building permits and spatial planning at the local level hampered market creation (Jacobsson and Bergek 2004: 226). Both large-scale and small-scale wind energy innovation systems existed in the Netherlands (Kamp 2008: 281). The development of large-scale wind turbines was almost entirely science-led with insufficient collaboration with the industry and electricity companies (Kamp 2008: 281). The focus of the Dutch wind policy was on energy utilities as main project developers (Wolsink 1996). The small-scale innovation system proved more successful but investment subsidies for promoting the demand were introduced only in 1986 (Kamp 2008: 283). Eventually, the progress was slow and ultimately hampered by the problems of local resistance, spatial planning, and the lack of willingness of central authorities to streamline administrative procedures (Jacobsson and Bergek 2004; Kamp 2008). All but one wind turbine manufacturer in the Netherlands disappeared from the market by 2000 (Kamp 2008: 283). In 1994, the budget for wind energy was significantly cut (Wolsink 1996: 1084) and reliance on a market-based approach gained prominence in the light of the privatization and deregulation reforms. The government adopted FiTs in 2003 aiming to reduce investment risks and promote the domestic supply of renewable energy (van Rooijen and van Wees 2006: 63). In 2010, the Netherlands had only 3.6 per cent of energy from renewables (Statistics Netherlands 2010), compared to the national target of 14 per cent by 2020. Following several changes to the support scheme, in 2011 a so-called SDE+ scheme was introduced with the sole purpose of meeting national targets in the most cost-efficient manner. In 2013 a society-wide consensus was reached on energy transition but it was acknowledged that national renewable energy targets for 2020 are not achievable (PLB 2014). By 2016, the Netherlands installed 3.4 GW of wind power (EWEA 2016b), of which 0.4 GW was in offshore wind (EWEA 2016a) (see Ćetković, Buzogány, and Schreurs 2016: 8). The Dutch wind energy sector is not of national significance but it is relatively well positioned internationally, particularly in operation and maintenance as well as manufacturing of small wind turbines (IEA WIND 2014: 134).

All surveyed simple CME countries, except Denmark, have failed to promote innovations in and capture direct economic benefits from onshore wind. The growth of the domestic market was notable initially, followed by a period of stagnation and relatively recent revival, especially in Sweden. How can the analysis of the national political-economic setting account for this development? In terms of the plurality of their political systems, simple CMEs are positioned between compound CMEs and LMEs. They have a multi-party system but with a unitary state structure and a strong tradition of a consensus-seeking policy style. It is precisely this overly consensus-based decision-making process that has hampered wind energy innovations. Both Sweden and the Netherlands are characterized by powerful national energy actors in hydropower (Sweden) and natural gas (Netherlands) with a rather hierarchical structure and strong state involvement, particularly characteristic for Sweden (Pettersson et al. 2010). It thus does not come as a surprise that early efforts to promote wind energy were crafted along the existing policy paradigm with energy utilities as natural partners. Such strategies not only proved inadequate for promoting diverse wind energy innovations, but were also ill-equipped to ensure market creation due to the reluctance of power utilities to rethink their business models and invest in new and financially risky wind energy. The Danish situation was the exception due to the lack of conventional domestic energy resources and a long tradition of distributed energy generation and cooperatives (van der Vleuten and Raven 2006). This created the space for domestic manufacturers to engage in developing and incrementally expanding the production of wind turbines. All three countries have strong systems of strategic state–industry–science–society coordination, but only in Denmark could the wind energy industry mobilize the necessary support.

6.3.2 Offshore Wind

6.3.2.1 Compound CME (Germany)

Offshore wind energy was not the focus of decision makers in Germany for a long time. There was insufficient knowledge about the challenges and costs of offshore wind technology. The initial FiTs and regulations for connecting renewable energy power plants to the grid were not suitable for driving the expansion of offshore wind projects. In 2002, the government published a strategy for the use of offshore wind in Germany (German Government 2002). It set the objective of 2–3 GW of offshore wind capacity to be installed by 2010, followed by 20–25 GW by 2030. These targets soon proved overly ambitious. Gradually, the regulatory and institutional changes were put forward to streamline administrative procedures, connect important actors in the field, and make offshore wind energy projects more attractive for investors. In 2006, amendments to the regulatory framework were made to shift the responsibility of providing grid connections from project developers to a Transmission System Operator (TSO) (Fitch-Roy 2015: 7). In 2011, the responsible ministry and state-owned development bank (KfW) launched the 'KfW Offshore Wind Energy Programme' to facilitate the financing of offshore wind energy projects (KfW 2015). Following the disputes about the grid connections between project developers and Tennet as the main TSO in 2012, the Federal Minister of Economy played a pivotal role through the so called '*AG Beschleunigung*' (Acceleration) initiative in bringing together the actors from the industry and the government and resolving open questions. This resulted in the revisions to the Renewable Energy Law in 2012 and 2014 which specified remuneration FiT models for offshore wind parks. The revisions from 2014 also define more realistic long-term targets for offshore wind energy development including 6.5–10 GW by 2020 and 15–25 GW by 2030 (Anzinger and Kostka 2015: 15).

The first commercial offshore wind park in Germany was built only in 2011. However, over the last two years, the German offshore wind market has been the most dynamic in Europe, alongside the UK. In 2015, Germany installed the highest amount of offshore wind power in Europe (2.3 GW) making the total offshore wind power capacity of 3.3 GW (EWEA 2016a). The expansion of the domestic offshore wind market has provided a boost for the already highly internationally competitive national wind energy clusters located in North-West Germany. The estimates from 2012 show that 18,000 people are directly employed in the offshore wind energy sector in Germany (GTAI 2014: 7) (see Ćetković, Buzogány, and Schreurs 2016: 8).

The recent success and the relatively certain long-term prospect of offshore wind energy in Germany represent a continuation of the German renewable energy market and industrial policy, building on the previously secured legitimacy of wind energy technology, established industrial networks and know-how, and the political consensus on the need for energy transition. The strategic state–market–science coordination proved even more important than in onshore wind, given the large scale, technology uncertainty, and high capital costs of offshore wind energy projects.

6.3.2.2 Simple LME (UK)

Early interest and R&D initiatives in offshore wind energy in the UK started in the mid-1990s, but the growth of the offshore wind sector was inhibited by two main factors: 1) liberalization of the electricity market in 1990 which put pressure on energy utilities to reduce costs, and 2) a technology-neutral support scheme for renewable energy sources which provided little incentive for costly and risky technology like offshore wind (Smit, Junginger, and Smits 2007: 6438). The first offshore wind park in the UK was built in 2001 but the rapid deployment of offshore wind only commenced in 2009 with the improvement in the regulatory framework and a more active government role (Kern et al. 2014). The RO system was changed to provide more generous support for infant renewable energy technologies (Kern et al. 2014: 638). The Crown Estate was given the mandate to manage the seabed, grant consents for offshore wind projects, and ensure profit maximization for the state. The role of the Crown Estate has been instrumental in improving the conditions for offshore wind energy investments in the UK and facilitating project implementation (Kern et al. 2014). The new support scheme CfD continues to provide necessary support for offshore wind projects. However, the government has made the support for offshore wind conditional on clear targets for cost-reduction. Consequently, the government has not yet set the targets for offshore wind development after 2020, implying high uncertainty for the sector.

Offshore wind is important technology for meeting the UK's renewable energy targets and promoting economic and industrial development (UK Government 2013). Unlike the locally embedded offshore wind energy sector in Denmark and Germany, offshore wind in the UK has largely been developed in an open-market international fashion under the leadership of foreign companies and weak science–industry collaboration (Smit, Junginger, and Smits 2007; Wieczorek et al. 2013). Since 2009, the government has directed efforts towards promoting innovations, developing manufacturing capacities, and involving domestic companies in the offshore wind supply-chain (Ćetković and Buzogány 2016). However, the majority of components for offshore wind farms in the UK are still produced in neighbouring countries, particularly Germany, Denmark, and the Netherlands (RenewableUK/The Crown Estate 2013; Wieczorek et al. 2013). The estimated number of direct jobs in the offshore wind sector in UK was 6,830 in 2013 (RenewableUK/The Crown Estate 2013: 4). In terms of the installed capacity in 2015 the UK had 5,098 MW of operational offshore wind power, which represents 45.9 per cent of the entire EU market (EWEA 2016a) (see Ćetković, Buzogány, and Schreurs 2016: 8).

The success of offshore wind market in the UK can be attributed to several factors including the abundant natural potential and the pressure for meeting national renewable energy targets. Perhaps the key explanatory variable is the 'fit' between the offshore wind sector and the national political-economic logic. Offshore wind development is associated with the construction of large-scale, concentrated, infrastructural projects. The planning and implementation of such projects resembles in many ways conventional energy projects. Such centralized energy planning is not only familiar to the national government in UK, but it also allows the government to directly monitor and manage the revenues from offshore wind investments (see also Kern et al. 2014). Nonetheless, the still prevailing market-based policy paradigm and the related lack of strategic state–industry–science collaboration have hampered the emergence of a genuine domestic offshore wind industry and failed to provide long-term certainty for sectoral growth.

6.3.2.3 *Simple CMEs (Denmark, Sweden, Netherlands)*

The first commercial offshore power plant was built in Denmark in 1991. Four years later, the second offshore wind farm was implemented (Bilgili et al. 2011: 907). The development of offshore wind in Denmark was closely tied to the onshore wind industry. Smit, Junginger, and Smits (2007: 6436) argue that during this early period only the established Danish wind manufacturers and project developers took responsibility for developing offshore wind technologies and projects. In the subsequent phase, due to the demonstrated feasibility of offshore wind, the offshore wind innovation system expanded to include government agencies, research centres, and component and service suppliers (Smit, Junginger, and Smits 2007: 6436–7). This led to the first action plan on offshore wind power in Denmark outlined in 1997 in cooperation between responsible ministries and the industry. The supportive policy instruments and the role of the Danish Energy Authority facilitated the construction of two large-scale offshore wind farms in 2002 and 2003 (Smit, Junginger, and Smits 2007: 6437). As noted by Smit, Junginger, and Smits (2007: 6441), the government's approach fostered interaction and learning across all important actors. The involvement of citizens and cooperatives was also significant and resulted in the implementation of fully of partly locally-owned offshore wind farms (Smit, Junginger, and Smits 2007: 6437). The regulatory changes and the transition towards a more competitive market-based support scheme led to pausing offshore wind deployment in the mid-2000s. In the energy strategy through 2050 adopted in 2011, the role of offshore wind is indicated as crucial (Danish Government 2011). Between 2009 and 2013, several new large-scale wind farms were connected to the grid leading to a total of 1.3 GW in 2015. This makes Denmark the third largest offshore wind market, behind the UK and Germany (EWEA 2016a) (see Ćetković, Buzogány, and Schreurs 2016: 8). The development of offshore wind is supported by considerable domestic investments in R&D, although the greatest share of R&D spending comes from private companies (Megavind 2010).

Sweden was the place where the first offshore wind turbine was constructed in 1990 (Bilgili et al. 2011: 207). Since then, several offshore wind farms have been built, mainly as demonstration projects for acquiring knowledge and testing the technology (Esteban et al. 2011). The policy incentive for constructing commercial offshore wind farms in Sweden has been weak. The government's focus snot been on promoting wider utilization of offshore wind and cost-reductions in the technology, but rather on enhancing knowledge accumulation and testing through pilot studies. This should provide a basis for implementing offshore wind once the technology becomes more mature and financially affordable (Söderholm and Petersson 2011: 522). The future of nuclear energy in Sweden is questionable and there is a debate whether the country should continue relying on the mature renewable energy technologies 'of the shelf' (e.g., onshore wind and biomass) for meeting energy needs or engage in actively exploring innovation and industrial opportunities in infant technologies (e.g., offshore wind) (Södeholm and Petersson 2011; 4C Offshore 2015). At the end of 2015, the size of the offshore wind market in Sweden was comparatively small amounting to a total of 0.2 GW (EWEA 2016a) (see Ćetković, Buzogány, and Schreurs 2016: 8).

The debate about offshore wind potential has a long tradition in the Netherlands, mainly in the context of diversifying the existing offshore oil and gas sectors (Verhees et al. 2015). Cooperation between government agencies, electricity companies, local manufacturers, and research centres was an important part of this process (Verhees et al. 2015). However, within the national wind energy research programmes during the 1980s, little funding was devoted to offshore wind, the focus being on more mature and less costly onshore wind installations (Verhees et al. 2015: 819). The implementation problems in onshore wind projects during the early 1990s renewed the interest in offshore wind. The new Wind Energy Programme in the period 1992–5, for the first time specified a goal of 200 MW of offshore wind capacity by 2010 (Verhees et al. 2015: 819). The Netherlands implemented the second commercial offshore wind power plant in the world in 1994 (Bilgili et al. 2011: 907). Although a clear legal framework was not in place, the government, based on a parliamentary consent and long-term energy planning, granted subsidies for two large offshore wind projects in 2001. They were connected to the grid in 2006 and 2008 (Verhees et al. 2015: 821). In 2002, an ambitious government target of 6000 MW in offshore wind power by 2020 was declared. Nevertheless, progress in developing offshore wind energy policy and a legal framework was slow (Verhees et al. 2015: 823). The potential of offshore wind for innovation and industrial development has been widely recognized, but the government has struggled to balance cost-efficiency with the active promotion of innovations and market creation. Since 2010, offshore wind has been defined as one of the priority economic sectors (Verhees et al. 2015: 825) and R&D funding for wind energy has almost entirely been directed to offshore wind (IEA WIND 2014:135). In 2015, the Netherlands had 426.5 MW in offshore wind power, or 3.9 per cent of the entire EU market (EWEA 2016a). The number of direct jobs in the offshore wind industry is higher than the small domestic market would suggest. In 2013, 1900 full-time employed people were registered in the offshore wind sector and this number increased to 2150 in 2014 (Ecofys 2014) (see Ćetković, Buzogány, and Schreurs 2016: 8).

All three countries have engaged early in exploring offshore wind energy potential relying on the state–industry–science cooperation. Similar to the situation in the onshore wind sector, a shift in the policy paradigm that would allow for enhancing the domestic market for offshore wind has been difficult to achieve. Where supportive policy conditions were present, Denmark and to a lesser extent the Netherlands, the offshore wind sector grew deeply embedded in the network of domestic energy utilities, suppliers, and research centres.

6.4 CONCLUSIONS

The analysis of onshore wind development has largely supported the theoretical assumption on the political-economic conditions for energy innovations. It has been demonstrated that the rise and growth of the German onshore wind energy sector can be attributed to the effective match between strategic state–industry–science–society coordination and the sufficiently plural and decentralized political environment. It is in this context that a variety of bottom-up wind energy solutions could thrive and mobilize long-term support at different levels of government. Although characterized as simple CME, Denmark proved capable of being a pioneer and one of the leading nations in the wind energy industry. This was the result of the inherited tradition of decentralized energy distribution. Some studies suggest that the Danish economy has evolved into a more decentralized coordinated market economy (Campbell and Pedersen 2007). This further supports the argument about the facilitating role of coordinated but decentral structures for energy innovations. It also suggests that that there are variations in the level of decentral ization and plurality among similar forms of capitalism and that the distinction between simple and compound polities should be refined. Other simple CMEs initially supported onshore wind development in a dominantly centralized large-scale manner but the willingness to promote domestic demand ceased quickly due to the resistance from established utilities, liberalization reforms, and the fact that Germany and Denmark were faster in capturing the largest economic benefits in the onshore wind industry. The latter emphasizes the importance of economic interests as drivers of energy transition and illustrates the barrier of the centralized consensus-seeking neo-corporatist relations for new energy technologies. Sweden has recently adopted onshore wind on a broader scale, as a result of the enhanced state role and international climate and energy commitments. Finally, the case of onshore wind in the UK provides evidence of how unitary market-led political economies tend to be laggards in providing stable support for new more decentralized energy technologies. Interestingly, and in support of the general argument, the political devolution in UK and the entrepreneurial role of the Scottish Government have facilitated onshore wind deployment, which draws attention to the dynamic character and institutional changes of national political economies.

The chapter has revealed similarities but also some differences between onshore and offshore wind development. Although the offshore wind sector does not represent simple diversification from onshore wind (Jacobsson and Karltorp 2013), the countries with the strong onshore wind industry and the secured legitimization of wind technology (Germany and Denmark) have also been among the leaders in the domestic offshore market and industrial competitiveness. Somewhat surprisingly, the UK has emerged recently as the most dynamic offshore wind market. This can be explained by the large-scale top-down character of offshore wind coupled with vast natural resources, industrial objectives, and the pressure for meeting national renewable energy targets. However, the long-term prospects for the offshore wind market have proved less stable than in Germany and Denmark. Furthermore, due to strategic coordination, CME countries have generated more locally embedded offshore wind sectors, than is the case in the UK (see Wieczorek et al. 2013). Another interesting finding is the increasing focus on promoting offshore wind market and innovations in the countries that were laggards in onshore wind (the UK but also the Netherlands). This reflects the motives to capture market share in the emerging offshore wind sector and further underlines the critical role of industrial and economic motives behind clean energy policies.

REFERENCES

Anzinger, N. and G. Kostka (2015). 'Offshore Wind Power Expansion in Germany: Scale, Patterns and Causes of Time Delays and Costs Overruns'. Large Infrastructure Projects in Germany between Ambition and Realities, Working Paper 4, Hertie School of Governance, Berlin.

Bergek, A., M. Hekkert, S. Jacobsson, J. Markard, B. Sandén, and B. Truffer (2015). 'Technological Innovation Systems in Contexts: Conceptualizing Contextual Structures and Interaction Dynamics'. *Environmental Innovation and Societal Transitions*, 16(1): 51–64.

BMWI/BMU (2010). 'Energy Concept for an Environmentally Sound, Reliable and Affordable Energy Supply'. Available at: <http://www.bmwi.de/English/Redaktion/Pdf/energy-concept,property=pdf,bereich=bmwi,sprache=en,rwb=true.pdf> (accessed 1 February 2016).

Bilgili, M., Y. Abdulkadir, and S. Erdogan (2011). 'Offshore Wind Power Development in Europe and Its Comparison with Onshore Counterpart'. *Renewable and Sustainable Energy Reviews*, 15(2): 905–15.

Boeckle, R., M. Dua, D. Henriques, P. Simon, and T. Francesco (2010). *The German Wind Energy Cluster: Microeconomics of Competitiveness*. Boston, MA: Harvard Business School/Harvard Kennedy School of Government.

Bruns, E., D. Ohlhorst, B. Wenzel, and J. Köppel (2011). *Renewable Energies in Germany's Electricity Markets. A Biography of the Innovation Process*. Dordrecht: Springer Netherlands.

Buen, J. (2006). 'Danish and Norwegian Wind Industry: The Relationship between Policy Instruments, Innovation and Diffusion'. *Energy Policy*, 34(18): 3887–97.

Bundestag (2014). 'Act on the Development of Renewable Energy Sources'. Available at: <https://www.erneuerbare-energien.de/EE/Redaktion/DE/Gesetze-Verordnungen/eeg_2014_engl.pdf?__blob=publicationFile&v=4> (accessed 1 February 2016).

BWE (Bundesverband Windenergie) (2016). 'Land-Based Wind Energy: Analysis of the German Market in 2015'. Available at: <https://www.wind-energie.de/en/press/press-releases/2016/land-based-wind-energy-analysis-german-market-2015> (accessed 1 February 2016).

4C Offshore (2015). 'Swedes Disapprove Offshore Wind'. Available at: <http://www.4coffshore.com/windfarms/swedes-dissaprove-offshore-wind-nid2298.html> (accessed 1 February 2016).

Campbell, J. L. and O. K. Pedersen (2007). 'The Varieties of Capitalism and Hybrid Success: Denmark in the Global Economy'. *Comparative Political Studies*, 40(3): 307–32.

Ćetković, S. and A. Buzogány (2016). 'Varieties of Capitalism and Clean Energy Transitions in the European Union'. *Climate Policy*. doi:10.1080/14693062.2015.1135778.

Ćetković, S., A. Buzogány, and M. Schreurs (2016). 'Varieties of Clean Energy Transitions in Europe: Political-Economic Foundations of Onshore and Offshore Wind Development', 2016/18. UNU-WIDER, Helsinki.

Colomer, J. M. (2012). 'The More Parties, the Greater Policy Stability'. *European Political Science*, 11(2): 229–43.

Danish Government (2011). 'Energy Strategy 2050: From Coal, Oil and Gas to Green Energy'. Available at: <http://dfcgreenfellows.net/Documents/EnergyStrategy2050_Summary.pdf> (accessed 16 November 2016).

DECC (Department of Energy & Climate Change) (2011). 'UK Renewable Energy Roadmap'. Available at: <https://www.gov.uk/government/uploads/system/uploads/attachment_data/file/48128/2167-uk-renewable-energy-roadmap.pdf> (accessed 1 February 2016).

DTI (Department of Trade and Industry) (2003). 'Energy White Paper: Our Energy Future Creating a Low-Carbon Economy'. Available at: <http://webarchive.nationalarchives.gov.uk/+/http:/www.berr.gov.uk/files/file10719.pdf> (accessed 1 February 2016).

Ecofys (2014). 'Dutch Offshore Wind Energy Cluster Generates 2150 Full-Time Jobs and over €900 Million Turnover'. Available at: <http://www.ecofys.com/en/news/dutch-offshore-wind-energy-cluster-generates-2150-full-time-jobs-and-over-9/> (accessed 1 February 2016).

Esping-Andersen, G. (1990). *The Three Worlds of Welfare Capitalism*. Hoboken, NJ: John Wiley & Sons.

Esteban, M. D., J. J. Diez, J. S. Lopez, and V. Negro (2011). 'Why Offshore Wind Energy?' *Renewable Energy*, 36(2): 444–50. doi:10.1016/j.renene.2010.07.009.

European Commission (2013). 'JRC Scientific and Policy Reports R & D Investment in the Technologies of the European Strategic Energy Technology Plan'. Available at: <https://ec.europa.eu/energy/sites/ener/files/swf_2013_0157_en.pdf> (accessed 1 February 2016).

EWEA (European Wind Energy Association) (2016a). 'The European Offshore Wind Industry: Key Trends and Statistics 2015'. Available at: <http://www.ewea.org/fileadmin/files/library/publications/statistics/EWEA-European-Offshore-Statistics-2015.pdf> (accessed 1 February 2016).

EWEA (European Wind Energy Association) (2016b). 'Wind in Power: 2015 European Statistics'. Available at: <http://www.ewea.org/fileadmin/files/library/publications/statistics/EWEA-Annual-Statistics-2015.pdf> (accessed 1 February 2016).

Fagerberg, J. (2015). 'Innovation Policy, National Innovation Systems and Economic Performance: In Search of a Useful Theoretical Framework'. TIK Centre for Technology, Innovation and Culture, Oslo.

Fitch-Roy, O. (2015). 'An Offshore Wind Union? Diversity and Convergence in European Offshore Wind Governance'. *Climate Policy.* doi: 10.1080/14693062.2015.1117958.

Garud, R. and P. Karnøe (2003). 'Bricolage versus Breakthrough: Distributed and Embedded Agency in Technology Entrepreneurship'. *Research Policy*, 32(2): 277–300.

German Government (2002). 'Strategy of the German Government on the Use of Off-Shore Wind Energy in the Context of its National Sustainability Strategy'. Available at: <http://www.offshorewindenergy.org/reports/report_033.pdf> (accessed 1 February 2016).

Giest, S. (2015). 'Comparative Analysis of Sweden's Wind Energy Policy: The Evolution of "Coordinated" Networks'. *Journal of Comparative Policy Analysis*, 17(4): 393–407.

GTAI (Germany Trade & Invest) (2014). '*Industry Overview: The Wind Energy Industry in Germany*', Berlin.

Hackley, R. (2015). 'Cameron's U.K. Conservatives Oppose Onshore Wind Energy Farms. *Bloomberg.* Available at: <http://www.bloomberg.com/news/articles/2015-04-14/cameron-s-u-k-conservatives-oppose-onshore-wind-farms> (accessed 1 February 2016).

Hall, P. A. and D. W. Soskice (2001). *Varieties of Capitalism: The Institutional Foundations of Comparative Advantage.* Oxford: Oxford University Press.

Hancké, B. (2009). *Debating Varieties of Capitalism: A Reader*: Oxford: Oxford University Press.

HM Revenue and Customs (2015). 'Climate Change Levy: Removal of Exemption for Electricity from Renewable Sources'. Available at: <https://www.gov.uk/government/publications/climate-change-levy-removal-of-exemption-for-electricity-from-renewable-sources/climate-change-levy-removal-of-exemption-for-electricity-from-renewable-sources> (accessed 1 February 2016).

Howard, R. and K. Drayson (2015). 'Powering Up: The Future of Onshore Wind in the UK'. Policy Exchange, London.

IEA (International Energy Agency WIND) (2014). '2013 IEA WIND Annual Report'. Available at: <http://www.ieawind.org/annual_reports_PDF/2013/2013%20AR_small_090114.pdf> (accessed 1 February 2016).

International Renewable Energy Agency (2012). '30 Years of Policies for Wind Energy: Lessons from 12 Wind Energy Markets'. IRENA, Abu Dhabi.

IWES (n.d.). 'Wind Monitor'. Available at: <http://windmonitor.iwes.fraunhofer.de/windmonitor_en/1_wind-im-strommix/1_energiewende-in-deutschland/5_Ausbau_der_Windenergie/> (accessed 1 February 2016).

Jacobsson, S. and A. Bergek (2004). 'Transforming the Energy Sector: The Evolution of Technological Systems in Renewable Energy Technology'. *Industrial and Corporate Change*, 13(5): 815–49.

Jacobsson, S. and K. Karltorp (2013). 'Mechanisms Blocking the Dynamics of the European Offshore Wind Energy Innovation System—Challenges for Policy Intervention'. *Energy Policy*, 63: 1182–95. doi:10.1016/j.enpol.2013.08.077.

Jacobsson, S. and V. Lauber (2006). 'The Politics and Policy of Energy System Transformation—Explaining the German Diffusion of Renewable Energy Technology'. *Energy Policy*, 34(3): 256–76.

Kamp, L. M. (2008). 'Socio-Technical Analysis of the Introduction of Wind Power in the Netherlands and Denmark'. *International Journal of Environmental Technology and Management*, 9(2/3): 276–93.

Kern, F., A. Smith, C. Shaw, R. Raven, and B. Verhees (2014). 'From Laggard to Leader: Explaining Offshore Wind Developments in the UK'. *Energy Policy*, 69(6): 635–46.

KfW (2015). 'KfW Programm Offshore Windenergie'. Available at: <https://www.kfw.de/Download-Center/F%C3%B6rderprogramme-%28Inlandsf%C3%B6rderung%29/PDF-Dokumente/6000001940-Merkblatt-273.pdf> (accessed 1 February 2016).

Lauber, V. and L. Metz (2004). 'Three Decades of Renewable Energy Policies in Germany'. *Energy and Environment*, 15(4): 599–623.

Lijphart, A. (1999). *Patterns of Democracy: Government Forms and Performance in Thirty-six Countries*. New Haven, CT: Yale University Press.

Mazzucato, M. (2015). *The Entrepreneurial State: Debunking Public vs. Private Sector Myths*. London: Anthem Press.

Megavind (2010). 'Denmark Supplier of Competitive Offshore Wind Solutions: Megavind's Strategy for Offshore Wind Research, Development and Demonstration'. Available at: <http://ipaper.ipapercms.dk/Windpower/Megavind/Denmark_supplier_of_competitive_offshore_wind_solutions/> (accessed 1 February 2016).

Pegels, A. and W. Lütkenhorst (2014). 'Is Germany's Energy Transition a Case of Successful Green Industrial Policy? Contrasting Wind and Solar PV'. *Energy Policy*, 74: 522–34.

Pettersson, M., K. Ek, K. Soderholm, and P. Soderholm (2010). 'Wind Power Planning and Permitting: Comparative Perspectives from the Nordic Countries'. *Renewable & Sustainable Energy Reviews*, 14(9): 3116–23.

PLB (Planbureau voor de Leefomgeving) (2014). 'Netherlands National Energy Outlook 2014 Summary'. Available at: <http://www.pbl.nl/sites/default/files/cms/publicaties/pbl-2014-netherlands-national-energy-outlook-2014.pdf> (accessed 1 February 2016).

Ragwitz, M. and A. Miola (2005). 'Evidence from RD&D Spending for Renewable Energy Sources in the EU'. *Renewable Energy*, 30(11): 1635–47.

RenewableUK (2014). 'Wind Energy in the UK: State of the Industry Report 2014'. Available at: <http://www.maritimeindustries.org/write/Uploads/UKMIA%20Uploads%20-%20DO%20NOT%20DELETE/State_of_the_Industry_-_Wind_Energy_Report_2014.pdf> (accessed 1 February 2016).

RenewableUK (2015). 'Wind Energy in the UK: State of the Industry Report 2015'. Available at: <http://c.ymcdn.com/sites/www.renewableuk.com/resource/resmgr/publications/reports/StateIndustryReport2015Full.pdf> (accessed 1 February 2016).

RenewableUK (2016). 'UK Wind Energy Database'. Available at: <http://ruk.pixl8-hosting.co.uk/en/renewable-energy/wind-energy/uk-wind-energy-database/index.cfm> (accessed 1 February 2016).

RenewableUK/The Crown Estate (2013). 'Building an Industry: Updated Scenarios for Industrial Development'. Available at: <https://hub.globalccsinstitute.com/sites/default/files/publications/115688/building-industry-updated-scenarios-industrial-development.pdf > (accessed 1 February 2016).

Sandén, B. A. and C. Azar (2005). 'Near-Term Technology Policies for Long-Term Climate Targets: Economy Wide Versus Technology Specific Approaches'. *Energy Policy*, 33(12): 1557–76.

Schmidt, V. (2012). 'The State and Political Economic Change: Beyond Rational Choice and Historical Institutionalism to Discursive Institutionalism'. In M. Telò (ed.), *State, Globalization and Multilateralism*, vol. 5. Dordrecht: Springer Netherlands, pp. 99–118.

Schneider, B. R. and D. Soskice (2009). 'Inequality in Developed Countries and Latin America: Coordinated, Liberal and Hierarchical Systems'. *Economy and Society*, 38(1): 17–52.

Simmie, J., R. Sternberg, and J. Carpenter (2014). 'New Technological Path Creation: Evidence from the British and German Wind Energy Industries'. *Journal of Evolutionary Economics*, 24(4): 875–904.

Smit, T., M. Junginger, and R. Smits (2007). 'Technological Learning in Offshore Wind Energy: Different Roles of the Government'. *Energy Policy*, 35(12): 6431–44.

Smith, A. and R. Raven (2012). 'What Is Protective Space? Reconsidering Niches in Transitions to Sustainability'. *Research Policy*, 41(6): 1025–36.

Söderholm, P., K. Ek, and M. Pettersson (2007). 'Wind Power Development in Sweden: Global Policies and Local Obstacles'. *Renewable and Sustainable Energy Reviews*, 11(3): 365–400.

Söderholm, P. and M. Pettersson (2011). 'Offshore Wind Power Policy and Planning in Sweden'. *Energy Policy*, 39(2): 518–25.

Statistics Netherlands (2010). 'Renewable Energy in the Netherlands 2010'. Available at: <http://www.cbs.nl/NR/rdonlyres/BED23760-23C0-47D0-8A2A-224402F055F3/0/2012c90pub.pdf> (accessed 1 February 2016).

Suurs, R. A. A. and M. P. Hekkert (2009). 'Competition between First and Second Generation Technologies: Lessons from the Formation of a Biofuels Innovation System in the Netherlands'. *Energy*, 34(5): 669–79.

UK Government (2013). *Offshore Wind Industrial Strategy: Business and Government Action*. Available at: <https://www.gov.uk/government/uploads/system/uploads/attachment_data/file/243987/bis-13-1092-offshore-wind-industrial-strategy.pdf> (accessed 1 February 2016).

van der Vleuten, E. and R. Raven (2006). 'Lock-In and Change: Distributed Generation in Denmark in a Long-Term Perspective'. *Energy Policy*, 34(18): 3739–48.

van Rooijen, S. N. M. and M. T. van Wees (2006). 'Green Electricity Policies in the Netherlands: An Analysis of Policy Decisions'. *Energy Policy*, 34(1): 60–71.

Verbong, G. and D. Loorbach (eds) (2012). *Governing the Energy Transition: Reality, Illusion or Necessity?* London: Routledge.

Verhees, B., R. Raven, F. Kern, and A. Smith (2015). 'The Role of Policy in Shielding, Nurturing and Enabling Offshore Wind in The Netherlands (1973–2013)'. *Renewable & Sustainable Energy Reviews*, 47(6): 816–29.

Wieczorek, A. J., S. O. Negro, R. Harmsen, G. J. Heimeriks, L. Luo, and M. P. Hekkert (2013). 'A Review of the European Offshore Wind Innovation System'. *Renewable and Sustainable Energy Review*, 26(2): 294–306.

Wolsink, M. (1996). 'Dutch Wind Power Policy—Stagnating Implementation of Renewables'. *Energy Policy*, 24(12): 1079–88.

7

The Political Economy of Energy Innovation

Shouro Dasgupta, Enrica de Cian, and Elena Verdolini

7.1 INTRODUCTION

Technological change directed towards more efficient and eco-friendly technologies is a priority for both developed and developing countries. Insights on past trends and determinants of energy innovation, including political economy factors and non-financial drivers, are important to set the basis for cost effective climate and energy policies in the coming years. Issues such as the role of institutions and lobbying as enabling or inhibiting factors are critical factors that have been only marginally examined by the existing literature.

This chapter uses two commonly used indicators of innovation, energy industrial R&D and energy patents for 20 countries for the years 1995–2010 to examine the influence of political economy factors on energy-related innovations using econometric analysis. Political economy factors can be broadly defined as those concerning the interactions and tensions between de jure and de facto power, including the distribution of resources, the rules for the exercise of power and the enforcement of contracts, the procedures and institutions for settling conflicts over these rules, or the physical and organizational infrastructure supporting economic activities, transactions, and collective actions.[1] We focus on four aspects: the types and stringency of government support to energy innovation (e.g., the various policy instruments implemented to this end such as environmental and R&D policies); the quality of governance (e.g., government effectiveness); the political orientation of the government; and the distribution of resources across interest groups.

[1] Stavins (2004) refers to political economy as the process through which political decisions are made.

The role of these factors in relation to energy innovation dynamics, energy transition, and sustainable development has been acknowledged by several contributions both in the policy and in the economics realms (Anadón 2012; Friedrichs and Inderwildi 2013; Hughes and Lipscy 2013; IPCC 2014; IEA 2015a). More than other sectors, energy can be dominated by large incumbent companies and utilities, which often seek to influence policy. Their investments, especially in new technologies are shaped by the incentives and regulations set by policy makers (Lockwood 2013a). Moreover, the actual impact of regulations and government policies is affected by the broader institutional settings (Stavins 2004). Good governance is particularly important as many of the government interventions are economic policies (Lockwood 2013a) and bureaucrats are the actors ultimately implementing these policies (Lockwood 2013b).

The role of governance quality, political orientation of the government, and distribution of resources across interest groups have received only marginal attention and have not been explored jointly in the empirical literature on energy innovation. In this specific domain, the role of public policies as drivers of innovation has received more attention than institutions and political economy factors. The multiple sources of market failures that characterize the energy sector and the recent debate regarding the actions governments should undertake to curb rising greenhouse gas emissions partly explains the focus of the current literature on the role of environmental, energy, and innovation policies. In the energy-environmental realm, state intervention is motivated by the presence of environmental externalities (a gap between private and social returns to pollution control), as well as of innovation externalities (a gap between private and social returns to innovation). Moreover, in comparison to other sectors, energy R&D often entails large-scale projects, which need public support (Anadón 2012).

This chapter contributes to the debate by jointly assessing the influence of environmental and R&D policies, governance quality, political orientation, and distribution of resources to energy intensive industries on energy innovation. The rest of the chapter is organized as follows. Section 7.2 discusses our measures of energy innovation. Section 7.3 provides the empirical framework and describes the main hypotheses explored in the empirical analysis. Section 7.4 discusses the results. Section 7.5 concludes, highlighting policy implications and future research needs.

7.2 MEASURING ENERGY INNOVATION TRENDS

Studying innovation systems and dynamics using an empirical approach is challenging, as innovation comprises both tangible and intangible outputs (e.g., new technologies, machines, products, patents but also ideas, process innovation, managerial, and organizational innovation). Following a large empirical innovation economics literature, we study the more tangible and measurable aspects of the innovation process. While these constitute only a part of the innovation output relevant for the energy system and sector, they nonetheless provide important insights that can complement those from qualitative and bottom-up case studies focusing on more intangible and less measureable aspects such as organizational innovation.

Here we focus on R&D expenditures and patent counts.[2] The former informs on the inputs of the innovation process, while the latter is a proxy of innovation outputs. Both indicators suffer from some specific shortcomings. R&D investments provide insights on innovation effort but not on innovation quality. Conversely, patent statistics provide a partial measure as not all innovations are patented, even though they can be weighted using information on several indicators to control for quality (for instance, claims or citations, see Griliches 1990). Furthermore, patents may increase due to changes in patent law or strategic reasons to signal in which companies to invest in, regardless of innovative activity (Mazzucato 2013).

In the case of energy innovation, matters are further complicated by the fact that it is unclear how clean innovation or even energy innovation are defined (Gallagher et al. 2011). A number of studies focus specifically on the energy supply sector (Salies 2010; Sterlacchini, 2012; Costa-Campi et al. 2014) but energy-saving R&D and innovation are pervasive. Energy is an input for nearly all sectors of the economy and the way in which energy is produced, transformed, and distributed depends on innovative activities well beyond those of the energy supply sector itself. All R&D expenditures are inputs into complex processes that ultimately lead to innovations that may or may not be clean. In order to proxy for industrial R&D investments in energy, we rely on the Analytical Business Enterprise Research and Development (ANBERD) database (OECD 2016), which provides information on the R&D expenditures at the sectoral level for 30 countries for the years 1990–2013.[3] We define energy R&D investments in two ways. First, we focus on R&D spending in the 'Electricity, water and gas distribution industry', which represents the downstream sector for energy production (power R&D). Second, we define energy investments as a combination of R&D expenditures from 'Electricity, water and gas distribution industry' and 'Mining', which capture the combined R&D effort in the upstream and downstream energy supply sector (energy R&D).

[2] Arguably, R&D investments and patents represent only part of the full innovation process, as they somewhat disregard the issue of technology diffusion. Specifically, patent data is an imperfect indicator of technology diffusion, but nonetheless widely used in the literature to proxy for the other, earlier stages of innovation (see, for instance, Hall and Rosenberg 2010).

[3] Our analysis focuses on the 20 countries between 1995 and 2010 for which both policy and institutional data are available.

These measures arguably represent a lower-bound estimate of energy-related innovation (Upstill and Hall 2006), as they only include the R&D directly performed by the energy supply sectors. Indeed, non-energy sectors indirectly contribute to energy-related innovation. For instance, improvements in the manufacturing of chemicals and chemical products, and in computer and electronics, contribute to the development of energy system technologies, such as solar power or smart grids. These are 'embedded' in the capital that is supplied to the energy supply sector. The sum of the direct and 'embedded' R&D can be considered an estimate of the upper bound of industrial energy-related R&D in a given country. Input–output data can be used to provide an estimate of energy-related R&D expenditures including the research performed in other economic sectors that are embedded in the capital purchased by the electricity and the mining sectors. Dasgupta, De Cian, and Verdolini (2016) provide a detailed description and application of this method.

While providing insights on the extent of energy innovation efforts, the ANBERD statistics have some shortcomings. For instance, they report R&D expenditure by sector of performance expenditure, regardless of whether funds were provided by the private or by the public sector. This means that industrial R&D reported by ANBERD statistics might include a fraction of R&D expenditure funded by the government and therefore reported in the government budget outlays as well. For this reason, we refer to the R&D reported in the ANBERD statistics as industrial rather than private R&D.

Another widely used proxy for innovation is patent counts, which is an indicator of the output of the industrial R&D process (Griliches 1990).[4] The temporal and country coverage of patent data is often broader than that of R&D statistics and makes it an attractive empirical proxy. In the specific case of energy-related innovation, a further advantage of using patent data is the possibility of assigning patents to specific energy technology classes in the energy sector, which also include renewables (Johnstone, Ivan Haščič, and Popp 2010) and efficient fossil-based technologies for electricity generation (Lanzi, Verdolini, and Haščič 2011). We collect patent statistics from the Organization for Economic Co-operation and Development (OECD) Patent Statistics Database (OECD 2015b) and count applications through the Patent Cooperation Treaty (PCT) by the inventor country and priority date. The technologies included in our patent counts are the following:[5]

[4] Indeed patents are positively correlated with power R&D (correlation coefficient for power patents is 0.41 and 0.62 for environmental patents) and with energy R&D (power patents 0.50 and environmental patents 0.44).

[5] Please refer to Haščič and Migotto (2015) and OECD (2015a) for more details about the technologies included.

(1) Power Patents: related to energy generation, they include both energy generations from renewable and non-fossil sources and technologies improving the efficiency of fossil fuels, such as Integrated Gasification Combined Cycle and improved burners. Both renewable and fossil-efficient technologies have significant mitigation potential (IEA 2014).

(2) Green Patents: include power patents as well as the patents in the technology domains of general environmental management, technologies specific to climate change mitigation, energy efficiency in buildings and lighting, technologies with potential or indirect contribution to emissions mitigation, emissions abatement, and fuel efficiency in transportation.

7.3 EMPIRICAL MODEL AND RESEARCH FRAMEWORK

Combining the data sources on our variables of interest described in Section 7.2, we build an unbalanced panel of 20 countries for the years 1995–2010. Please refer to Dasgupta, De Cian, and Verdolini (2016) for a detailed description of the data and descriptive statistics. The literature has used alternative methods to examine the role of political economy in the context of energy technology choices and in the context of innovation, such socio-technical transition studies (see Johnstone and Stirling (2015) for an example of such approach and Turnheim et al. (2015) for a review of the approach). Socio-technical transition studies can analyse multiple dimensions of change, including economic, political, and socio-cultural aspects at different levels and temporalities. Econometric analysis, specifically panel regression analysis, can complement this approach and isolate the influence of environmental policy, institutional quality, political orientation, and resource distribution conditional on each of the other factors, controlling for country-invariant and time-invariant characteristics.

7.3.1 Empirical Model

We use our data to estimate the following general reduced form equation:

$$y_{it} = \alpha_i + \gamma_t + \pi_{it}\beta_1 + x_2\varphi_{it} + \beta_3\rho_{it} + \beta_4\theta_{it} + Z_{it}\omega + \epsilon_{it}, \qquad (1)$$

where the subscripts i and t indicate respectively the country and the year, and:

- y_{it} is a variable measuring the energy innovation intensity of the economy. Specifically, we define y_{it} as the share of one of our innovation proxies discussed in Section 7.2 (i.e., industrial energy R&D, power

R&D, power patents, or environmental patents) over total value added. We scale all innovation proxies relative to the total value added to account for the heterogeneity in the countries included in our sample. This is in line with the general literature on this topic (see, for instance, Popp 2002).

- π_{it} is a vector of policy stringency measures, discussed in detail in Section 7.3.2 and includes both market-based and non-market-based instruments directly targeting the environmental externality, such as taxes or standards, as well as government R&D investments in energy innovation targeting the knowledge externality.
- φ_{it} is a proxy for institutional quality, measured either by government effectiveness or by an aggregate indicator of governance quality discussed in Section 7.4.
- ρ_{it} is a proxy of the political orientation of the government.
- θ_{it} is a proxy of the distribution of resources to the energy sector relative to the rest of the economy, which in our framework inform on two different aspects, market-size effect and the power of the energy lobby within each country.
- Z_{it} is a vector of other relevant control variables influencing innovation investments, including an index for industrial energy prices and trade openness. Higher energy prices are expected to increase innovation incentives, net of any political economy consideration (Popp 2002), whereas trade openness can have an ambiguous effect.
- α_i and γ_t are country and year fixed effects, while ϵ_{it} is a random error term. Country fixed effects control for time-invariant factors, including persistent institutional factors, such as the democratic/autocratic characteristics and system of government of countries. The time fixed effects control for inter-temporal trends that are uniform across countries, such as the economic cycle.

The expectations about the roles of the variables of interest, π_{it}, ρ_{it}, θ_{it}, is detailed in the research hypotheses presented in Section 7.3.2. The regressions are estimated using fixed effect linear models, as both our R&D and patent data are continuous variables.[6] Due to the different nature of R&D investments and patents, we use a different lag structure in our models. Specifically, we assume that R&D investments react faster to environmental policies than patents. This is due to the fact that patents measure the output of the innovation process. Applying for patent requires first to put the R&D investment to work and then develop and test new ideas. Thus, the

[6] The patents from the OECD database are computed using fractional counting and hence are continuous in nature.

R&D specifications use a one-year time lag, while the patent equation uses a two-year time lag.[7]

7.3.2 Research Hypotheses

We use the model presented in Section 7.3.1 to test a set of hypotheses inspired by the existing literature. The four hypotheses of interest are discussed in this section.

Hypothesis 1 (H1): *Environmental policy stringency* (π_{it}) *results in dynamic efficiency gains. Stringent regulations provide long-term incentives for innovation in energy-saving and pollution-reducing technologies.*

Overcoming environmental issues requires addressing two market failures. Since pollution is not priced appropriately, private firms tend to over-pollute with respect to the social optimum. The environmental externality can be directly targeted by using two different policy instruments: market-based policies, such as a tax on pollution, feed-in tariffs, or trading schemes, and non-market-based policies, such as standards or incentives for R&D investments in cleaner energy. Both instruments have been widely used in the countries in our sample.

The available literature provides evidence on both market-based and non-market-based instruments, together with innovation policies, supporting cleaner technologies, and affecting the rate and direction of technological change (Jaffe, Peterson, and Portney 1995; Popp 2002; Johnstone, Ivan Haščič, and Popp 2010) but a priori their effectiveness may be different (Fischer, Parry, and Pizer 2003; Newell 2010). Previous empirical studies on the inducement effect of environmental and energy policy on innovation have employed different measures of environmental stringency, policy instruments, and innovation indicators (Brunel and Levinson 2013). Here we rely on the recent environmental policy stringency (EPS) database of the OECD (Botta and Koźluk 2014), which provides detailed cross-country information on several instruments and on the International Energy Agency (IEA) Energy

[7] Note that, incidentally, allowing for a one (two)-year lag structure in the R&D (patent) equation also partly addresses concerns regarding the endogeneity of the explanatory variables. Regarding country-level variables such as good governance, the political orientation of the government and the lobbying power of the energy sector, endogeneity corners are weak to non-existent, since environmental innovation represents only a fraction of the innovative capacity of the countries in our sample during the time period explored. Hence, it is unlikely to be a major driver of country-level variables. Conversely, there may be concerns regarding the endogeneity of the EPS policy indicators, as the availability of cleaner and more efficient energy technologies may be influencing the ability of countries to propose, pass, and adopt environmental policies (Carrion-Flores and Innes 2010). Allowing for a time lag reduces concerns in this respect.

Technologies R&D database (IEA 2015b). The EPS aggregate policy indicator (EPS-Total score) is constructed using information on both market-based and non-market policies.[8] For each policy instrument, countries are scored on a scale from 0 to 6 depending on the stringency of the policy they implement. Such scores are then weighted and aggregated to construct the aggregate policy indicator.[9]

Hypothesis 2 (H2): *Institutional quality, measured as good governance, increases the incentives to invest in energy-related innovation.*

The role of governance quality has been widely examined in the context of investments by the literature on Foreign Direct Investment (FDI) drivers and to a lesser extent, by the general literature on innovation.[10] The dominant view is that good governance aids FDI (Ayal and Karras 1996; Globerman and Shapiro 2003; Biglaiser and DeRouen 2006; Gani 2007; Staats and Biglaiser 2012).[11] However, poor governance can also lead to more foreign investments if combined with high levels of corruption (Bellos and Subasat 2012). The literature on general innovation states that better institutions are likely to promote general innovation and investments (Habiyaremye and Raymond 2013; Tebaldi and Elmslie 2013; Silve and Plekhanov 2015). We test whether government effectiveness and more broadly good governance as measured by the World Governance Indicators (WGI) (Kaufman, Kraay, and Mastruzzi 2010)[12] affect energy-related innovation. WGI institutional quality indicators are measured on a normalized scale from (−)2.5 to (+)2.5, where the highest value indicates better governance. We focus on government effectiveness, which is an indicator of bureaucratic quality and speed. Low levels of government effectiveness can be associated with excessive regulations, lengthy processes, and lower transparency in the form of flow of information.

Hypothesis 3 (H3): *The political orientation of the government influences investments in energy R&D and patents. On the one hand, left-leaning*

[8] Market-based policies include feed-in tariffs (FITs—solar and wind), taxes (on CO2, SOx, NOx, and diesel), certificates (White, Green, and CO2), and the presence of deposit and refund schemes (DRS). Non-market-based policies include standards, such as emission limits for SOx, NOx, and SO2 and on the sulphur content of diesel, as well as public R&D investment in energy technologies.

[9] We refer the interested reader to Botta and Koźluk (2014) for details about the indicators' construction.

[10] The general literature on the determinants of innovation in the manufacturing sector is broad (Becheikh et al. 2006; Hall and Rosenberg 2010) and focuses on several key internal factors, such as size, firm age, skills, and qualified personnel.

[11] Good governance is defined as a government that entails an independent judiciary and legislation, fair and transparent laws with impartial enforcement, reliable public financial information, and high public trust (Subasat and Bellos 2011).

[12] For more detailed information on the WGI, please see <http://info.worldbank.org/governance/wgi/index.aspx#home> (accessed 12 October 2016).

governments are more likely to implement regulations that attract innovation and investment in energy-related R&D. On the other hand, right-wing-oriented governments are more likely to take a laissez faire approach. Therefore, the impact of political orientation can be ambiguous.

The role of government political orientation has been examined in the context of environmental policy adoption (Fankhauser, Gennaioli, and Collins 2014; Folke 2014) and to some extent, in the context of private investment and FDI. In the former, the consensus seems to be that right-wing governments generally oppose laws to support climate regulations (McCright and Dunlap 2011; Painter and Ashe 2012), while left-wing governments are more likely to pass them (Neumayer 2003 and Fankhauser, Gennaioli, and Collins 2014). On the contrary, right-leaning governments are more inclined to allow the market forces to stimulate investment efforts (Esping-Andersen 1990; Boix 1998). The FDI literature, however, provides somewhat contrasting insights. Shleifer (1998) states that right-wing governments consider the private sector to be more conducive in terms of innovation and therefore tend not to intervene in the market while Hawkins, Mintz, and Provissiero (1976) and Jensen (2006) mention that left-leaning governments are more likely to expropriate foreign assets, which discourages FDI. We use political orientation of governments from the Database of Political Institutions 2012 (DPI) (Beck et al. 2001) as proxies for political institutions, specifically, the political orientation of the executive party with respect to economic policy. This is a categorical variable that takes three values: right (1), centre (2), and left (3) orientations.

Hypothesis 4 (H4): A higher share of energy intensive sectors will (a) give rise to a market-size effect (i.e., higher demand for energy), (b) lead to more lobbying power of the energy intensive sectors towards the government, and (c) increase the coordination costs of such lobbying activities. Therefore, the impact of resource distribution on energy-related innovation is not clear a priori.

The distribution of resources across interest groups may give rise to several dynamics. First, if an economy relies more on energy intensive sectors, the market for new energy inventions will be larger. As a result, the value associated with any innovation relative to energy goods will be higher, as it is more profitable to develop technologies that have a larger market (see, for instance, the discussion of market-size effects in the directed technical change literature in Acemoglu (2002)). This would suggest that larger energy-intensive sectors will likely result in more energy-related innovation. Second, energy intensive incumbent industries with access to significant resources tend to engage in lobbying for government support and seek to influence policy decisions. For instance, Fredriksson and Svensson (2003) argue that strong industry lobbies may engage in corruption to reduce environmental policy

Table 7.1. Political economy factors: hypothesis and proxy variables

Hypothesis	Proxy Variables
H1: Environmental policy	EPS-Market, EPS-Non-market, EPS-Total
H2: Governance	Governance effectiveness, Governance Average WGI indicator, Governance x EPS-Total
H3: Left-wing political orientation	Political orientation
H4: Lobbying	Value added share of energy-intensive industries Value added share of carbon-intensive industries Value added share of electricity

Source: Authors' conceptualization.

stringency while Fredriksson, Vollebergh, and Dijkgraaf (2004) remark that incumbent industries utilize their lobbying power to oppose structural transformation. Third, larger sectors may imply more firms/actors, and this would result in higher coordination costs of such lobbying activities (Olson 1965; Fredriksson, Vollebergh, and Dijkgraaf 2004).

In line with previous literature (Fredriksson, Vollebergh, and Dijkgraaf 2004; Costa-Campi et al. 2014), we use the value-added share of energy-intensive industries (Coke, Refined Petroleum and Nuclear Fuel, Chemicals and Chemical Products, Rubber and Plastics, Water and Air Transport, Electricity, Basic Metals and Fabricated Metal Mining)[13] in the economy computed using industrial value added data from the WIOD database (Timmer et al. 2015) as an indicator of market-size, lobbying power, and coordination costs.

Table 7.1 summarizes the main proxy variables that are used to measure the key drivers behind the four hypotheses described in this section.

7.4 RESULTS

The empirical results of our analysis using the two main indicators of innovation described in Section 7.2 are provided in Tables 7.2 and 7.3. Table 7.2

[13] Energy-intensive sectors have been defined as the sectors with energy intensity above the 75th percentile. As a robustness test, two other proxy variables have been considered; the value-added share of carbon-intensive industries (other non-metallic mineral, inland, water, and air transport, electricity, mining), and value-added share of the electricity or energy (electricity +mining) sector. Carbon-intensive sectors have been defined as the sectors with carbon intensity above the 75th percentile.

Table 7.2. Regression results using R&D intensity over value added as innovation proxy: one-year lag for all independent variables

		1	2	3	4	5	6	7	8	9
	Dependent Variable	Log of R&D Intensity—Power					Log of R&D Intensity—Energy			
	EPS Market Score	0.198+	0.189+	0.165			−0.006	−0.018		
H1		(0.125)	(0.121)	(0.122)			(0.111)	(0.111)		
	EPS Non-market Score	−0.089	−0.014	0.018			−0.058	−0.043		
		(0.108)	(0.107)	(0.108)			(0.098)	(0.098)		
	EPS Total Score				0.135	0.164			−0.079	0.006
					(0.110)	(0.162)			(0.101)	(0.148)
	Govt. Effectiveness	0.964***	0.769**		0.619**	0.666*	0.399		0.409	0.549+
H2		(0.323)	(0.317)		(0.312)	(0.367)	(0.294)		(0.288)	(0.338)
	WGI			0.754				0.418		
				(0.538)				(0.498)		
	Govt. Effectiveness*EPS Interaction					−0.033				−0.098
						(0.135)				(0.123)
H3	Political orientation	0.222***	0.211***	0.202***	0.200***	0.199***	0.112*	0.107*	0.111*	0.106*
		(0.065)	(0.063)	(0.064)	(0.063)	(0.063)	(0.058)	(0.058)	(0.057)	(0.058)
	VA Share Energy-intensive industries	0.710**	0.827**	0.814**	0.815**	0.822**	0.537+	0.525+	0.542+	0.553+
H4		(0.356)	(0.356)	(0.363)	(0.356)	(0.358)	(0.342)	(0.348)	(0.341)	(0.341)
	Energy price index		−3.053***	−3.203***	−3.193***	−3.202***	−8.309***	−8.629***	−8.286***	−8.388***
			(0.756)	(0.760)	(0.753)	(0.756)	(3.034)	(3.029)	(3.012)	(3.017)
	Trade openness		0.005	0.005	0.004	0.004	−0.027***	−0.027***	−0.027***	−0.027***
			(0.008)	(0.008)	(0.008)	(0.008)	(0.007)	(0.007)	(0.007)	(0.007)
	Observations	256	256	256	256	256	256	256	256	256
	R-squared	0.200	0.257	0.244	0.254	0.254	0.235	0.231	0.236	0.239
	Number of countries	20	20	20	20	20	20	20	20	20

Note: Standard errors in parentheses; *** p<0.01, ** p<0.05, * p<0.1, + p<0.15.
Source: Authors' estimations.

focuses on power and energy R&D intensity while Table 7.3 presents the results for the power and environmental patent intensity specifications.

7.4.1 Role of Environmental Policy Stringency

Our results generally confirm previous findings on the inducement effect of environmental policies with respect to energy-related innovation activities. We find that the effect is weaker in the case of energy-related R&D and stronger in the case of energy-related patents.

Focusing on the R&D specification (Table 7.2), the coefficient for EPS variables is positive only if we consider investments in the power sector alone (hence, electricity) and market-based policy instruments. Furthermore, the coefficient is only significant at the 15 per cent level. Non-market-based policies instead do not have any significant effect on R&D, in line with Ulph and Katsoulacos (1998) and Fischer, Parry, and Pizer (2003), who suggest that stricter regulations fail to have any significant effect on R&D.

Conversely, stronger results emerge when patent intensity (Table 7.3) is used as the indicator for innovation. Both market and non-market-based environmental policy stringency are positive and significant for both types of patents. The effect of market-based instruments is stronger in most specifications and the inducement effect is larger when the broader definition based on environmental patents is considered. Our results suggest that one unit increase in the market-based score (corresponding approximately to one interquartile range (IQR) change)[14] increases power patents intensity by between 1.3 and 1.4 per cent and environmental patent intensity by between 3 and 3.2 per cent. In the case of non-market-based policies, a similar change increases power patents intensity by between 1.2 and 1.5 per cent,[15] and environmental patents intensity by 2.3 per cent. It should be noted that the median improvement in policy stringency between 1995 and 2010 across the 20 countries has been approximately 1 unit for EPS market-based score and 2 units on a scale of 0–6 for EPS non-market-based score.[16]

These findings are in line with Johnstone, Ivan Haščič, and Popp (2010), who show that increasing number of international climate policies have resulted in an increase in renewable energy patents. These results are also in

[14] In the case of EPS market score, moving from the 25th quartile (1.1) to the 75th quartile (2.3) is equivalent to a one-unit increase in the EPS and is equivalent to moving from the policy stringency of Belgium to that of Finland in 2010.

[15] In the case of EPS non-market score, the IQR is larger than one. Moving from the 25th quartile (1.1) to the 75th quartile (2.6) is equivalent to the increase in policy stringency observed in Portugal between 1995 and 2010.

[16] During 1995–2010, modest increases of 1 unit have been achieved in Italy, Australia, Portugal, while more ambitious increases of 3–4 units have been achieved in South Korea and The Netherlands, while Germany has achieved increase of about 2 units.

Table 7.3. Regression results using patent intensity over value added as innovation proxy: two-year lag for all independent variables

		1	2	3	4	5	6	7	8	9	10
	Dependent Variable	Log of Patent intensity—Power					Log of Patent intensity—Environment				
H1	EPS Market Score	0.013***	0.014***	0.013***			0.031***	0.032***	0.029**		
		(0.005)	(0.003)	(0.007)			(0.012)	(0.007)	(0.016)		
	EPS Non-market Score	0.012***	0.013***	0.015***			0.018+	0.018+	0.023**		
		(0.004)	(0.005)	(0.001)			(0.011)	(0.110)	(0.046)		
	EPS Total Score				0.017***	0.004			−0.007	0.030**	0.020
					(0.000)	(0.587)				(0.011)	(0.277)
H2	Govt. Effectiveness	0.069***	0.070***		0.065***	0.045***	0.211***	0.212***		0.199***	0.183***
		(0.013)	(0.000)		(0.000)	(0.003)	(0.033)	(0.000)		(0.000)	(0.000)
	WGI			0.095***					0.313***		
				(0.000)					(0.000)		
	Govt. Effectiveness*EPS Interaction					0.015**					0.012
						(0.010)					(0.416)
H3	Political Orientation	−0.002	−0.002	−0.002	−0.002	−0.002	−0.006	−0.007	−0.007	−0.008	−0.007
		(0.003)	(0.395)	(0.379)	(0.330)	(0.514)	(0.006)	(0.286)	(0.314)	(0.212)	(0.256)
H4	VA Share Energy-intensive industries	−0.003	−0.009	−0.009	−0.012	−0.012	−0.048	−0.061*	−0.058+	−0.068*	−0.069*
		(0.014)	(0.546)	(0.557)	(0.419)	(0.380)	(0.035)	(0.089)	(0.114)	(0.061)	(0.059)
	Energy price index		0.025	0.016	0.024	0.027		0.058	0.032	0.053	0.056
			(0.338)	(0.544)	(0.367)	(0.309)		(0.390)	(0.640)	(0.431)	(0.413)
	Trade Openness		−0.000	−0.000+	−0.000	−0.000		−0.001	−0.001+	−0.001	−0.001
			(0.170)	(0.130)	(0.175)	(0.277)		(0.180)	(0.127)	(0.189)	(0.224)
	Observations	256	256	256	256	256	256	256	256	256	256
	R-squared	0.662	0.666	0.651	0.657	0.663	0.634	0.638	0.623	0.630	0.630
	Number of countries	20	20	20	20	20	20	20	20	20	20

Note: Standard errors in parentheses; *** $p<0.01$, ** $p<0.05$, * $p<0.1$, + $p<0.15$.

Source: Authors' estimations.

line with findings from some of the previous literature including Lanjouw and Mody (1996) and Popp (2002), namely that the number of environmental patents tends to increase as the cost of pollution abatement rises. Finally, the apparently stronger results in the case of the patent specification than in the R&D specification are in line with the evidence presented by Rubashkina, Galeotti, and Verdolini (2015), who focus on overall patenting within different sectors of the economy. A reason for the stronger evidence of induced innovation when using patents as opposed to R&D in the present work might be due to the different ways patents and R&D are defined. Patents explicitly refer to clean and energy-saving innovations while the definition of energy R&D does not specify the purpose of the expenditure. Overall, with respect to Hypothesis 1, our regression results suggest that more stringent environmental policies provide dynamic efficiency gains and incentives for innovation in energy-saving and pollution-reducing technologies.

7.4.2 Role of Good Governance

Good governance appears to be an important driver of innovation. Depending on the governance proxy used, a one-unit increase in government effectiveness is associated with between 62 per cent and 96.4 per cent increase in power R&D intensity (Table 7.2) and between 6.5 per cent and 31.3 per cent increase in patent intensity (Table 7.3). This suggests that stronger economic institutions promote innovation and are in line with the existing literature (Ayal and Karras 1996; Habiyaremye and Raymond 2013; Tebaldi and Elmslie 2013; Silve and Plekhanov 2015).

The marginal effect of governance might appear substantial given the coefficient interpretation provided in the paragraph above. However, a one-unit increase in the governance proxy is a rather significant change. It is comparable to moving from the governance quality of a country such as Portugal (1.02) or Slovenia (1.03) to that of countries such as Sweden or Finland (2.01 and 2.25) in 2010. Historically, the biggest improvements in governance quality have been achieved by South Korea and Estonia, where the governance WGI score increased by 0.6 and 0.5 between 1995 and 2010, respectively.

Overall, with respect to Hypothesis 2, our regression results suggest that improvements in governance and government effectiveness provide incentives for energy-related innovation.

7.4.3 Role of Political Orientation

Political orientation seems to be a more important factor for the input rather than the output of innovation, as the variable has a statistically significant effect only in the case of power and energy R&D intensity. A change in the political orientation of the government from right towards a left-leaning position, which corresponds to an IQR change in our sample, is associated with an increase in industrial R&D of 11 per cent (power) and 22 per cent (energy), respectively. To put these effects in perspective, countries such as Portugal in our sample moved from a right-leaning orientation in 1995 to a left-leaning government in 2010, while countries such as Canada, The Netherlands, and Sweden underwent the opposite change.

Overall, with respect to Hypothesis 3, left-leaning governments are more likely to implement regulations that attract energy R&D investments, but this does not translate into higher patent intensity.

7.4.4 Role of Resource Distribution, Market-Size Effect, and Lobbying

The size of the energy sector, measured as the value-added share of energy-intensive industries, has a positive impact on R&D intensity, suggesting that either industries will allocate more resources towards R&D due to the larger size of the potential market for energy innovations, or a larger energy sector will be able to lobby for more resources to be allocated to energy R&D. A 1 per cent increase in the value added share of energy intensive industries, approximately corresponding to an IQR change, increases power R&D intensity by between 0.54 and 0.83 per cent. It should be noted that a 1 per cent increase is a rather modest increase in this case. Between 1995 and 2010, changes in the share of energy intensive industries in our sample varied between (−) 62 per cent to (+) 28 per cent in France and Australia, respectively.

The smaller marginal effect on energy R&D intensity might reflect a different relevance of political economy factors within the energy sector itself.[17] As explained by Hughes and Lipscy (2013), power markets tend to be more concentrated within domestic markets whereas many oil and gas companies are vertically integrated and international in scope. Therefore, the political economy factors that matter for electricity are likely to differ from those relevant for the oil and gas industry, which are included in our definition of energy R&D. Factors such as lobbying are therefore more relevant for the more inward-oriented sectors, such as power. Since the size of the energy sector is a proxy of the lobbying power of energy-intensive industries, it has the opposite effect on patent intensity, indicating that a larger energy sector reduces the incentive to carry out energy-saving and clean innovation.

[17] Including the mining sector (oil and gas extraction).

Overall, with respect to Hypothesis 4, the larger the size of the potential markets for energy innovation, the larger the inducement effect for industries to invest in energy R&D. At the same time, larger energy sector has power to lobby for more resources to be allocated to energy R&D. These effects seem to prevail over coordination costs, however, market-size effects or lobbying from the energy sector do not result in a larger number of cleaner patents. This could mean that R&D investments are either used less effectively, or that they are used to improve other aspects of the technologies, which are more intangible and which are not codified in patents.

7.4.5 Role of Other Factors

We briefly comment here on the coefficients associated with our additional control variables, namely the energy price index and trade openness. The energy price index has a negative and statistically significant effect on both power and energy R&D intensities. A possible explanation in this respect is that higher energy prices increase energy expenditure, both in the private and public sectors, reducing resources available for other uses, including R&D. Energy prices provide a positive incentive for patents, but the coefficients are not statistically significant. Although the evidence is only imprecisely estimated, it suggests that even though fewer resources are allocated as input to innovation, the innovation process is more efficient at delivering new inventions.

Trade openness has a negative and significant effect on energy R&D intensity, suggesting that countries with developed trade relationships have fewer incentives to allocate resources to power and mining R&D and that technology adoption and imitation displace domestic innovation. Note that the effect is only significant when the definition of energy R&D include the mining sector, which is more outward-oriented than power, making the energy aggregate sensitive to changes in trade exposure.[18]

7.5 CONCLUSION

This chapter empirically investigates the impact of political economy and institutional factors on the incentives to innovate in the energy sector. We propose four empirical proxies that can measure energy-related innovation, namely power R&D, energy R&D (consisting of the investment of the power

[18] Refer to Dasgupta et al. (2016) for additional regressions and robustness tests.

and mining sector), power patents (related to renewable and energy efficient technologies for power production), and environmental patents (including energy patents as well as patents generally aimed at environmental protection). We focus on the empirical analysis of the role of four political economy factors, namely environmental policy, good governance, political orientation, and the distribution of resources to energy intensive industries that can induce effects of both market-size and lobbying.

The analysis suggests that all abovementioned factors affect the incentives to devote resources to energy R&D and to create new clean and energy efficient technologies. Specifically, market-based incentives, and to some extent also non-market based incentives, results in dynamic efficiency gains. Countries with better governance are characterized by higher levels of energy-related R&D, while left-wing governments are more likely to devote R&D resources to the energy sector but this does not translate into higher power-related patent intensity. A larger distribution of resources towards energy-intensive sectors can induce market-size effects and have more power to lobby for more resources to be allocated to energy R&D but this does not translate into higher patent intensity.

The empirical analysis described in this chapter shows that political economy factors can act as barriers even in the presence of stringent environmental policy. This implies that in order to favour changes towards a greener economy, countries should combine environmental policy with a general strengthening of institutional quality, consider the influence of government's political orientation on environmental policy, as well as the size of energy intensive sectors in the economy, which affect both the lobbying structure and the demand for energy innovations. These results point to the need to move the literature on the determinants of energy-related innovation beyond the focus on environmental policy instruments that has dominated the environmental economics literature in recent years.

REFERENCES

Acemoglu, D. (2002). 'Directed Technical Change'. *Review of Economic Studies*, 69(4): 781–809.

Anadón, L. D. (2012). 'Missions-Oriented R&D Institutions in Energy between 2000 and 2010: A Comparative Analysis of China, the United Kingdom, and the United States'. *Research Policy*, 41(10): 1742–56. doi:10.1016/j.respol.2012.02.015.

Ayal, E. and G. Karras (1996). 'Bureaucracy, Investment, and Growth'. *Economics Letters*, 51(2): 233–59.

Becheikh, N., R. Landry, and N. Amara (2006). 'Lessons from Innovation Empirical Studies in the Manufacturing Sector: A Systematic Review of the Literature from 1993–2003'. *Technovation*, 26(5/6): 644–64. doi:10.1016/j.technovation.2005.06.016.

Beck, T., G. Clarke, A. Groff, P. Keefer, and P. Walsh (2001). 'New tools in Comparative Political Economy: The Database of Political Institutions'. *World Bank Economic Review*, 15(1) (September): 165–76.

Bellos, S. and T. Subasat (2012). 'Corruption and Foreign Direct Investment: A Panel Gravity Model Approach'. *Bulletin of Economic Research* 64(4): 565–75.

Biglaiser, G. and K. DeRouen (2006). 'Economic Reforms and Inflows of Foreign Direct Investment in Latin America'. *Latin American Research Review*, 41(1): 51–75.

Boix, C. (1998). *Political Parties, Growth and Equality: Conservative and Social Democratic Economic Strategies in the World Economy*. Cambridge: Cambridge University Press.

Botta, E. and T. Koźluk (2014). 'Measuring Environmental Policy Stringency in OECD Countries—A Composite Index Approach'. OECD Economics Department Working Paper 1177, OECD Publishing, Paris.

Brunel, C. and A. Levinson (2013). 'Measuring Environmental Regulatory Stringency'. OECD Trade and Environment Working Papers 2013/05. Paris: OECD Publishing. doi: <http://dx.doi.org/10.1787/5k41t69f6f6d-en>.

Carrión-Flores, C. and R. Innes (2010). 'Environmental Innovation and Environmental Performance'. *Journal of Environmental Economics and Management*, 59(1): 27–42.

Costa-Campi, M. T., N. Duch-Brown, and J. García-Quevedo (2014). 'R&D Drivers and Obstacles to Innovation in the Energy Industry'. *Energy Economics*, 46: 20–30. doi:10.1016/j.eneco.2014.09.003.

Dasgupta, S., E. De Cian, and E. Verdolini (2016). 'The Political Economy of Energy Innovation'. WIDER Working Paper 2016/17. UNU-WIDER, Helsinki.

Esping-Andersen, G. (1990). *The Three Worlds of Welfare Capitalism*. Princeton, NJ: Princeton University Press.

Fankhauser, S., C. Gennaioli, and M. Collins (2014). 'Domestic Dynamics and International Influence: What Explains the Passage of Climate Change Legislation?' Grantham Research Institute on Climate Change and the Environment Working Paper 156, London School of Economics and Political Science, London.

Fischer, C., I. Parry, and W. Pizer (2003). 'Instrument Choice for Environmental Protection When Technological Innovation Is Endogenous'. *Journal of Environmental Economics and Management*, 45(3): 523–45.

Folke, O. (2014). 'Shades of Brown and Green: Party Effects in Proportional Election Systems'. *Journal of European Economic Association*, 12(5): 1361–95.

Fredriksson, P. G. and J. Svensson (2003). 'Political Instability, Corruption and Policy Formation: The Case of Environmental Policy'. *Journal of Public Economics*, 87(8): 1383–405.

Fredriksson, P. G., H. R. J. Vollebergh, and E. Dijkgraaf (2004). 'Corruption and Energy Efficiency in OECD Countries: Theory and Evidence'. *Journal of Environmental Economics and Management*, 47(2): 207–31. doi:10.1016/j.jeem.2003.08.001.

Friedrichs, J. and O. R. Inderwildi (2013). 'The Carbon Curse: Are Fuel Rich Countries Doomed to High CO_2 Intensities?' *Energy Policy*, 62: 1356–65. doi:10.1016/j.enpol.2013.07.076.

Gallagher, K. S., L. D. Anadón, R. Kempener, and C. Wilson (2011). 'Trends in Investments in Global Energy Research, Development, and Demonstration'. *Wiley Interdisciplinary Reviews: Climate Change*, 2(3): 373–96. doi:10.1002/wcc.112.

Gani, A. (2007). 'Governance and Foreign Direct Investment Links: Evidence from Panel Data Estimations'. *Applied Economics Letters*, 14(10): 753–6.

Globerman, S. and D. Shapiro (2003). 'Assessing Recent Patterns of Foreign Direct Investment in Canada and the United States'. In R. Harris (ed.), *North American Linkages: Opportunities and Challenges for Canada*. Calgary: University of Calgary Press, pp. 281–312.

Griliches, Z. (1990). 'Patent Statistics as Economic Indicator: A Survey'. *Journal of Economic Literature*, 28(4): 1661–707.

Habiyaremye, A. and W. Raymond (2013). 'Transnational Corruption and Innovation in Transition Economies'. UNU-Merit Working Paper Series 2013–050, UNU-MERIT, Maastricht.

Hall B. H. and N. Rosenberg (2010). *Handbook of the Economics of Innovation*. Amsterdam: Elsevier.

Haščič, I. and M. Migotto (2015). 'Measuring Environmental Innovation Using Patent Data'. OECD Environment Working Papers 89, OECD Publishing, Paris. Available at: <http://dx.doi.org/10.1787/5js009kf48xw-en>.

Hawkins, R. G., N. Mintz, and M. Provissiero (1976). 'Government Takeovers of U.S. Foreign Affiliates'. *Journal of International Business Studies*, 7(1): 3–16. doi:10.1057/palgrave.jibs.8490690.

Hughes, L. and P. Y. Lipscy (2013). 'The Politics of Energy'. *Annual Review of Political Science*, 16(1): 449–69. doi:10.1146/annurev-polisci-072211-143240.

IEA (International Energy Agency) (2014). 'Energy Technology Perspectives 2014', International Energy Agency, Paris. doi: <http://dx.doi.org/10.1787/energy_tech-2014-en>.

IEA (International Energy Agency) (2015a). 'Energy Technology Perspectives 2015: Energy Technology Perspectives 2015—Mobilising Innovation to Accelerate Climate Action'. International Energy Agency, Paris.

IEA (International Energy Agency) (2015b). 'RD&D Budget'. IEA Energy Technology RD&D Statistics database. doi: <http://dx.doi.org/10.1787/data-00488-en>.

IPCC (2014). 'Climate Change 2014: Mitigation of Climate Change'. Contribution of Working Group III to the Fifth Assessment Report of the Intergovernmental Panel on Climate Change. New York and Cambridge: Cambridge University Press.

Jaffe, A. B., S. R. Peterson, and P. R. Portney (1995). 'Environmental Regulation and the Competitiveness of U.S. Manufacturing: What Does the Evidence Tell Us?' *Journal of Economic Literature*, 33(1): 132–63.

Jensen, N. M. (2006). *Nation-States and the Multinational Corporation: A Political Economy of Foreign Direct Investment*. Princeton, NJ: Princeton University Press.

Johnstone, N., I. Ivan Haščič, and D. Popp (2010). 'Renewable Energy Policies and Technological Innovation: Evidence Based on Patent Counts'. Environmental Resource Economics, 45(1): 133–55.

Johnstone, P. and A. Stirling (2015). 'Democracies' in Sociotechnical Transitions and Discontinuities'. SPRU Working Paper Series, University of Sussex, Brighton.

Kaufman, D., A. Kraay, and M. Mastruzzi (2010). 'The Worldwide Governance Indicators: Methodology and Analytical Issues'. World Bank Policy Research Working Paper 5430. Washington, DC: World Bank.

Lanjouw, J. O. and A. Mody (1996). 'Innovation and the International Diffusion of Environmentally Responsive Technology'. *Research Policy*, 25(4): 549–71.

Lanzi, E., E. Verdolini, and I. Haščič (2011). 'Efficiency-improving Fossil Fuel Technologies for Electricity Generation: Data Selection and Trends'. *Energy Policy*, 39(11): 7000–14.

Lockwood, M. (2013a). 'The Political Dynamics of Green Transformations: The Roles of Policy Feedback and Institutional Context', EPG Working Paper 1403. Energy Policy Group, University of Exeter, Exeter.

Lockwood, M. (2013b). 'The Political Sustainability of Climate Policy: The Case of the UK Climate Change Act'. *Global Environmental Change*, 23(5): 1339–48.

McCright, A. and R. Dunlap (2011). 'The Politicization of Climate Change and Polarization in the American Public's Views of Global Warming, 2001–2010'. *The Sociological Quarterly*, 52(2): 155–94.

Mazzucato, M. (2013). *The Entrepreneurial State. Debunking Public vs. Private Sector Myths*. London: Anthem Press.

Neumayer, E. (2003). 'Are Left-Wing Party Strength and Corporatism Good for the Environment?' *Ecological Economics*, 45(2): 203–20.

Newell, R. G. (2010). 'The Role of Markets and Policies in Delivering Innovation for Climate Change Mitigation'. *Oxford Review of Economic Policy*, 26(2): 253–69. doi:10.1093/oxrep/grq009.

OECD (2015a). 'Patent Search Strategies for the Identification of Selected Environment-Related Technologies'. OECE Environment Directorate, Paris. Available at: <http://www.oecd.org/environment/consumption-innovation/env-tech-search-strategies.pdf> (accessed 7 March 2016).

OECD (2015b). 'Patents by Main Technology and by International Patent Classification (IPC)'. OECD Patent Statistics (database). doi: <http://dx.doi.org/10.1787/data-00508-en>.

OECD (2016). 'ANBERD database'. Available at: <http://stats.oecd.org/Index.aspx?DataSetCode=ANBERD_REV4> (accessed 22 March 2016).

Olson, M. (1965). *The Logic of Collective Action*. Cambridge, MA: Harvard University Press.

Painter, J. and T. Ashe (2012). 'Cross-National Comparison of the Presence of Climate Scepticism in the Print Media In Six Countries, 2007–10'. *Environmental Research Letters*, 7(4): 1–8.

Popp, D. (2002). 'Induced Innovation and Energy Prices'. *American Economic Review*, 92(1): 160–80.

Rubashkina, Y., M. Galeotti, and E. Verdolini (2015). 'Environmental Regulation and Competitiveness: Empirical Evidence on the Porter Hypothesis from European Manufacturing Sectors'. *Energy Policy*, 83: 288–300. Available at: <http://dx.doi.org/10.1016/j.enpol.2015.02.014> (accessed 7 March 2016).

Salies, E. (2010). 'A Test of the Schumpeterian Hypothesis in a Panel of European Electric Utilities'. In J. L. Gaffard and E. Salies (eds), *Innovation, Economic Growth and the Firm*. London: Edward Elgar Publishing.

Shleifer, A. (1998). 'State versus Private Ownership'. *Journal of Economic Perspectives*, 12(4): 133–50.

Silve F. and A. Plekhanov (2015). 'Institutions, Innovation and Growth: Cross-Country Evidence'. European Bank for Reconstruction and Development, Working paper 177, EBRD, London.

Staats, J. L. and G. Biglaiser (2012). 'Foreign Direct Investment in Latin America: The Importance of Judicial Strength and Rule of Law'. *International Studies Quarterly*, 56(1): 193–202.

Stavins, R. N. (2004). 'Introduction to the Political Economy of Environmental Regulations'. RFF Discussion Paper 04–12, Resources for the Future, Washington, DC.

Sterlacchini, A. (2012). 'Energy R&D in Private and State-Owned Utilities: An Analysis of the Major World Electric Companies'. *Energy Policy*, 41: 494–506. doi:10.1016/j.enpol.2011.11.010.

Subasat, T. and S. Bellos (2011). 'Economic Freedom and Foreign Direct Investment in Latin America: A Panel Gravity Model Approach'. *Economics Bulletin*, 31(3): 2053–65.

Tebaldi, E. and B. Elmslie (2013). 'Does Institutional Quality Impact Innovation? Evidence from Cross-Country Patent Grant Data'. *Applied Economics*, 45(7): 887–900.

Timmer, M. P., E. Dietzenbacher, B. Los, R. Stehrer, and G. J. de Vries (2015). 'An Illustrated User Guide to the World Input–Output Database: The Case of Global Automotive Production'. *Review of International Economics*, 23(3): 575–605.

Turnheim, B., F. Berkhout, F. W. Geels, A. Hof, A. McMeekin, B. Nykvist, and D. van Vuuren (2015). 'Evaluating Sustainability Transitions Pathways: Bridging Analytical Approaches to Address Governance Challenges'. *Global Environmental Change*, 35: 239–53.

Ulph, D. and Y. Katsoulacos (1998). 'Endogenous Spillovers and the Welfare Performance of Research Joint Ventures', Mimeo, University College London, London.

Upstill, G. and P. Hall (2006). 'Innovation in the Minerals Industry: Australia in a Global Context'. *Resources Policy*, 31(3): 137–45.

8

Is Feed-In-Tariff Policy Effective for Increasing Deployment of Renewable Energy in Indonesia?

Dewi Yuliani

8.1 INTRODUCTION

Indonesia is an archipelago rich in renewable energy resources including geothermal, hydro, biomass, solar, and wind energy. Due to its location in the 'Ring of Fire', geothermal energy is its main potential source of renewable energy, amounting to 29 GW—40 per cent of the world's total. Of the total potential, up till now only 1,340 MW (less than 5 per cent) has been installed for electricity generation, with another 1,500 MW at various stages of development (DGNREEC 2015). Hydropower is another major source of potential in Indonesia and is estimated to be around 75,000 MW. At present, some 6,000 MW has been installed as large Hydro Electric Power, mostly in Java. Included in this category is small hydro (up to 10 MW). 217.16 MW of small hydropower plant has been installed in 2015, a large portion of which is used not for profit but for electrifying remote areas (DGNREEC 2015).

For a country on the equator, solar energy is considered the most underutilized source of energy in Indonesia. Only around 71 MW is installed but mostly belongs to the state-owned electricity company (SEC or Indonesia's PLN) for electrifying remote islands in the eastern parts of Indonesia. At present, the imported amount of solar photovoltaics (PV) is still considerably high, so the Government of Indonesia (GOI) has attempted to boost the local production of PV systems and encourage investors to build PV manufacturers. As an agricultural country, biomass is the other major potential source of Indonesia's renewable energy. This is thought to have a potential 32,000 MW, of which 1,740 MW is planned to be installed in the near future (DGNREEC 2015). Biomass sources include palm oil, cassava, sugar cane, and so on, which can be the source of bioethanol and biofuel production.

Despite this abundant potential, the development of renewable energy in Indonesia still faces a hard battle, mostly because of the long legacy of the government's energy subsidies for fossil fuels. According to the Ministry of Energy and Mineral Resources (MEMR), Indonesia's energy demand has been increasing faster than its economic development growth, which amounted to 7 per cent in 2014 (DGNREEC 2014). This demand is currently met largely by fossil fuels, which in turn bleeds the country's finances as Indonesia has become a net importer of both crude oil and refined products since 2004.

As a response to the increasing demand for energy and for cleaner energy due to climate issues, in January 2014 The National Energy Policy was adopted by the House of Representatives in the form of Government Regulation No 79/2014 (Government of Indonesia 2014), in which the GOI projected that renewable sources would supply at least 23 per cent of Indonesia's energy needs in 2025. The GOI is also committed to reducing greenhouse gas emissions by 26 per cent by 2020 to combat global climate change (MEMR 2015).

Many efforts have been undertaken to increase the deployment of renewable energy (RE), especially to support the increasing electricity demand. These include: obligating the SEC to buy generated electricity from small-scale producers, prioritizing the use of RE, tax incentives, tax holidays for exported equipment, and simplifying the procedure for licences. Lastly, since around 2010 the GOI has implemented its feed-in-tariff (FIT) policy in order to encourage the private sector to develop renewable electricity. The FIT policy ensures that developers get a relatively fair price for electricity generation from renewable energy resources, which are now being used widely in many countries.

This study undertakes exploratory research that aims to evaluate whether the implementation of the FIT policy is effective for the purpose of increasing deployment of renewable energy in Indonesia. A number of official documents and supporting policies set out the plans for renewable energy development in Indonesia. However, no critical study has been carried out into the real outcome of policy implementation in the field. Data analysis is expected to show whether the FIT policy is sufficient, and if not, what are the reasons for this and what can be done to improve it.

This chapter is organized as follows. After the introduction, which describes Indonesia's renewable energy situation and outlines the purpose of the study, Section 8.2 briefly reviews the FIT policy, its principles, case studies, and barriers to implementation. Section 8.3 outlines the methodology used for data collection and the locations covered by the site survey for the purpose of data verification. Section 8.4 provides the data findings and undertakes data analysis to identify factors that hinder the deployment of renewable energy. Section 8.5 offers conclusions and recommendations for better implementation of FIT policy in the future.

8.2 A BRIEF REVIEW OF THE FEED-IN-TARIFF POLICY

8.2.1 What is FIT Policy?

Historically, to spur the growth of renewable electricity, European countries set a price for electricity that is attractive to investors. This policy, which was subsequently called feed-in-tariff (FIT) policy, has three components: a fixed price which is sufficient for a return on investment, a guaranteed grid connection, and a long-term contract to ensure the stability of the investment (Mendonça 2007). By using this policy, investors do not have to deal with the cumbersome process of various incentives and tax exemptions, as applied in America (Farrel 2009: 6). This price is reviewed annually and it can be lowered if it attracts too many investors or creates windfall profit, and vice versa. Later on, this fixed price policy is accompanied by a mandate for utility companies to connect renewable electricity to the grid. In Europe, the cost of connection to the higher voltage is borne by the utility company, and the contract for selling electricity is short and uniform (Farrel 2009: 6).

According to IRENA (2012), there are four categories of deployment policies, namely: fiscal incentives, public finance, regulations, and access policies (Mitchell et al. 2011). The FIT policy falls into the price-driven public finance policy category (IRENA 2012). Deployment, which is defined by IRENA (2012) as the stage of commercial roll-out, is a stage preceding full competition and maturity in the development of a new technology. In the commercial roll-out stage or deployment, economic support is needed to correct market failure and/or inadequate policy requirements. In the renewable energy market, failure comes from the discrepancy between the average production costs of electricity from fossil fuels or conventional energy and the production costs using renewables. Without consideration of the cost of research and development, externalities related to climate change, and other barriers to mass production, there is no way that renewable energy technology can compete with the energy price produced by conventional technology.

Another benefit of FIT policy implementation includes support for local ownership of renewable plants because of its simplicity, and thus it creates more jobs and increases the economic impact for the community. If we look at the experience of European countries, about half of wind turbines in Germany and about 80 per cent in Denmark are owned by local residents (Farrel 2009).

8.2.1.1 Examples of FIT Implementation

The most notable success story of FIT policy is Germany's and Denmark's experiences in FIT implementation. Germany started to apply FIT policy in

1990, when the law required electricity utilities to connect renewable electricity to the grid and to buy the electricity at a price ranging from 65 per cent to 90 per cent of the average tariff for consumers (Mez 2012: 24). The implementation of FIT since then has spurred the expansion of wind turbine power plants in particular, from 20 MW in 1989 to more than 1,100 MW in 1995 (Lauber and Mez 2006 in Mez 2012: 24). In addition, Germany has also implemented a wind power programme and provides several kinds of subsidies and financial incentives for renewable energy investors. This policy, along with the implementation of the Renewable Energy Sources Act (RESA) in 2000, which was passed in order to achieve the targeted 25 per cent reduction of carbon-dioxide emissions, has resulted in an almost tenfold expansion in renewables from 1.3 per cent in 1990 to nearly 11 per cent in 2011 (Mez 2012: 26).

In 1988, Denmark applied a price policy for renewable electricity that paved a way for FIT policy, in which utility companies are required to set a fair price, purchase, and connect to the renewable electricity produced. This policy was successful in triggering the establishment of wind turbines, reaching 300 MW of capacity. The FIT policy was formally applied in Denmark in 1993, when the purchase price of renewable electricity was set at 85 per cent of the average production and distribution cost and connection to the grid and purchase of power was guaranteed. This policy, along with an exemption for cooperatives from paying, has resulted in 28 per cent of Danish energy coming from renewables with wind turbines producing almost 3,000 MW of electricity. However, in 2004 the Danish ended the FIT policy and nowadays investors in renewable energy have to follow the renewable energy portfolio standard with tradable credits, such as applied in the USA (Farrel 2009).

8.2.1.2 Barriers to FIT Implementation

Klein et al. (2010) warned that FIT policy should be kept transparent and not too complex, and their study evaluated the structure of electricity tariffs in several countries that could act as barriers to FIT implementation. In the USA, Farrel (2009) complains that the contract for electricity purchase can exceed 100 pages, while in Germany, for instance it is kept short to 4–5 pages. Another potential barrier is the condition of the electricity market, such as liberalization of electricity in New Zealand (White, Lloyd, and Wakes 2013).

However, another study in the European Union has suggested that the poor effectiveness of FIT policy is mainly caused by failure to tend to non-economic barriers, such as uncertain and lengthy administrative processes, and difficulties in gaining access to the grid (IRENA 2012: 13). Research by ECORYS (2008) into European Union member states showed that there are some non-cost barriers that negatively impact the deployment of renewables to a different degree. One of the barriers most impacted is the administrative hurdles which include many issues, such as lack of coordination between institutions, length of time for obtaining authorization, the incoherence of procedures and regulations, big overhead costs, insufficient spatial planning, and social opposition. Other significant barriers are difficulty in getting grid connection and access, and limited information and awareness related to renewable energy among civil servants and the community in general (ECORYS 2008).

8.2.2 FIT Policy in Indonesia

In Indonesia, a policy, which is similar to FIT, was issued in 2002 for small-scale hydropower plants. The MEMR issued a requirement for the SEC to buy electricity at 60 per cent of regional electricity production costs for low-voltage, and 80 per cent for middle-voltage connection. However, there have been some obstacles encountered in its implementation, one being the difficulty of determining the regional standard of production cost, and another being because energy production (especially of petroleum) was still heavily subsidized by the GOI, so that the utility company did not need to prioritize the use of renewable energy. The policy was improved in 2006 to include higher-scale hydropower plants (up to 10 MW), and then underwent some revisions in 2009 to include a more attractive tariff. Hasan and Wahjosudibjo (2014) argue that the obstacles to FIT implementation are mainly because there is no guarantee of long-term contracts and nor that the utility company will accept the contract.

A more formal kind of FIT policy was issued in 2011 and 2012, when the MEMR required the PLN to buy electricity from geothermal power plants at a fixed tariff of US$0.97/kWh. Subsequently, the FIT for other renewable energy sources, such as solar and biomass, was issued in 2013 and 2014. In general, FIT regulation is targeted at reducing the time required for negotiating a price with the SEC (which, at times, can be several years), and making investment more attractive and long-term planning possible, thereby making RE investment viable. Table 8.1 shows the different FIT regulations for various RE sources, their original version, and the revisions.

It can be seen from Table 8.1 that the GOI made several attempts to adjust and readjust the RE selling price; in some cases revisions were made more than once. In small hydropower plants, for instance, there were at least four MEMR regulations with different electricity prices which had been revised due to inputs from related parties, adjustment for inflation, and exchange rates. From the many revisions of FIT regulation, it can be concluded that the GOI puts in a lot of effort, and thus relies on this policy, to accelerate deployment of renewable energy.

Table 8.1. Regulation of feed-in tariff in Indonesia for various RE sources

No	RENEWABLE ENERGY SOURCE	FEED-IN TARIFF REGULATION				NOTE
		Original Version	Electricity Price (US$/kWh)	Revised Version	Electricity Price (US $/kWh)	
1	Micro/Mini Hydro	MEMR Regulation No. 12 year 2014	<10 MW = 8 cent	MEMR Regulation No. 22 year 2014	<10 MW = 8 cent	7.2 cent if using public infrast, i.e., irrigation
			>250 kWh = 9.4 cent		>250 kWh = 9.4 cent	8.4 cent if using public infrast, i.e., irrigation
				MEMR Regulation No. 19 year 2015	<10 MW = 12 cent	10.8 cent if using public infrast, i.e., irrigation
					>250 kWh = 14.4 cent	13 cent if using public infrast, i.e., irrigation
2	Geothermal	MEMR Regulation No. 02 year 2011	on high volt. only 9.7 cent	MEMR Regulation No. 22 year 2012	on high volt. 10–17 cent on med. Volt. 11.5–18.5 cents	depend on location
3	Biomass	MEMR Regulation No. 04 year 2012	9.75 cent 13.5 cent	MEMR Regulation No. 27 year 2014	8.5 cent 11.1 cent	med volt. Interconn. low volt. Interconn.
	Biogass		9.75 cent 13.5 cent		8.5 cent 10.4 cent	med volt. Interconn. low volt. Interconn.
	Municipal Waste		10.5 cent 13.98 cent	MEMR Regulation No. 44 year 2015	16.55 cent 20.16 cent	high & med volt. Intercn. low volt. Interconn.
4	SOLAR PV	MEMR Regulation No. 17 year 2013	up to 25 cent up to 30 cent			local content < 40% Local content > 40%

Note: Prices converted from rupiahs rely on its annual exchange rate.

Source: Based on data from MEMR (2015).

8.3 DATA ACQUISITION

Effectiveness is defined as 'the extent to which intended objectives are met, for instance, the actual increase in the output of renewable electricity generated or shares of renewable energy in total energy supplies within a specified time period' (IRENA 2012: 12). In this study, the effectiveness of the FIT policy is measured by the increase in deployment of renewable energy power plants over the period since the policy was made in 2010 to date (a five-year period), compared with the GOI's target for achieving its share of the national energy mix. As suggested by IRENA (2012), in analysing the effectiveness of a policy, it is necessary to conduct an in-depth analysis to elaborate factors that support or hinder the outcome of the policy, such as external factors or indirectly related barriers to the success of policy implementation.

8.3.1 Study Location

Indonesia is made up of 17,000 islands and has one of the longest coastlines in the world. Most of the regions in Indonesia, especially in the eastern part, still have very poor infrastructure conditions. However, with a population of about 300 million and relatively stable economic growth (4–6 per cent per annum over the last five years), it is a country with many interesting prospects.

Politically speaking, Indonesia can be categorized as a country in transitional democracy, because, after its independence on 17 August 1945, democratic transition did not start to happen until 1998 when the long reign of General Suharto was toppled by students' demonstrations after 30 years of authoritative governance. After the so-called reformation era, Indonesian society has undergone a drastic change, including a shift from centralized to decentralized governance. Nowadays the people of Indonesia are enjoying a new sense of democracy. This new-found freedom allows them to freely question or even disobey what the government wants. This understanding of the political situation is important in comprehending the real condition of governance related to the deployment of renewable energy power plants.

This study uses a quantitative and qualitative research approach. The quantitative study is conducted by collecting, comparing, and analysing data from various institutions related to the targeted deployment of renewable energy power plants and its realization in the field. The comparison is followed by site visits to several locations for the purpose of confirmation. Due to limited time and budget, the confirmation field surveys were only conducted in West Java, a province adjacent to Jakarta, to gain a deeper perspective from players in renewable energy in real circumstances.

One reason for choosing West Java for site confirmation is because it is the province where the development of RE is considered to be the highest, and it has been a pioneer in RE development in Indonesia, such as in geothermal, micro-hydro, biogas, biofuel, and so on. It is densely populated with about 46 million people inhabiting some 35,000 square kilometres of land. With 6.5 per cent annual economic growth in 2012, its energy consumption has continued increasing to around 8–11 per cent for all sectors, implying that the usage of energy is still inefficient. Thus, serious efforts to lower energy consumption, as well as efforts to diversify fossil-fuel energy with more renewable kinds of energy have been conducted since 2001. Some renewable energy technologies, such as micro-hydropower plants, are considered to be relatively successful.

8.3.2 Quantitative Data

Data analysis from several data resources (MEMR, SEC, local government) resulted in a wide discrepancy between targeted installed capacity and its realization. Most commercial RE projects have been delayed by one to five years. As has been stated before, the comparison between projected and real data for small hydro power plants is only conducted for West Java, which is considered to mirror Indonesia's condition. As per IRENA's definition (2012), deployment of renewable energy power plants includes the commercial operation of the plant, not just installation. The financial stage includes preparation of funding sources, negotiating and signing of a Power Purchasing Agreement (PPA) with the SEC, the process of obtaining various kinds of permits (from central government as well as from local government), and the construction stage, which includes land acquisition, community preparation, mobilization, and other construction activities. Meanwhile, number of permits refers to the sum of permits related with small hydro power plants, issued either by central government (SEC, DGNRE, and DGE), or local government in the form of location permits. Figure 8.1 shows the actual installed capacity and the various stages of renewable energy power plant deployment from 2010, or since the FIT implementation, until December 2015.

Official data issued by central government institutions are not consistent. For instance, for small hydropower, one published report says that there are only 19 small hydropower licences in West Java, while another institution states there are 31, and yet another says 34 licences. When these data are cross-checked with data from local government (regency), the compilation from regencies which have small hydropower sites suggests there are 61 small hydropower licences in West Java.

Another interesting result of the data analysis is that, on average, only 10 per cent of permits issued for power generation from the renewable energy source, are already at the commercial operation stage or deployed. Some of the permits had been issued for more than five years. Thus, there must be some problems in the

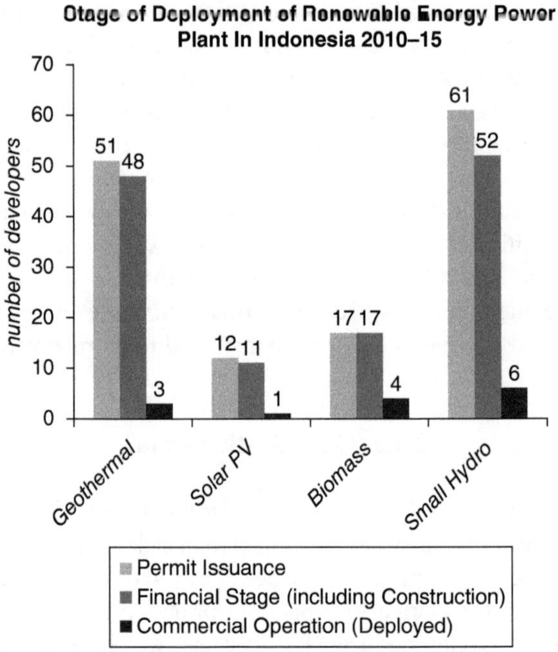

Figure 8.1. Deployment status of renewable energy power plants after FIT policy.

Source: Author's data compilation from various institutions and field surveys.

Note: For Small Hydro, data only from West Java Province.

field that prevent the FIT policy from having a positive effect on the deployment of renewable energy power plants. To delve deeper into this discrepancy, qualitative surveys were conducted using semi-structured interviews. The identities of all those interviewed in this study, including experts, developers, and public authorities, are confidential due to the sensitivity of the issue.

8.3.3 Qualitative Data

From the collected data and results of the analysis, an attempt to elaborate the causes of the discrepancy between projected goals and the real situation is conducted using semi-structured interviews. It is worth noting that not everyone contacted who was directly involved in the deployment of RE was willing to be interviewed, especially government officials and developers. Some were eventually willing to give information when the interview was conducted outside of their offices, some objected to the interview being recorded, and most of them agreed to be interviewed with the promise that their names would not be revealed. On average, the interviews lasted between about 15 minutes and 2 hours; some were recorded while others were field-noted.

Table 8.2. List of guidance questions for key respondents

Guidance questions for investors/developers:
1. How has the introduction of FIT policy affected your decision to invest in RE power plant?
2. Is the current structure of FIT policy sufficient to foster deployment of RE? if not, why?
3. What obstacles do you find in the deployment of RE power plants?
4. Is there any suggestion you want to propose to overcome those obstacles?
Guidance questions for government officials (central, regional, and local government):
1. Is there any special treatment for the issuance of permits for RE investment, compared with other kinds of permits?
2. Do you know the government's target for increasing the usage of RE sources?
3. What obstacles do you find in the deployment of RE power plants?
4. Is there any suggestion you want to propose to overcome those obstacles?
Guidance questions for government officials (expert, practitioner, and utility company):
1. Is the recent structure of FIT policy sufficient to foster deployment of RE? if not, why?
2. What are the obstacles to implementing FIT policy in Indonesia?
3. Is there any suggestion you want to propose to overcome those obstacles?

Source: Author's illustration.

In total, there were 22 key respondents, consisting of 5 investors (4 small hydropower developers, 1 geothermal), 14 government officials (5 from central government, 2 from provincial government, and 7 from regency (local) government), and 3 respondents from related agencies (1 SEC, 1 expert, 1 practitioner). Table 8.2 shows a general guidance questions to semi-structure interviews conducted, which are slightly different for each group of respondents.

The analysis is focused on the evaluation of FIT policy in Indonesia: whether it is effective in increasing the usage of RE, whether there is any room for improvement, and if not effective, then what are the alternatives. For definition purposes, deployment of renewable energy means the power plant is not just already installed and producing electricity, but it also has generated income from selling electricity.

8.4 ANALYSIS

8.4.1 Effectiveness of FIT Policy

It is interesting to note that since around 2011 when the FIT policy was first made, investment in renewable energy has been increasing significantly. However, the study results showed a totally different outcome regarding real deployment of renewable energy power plants. Take geothermal energy, for instance: from 51 licences issued by the MEMR, only 3 are operational commercially in 2015. This means that the amount of delayed deployment of geothermal power plants exceeds 90 per cent. For solar PV and biomass, data for real deployment of power plants are not very different. For small hydropower plants, the data, which are taken only from West Java, show that the rate of successful deployment is no better, with only 6 out of 61 licences having been fully deployed commercially.

Those at the construction phase (representing a total 56.69 MW from nine companies) report that the delayed construction stage is mostly caused by difficulties in obtaining a location permit or a building permit from local (regency) government, hard negotiations of land price, and the demands of local communities. In West Java, there are no biomass (domestic waste to energy power plant), solar PV, or wind turbine power plants which are fully operational in a commercial sense. Some small MWs have been installed (solar PV and biomass), but negotiations of price and other related processes are still underway.

The 19 small hydropowers that are in the finance phase (total 88,174 MW) report several problems including difficulties in obtaining PPA with the SEC and the long process of permit issuance for surface water usage, in which the two processes are intermingled. It is interesting that policies for river usage permits, which depend on where the river is located, are different from one locality to another, and from local government to central government.

One unique example is a small hydropower plant of 600 kW in a region of Central Java, which has been fully constructed but has not been operational because it has not obtained the PPA from the SEC. And the SEC cannot issue the PPA because the related institution has not been issued with a permit for water resource utilization. There has been much confusion over who is authorized to issue the permit: initially it was thought to be the responsibility of local government, but, then it was clear that the specific watershed's authority belongs to central government (Ministry of Public Works). However, up until now the permit has not been issued by the minister, with no clear explanation.

When asked whether the structure of FIT policy has met their expectations, all respondents from investors of RE power plants admitted that the structure and the selling price have been sufficient to make the project (on paper) seem viable, and this has greatly affected their decision to invest in renewable energy. Especially for small hydropower investors, who enjoy a high price of electricity and for whom the domestic content of the technology is relatively high, the structure of small hydropower has caused a surge of domestic investors who are interested in investing. This attractiveness of RE investment has been confirmed by the US Department of Commerce report (US Department of Commerce 2010) and the UNDP (United Nations Development Programme) report (UNDP 2015). However, the investors convey that, in the process of development, the many additional costs incurred by the many kinds of permits that have to be obtained before they are fully operational, and the long delays caused by them, have eroded the profitability of RE projects.

8.4.2 Procedure for Obtaining a Renewable Energy Licence

In delving deeper into the permit procedure, the research is focused on small hydro development in West Java only, due to limited time and budget. There is no coherent data between institutions and no transparency of information. At the lower level of government, which is provincial and regency government, this is even worse. That is why interviews were conducted covertly and some of those interviewed agreed to be interviewed only on condition that their names would not be revealed. Figure 8.2 displays the general procedure to be conducted in order to obtain a small hydropower plant licence. The flows of procedures at the lower levels of government (province and regency government) are simplified and compiled from practices existing in different regencies.

The upper part of the procedure illustrates the flow of applications for power plant licences according to MEMR Regulations No 22/2010, where the developers of on-grid power generation must submit an application to the Directorate General of Electricity and must previously have had a memorandum of understanding with the SEC that the planned location of the development is already included in the Master Plan. Then, the developer must submit a deposit of 5 per cent of total investment to the Directorate General of New and Renewable Energy, in order to obtain Approval for Hydro Utilization. Thereafter, a temporary permit for small hydropower generation is issued. Based on this temporary permit, the developer must then go through

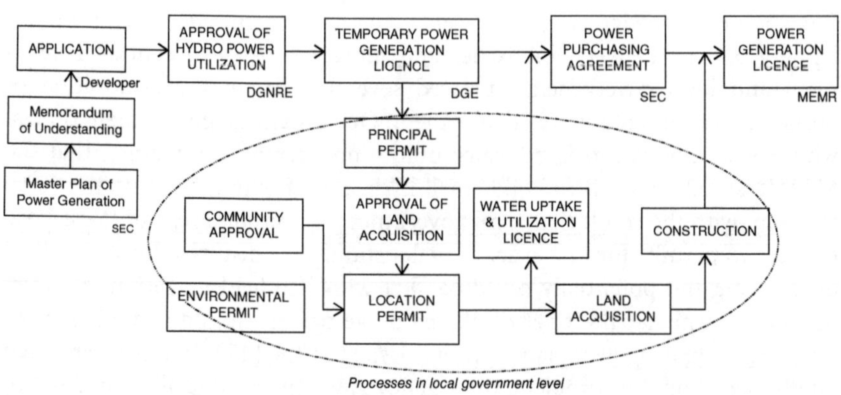

Figure 8.2. Procedure for the issuance of licences for small hydropower generation.
Source: Author's illustration.

the procedure at the lower level of government (province government and related regency government).

This is where matters get more complicated, because, in order to conduct land acquisition, the developer must go through several processes, often very tiring, with no time limit, and a lot of additional (and unofficial) cost. To make matters worse, there is no standard procedure for this land acquisition between regencies, and the process may vary greatly from one place to another. When there is forest land included in the design, the matter is even more complicated because the timeline of the procedure to obtain permission to use the land often exceeds the length of the permit.

In general, there are at least 14 kinds of licences and permits (and other permit-like processes) that have to be obtained before a developer can start the construction of a RE power plant: four from central government, one from provincial government, and nine from regency government. There are also some associated recommendations from related bodies which are a must-have requirement before a permit can be issued (this includes community approval). For geothermal investment, this number may well be higher, especially for foreign investors. It is worth noting that the costs associated with permit issuance are generally not stated clearly; sometimes they need to be negotiated, and sometimes they require much higher costs than expected. Although some RE investors admit that they have prepared for such excess costs, nevertheless the negotiation process requires time, delays schedules, and lowers the profitability of RE investments.

8.4.3 Obstacles to Deployment of Renewable Energy

8.4.3.1 Incoherent Regulations between Related Institutions

Incoherent knowledge about renewable energy, different agendas, and overlapping policies between ministries and moreover between central government and local government, surfaced several times in the elaboration of obstacles to renewable energy deployment. One provincial government official who was authorized to issue water utilization permits complained that the MEMR issued a permit to build small hydropower without stating the exact location, with the result that local government, in certain areas of West Java, issued two permits for one exact same location. This disturbs the steady flow of the river and potentially could conflict with rice field irrigation. He then questioned whether the river discharge usage permit should come before or after a small hydropower plant permit is issued by the MEMR. However, when questioned about this problem, the regency government official seemed not to recognize this problem, stating that they only process location permits as they are proposed, and that there is no clarification process with the higher level of government or with related institutions whatsoever. Thus, if there is overlapping of location permits, it is not their problem.

The institutional problem becomes worse when an RE power plant is to be built in a conservation or protected forest area because the clearance and permit to use the forest area must come from the Minister of Forestry. Even though they know that renewable energy plants would not damage the forest (unlike mining, for instance), still there is no special treatment for the so-called sustainable energy and non-sustainable usage of the forest. Even though, recently, there has been a revision for geothermal energy that enables protected forest to be used for geothermal power plants, a geothermal investor respondent still complained that the issuance of the permit is long and complex.

8.4.3.2 Long Process of Electricity Purchasing

Investors also convey that the signing of a PPA with the SEC is extremely complex and technical and bargaining is tough. Sometimes the process took longer than a year. A geothermal power plant investor stated that he has to wait four years before the agreement is reached. Another investor even suspects that the SEC is reluctant to buy electricity from renewable energy sources because it then has to cover additional costs compared to when it uses coal-fired power plants (personal interviews, August–September 2015).

Meanwhile, in reply to questions about the many difficulties experienced in obtaining PPAs, the Director of Business Ventures of the SEC (personal interview, September 2015) denied that the SEC is reluctant to buy electricity from RE sources, stating that it is mandatory for the SEC to buy in accordance with the MEMR regulations. However, as a state-owned company, there are several requirements that have to be met before issuance of a PPA including, among others, obtaining all related permits, land clearance, and agreement with the local community. In reply to questions about the profitability of trading electricity from RE sources, she admitted that the price is considerably higher than coal-generated electricity (average production cost of around US$0.5). At present, the buying price of electricity from RE varies between US$0.10 and US$0.17, which is sold to the public at an average price of around US$0.8 to US$0.9, and this makes the discrepancy in cost a burden for the SEC to cover from elsewhere.

Another complaint from the SEC is that while they have to pay instantly the electricity price as generated, reimbursement from the government is dependent on the national budgeting agenda. Reimbursement is sometimes delayed for several months and sometimes is not provided at the exact amount that has been disbursed by the SEC. Thus, she proposed that there should be an independent agency to manage the financing of electricity from RE sources which deal with the electricity price negotiations, payments of power purchasing, and all related matters. Learning from the case of Malaysia and Thailand, this agency would ensure that the discrepancy in cost between the average production cost and the selling price from conventional power plants, which is reimbursed in the form of government subsidy, can be managed effectively without depending on political processes in the annual national budgeting agenda.

8.4.3.3 Problems with Community and Land Acquisition

Another factor that hinders speedy deployment of RE power plants is that in some locations the community does not wish to sell its land due to lack of knowledge about renewable energy. This is especially true for geothermal energy, where in West Java at least, there are four locations where geothermal plants are delayed due to various kinds of conflict. Yuliani (2013) stated that community resistance in many locations of geothermal development in Indonesia represents the major obstacle to renewable energy deployment. Similarly, Hasan and Wahjosudibjo (2014) suspected that there are several issues other than the electricity tariff that have to be addressed by the government in order to improve geothermal power plants.

It is worth noting that, after some 30 years of relatively authoritarian governance, since around the 2000s, the Indonesian people have been enjoying a more democratic atmosphere in what is called the reformation era, where governance is more decentralized, public policy is more bottom-up and participatory, and there is a new-found freedom of speech among the people. Yuliani (2013) noted that this new condition in a community sometimes brings problems for the development of infrastructure, especially when there is no sufficient preparation for the community to accept new things, such as renewable energy technology. Even respondents from local government officials (regency government) show that this lack of information is not only true for the layman but also for bureaucrats.

In the deployment of small hydropower plants, community conflicts are usually about the amount of water used for irrigation, which mostly can be solved by good communication. However, unavailability of a standard price for land, especially in remote areas, often delays the land acquisition process and if the price wanted by the community exceeds the expected figure this can lower the economic viability of the project. It is worth noting that, for the foreign investor, this problem of the price of land skyrocketing may worsen, and needs to be carefully considered.

8.4.4 Miscellaneous: Unanticipated Impacts of FIT Policy

The significant increase in renewable energy investment, especially for small hydropower, creates several unanticipated impacts which are not always positive.

Government officials, as well as developers, report that there is some kind of rivalry between institutions related to the importance of their role in permit processes. The struggle to acquire the most profitable sites for hydropower plants has created a situation where power plants are located very near each other. On the supply side, it is a worrying situation, because there is no capacity cap that limits how much of the generated electricity the electricity company is willing to buy.

Related to land acquisition, a growing number of people play the role of a land realtor, hoping that they can benefit from discrepancies in land prices. This, in turn, makes the price of land uncontrollable in some regions. One developer complained that government officials are also players in land acquisition deals.

Before the implementation of FIT, small hydropower was used to electrify communities in remote regions, usually with aid from the government. However, with the massive emergence of hydro investors and the fight for the best locations, these community-based small hydropower plants are left behind. In West Java, there are at least two locations where micro-hydro plants were ready to be constructed but eventually failed because the power plant licence for the exact location had been issued by the MEMR.

8.5 CONCLUSION AND RECOMMENDATIONS

8.5.1 Conclusion

Indonesia is an example of a country where FIT policy, which is considered quite successful in developed countries, is duplicated without careful considerations of the existing sociopolitical conditions. The FIT policy in Indonesia, which has been formulated in eight ministry regulations (four original versions and five revised versions) has not significantly improved deployment of its renewable energy. It is true that investment interest in renewable energy has soared to a level which has never been seen since its implementation around the 2010s. However, data confirmed through field reports shows that about 90 per cent of scheduled deployments of renewable energy power plants are delayed, and targets are far from realized.

Many complaints related to difficulties in obtaining a PPA, lack of a guarantee of grid connection, and complex contract structure, indicating that some of the FIT policy principles are not being met, although the price of electricity is satisfactory. The study also reveals several other issues that pose obstacles to renewable energy deployment, which can be grouped into three big problems: unclear and inconsistent policies both horizontally (between ministries) and vertically (between levels of government), the complex requirements of many permits, and difficulties related to land acquisition which includes conflict with the community. Adding to those problems in the field is the corrupt behaviour in permit issuance processes, which has made the calculation of the economic scale of a project difficult.

To sum up, the FIT policy when applied in a less developed country such as Indonesia must work in tandem with other efforts. The complex structure of governance, lack of community knowledge about renewable energy, and incoherent coordination between related institutions, all add up to difficulties of implementation in the field. If Indonesia is serious about achieving its projected goals for renewable energy deployment, then efforts must be strengthened and calculated measures must be taken. On the other hand, the study also underlines the importance of narrowing the wide gap in transitioning to cleaner energy, where less developed and developed countries both face very different challenges.

8.5.2 Recommendation

In order to enhance the effectiveness of FIT policy in increasing the success of renewable energy deployment, it is recommended that the GOI simplifies the licence procedures, either at the central government level or most especially at the local government level. This will require an exceptionally strong commitment from all parties, because every institution holds to its own laws and regulations, and may not have the same agenda for the development of renewable energy.

The existence of specific budgets allocated for renewable energy subsidies, as implemented by many countries, can play a significant role in lessening the burden of developers. The subsidy which is intended to pay for the discrepancy of production costs between renewable energy and conventional power plants has to be independently managed, and not be dependent on the political process in the national budget planning which is decided annually. One realistic option is to form an independent body which is given the mandate to facilitate the permit process, formulate and finalize PPA, and manage various aspects related to pricing and payments of the government subsidy for renewable energy.

It is also recommended that the GOI should not abandon community-based efforts in renewable energy deployment. The excessive focus on investment to boost renewable energy deployment is not sustainable because it does not create a sense of belonging from the surrounding community. On the other hand, there is a lot of under-utilized renewable energy potential which is most suitable to be used for domestic energy, such as biomass.

8.5.3 Suggestions for Further Research

Due to the limited time and budget for conducting this field survey, in-depth investigation of locations for the deployment of renewable energy power plants has only been conducted at a limited number of sites. It is suggested that a thorough assessment is conducted in other parts of Indonesia, including of all kinds of renewable technologies. Field practices at the lower level of government may vary from place to place. Thus, a comparison between regions is recommended in order to obtain a general description of how permit procedures are carried out at local government level, and eventually, to suggest the best option for simplifying procedures and increasing the effectiveness of deployment. There are at least two small hydropower plants which are said to have been operational in published reports by related institutions; in reality, they are still in the process of becoming fully operational commercially. Thus, it is best to check the location of development on the site to assess the real situation, especially for solar PV and biomass power plants which are located in remote areas of Indonesia.

The availability of low-interest loans dedicated to renewable energy development from various banking and financial institution, both national and international, must also be scrutinized. The fact that about 90 per cent of power plant licences have not shown any progress after several years, raised some concerns about prejudice from several of those surveyed that the licences are used to obtain loans to be invested elsewhere. In this study, the researcher does not have access to any banking personnel and, under the bank privacy principle, it could be difficult to get any information. However, in order to create a healthy investment climate for renewable energy, it is suggested that measures are taken to ensure that renewable energy loans are used appropriately.

ACKNOWLEDGEMENTS

This research was made possible by UNU-WIDER through the project on 'Political Economy of Clean Energy Transitions'. The author would like to thank UNU-WIDER for the opportunity to conduct the research, and to meet and network with other researchers.

REFERENCES

DGNREEC (Direktorat Jenderal Energi Baru Terbarukan dan Konservasi Energi) (2014). 'Role and Strategy for Development of Renewable Energy and Energy Conservation'. Unpublished Presentation.

DGNREEC (2015). 'Statistik EBTKE 2015'. DGNREEC, Jakarta. Available at: <http://www.ebtke.esdm.go.id/>(accessed 30 September 2015).

ECORYS (2008). 'Assessment of Non-Cost Barriers to Renewable Energy Growth in EU Member States—AEON'. DG TREN No. TREN/D1/48–2008. ECORYS,

Rotterdam. Available at: <https://ec.europa.eu/energy/sites/ener/files/documents/2010_non_cost_barriers.pdf> (accessed 19 February 2016).

Farrel, J. (2009). 'Feed-in Tariffs in America: Driving the Economy with Renewable Energy Policy that Works'. Minneapolis. Available at: <https://ilsr.org/wp-content/uploads/files/feed-in%20tariffs%20in%20america.pdf> (accessed 1 March 2016).

Government of Indonesia (2014). 'Government Regulation No.79/2014 on National Energy Policy'. Available at: <http://prokum.esdm.go.id/pp/2014/PP%20Nomor%2079%202014.pdf> (accessed 30 September 2015).

Hasan, M. and A. S. Wahjosudibjo (2014). 'Feed-In Tariff for Indonesia's Geothermal Energy Development, Current Status and Challenges'. Proceedings of Thirty-Ninth Workshop on Geothermal Reservoir Engineering Stanford University, 24–26 February, Stanford, California.

IRENA (International Renewable Energy Agency) (2012). 'Evaluating Policies in Support of the Deployment of Renewable Power'. Available at: <http://www.irena.org/menu/index.aspx?mnu=Subcat&PriMenuID=36&CatID=141&SubcatID=251> (accessed 19 September 2014).

Klein, A., E. Merkel, B. Pfluger, A. Held, M. Ragwitz, G. Resch, and S. Busch (2010). 'Evaluation of Different Feed-in-Tariff Design Options—Best Practice Paper for the International Feed-In Cooperation'. Third Edition. A Joint Research Project between Energy Economics Group (EEG) and Fraunhofer ISI. Available at: <file:///C:/Users/George/Downloads/Best_practice_Paper_3rd_edition.pdf> (accessed 1 March 2016).

Lauber, V. and L. Mez (2006). 'Renewable Electricity Policy in Germany, 1974 to 2005'. *Bulletin of Science, Technology & Society*, 26(2): 105–20.

MEMR (Ministry of Energy and Mineral Resources) (2015). 'Road Map for Accelerated Development of New and Renewable Energy 2015–2025'. MEMR, Jakarta.

Mendonça (2007). *Feed-in Tariffs: Accelerating the Deployment of Renewable Energy*. London: Earth Scan.

Mez, L. (2012). 'Germany's Merger of Energy and Climate Change Policy'. *Bulletin of the Atomic Scientists*, 68(6): 22–9.

Mitchell, C. et al. (2011). 'Policy, Financing and Implementation'. In *IPCC Special Report on Renewable Energy Sources and Climate Change Mitigation*. Cambridge and New York: Cambridge University Press, pp. 865–950.

UNDP (United Nations Development Programme) (2015). 'The State of Indonesia's Renewable Energy'. Available at: <http://www.id.undp.org/content/indonesia/en/home/presscenter/articles/2015/07/22/the-state-of-indonesia-s-renewable-energy.html> (accessed 30 September 2015).

US Department of Commerce (2010). 'Renewable Energy Market Assessment Report: Indonesia'. Available at: <http://www.iea.org/publications/freepublications/publication/indonesia2008.pdf> (accessed 30 September 2015).

White, L. V., B. Lloyd, and S. J. Wakes (2013). 'Are Feed-in Tariffs Suitable for Promoting Solar PV in New Zealand Cities?' *Energy Policy*, 60: 167–78. Available at: <http://dx.doi.org/10.1016/j.enpol.2013.04.079> (accessed 16 April 2015).

Yuliani, D. (2013). 'Dynamics of Community Resistance in Planning (Case Study: Development of Geothermal Power Plant in Mount Tampomas, Sumedang, Indonesia)'. Unpublished Dissertation, Institute of Technology of Bandung, Indonesia.

9

Do Political Economy Factors Matter in Explaining the Increase in the Production of Bioenergy?

Éric Nazindigouba Kere

9.1 INTRODUCTION

The international political and economic context is characterized by a growing awareness of the need to fight against global warming (due largely to the increase in greenhouse gas emissions, according to the Intergovernmental Panel on Climate Change (IPCC 2014) and to find alternatives to fossil fuels. Indeed, the increase in temperature, the multiplication of natural disasters (storms, droughts, floods, etc.), and the volatility of oil prices are signals that should encourage states to act against global warming.

In such a context, the 'bioenergy with carbon capture and storage' (BECCS) technology appears as an alternative to fossil fuels. Indeed, these energies have a very favourable carbon footprint because they are produced from agriculture and forestry biomass.[1] They are mainly used to produce heat, biofuels, and electricity. For these reasons, many countries have chosen to respond to environmental and energy challenges through increased production of bioenergy. For example, the United States, with the Renewable Fuel Standard Program (RFS2) and the Energy Independence and Security Act of 2007, set the goal of incorporating a minimum of 36 billion gallons of biofuels into the fuel market by 2022 (EPA 2010). According to data from the United States Energy Information Administration (EIA 2015a), illustrated in Figure 9.1, world production of bioenergy (ethanol and biodiesel) has increased dramatically over the last decade. Global bioenergy production increased from 300,000 barrels per day (BPD) in 2000 to 1.9 million BPD in 2012. This increase was primarily

[1] If they do not generate indirect land use changes.

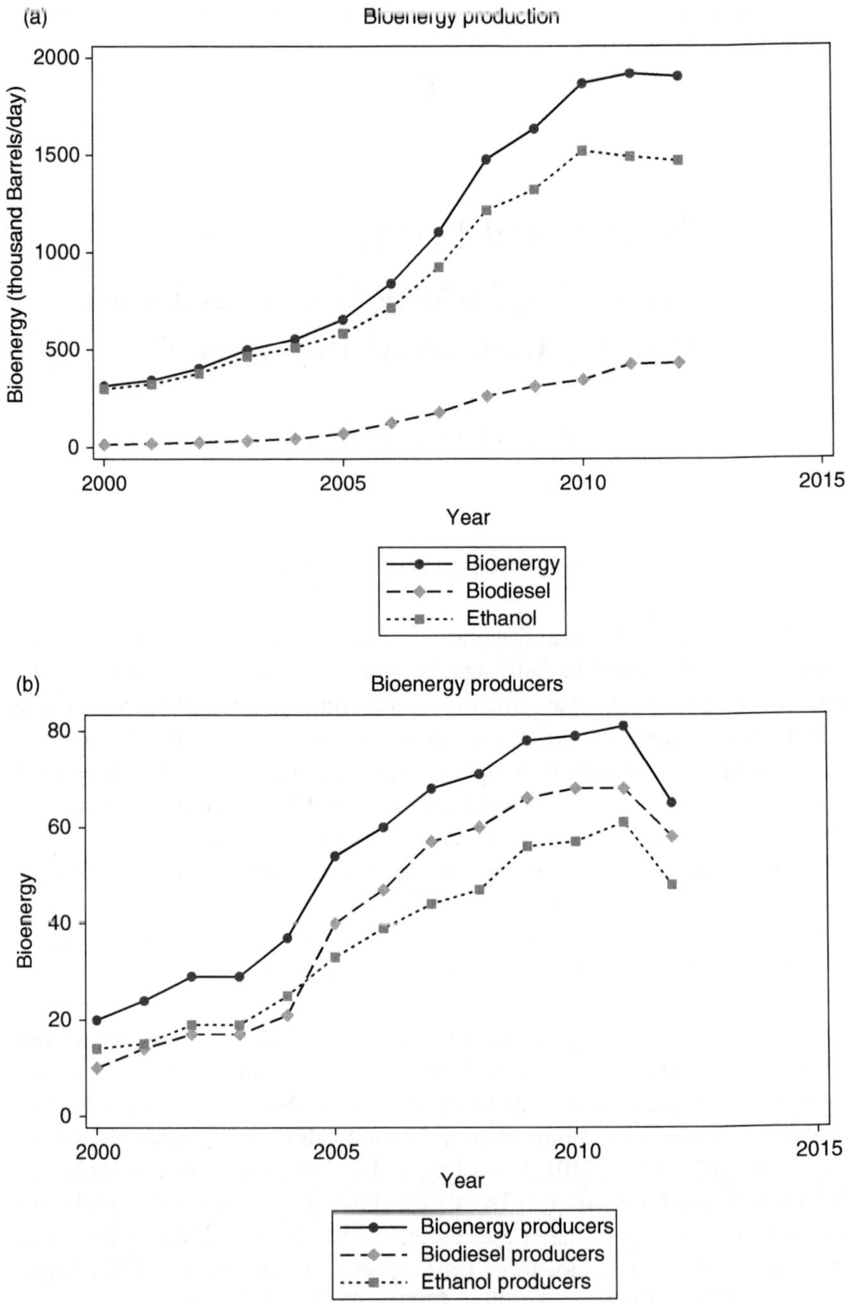

Figure 9.1. Evolution of bioenergy production and number of producers.
Source: Author's calculations based on data from EIA (2015a).

due to strong growth in ethanol production: 1.47 million BPD in 2012 versus 299,000 BPD in 2000. Biodiesel production increased from 15,000 to 431,000 BPD between 2000 and 2012. At the same time, the number of bioenergy-producing countries quadrupled between 2000 and 2011, followed by a slight decline in 2012.[2] Despite this progress, the IPCC (2014: 82) recommends a fourfold increase in investments in BECCS in order to limit global warming to 2°C. Therefore, a better understanding of the determinants of bioenergy production is essential to promote the transition to clean energy.

To the best of our knowledge, only Gan and Smith (2011) have empirically analysed the macroeconomic determinants of bioenergy. In the OECD (Organization for Economic Cooperation and Development) countries, they showed that gross national product along with bioenergy market deployment policies have significant and positive impacts on the per capita supply of renewable energy and bioenergy. Surprisingly, the impact of political economy factors was ignored in their study, even though several papers have shown that the quality of political and economic governance influences economic activity (including bioenergy production) through its impact on investment and entrepreneurship (Acemoglu, Johnson, and Robinson 2001; Baum and Lake 2003; Christopoulos and Tsionas 2004; La Porta et al. 2008; Nelson and Singh 1998).

We classify political economy factors into two categories: governance quality and macroeconomic policies. Quality of governance reflects the quality of political and economic institutions, and can be captured by the following indicators: environmental policy stringency (EPS) instruments, bureaucracy quality, corruption, investment profile, democratic accountability, government stability, and law and order. Furthermore, according to La Porta, de Silanes, and Shleifer, 'the historical origin of a country's laws is highly correlated with a broad range of its legal rules and regulations, as well as with economic outcomes' (2008: 285), and thus, indirectly, with the economic and political conditions in which bioenergy is produced. High-quality governance helps to create institutional and political dynamics, a transparent and predictable framework that encourages economic actors to invest in growth sectors of the future, including bioenergy. Macroeconomic policies (financial development, interest rates, trade openness, and oil scarcity) may also play an important role in increasing the production of bioenergy. Indeed, financial development facilitates the financing of future projects at low cost. It results in low interest rates and high credit volume. Open trade facilitates access to technology, increases competition, and provides new market opportunities for

[2] Countries (Barbados, Trinidad and Tobago, Kazakhstan, Vietnam, Jamaica, Honduras, Nicaragua, El Salvador, Singapore, Hong Kong, Pakistan, Serbia, Costa Rica, Cambodia, Switzerland, Bosnia and Herzegovina, and Norway) that have stopped production of bioenergy between 2011 and 2012 account for only 0.4 per cent of production; hence the small effect on global production. This phenomenon can be explained by the increase in agricultural prices in 2012.

bioenergy, whereas the increasing scarcity of physical oil (high prices and low reserves) stimulates the transition to clean energy (Grafton et al. 2012; Heun and de Wit 2012). Finally, the amount of renewable energy produced is an important determinant of biofuels production. Energy transition can be achieved through specialization in the production of bioenergy and/or renewable energy based on the comparative advantages of each country.

This study tries to fill the gap in the literature on the macroeconomic drivers of bioenergy by examining the impact of political economy factors, oil production, renewable energies, and macroeconomic factors. First, we present a simple theoretical model in which oil, renewable energy, and bioenergy are produced simultaneously. This formalization allows us to highlight the theoretical connections between oil production, supply of bioenergy, and political economy factors. We show that the supply of bioenergy depends positively on governance quality, financial development, land yields, and market conditions (price, income). On the negative side, oil reserves and renewable energies tend to reduce bioenergy supply. Second, we empirically highlight the determinants of the supply of bioenergy using an unbalanced panel dataset of 112 countries for the period between 2000 and 2012. Motivated by the fact that 'zeros' represent a large fraction of the bioenergy data (40 per cent), we use a fixed effects Tobit model (Honoré 1992) to address the censoring problem, as well as unobservable heterogeneity specific to each country. Third, we make separate estimates of the supply function of bioenergy for all the countries in our sample of developed and developing countries. This allows us to analyse and compare the effects of political economy factors on bioenergy production, depending on the level of development. Finally, given that most countries do not produce bioenergy, we analyse the determinants leading to the decision of whether or not to produce bioenergy by using a random effects probit model.[3] We show that political economy factors (governance quality and macroeconomic policies) are central in deciding whether or not to produce bioenergy, but their impact on production size is limited. Indeed, when the decision to produce is made, market size and production conditions have a greater influence on the amount of bioenergy. Using a long-term analysis, we find that the countries whose laws are of Germanic, Scandinavian, and French origin produce relatively more bioenergy, whereas those whose laws have a Socialist origin produce less than other countries.

The remainder of the chapter is organized as follows. Section 9.2 outlines the theoretical model, the empirical strategy, and data. Section 9.3 discusses the results and their implications. Section 9.4 concludes by indicating how political economy factors can facilitate the transition to clean energy.

[3] There are several countries that produce bioenergy throughout the period and others that do not produce anything. Using a discrete choice model with fixed effects removes these countries from the analysis. This is why we choose a random effects model.

9.2 ANALYTICAL FRAMEWORK

9.2.1 A Simple Theoretical Model

To motivate the empirical analysis in this section, we develop a function of bioenergy supply taking into account the governance quality, macroeconomic policies, energy demand, and supply of oil and renewable energy. First, we assume that bioenergy, oil, and renewable energy are perfect substitutes in producing energy. The total energy demand D_E is decreasing with the price of energy P and increasing with national income Y: $D_E = (P,Y)$. Based on the Hotelling (1931) rule, we define the function of oil supply S_O as an increasing function of price P, interest rates r, and oil reserves R: $S_O = (P,r,R)$. If the country is not an oil producer, oil supply will be provided by imports and will depend on world oil reserves and international prices. Significant global reserves of oil and low international prices will make imports cheaper. However, a non-oil-producer is more likely to promote the development of bioenergy for reasons of energy sovereignty. Second, we suppose that for each unit of bioenergy sold, the producer receives $P_B = P + \tau$, where $\tau \geq 0$, a premium price for bioenergy. This prime can be the willingness of consumers to pay more for, or a state to subsidize, bioenergy. Finally, we assume that bioenergy is produced by a representative firm whose profit function is:

$$\Pi = (P + \tau)BioE - CT(BioE, L, Gov, MP), \quad (1)$$

where *BioE* represents the quantity of bioenergy supplied, *L* the productivity of the land, *Gov* the indicator of the quality of governance, *MP* the macroeconomic policies, and *CT(BioE,L,Gov,MP)* the total cost of production. We assume that the marginal cost is positive ($CT_{BioE} > 0$) and decreases with the quantity produced ($CT_{BioE,BioE} < 0$). In addition, the marginal cost of production decreases with productivity of land ($CT_L < 0$), governance quality ($CT_{Gov} < 0$), and macroeconomic policies ($CT_{MP} < 0$). Indeed, the better the governance and macroeconomic policies, the more conducive the conditions for the development of economic activities, including the production of bioenergy. In addition, good governance and macroeconomic policies can lead a country to invest more in the energy transition.

Let $D_E(P,Y)$ represent the demand of energy and *Renew* the amount of renewable energy produced. At equilibrium, demand equals supply: $D_E(P,Y) = S_O(P,r,R) + Renew + BioE$. This means that $BioE = D_E(P,Y) - S_O(P,r,R) - Renew$. As a result, the demand for bioenergy can be defined as the difference between the total energy demand and the supply of oil:

$$D_{BioE}(P, Y, r, R) = D_E(P, Y) - S_o(P, r, R) - Renew. \quad (2)$$

By inserting the inverse demand for bioenergy, $P(BioE,Renew,Y,r,R) = P(D_{BioE})$, in the profit equation (1) we get:

$$\max_{BioE} \Pi = [P(BioE, Renew, Y, r, R) + \tau]BioE - CT(BioE, L, Gov, MP). \quad (3)$$

On the basis of the first-order conditions of this maximization programme, we have:

$$[\left(\partial P(BioE, Renew, Y, r, R)\right)/\partial BioE]BioE + P(BioE, Renew, Y, r, R) + \tau = \mathrm{Cm}(BioE, L, Gov, MP). \quad (4)$$

At the optimum, the marginal benefit of an additional unit of bioenergy is equal to the marginal cost of the unit. The optimal bioenergy supply can be expressed as:

$$BioE^* = BioE(P, Renew, Y, r, R, \tau, L, Gov, MP). \quad (5)$$

Using simple linear functions, a comparative static analysis allows us to better understand the insights of this result.

Let $S_O(P,r,R) = \delta_1 P + \delta_2 R + \delta_3 r$, $D_E(P,Y) = \gamma_1 Y - \gamma_2 P$, $CT(BioE,L,Gov,MP) = (\nu_1 - \nu_2 L - \nu_3 Gov + \nu_4 MP)BioE$. Therefore, the demand for bioenergy equals $D_{BioE}(P,Y,r,R) = \gamma_1 Y - \gamma_2 P - \delta_1 P - \delta_2 R - \delta_3 r - n_1 Renew$ and the optimal bioenergy supply can be expressed as follows:

$$BioE^* = \frac{1}{2}\left(\gamma_1 Y - n_1 Renew - \delta_2 R - \delta_3 r + (\gamma_2 + \delta_1)(\tau - \nu_1 + \nu_2 L + \nu_3 Gov + \nu_4 MP)\right) \quad (6)$$

A comparative static analysis can be conducted by differentiation of Equation (6) with respect to each of the explanatory variables. Differentiation yields:

$$\frac{\partial BioE^*}{\partial Renew} = -\frac{n_1}{2} < 0$$

$$\frac{\partial BioE^*}{\partial Gov} = \frac{(\gamma_2 + \delta_1)\nu_3}{2} > 0$$

$$\frac{\partial BioE^*}{\partial \tau} = \frac{(\gamma_2 + \delta_1)}{2} > 0$$

$$\frac{\partial BioE^*}{\partial MP} = \frac{(\gamma_2 + \delta_1)\nu_4}{2} > 0$$

$$\frac{\partial BioE^*}{\partial Y} = \frac{\gamma_1}{2} > 0$$

$$\frac{\partial BioE^*}{\partial R} = -\frac{\delta_2}{2} > 0$$

$$\frac{\partial BioE^*}{\partial r} = -\frac{\delta_3}{2} > 0$$

$$\frac{\partial BioE^*}{\partial L} = \frac{(\gamma_2 + \delta_1)\nu_2}{2} > 0.$$

The political economy factors (governance quality and macroeconomic policies) have a positive impact on bioenergy supply. These factors will promote the establishment of the necessary conditions for investment in growth sectors of the future, including bioenergy. The premium price for bioenergy favours the production of bioenergy. Indeed, the higher the premium price, the higher the country internalizes the environmental and social benefits of bioenergy. It also appears that biofuels and renewable energy are substitutable. Oil reserves have a negative effect on the supply of bioenergy. In addition, the interest rate has a negative effect on the production of bioenergy. According to the Hotelling (1931) rule, a high interest rate increases the incentive to exploit the oil reserves, investing the income from the sale of oil in a more profitable alternative. In addition, a higher interest rate discourages new investments. In the long term, however, the extraction of fossil fuels may cause oil scarcity. Then prices will rise making profitable renewable energy as in the case of low interest rates. The supply of bioenergy also depends positively on income. According to the environmental Kuznets curve, income growth leads to higher preferences for environmental goods. Finally, agricultural productivity has a positive effect on the production of bioenergy.

9.2.2 Econometric Model

One of the aims of this chapter is to estimate the supply equation of bioenergy obtained in the theoretical model (Equation (5)). However, there are some challenges, particularly regarding censoring of the dependent variable and how the unobserved heterogeneity is taken into account. Bioenergy production ($BioE_{it}$) is a latent variable because it is observable only if production is positive. If the optimal production is negative, the country does not produce. The dependent variable is limited by a positivity constraint. In our sample, only 11 per cent of all countries were producing bioenergy in 2001 and 48 per cent in 2012. Censoring is important and must be taken into account. It may therefore be argued that a Tobit model is the preferred estimator (Greene 2002; Wooldridge 2010). The model to be estimated can be written as follows:

$$BioE_{it} = \beta X_{it} + \alpha_i + \epsilon_{it}, \tag{7}$$

whereas the observed value of the bioenergy variable, $BioE_{it}$, is given by:

$$BioE_{it} = \begin{cases} BioE_{it}^{*} & if\ BioE_{it}^{*} > 0 \\ 0 & if\ BioE_{it}^{*} \leq 0, \end{cases} \tag{8}$$

where the $BioE_{it}$ variable is greater than zero only when the latent

$BioE^*_{it}$ variable exceeds zero (which represents cases where a country's willingness to produce bioenergy is positive), X represents the matrix of explanatory variables, β the vector of the coefficients associated with X, α_i country fixed effects, and ϵ_{it} the term error, which is independently and identically distributed.

Equation (8) can be estimated using a Tobit model with random effects, if the country-specific effects are not correlated with the explanatory variables. Otherwise, a fixed effects model must be used. In this case study, fixed effects are essential to control for unobserved heterogeneity specific to each country that does not vary over time (at least in the short term), such as institutional framework or preferences for bioenergy. The introduction of fixed effects allows us to take these factors into account in our regressions. To avoid the 'incidental parameters problem', we use the fixed effects Tobit model suggested by Honoré (1992). This method is based on the work of Powell (1984, 1986), on least absolute deviations estimation for the censored regression model and symmetrically trimmed least squares estimation for Tobit models.[4] To determine whether the fixed effects or random effects model is appropriate, we use a Hausman specification test. For all specifications, the test statistic χ^2 allows us to reject the null hypothesis of independence between errors and explanatory variables, and accordingly opt for a fixed effects model.

9.2.3 Data Description

This study examines the drivers of per capita supply of bioenergy (see Appendix Table 9.A1 for a description and Kere (2016: appendix D1) for summary statistics of the data).[5] Owing to the availability of data,[6] we use ethanol, biodiesel, and total bioenergy (ethanol + biodiesel) as a proxy of bioenergy. So this assumption excludes mainly fuelwood and biogas. The study sample contains 112 countries over the period between 2000 and 2012. Data on bioenergy, renewable energy, international price of oil, and oil reserves are from EIA (Energy Information Administration) (2015b).

The other explanatory variables are from World Development Indicators (WDI 2015): gross domestic product (*GDP*), population density (*Pop_dens*), percentage of urban population (*Urban*) and cereal production (*Cereal_yield*), agricultural land (*Agri_land*), and financial development (*Financial*). In this study, we do not use the production of sugarcane as an explanatory variable

[4] Fixed effects Tobit models developed by Honoré (1992) was implemented with STATA 'pantob' and is available at: <http://www.princeton.edu/~honore/stata/#1._Pantob_version_0.6> (accessed 15 October 2016).

[5] The list of countries represented in the estimates is available in Kere (2016: appendix A3).

[6] For example, it is difficult to find homogeneous data on fuelwood production by country.

because there are too many missing data. North America is the largest bioenergy producer, followed by Latin America, Europe, and Central Asia. Latin America and North America are the largest producers of ethanol whereas Europe is the largest producer of biodiesel.

Governance is an index of the overall quality of governance, constructed using six dimensions of governance of the International Country Risk Guide (ICRG 2015): bureaucracy quality, corruption, investment profile, democratic accountability, government stability, and law and order. We combined these indicators using the first two principal components (81 per cent of the overall variance) of the vector of six indicators of governance. We standardized the index by the following method: $Governance = [Index - \max(Index)]/[\max(Index) - \min(Index)]$. The results of the principal component analysis are available in Kere (2016: appendix table B1). The renewable energy variable and governance index are lagged by five years to avoid endogeneity problems.

Biofuel mandates are a prime targeted mechanism to stimulate the production of biofuel. Unfortunately, there are very few panel data on these policies. Moreover, these mandates do not vary much over time; therefore, they are captured by fixed effects of our empirical model. For these reasons and as a proxy, we use the composite indexes of EPS elaborated by Botta and Koźluk (2014): market and non-market instruments. These policies are the instruments used to achieve the mandates. The market component of the EPS includes taxes and charges applied on input or output of a production process (diesel) or pollution source (CO_2, NO_x, and SO_x); trading schemes (green certificate, white certificate, and CO_2 trading); feed-in tariffs (solar and wind); and deposit refund scheme (waste). The non-market component of the EPS includes standards (emission limit values: NO_x, SO_x, PM_x, and sulphur in diesel content limit) and research and development subsidies. Both indices can be used as a proxy of premium price because they allow internalizing the market and environmental externalities. Unfortunately, these variables exist only for OECD member countries. In developing countries, energy policies include policy statements on the development of bioenergy, often without concrete strategies or the institutional framework necessary for implementation (Jumbe et al. 2009).

Our measurement of financial development is the ratio of domestic credit allocated to private sector to GDP (excluding credit to central government and public enterprises and credit issued by central banks). This measure goes beyond the market interest rates because it allows us to account for the quality and capacity of the domestic financial system to finance the private sector.

Oil_reserve, representing the crude oil proved reserves, is an exogenous variable that does not vary much over time. The international price of oil (*Inter_Oil_price*) is the result of the interaction between global oil supply and demand. It is exogenous but does not vary across countries. Using these terms, we generate a new variable dividing oil reserves by the international oil prices

Table 9.1. Expected signs of the explanatory variables

Explanatory variables	Indicators	Expected sign
Quality of governance	*Governance* (index of overall governance quality)	+
EPS	*Market-based instruments*	+
	Non-market-based instruments	+
Legal origins	*British*	Reference
	German	+
	Scandinavian	+
	French	+
	Socialist	−
Substitute products	*Oil scarcity* (oil reserve to oil price ratio)	−
	Renewable energy production (hydroelectric, geothermal, wind, solar, waste, tide, and wave)	−
Macroeconomic policies	*Financial development*	+
	Trade openness (trade to GDP ratio)	+
Agricultural productivity	*Cereal yield*	+
Market size	*GDP per capita*	+
	Population density	+
	Urban (%)	+

Source: Author's compilation.

(*Oil* = *Oil_reserves*/*Inter_Oil_price*). *Oil* is the relative value of oil reserves compared to international oil prices. This new variable is exogenous and varies over time and between countries.

Our definition of the legal origin of laws is based on the distinction made by La Porta et al. (1999). According to the historic and legal tradition of company law, or commercial code, countries are classified as: British law (or common law), French civil law, German civil law, Scandinavian law, and socialist law. These traditions historically come from Great Britain, France, Germany, Scandinavia, and the Soviet Union, respectively. For our purposes, the legal origins are represented by binary variables that identify the legal origin of a country's laws as French, German, Scandinavian, or socialist, with British law as the omitted group.[7]

Based on the results of the theoretical model, the literature on the determinants of bioenergy production, and political economy literature, we classify the determinants of bioenergy production into seven sets of explanatory variables: indicators of governance quality, EPS instruments, legal origins of law, macroeconomic policies, substitute products, agricultural policy factors (supply factors), and market size (demand factors) (Table 9.1).

[7] See Section 9.3.4 for further explanation of the expected signs of legal origin.

9.3 DETERMINANTS OF THE PRODUCTION OF BIOENERGY

9.3.1 Bioenergy and Political Economy Factors: Naïve Evidence

In this section, we give an insight into the relationships between our variables of interest. Figure 9.2 provides a combined visual analysis of the relationship between bioenergy and political economy factors over the entire sample of developing and developed countries. The graphs (Figures 9.2a–f) suggest that there is a strong positive effect of governance and EPS instruments on bioenergy production. They also show that the United States and Brazil have levels of bioenergy production higher than other countries. These countries can be considered outliers. Indeed, the United States and Brazil are the world's largest producers of bioenergy, and have developed proactive policies of bioenergy production. For example, Brazil has committed to increase the share of ethanol to 27.5 per cent in its overall energy balance, and, as mentioned in the introduction, the United States set the goal of incorporating a minimum of 36 billion gallons of biofuels into the fuel market by 2022 (EPA 2010). For robustness, we will test the sensitivity of our results to the presence of these countries in our regressions.

Figure 9.2d analyses the average per capita production of bioenergy by legal origin.[8] This graph shows that countries with laws of Scandinavian and Germanic origin produce more bioenergy than others, whereas those of French and Socialist origin produce relatively less bioenergy. Taking into account the characteristics of the origin of a country's laws, highlighted by La Porta et al. (1999; La Porta, de Silanes, and Shleifer 2008), these statistics are consistent with our expectations. Indeed, we think that countries following Scandinavian and German civil law are most likely to promote the production of bioenergy. The interventionist stance of their laws, and efficiency of their bureaucracy, will allow them to opt for clean energy, despite pressure. We expect countries subject to the French civil law to be intermediate regarding the adoption of clean energy. In these countries the bureaucracy is powerful and largely unconstrained, which helps in the implementation of environmental policies. However, this system typically generates more corruption, low financial development, and unemployment, which tends to reduce the effectiveness of environmental policies. However, laws of British origin rely on strong private property rights and low regulation, which promotes the development of private activity with low regulation. Therefore, binding environmental policies will be difficult to implement. Finally, countries with laws of

[8] This analysis is a simple correlation that does not take into account other factors, such as mandates on both ethanol and biodiesel production, which can have a significant impact on the relationship between bioenergies and the legal origin of laws.

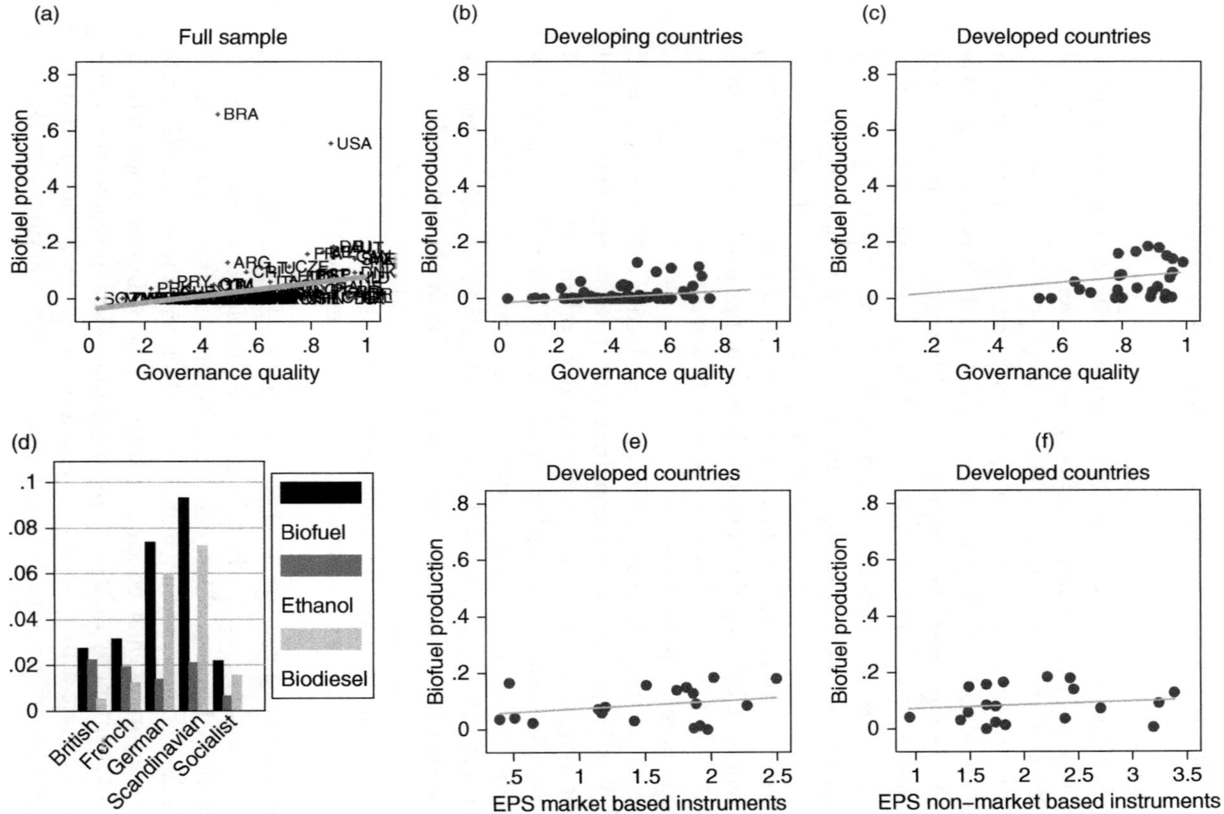

Figure 9.2. Correlation between bioenergy, governance quality, and environmental policy stringency (EPS).

Note: Biofuel is measured in 1,000 barrels per year; governance quality and EPS are index.

Source: Author's calculations based on data from EIA (2015a), ICRG (2015), and Botta and Koźluk (2014).

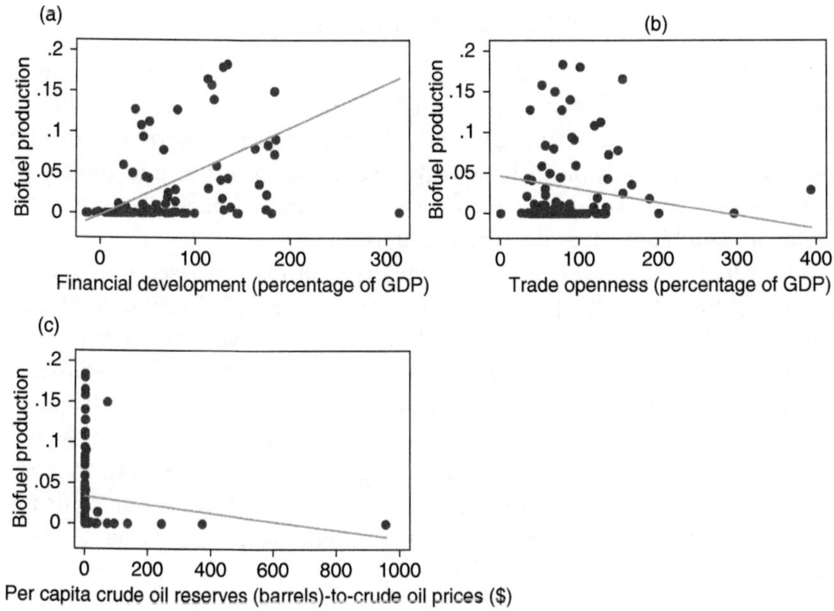

Figure 9.3. Correlation between bioenergy and macroeconomic policies of (a) financial development, (b) trade openness, and (c) oil scarcity.

Note: Biofuel is measured in 1,000 barrels per year.

Source: Author's calculations based on data from EIA (2015a) and WDI (2015).

Socialist origin have the least effective environmental policies because of the high probability of corruption associated with them.

Figure 9.3 shows the impact of macroeconomic policies on bioenergy. As expected, these results show that financial development and oil reserves influence the production of bioenergy positively and negatively, respectively, whereas the impact of trade openness is negative. Indeed, the least competitive countries may choose to import bioenergy from the most competitive countries and therefore reduce their own productions. However, this result seems counterintuitive because bioenergy trade is not very developed.

9.3.2 Results of Tobit Model with Fixed Effects

The estimation results using the Tobit model with fixed effects are presented in Table 9.2 and random effects results are available in Kere (2016: appendix table C1). The drivers of bioenergy production can be classified into four categories: governance quality, macroeconomic policies, size of market, and agricultural factors.

9.3.2.1 Governance Quality

Across the entire sample, the index of the quality of governance has a significant positive effect at 5 per cent level on the production of biodiesel. When looking at the results by level of development, governance quality promotes the production of bioenergy. According to our results, quality of governance significantly influences the production of ethanol in developing countries and biodiesel in developed countries.

Regarding the impact of EPS instruments, non-market-based instruments positively affect the total production of bioenergy and biodiesel in developed countries whereas market-based instruments do not have a significant effect. Such results are not surprising. Indeed, non-market-based instruments (standards and research and development subsidies) provide research support for the introduction of new technologies and improve the profitability and social acceptance of clean energy including bioenergy. In contrast, market-based instruments (taxes on carbon, diesel, or sulphide and feed-in tariffs for solar and wind) are intended for promoting renewable energy such as solar and wind.

9.3.2.2 Other Energy

Oil reserves have a negative effect on the production of bioenergy in the overall sample and on biodiesel in developing countries. Large oil reserves represent a guarantee of availability of oil at low cost, which tends to discourage the production of bioenergy. However, oil reserves significantly promote the production of ethanol and biodiesel in developed countries. In fact, the rich oil-producing countries tend to invest more in clean energy.

Renewable energy has a significant negative impact only on the total bioenergy production in developing countries. Unlike developed countries, these countries lack the financial capacities to invest in both renewable energy and biofuels.

9.3.2.3 Macroeconomic Policies

The impact of macroeconomic policies is contrasted with the level of development. Our results show that financial development positively influences the production of bioenergy in developing countries. These countries face enormous difficulties financing their economies, yet they have vast unexploited fertile farmland. A structured and efficient financial system would allow projects for bioenergy production to be financed at a lower cost. Financial development has no impact on the production of bioenergy in developed countries because they have little problem financing promising projects. Finally, as expected, trade openness has a positive impact on bioenergy and ethanol, but only in the subsample of developing countries.

9.3.2.4 Market Size

In developed countries, the size of the market plays a key role in the development of bioenergy. Per capita income, urbanization, and population density promote energy demand. Given the relatively high income in these countries, according to the environmental Kuznets curve, a part of the energy demand will be transferred to clean energy, including bioenergy.

Market size factors do not significantly influence the production of bioenergy in developing countries. This result can be explained by the relatively higher cost of bioenergy; they are mainly produced using subsidies. It is difficult for developing countries, which have other priorities such as food security or fighting against poverty, to engage in subsidized bioenergy production.

9.3.2.5 Agricultural Factors

Cereal yield is an important determinant of bioenergy production. As expected, it has a positive and significant influence on the production of bioenergy, ethanol, and biodiesel in the total sample of developing and developed countries.

9.3.2.6 Estimation Results without the United States and Brazil

As a test of robustness, we excluded Brazil and the United States in our sample because their production of bioenergy represents a large portion of total world production. For example, in 2012, these two countries accounted for 30 per cent of the world's production (EIA 2015a). In addition, as shown in Figure 9.2a, these two countries can be considered outliers. The results are in line with the previous results.[9] Therefore, these countries have no influence on the quality of our results.

9.3.3 Decision to Produce Bioenergy: Results of Random Effects Probit Model

To identify the influence of political economy factors on the decision of whether or not to produce bioenergy, we use a random effects probit model.

[9] The results are available in Kere (2016: appendix D1).

Table 9.2. Tobit model with fixed effect of bioenergy supply

Variable	All countries			Developing countries			Developed countries		
	Biofuel	Ethanol	Biodiesel	Biofuel	Ethanol	Biodiesel	Biofuel	Ethanol	Biodiesel
Governance quality									
Governance	0.4132 (0.2684)	0.2917 (0.2450)	0.3858** (0.1540)	0.3593** (0.1747)	0.3124** (0.1400)	0.0924 (0.0970)	−0.1823 (0.4606)	−0.4648 (0.3381)	0.3385* (0.1948)
EPS market							−0.0323 (0.0304)	−0.0782* (0.0426)	0.0107 (0.0111)
EPS non-market							0.0464* (0.0273)	−0.0023 (0.0238)	0.0555*** (0.0149)
Other energy									
Renewable	−0.0280 (0.0310)	−0.0532 (0.0564)	0.0126 (0.0214)	−0.0408*** (0.0125)	−0.0261 (0.0238)	−0.0200 (0.0163)	−0.0403 (0.0421)	−0.0147 (0.0301)	−0.0034 (0.0221)
Oil	−0.0009*** (0.0002)	−0.0006 (0.0004)	0.0003 (0.0004)	−0.1230 (0.0897)	−0.0693 (0.1380)	−0.0380** (0.0176)	−0.0003 (0.0003)	0.0009 (0.0006)	0.0009** (0.0004)
Macroeconomic policies									
Financial	0.0028 (0.0018)	0.0027 (0.0018)	0.0013 (0.0009)	0.0056*** (0.0017)	0.0046* (0.0025)	0.0029*** (0.0010)	−0.0009 (0.0015)	−0.0012 (0.0009)	0.0000 (0.0010)
Openness	0.0025 (0.0018)	0.0064 (0.0044)	0.0002 (0.0008)	0.0023* (0.0012)	0.0040** (0.0020)	0.0003 (0.0005)	−0.0022 (0.0033)	0.0024 (0.0050)	−0.0003 (0.0015)
Market size									
GDP	0.0342*** (0.0109)	0.0330*** (0.0085)	0.0209*** (0.0056)	−0.0062 (0.0111)	0.0412 (0.0456)	−0.0008 (0.0067)	0.0481*** (0.0148)	0.0114 (0.0417)	0.0177** (0.0078)
Population density	0.0881** (0.0449)	0.0572** (0.0287)	0.0568* (0.0313)	0.0097 (0.0573)	0.0271 (0.0653)	0.0026 (0.0352)	0.1789*** (0.0641)	0.1150 (0.1796)	0.0523** (0.0259)
Urban	0.0193 (0.0376)	0.0056 (0.0116)	0.0285*** (0.0110)	0.0062 (0.0090)	−0.0026 (0.0139)	0.0063 (0.0068)	0.1381 (0.1267)	0.3337** (0.1438)	0.0169 (0.0146)
Agricultural productivity									
Cereal yield	0.1099** (0.0450)	0.1745** (0.0772)	0.0307*** (0.0080)	0.0575*** (0.0189)	0.0616*** (0.0182)	0.0181 (0.0133)	0.1129** (0.0488)	0.1467*** (0.0277)	0.0236*** (0.0083)
Observations	1338	1338	1338	1021	1021	1021	243	243	243
(Hausman test)									
χ^2	59.827	389.80	40.151	90.113	146.83	15.421	279.42	165.83	221.30
Probability	0.0000	0.0000	0.0000	0.0000	0.0000	0.2013	0.0000	0.0000	0.0000

Note: Standard errors in parentheses; *p<0.10, **p<0.05, ***p<0.01.

Source: Author's calculations based on data from EIA (2015a); ICRG (2015); Botta and Koźluk (2014); and WDI (2015).

Table 9.3. Random effects probit model of the decision to produce bioenergy

Variable	All countries			Developing countries			Developed countries		
	Biofuel	Ethanol	Biodiesel	Biofuel	Ethanol	Biodiesel	Biofuel	Ethanol	Biodiesel
Governance quality									
Governance	2.2511*	1.7955	3.0700**	0.6322	−1.2634	0.8945	18.622***	12.840***	16.779***
	(1.1825)	(1.1439)	(1.2867)	(1.3074)	(1.3792)	(2.3536)	(5.6353)	(3.7821)	(5.8535)
EPS market							0.8480	0.1780	2.5593**
							(1.0153)	(0.2827)	(1.1744)
EPS non-market							3.1558***	0.7220***	4.8830***
							(0.8699)	(0.2601)	(0.9087)
Other energy									
Renewable	0.3149**	0.1021	0.0828	0.4512**	0.3105	0.2981*	0.2232	−0.0584	−0.1196
	(0.1571)	(0.1048)	(0.1570)	(0.2102)	(0.2392)	(0.1710)	(0.3653)	(0.1603)	(0.2798)
Oil	−0.1013***	−0.0322**	−0.0166**	−0.0880**	−0.0731**	−9.9969***	−0.0748**	−0.0160	−0.0186
	(0.0158)	(0.0163)	(0.0067)	(0.0345)	(0.0298)	(2.4561)	(0.0318)	(0.0169)	(0.0136)
Macroeconomic policies									
Financial	0.0353***	0.0297***	0.0351***	0.0182***	0.0189***	0.0221**	0.1022***	0.0233***	0.0492***
	(0.0051)	(0.0047)	(0.0054)	(0.0069)	(0.0073)	(0.0107)	(0.0265)	(0.0086)	(0.0189)
Openness	0.0210***	0.0126**	0.0150**	0.0158**	0.0081	0.0208*	0.0762	0.0093	−0.0020
	(0.0064)	(0.0064)	(0.0062)	(0.0067)	(0.0077)	(0.0119)	(0.0483)	(0.0153)	(0.0338)
Market size									
GDP	0.0831**	−0.0117	0.0386	0.9492***	0.7657***	1.3620***	0.0760	0.0032	0.4271***
	(0.0381)	(0.0371)	(0.0426)	(0.1796)	(0.1378)	(0.2674)	(0.1658)	(0.0824)	(0.1517)
Population density	0.0433*	0.0198	0.0329	0.0480**	0.0420**	0.0744***	−0.5121***	−0.1143*	−0.2221*
	(0.0255)	(0.0328)	(0.0227)	(0.0200)	(0.0199)	(0.0269)	(0.1645)	(0.0628)	(0.1335)
Urban	0.1437***	0.0707***	0.1555***	0.0253	0.0021	0.2007***	0.1818	0.1227*	0.1799
	(0.0167)	(0.0244)	(0.0283)	(0.0255)	(0.0302)	(0.0472)	(0.1821)	(0.0678)	(0.1185)
Agricultural productivity									
Cereal yield	1.0549***	0.7913***	0.9557***	1.0669***	0.9406***	0.9855***	0.7682	0.1918	0.4427
	(0.1592)	(0.1655)	(0.1913)	(0.1928)	(0.2337)	(0.3672)	(0.5729)	(0.3176)	(0.5389)
$\ln\sigma^2_\mu$	4.0946***	3.3494***	3.7872***	3.5999***	3.5530***	4.8731***	4.3027***	2.0114***	4.7753***
	(0.2055)	(0.2250)	(0.2097)	(0.2561)	(0.2605)	(0.3027)	(0.4997)	(0.5537)	(0.4480)
Constant	−21.00***	−14.74***	−23.25***	−12.88***	−10.37***	−27.23***	−49.60***	−25.46***	−55.44***
	(1.0763)	(1.1142)	(1.1590)	(1.0807)	(1.2486)	(2.5000)	(13.4772)	(6.4594)	(9.8595)
Observations	1338	1338	1338	1021	1021	1021	243	243	243

Note: Standard errors in parentheses; $^*p<0.10$, $^{**}p<0.05$, $^{***}p<0.01$.

Source: Author's calculations based on data from EIA (2015a); ICRG (2015); Botta and Koźluk (2014); and WDI (2015).

More than half the numbers of countries in our estimation sample do not produce bioenergy; therefore, it is important to study the determinant of this decision to better target energy policies.

The results presented in Table 9.3 show that governance quality is an important factor in the decision to produce bioenergy in developed countries. Indeed, governance has a positive effect on all types bioenergy production (total bioenergy, ethanol, and biodiesel) in developed countries, bioenergy and ethanol in the total sample, and no effect in developing countries. EPS instruments favour the production of bioenergy in developed countries. Non-market-based instruments have a positive impact on all types of bioenergy whereas market-based instruments only favour biodiesel production. The impact of governance quality and EPS is more important in the decision to produce bioenergy in developed countries because they create favourable conditions that encourage the development of business and investments. The high level of income and subsidies increase demand and profitability of bioenergy.

Among factors of macroeconomic policies, financial development promotes the production decision of all bioenergy types, whatever the level of development. Our results also show that trade openness promotes all types of bioenergy in the total sample and developing countries.

Oil reserves reduce the production of all types of bioenergy in the overall sample and the developing countries sample, but only bioenergy production in developed countries. In contrast, the production of renewable energy favours the decision to produce bioenergy in the total sample. It promotes bioenergy and biodiesel production in developing countries, but it has no significant effect in developed countries.

Finally, both the size of the market and agricultural factors increase the likelihood of producing bioenergy, excluding the density of the population that negatively influences bioenergy developed in the country.

9.3.4 Long-Run Analysis: Impact of Legal Origins on Bioenergy Production

The energy transition generates a significant additional cost. The production of bioenergy is currently more expensive than that of gasoline or diesel. To be competitive, bioenergy needs to be widely supported and provided tax exemptions, duty incorporation in fuels, and so on. However, the cost of inaction with regard to climate change is much more important. Therefore, the development of clean energy requires significant investment, a political will, along with the state's willingness to impose binding environmental standards on its industries, as well as capacity for the economy to absorb additional costs in the short term (Table 9.4).

Table 9.4. Long-run analysis: impact of legal origins on bioenergy production

Variable	All countries			Developing countries			Developed countries		
	Biofuel	Ethanol	Biodiesel	Biofuel	Ethanol	Biodiesel	Biofuel	Ethanol	Biodiesel
Legal origins[a]									
French	-0.0049 (0.0329)	-0.0139 (0.0361)	0.0093 (0.0124)	0.0189 (0.0423)	0.0370 (0.0493)	-0.0021 (0.0132)	0.0165 (0.3089)	-0.0802 (0.2983)	0.0962** (0.0345)
Socialist	0.0265 (0.0411)	0.0125 (0.0458)	0.0175 (0.0149)	0.0502 (0.0480)	0.0602 (0.0566)	0.0142 (0.0140)	0.0000 (.)	0.0000 (.)	0.0000 (.)
German	-0.0663 (0.0624)	-0.1013 (0.0644)	0.0266 (0.0216)				-0.0899 (0.3219)	-0.2259 (0.3111)	0.1349** (0.0356)
Scandinavian	-0.0341 (0.0764)	-0.0742 (0.0786)	0.0521* (0.0266)				0.0176 (0.2907)	-0.1790 (0.2812)	0.1959*** (0.0404)
Governance quality									
Governance	0.0629 (0.1458)	-0.1022 (0.1548)	0.1169** (0.0538)	-0.0477 (0.1854)	-0.1479 (0.2051)	-0.0010 (0.0542)	1.0546 (0.9761)	0.4204 (0.9307)	0.6348*** (0.1529)
EPS market							0.8480 (1.0153)	0.1780 (0.2827)	2.5593** (1.1744)
EPS non-market							3.1558*** (0.8699)	0.7220*** (0.2601)	4.8830*** (0.9087)
Other energy									
Renewable	0.0043 (0.0057)	0.0064 (0.0058)	-0.0002 (0.0021)	0.0197 (0.0132)	0.0205 (0.0138)	0.0008 (0.0037)	-0.0090 (0.0189)	0.0023 (0.0181)	-0.0113** (0.0033)
Oil	-0.0019 (0.0014)	-0.0016 (0.0013)	-0.0007 (0.0005)	-0.0044 (0.0030)	-0.0035 (0.0026)	-0.0049* (0.0029)	-0.0003 (0.0041)	-0.0020 (0.0040)	0.0017** (0.0006)
Macroeconomic policies									
Financial	0.0005 (0.0003)	0.0006* (0.0003)	0.0000 (0.0001)	0.0007 (0.0005)	0.0009* (0.0005)	-0.0000 (0.0001)	-0.0002 (0.0008)	0.0003 (0.0007)	-0.0005** (0.0001)
Openness	-0.0007** (0.0003)	-0.0010** (0.0004)	-0.0000 (0.0001)	-0.0006 (0.0004)	-0.0010* (0.0005)	0.0000 (0.0001)	-0.0001 (0.0020)	-0.0003 (0.0019)	0.0002 (0.0002)
Market size									
GDP	0.0002 (0.0018)	0.0009 (0.0020)	-0.0005 (0.0006)	0.0057 (0.0072)	0.0053 (0.0080)	0.0018 (0.0020)	-0.0002 (0.0082)	0.0003 (0.0078)	-0.0005 (0.0013)
Population density	-0.0020 (0.0013)	-0.0015 (0.0013)	-0.0004 (0.0004)	-0.0009 (0.0015)	-0.0000 (0.0014)	-0.0004 (0.0005)	-0.0033 (0.0062)	-0.0030 (0.0061)	-0.0002 (0.0009)
Urban	0.0012 (0.0008)	0.0012 (0.0009)	0.0004 (0.0003)	0.0010 (0.0011)	0.0011 (0.0012)	0.0005 (0.0004)	-0.0011 (0.0051)	0.0018 (0.0046)	-0.0029** (0.0009)
Agricultural productivity									
Cereal yield	0.0666*** (0.0093)	0.0627*** (0.0100)	0.0291*** (0.0054)	0.0360*** (0.0083)	0.0404*** (0.0084)	0.0122*** (0.0043)	0.0625*** (0.0169)	0.0659*** (0.0178)	0.0170** (0.0074)
σ	4.0946*** (0.2055)	3.3494*** (0.2250)	3.7872*** (0.2097)	3.5999*** (0.2561)	3.5530*** (0.2605)	4.8731*** (0.3027)	4.3027*** (0.4997)	2.0114*** (0.5537)	4.7753*** (0.4480)
Constant	0.1065*** (0.0097)	0.1082*** (0.0110)	0.0368*** (0.0037)	0.1042*** (0.0121)	0.1082*** (0.0140)	0.0286*** (0.0039)	-0.7818 (0.7648)	-0.6046 (0.7333)	-0.1786 (0.1098)
Observations	112	112	112	86	86	86	20	20	20

Note: [a] British civil law is excluded. Standard errors in parentheses; $^{*}p<0.10$, $^{**}p<0.05$, $^{***}p<0.01$.

Source: Author's calculations based on data from EIA (2015a); ICRG (2015); Botta and Koźluk (2014); La Porta et al. (1999); and WDI (2015).

Since legal origin does not vary over time, we analyse its impact on the production of bioenergy in the long term. Moreover, we believe that the impact of these historical factors should be more noticeable over the long term. To do this, we take the average of all the variables used in the estimation across the reporting period. This produces a new bioenergy variable that is censored (44 and 54 per cent of countries do not produce bioenergy in the overall sample and the developing countries sample, respectively). We use a Tobit model in cross-section to estimate the impact of legal origin in these two samples and ordinary least square of developed countries.

Legal origin only affects the production of biodiesel in the total sample. It has no effect on biodiesel in developing countries, but it affects the production of bioenergy in developed countries. As expected, countries whose laws are of Scandinavian origin produce relatively more biodiesel than countries whose laws are of British origin. Other legal origins have no significant effects on the production of bioenergy.

9.4 SUMMARY AND POLICY IMPLICATIONS

The adverse effects of climate change and the need to find an alternative to fossil fuels have significantly increased the interest of states and investors in bioenergy. The objective of this study was to analyse the impact of political economy factors on bioenergy production. Our methodological contribution is a theoretical and empirical analysis of the determinants of bioenergy production.

First, we showed theoretically that governance quality and premium price (EPS instruments) promote the development of bioenergy production. We also showed that the factors favouring oil production (oil reserves and low price of oil) negatively influence the development of bioenergy production. This result is due to these products being substitutable. The conditions of production (cereal yield) and demand (GDP, population density, and urbanization) tend to favour the production of bioenergy.

Second, we showed empirically that the political economy factors (governance quality, EPS instruments, and macroeconomic policies) create the necessary conditions (subsidies, standards, less corruption, strong properties rights, effective governance, etc.) for development and investment in bioenergy. However, once the investment decision is made, factors of supply and demand determine the quantity produced. Indeed, cereal yield increases production by reducing production costs when the purchasing power of the population (per capita income), urbanization, and population density increase demand.

Third, we showed that bioenergy production is guided by the factors of demand (market size) in developed countries, but in developing countries the impact of supply factors (cereal yield, financial development, and governance quality) is more important.

Bioenergies are not a magic solution to the fight against climate change due to the relatively high costs of these energies and indirect changes in land uses that could cause increase in greenhouse gas emissions. However, they can contribute to the fight against global warming and the fight against poverty in developing countries if technical innovations reduce their production costs and measures taken to minimize indirect land use changes. Therefore, in developing countries, global and national efforts should be made to attract more investors in this area, including: (i) research and innovation to reduce production costs, (ii) adding bioenergy to national energy policies, (iii) creating a legal framework for their development, (iv) promoting the development of the application, and (v) providing a basis for learning more about the development of viable and sustainable bioenergy models. In this context, sub-Saharan Africa (SSA) has an important role to play. It is lagging in the production of bioenergy, despite its enormous potential (<1 per cent of world production). According to Fischer and Shah (2010), 445 million hectares of land area is available for agriculture worldwide, 201 million hectares of which is in SSA. These are not protected areas or forests, they are not yet being used for agriculture, and they have a low population density (less than 25 inhabitants per square kilometre). The development of these lands could significantly increase the production of bioenergy and contribute to the growth of these countries. For example, using a computable general equilibrium model, Arndt, Msangi, and Thurlow (2010) showed that the expansion of biofuels production in Mozambique and Tanzania can contribute to boost growth, reduce the dependence on imports of fossil fuels, increase investment, and positively influence the agricultural sector and processors downstream. However, as pointed out by Jumbe, Msiska, and Madjera (2009), bioenergy development in SSA could have a negative impact on food security. Furthermore, no regulatory framework should come at the expense of small rural farmers. In this context, it is important for each country to develop its own bioenergy policies, taking into account the need to provide food at acceptable prices for its citizens. Such policies should encourage local and foreign investments that benefit the entire agricultural sector, while protecting small farmers.

APPENDIX A: DESCRIPTION AND SOURCES OF DATA, SUMMARY STATISTICS, AND LIST OF COUNTRIES

Table 9.A1. Description and sources of data

Variable	Description	Source
Bioenergy	Bioenergy per capita (1000 barrels per year)	EIA (2015a)
Ethanol	Fuel ethanol per capita (1000 barrels per year)	EIA (2015a)
Biodiesel	Biodiesel per capita (1000 barrels per year)	EIA (2015a)
Renewable	Renewable energy per capita (billion kilowatt-hours)	EIA (2015a)
Governance	Aggregated governance index	Author's calculations using ICRG (2015) data
Bureaucracy quality	Bureaucracy quality	ICRG (2015)
Government stability	Government stability	ICRG (2015)
Corruption	Corruption	ICRG (2015)
Investment profile	Investment profile	ICRG (2015)
Law and order	Law and order	ICRG (2015)
Democratic accountability	Democratic accountability	ICRG (2015)
EPS market	Market-based instrument of environmental policy stringency (EPS)	Botta and Koźluk (2014)
EPS non-market	Non-market-based instrument of EPS	Botta and Koźluk (2014)
Legal origin	Dummy variables identify the legal origin of law of a country as British, French, German, Scandinavian, or Socialist	La Porta et al. (1999)
Financial	Domestic credit to private sector by banks (percentage of gross domestic product, GDP)	WDI (2015)
Openness	Trade openness: imports plus exports in percentage of GDP	WDI (2015)
Oil	Per capita crude oil proved reserves (barrels)-to-crude oil prices ratio	EIA (2015a)
GDP	Real GDP per capita (USD 1000, constant 2005)	WDI (2015)
Pop_dens	Population density (1000 people per square kilometre of land area)	WDI (2015)
Urban	Urban population (percentage of total population)	WDI (2015)
Cereal_yield	Cereal yield (1000 kilogrammes per hectare)	WDI (2015)

EIA: Energy Information Administration (United States); ICRG: International Country Risk Guide; WDI: World Development Indicators; GDP: gross domestic product.

Source: Author's compilation.

REFERENCES

Acemoglu, D., S. Johnson, and J. A. Robinson (2001). 'The Colonial Origins of Comparative Development: An Empirical Investigation'. *The American Economic Review*, 91(5): 1369–401.

Arndt, C., S. Msangi, and J. Thurlow (2010). 'Are Biofuels Good for African Development? An Analytical Framework with Evidence from Mozambique and Tanzania'. WIDER Working Paper 2010/110, UNU-WIDER, Helsinki.

Baum, M. A. and D. A. Lake (2003). 'The Political Economy of Growth: Democracy and Human Capital'. *American Journal of Political Science*, 47(2): 333–47.

Botta, E. and T. Koźluk (2014). 'Measuring Environmental Policy Stringency in OECD Countries: A Composite Index Approach'. OECD Economics Department Working Paper 1177, OECD Publishing, Paris.

Christopoulos, D. K. and E. G. Tsionas (2004). 'Financial Development and Economic Growth: Evidence from Panel Unit Root and Cointegration Tests'. *Journal of Development Economics*, 73(1): 55–74.

EIA (Energy Information Administration) (2015a). 'International Energy Statistics (IES)'. US Department of Energy, US (EIA), Washington, DC. Available at: <http://www.eia.gov/cfapps/ipdbproject/IEDIndex3.cfm> (accessed 15 February 2015).

EIA (2015b). 'International Energy Statistics (IES), Renewables, Biofuels Production, 2008–2012'. US Department of Energy, US Energy Information Administration (EIA), Washington, DC. Available at: <http://www.eia.gov/cfapps/ipdbproject/IEDIndex3.cfm?tid=79&pid=79&aid=1> (accessed 22 May 2015).

EPA (Environmental Protection Agency) (2010). 'Renewable Fuel Standard Program (RFS2) Regulatory Impact Analysis'. US Environmental Protection Agency (EPA), Washington, DC. Available at: <http://www.epa.gov/sites/production/files/2015-08/documents/420r10006.pdf> (accessed 5 March 2016).

Fischer, G. and M. Shah (2010). 'Part 2: Statistical Annex'. *Farmland Investments and Food Security*. World Bank, Washington, DC. Available at: <http://documents.worldbank.org/curated/en/2010/08/14999367/farmland-investments-food-security> (accessed 8 March 2016).

Gan, J. and C. T. Smith (2011). 'Drivers for Renewable Energy: A Comparison among OECD Countries'. *Biomass and Bioenergy*, 35: 4497–503.

Grafton, R. Q., T. Kompas, and N. V. Long (2012). 'Substitution between Biofuels and Fossil Fuels: Is There a Green Paradox?' *Journal of Environmental Economics and Management*, 64(3): 328–41.

Greene, W. H. (2002). *Econometric Analysis*. Fifth Edition. Upper Saddle River, NJ: Prentice Hall.

Heun, M. K. and M. de Wit (2012). 'Energy Return on (Energy) Invested (EROI), Oil Prices, and Energy Transitions'. *Energy Policy*, 40(C): 147–58.

Honoré, E. B. (1992). 'Trimmed Lad and Least Squares Estimation of Truncated and Censored Regression Models with Fixed Effects'. *Econometrica*, 60(3): 533–65.

Hotelling, H. (1931). 'The Economics of Exhaustible Resources'. *Journal of Political Economy*, 39(2): 137–75.

ICRG (2015). 'International Country Risk Guide'. The PRS Group, New York. Available at: <http://www.prsgroup.com/about-us/our-two-methodologies/icrg> (accessed 22 May 2015).

IPCC (Intergovernmental Panel on Climate Change) (2014). 'Climate Change 2014: Synthesis Report' (Contribution of Working Groups I, II, and III to the Fifth Assessment Report of the Intergovernmental Panel on Climate Change, edited by R. K. Pachauri and L. A. Meyer). IPCC, Geneva. Available at: <http://www.ipcc.ch/pdf/assessment-report/ar5/syr/AR5_SYR_FINAL_All_Topics.pdf> (accessed 5 March 2016).

Jumbe, C. B., F. B. Msiska, and M. Madjera (2009). 'Biofuels Development in Sub-Saharan Africa: Are the Policies Conducive?' *Energy Policy*, 37(11): 4980–6.

Kere, E. N. (2016). 'Do Political Economy Factors Matter in Explaining the Increase in the Production of Bioenergy?' WIDER Working Paper 2016/25, UNU-WIDER, Helsinki.

La Porta, R., F. L. de Silanes, and A. Shleifer (2008). 'The Economic Consequences of Legal Origins'. *Journal of Economic Literature*, 46(2): 285–332.

La Porta, R., F. L. de Silanes, A. Shleifer, and R. W. Vishny (1999). 'The Quality of Government'. *Journal of Law, Economics and Organization*, 15(1): 222–79.

Nelson, M. A. and R. D. Singh (1998). 'Democracy, Economic Freedom, Fiscal Policy, and Growth in LDCs: A Fresh Look'. *Economic Development and Cultural Change*, 46(4): 677–96.

Powell, J. L. (1984). 'Least Absolute Deviations Estimation for the Censored Regression Model'. *Journal of Econometrics*, 25(3): 303–25.

Powell, J. L. (1986). 'Symmetrically Trimmed Least Squares Estimation for Tobit Models'. *Econometrica*, 54(6): 1435–60.

WDI (2015). 'World Development Indicators, World Databank'. World Bank, Washington, DC. Available at: <http://databank.worldbank.org/data/reports.aspx?source=world-development-indicators> (accessed 22 May 2015).

Wooldridge, J. M. (2010). *Econometric Analysis of Cross Section and Panel Data*. Second Edition. Cambridge, MA: The MIT Press.

10

Understanding Indicator Choice for the Assessment of RD&D Financing of Low-Carbon Energy Technologies

Lessons from the Nordic Countries

Jonas Sonnenschein

10.1 INTRODUCTION

Many climate scenarios show potential pathways to limit global warming to two degrees as stipulated in the Paris Agreement of 2015 (Edenhofer et al. 2014). These scenarios have in common that in order to decarbonize industrialized economies, further research, development, and demonstration (RD&D) of low-carbon energy technologies (LCET) and of technologies inducing 'negative emissions' are urgently needed (Clarke et al. 2014; Anderson 2015). In addition to new technological solutions, the speed of deployment and the integration of solutions into the energy system are critical factors in climate change mitigation.

When decarbonization scenarios go beyond technological feasibility and economic factors are accounted for, the focus is often on costs and additional investment needs (Gupta et al. 2014). The importance of low abatement costs is well-reflected in some of the main climate policy instruments, such as carbon-energy taxation, carbon trading, and green and white certificate schemes (Somanathan et al. 2014), which induce marginal changes in price structures. From a national perspective, the focus on low-cost abatement is justified as domestic climate change-related benefits do not outweigh the costs of the unilateral adoption of more expensive abatement options (Stavins 2014). Weak (carbon) price signals and—in the case of trading schemes—price fluctuations do not create sufficient incentives to fix the market failures in the generation of LCET change (Jaffe, Newell, and Stavins 2005), leading to underinvestment in RD&D and innovation. Due to spill-overs the social rate of return of RD&D investments is often higher than the commercial return rate (Griliches 1992). Moreover, it is in many cases particularly difficult to finance LCET as it has high capital requirements and a long time to market (Ghosh and Nanda 2010).

In order to scale up RD&D activity in this area, it is critical to know if governmental intervention can correctly identify RD&D initiatives with high social returns that are under-supplied with financing from the market. Thus, it is relevant to understand both the motivation for setting up new public RD&D support instruments and how their success is assessed and measured. As success is a normative concept, different stakeholders may have their own specific criteria or indicators for success of public RD&D in this context.

While there are various methods for evaluating the performance of RD&D support policies, many of them rest on few aggregated indicators, such as public and private RD&D expenditure as well as patent counts (Bozeman and Melkers 1993). These indicators alone do not reflect the complexity and dynamics of public RD&D, let alone innovation processes (Gallagher, Holdren, and Sagar 2006). The quantitative estimation of innovation policy indicators has been frequently criticized for rarely coming to conclusions with high policy relevance (Bergek et al. 2008).

The approach of *evaluating* indicators addresses this criticism without completely abolishing the indictor-based method. Indicator evaluation in the field of LCET RD&D is neither very far developed nor tested. Notable attempts are: Gallagher, Holdren, and Sagar (2006), who discuss the merits of various input, output, and outcome metrics but do not apply a uniform indicator evaluation framework; Wilson et al. (2012: 781), who roughly estimate the suitability of various indicators to research 'directed innovation efforts in response to climate change mitigation'; and Carley, Brown, and Lawrence (2012), who propose an evaluation framework for 'energy-based economic development' which includes the categorization of relevant indicators but not an actual indicator evaluation.

The purpose of this study is to assess the performance of indicator-based evaluation in the context of LCET support policies and to contribute to the structured assessment of potential indicators.

The Nordic countries have been chosen as a geographic area of study. Norway, Sweden, Finland, and Denmark have innovation ecosystems in place which provide dedicated support to LCET. They perform well on indexes related to eco innovation, such as the Global Green Economy Index 2016 (Dual Citizen LLC 2016), the 2014 Global Cleantech Innovation Index (WWF and Cleantech Group 2014), and the EU Eco-Innovation index (European Commission 2015). However, in Finland and Sweden in particular, the gaps between the evidence of emerging cleantech innovation and the evidence of commercialized cleantech innovation are large (WWF and Cleantech Group 2014). Within cleantech the sub-sector of LCET is particularly challenging due to long times to market and high capital requirements (Ghosh and Nanda 2010). Both the success of their cleantech industry and the remaining challenges in commercialization make the Nordic countries a suitable case study to identify and analyse indicators for the assessment of RD&D support policies.

Section 10.2 in this chapter includes the research design. In Section 10.3 the analysis of the indicator-based evaluation framework is presented. Section 10.4 discusses policy implications of indicator choice and Section 10.5 concludes.

10.2 RESEARCH DESIGN: THE INDICATOR-BASED EVALUATION FRAMEWORK

The research was framed as an exploratory case study of public RD&D financing of LCET in the Nordic countries. The study is constructed around indicator-based evaluation, confronting a literature review of indicators in RD&D policy evaluation with the actual usage of indicators in the Nordic countries. In order to enhance the understanding of indicator choice, an assessment of the indicator-based evaluation method was performed. Both primary and secondary data were collected to understand the respective funding instruments, their performance, and indicators used for their evaluation.[1]

10.2.1 Conceptualization of Indicator-Based RD&D Policy Evaluation

The multitude of indicators that is used in the assessment of RD&D policy can be categorized in different ways. A common differentiation is made between input, outcome, and impact indicators (Fischer 1995; Guedes et al. 2001; Neij and Åstrand 2006; Miedzinski et al. 2013). Another (complementary) approach to conceptualize the use of indicators is to view them as a way to operationalize criteria for policy evaluation (Mickwitz 2003). Relevant criteria that were used to structure this study are administrative capacity, effectiveness, and additionality.

It is debatable whether *administrative capacity* should be seen as an evaluation criterion as such or as a 'determinant of implementation' (Vedung 2000: 226). Following the IPCC (Intergovernmental Panel on Climate Change) (Kolstad et al. 2014), it was used as a criterion in this study. *Effectiveness* refers

[1] Further information about the case study, data collection and limitations of this study can be found in Sonnenschein (2016).

to the degree to which 'achieved outcomes correspond to the intended goals of the policy instrument' (Mickwitz 2003: 426). For this study also indicators were included, which were *de facto* used to assess the effects of a financing instrument but for which no explicit goals were formulated. Due to the large number of potential indicators for effectiveness this criterion is frequently subdivided into environmental effectiveness, technological progress, and commercial effectiveness (Carlsson et al. 2002: 243; Carley, Brown, and Lawrence 2012: figure 2). The *additionality* criterion complements effectiveness. It is the degree to which achieved outcomes differ from a baseline development that assumes the absence of the respective policy instrument. The challenge of attributing specific developments to individual policy instruments is large (Scriven 1991). Still, additionality is a core criterion to establish accountability for the success or failure of RD&D support policies.

10.2.2 Indicators Used in the Evaluation of LCET Support Policy

A comprehensive review of potential indicators used to assess LCET RD&D support is presented in Table 10.1. The table excludes social indicators and environmental indicators other than the ones related to greenhouse gas emissions. It differentiates between national level indicators and programme-level indicators and is structured according to the evaluation criteria presented in Section 10.2.1.

10.2.3 Key Indicators in the Context of Public RD&D Financing of LCET in the Nordic Countries

In order to reduce the scope of this study and increase its relevance, only the most salient indicators in the case study of LCET RD&D support in the Nordic countries were analysed. Moreover, only numeric indicators were chosen; and indicators included in the analysis had to be relevant at both national and programme level. Selected indicators included RD&D spending, CO_2 emissions, patents, commercial indicators (turnover, exports and jobs), return on investment (ROI), and the ratio of public and private RD&D. Further clarification about indicator choice follows (while the actual analysis of these indicators is presented in Section 10.3):

- The indicator 'CO_2 emissions', in this case, refers to CO_2 emissions from fossil fuel combustion determined with a production-based approach.
- The commercial indicators 'turnover', 'exports', and 'jobs' were grouped together as they are typically part of the same accounting system at the

Table 10.1. Indicators used in the literature on LCET support policies

	National level	Programme level
Administrative Capacity	*RD&D spending* *RD&D staff (and their formal qualification)*	*RD&D spending* *RD&D staff (and their formal qualification)*
	Effectiveness	
Environmental effectiveness	CO_2 emissions	CO_2 emissions
	CO_2 intensity of energy supply CO_2 intensity of the economy	
Technological progress	Patents (filed, granted, cited)	Patents (filed, granted, cited)
	Scientific papers (incl. PhD theses)	Scientific papers (incl. PhD theses)
	Learning rates	
	Technology/abatement costs	Technology/abatement costs
	Technology performance/efficiency	Technology performance/efficiency
	Energy efficiency/intensity of the economy	
Commercial effectiveness	Jobs	Jobs
	Exports	Exports
	Turnover	Turnover
		Turnover/employee (productivity)
	Profits	Profits
	Return on investment	Return on investment
	Number of enterprises	
	Energy cost savings	Energy cost savings
Other	Energy self-sufficiency Share of renewable energy in energy supply	
	Additionality	
	Ratio of public and private RD&D spending	Ratio of public and private RD&D spending
	Jobs per energy output Net employment effect Macroeconomic multipliers	
		Scale and timing of private sector RD&D activity

Source: Author's compilation based on Stosic et al. (2016); Wilson et al. (2012); Carley et al. (2011) ; Carley et al. (2012); Gallagher et al. (2006); Neij and Åstrand (2006); Jacobsson and Rickne (2004); Spangenberg (2004); Kleinknecht, Van Montfort, and Brouwer (2002); Schoenecker and Swanson (2002); and Grupp (2000).

national level; also at the programme level they are often measured and presented together.

- As well as RD&D budgets, administrative capacity, in terms of knowledge and skills, was highlighted in the interviews as a key input factor for the success of public interventions, but did not seem to be reflected in evaluations. Input indicators related to administrative capacity, such as the number and qualification of fund managers and public officers in RD&D schemes (Gallagher, Holdren, and Sagar 2006), were not frequently used.
- Technological output indicators, other than patent counts, are bibliometric indicators and the number of supported PhDs, neither of which were explicitly used in this specific case and are potentially subject to large biases (Jacobsson and Rickne 2004).
- Only one additionality indicator was chosen to be part of this study, since no data on output and outcome additionality of publically financed RD&D programmes could be obtained.

10.2.4 Assessment of the Indicator-Based Method

Once indicators used to assess RD&D support to LCET in the Nordic countries were identified, categorized, and selected, they were analysed in order to assess the indicator-based evaluation method. The analysis focused on the *acceptance* of relevant stakeholders, on the *ease of monitoring* an indicator, including measurability and data availability, and on an indicator's *robustness* against manipulation. This evaluation approach was inspired by the 'RACER framework' for indicator choice in impact assessments (European Commission 2005). RACER stands for relevant, accepted, credible, easy to monitor, and robust. Both relevance and measurability were also suggested as criteria for the assessment and selection of green growth indicators (GGKP 2013).

The acceptance of an indicator was included in the analysis, since the results of an assessment that is based on poorly accepted indicators is not likely to resonate with key stakeholders and tends to have less policy impact. Moreover, indicators that are difficult to monitor or can only be monitored at very high costs are less likely to be applied in evaluations. The more expensive it gets to monitor the development of indicators, the harder it gets to justify resource use for evaluation. In contrast, the robustness of an indicator does not have immediate influence on programme evaluation, as less robust indicators can still be influential if they are widely accepted and monitored. Still, robustness is crucial from the academic perspective as indicators that are not robust may not provide conclusive indications for the (re-)design of LCET support schemes. Moreover, manipulation of indicators may eventually erode acceptance.

10.3 ANALYSIS: INDICATORS FOR PUBLIC RD&D FINANCING OF LCET IN THE NORDIC COUNTRIES

As outlined in Section 10.1, the development of the LCET sector in the Nordic countries is generally perceived as a success story. In contrast, the role of public RD&D financing in this story is more difficult to grasp as it has not been comprehensively researched. This study contributes to the evaluation of RD&D financing of LCET in the Nordic countries by scrutinizing the use of indicators rather than by presenting a comprehensive indicator-based evaluation as such. Hence, specific performance data from the case study is merely used to illustrate the use of indicators and their assessment.[2]

Indicators are analysed and ranked according to the criteria acceptance, ease of monitoring, and robustness (see Table 10.2). The estimation of indicators is presented on an ordinal three-point scale (zero, one, or two stars). The results represent the specific case of RD&D financing of LCET in the Nordic countries. Generalizability beyond LCET in the Nordic countries is particularly limited in the case of acceptance, while similar results can be expected for the criteria ease of monitoring and robustness if the study is repeated in a different context.

While most of the results are indicator-specific, there are some cross-cutting results, in particular with respect to robustness. First, the assessed performance may vary significantly depending on the definition of LCET, which is sometimes also referred to as green energy or clean energy technology. The decision to include controversial and capital-intensive technologies such as carbon capture and storage (CCS) or nuclear energy in the definition can make a large difference. Time-lags are another aspect that influences the robustness of indicators. While inputs into LCET RD&D are visible right away, outcomes and impacts of RD&D support programmes manifest themselves only after several years. Finally, for all aggregated indicators there is the challenge of attribution. It is virtually impossible to separate the effects induced by individual support schemes from other factors such as larger business cycles and general technological progress. In Sections 10.3.1–10.3.6, the schematic overview of results (Table 10.2) is substantiated for each of the six indicators.

10.3.1 RD&D Spending

10.3.1.1 Acceptance

RD&D spending is a widely accepted indicator in the Nordic countries. Policy makers have stressed the leading role of the Nordics in LCET RD&D by

[2] Further case-specific data and figures are included in Sonnenschein (2016).

Table 10.2. Overview of assessment indicators for public RD&D support to LCET, their acceptance, ease of monitoring, and robustness

Indicator	Acceptance	Ease of monitoring	Robustness
		Administrative Capacity	
RD&D budgets	**	**	*
		Effectiveness	
CO_2 emissions	*	*/**	*
Patents	*	**	**
Turnover, exports, jobs	**	*	o
ROI	o	*	o
		Additionality	
Ratio of public & private RD&D	**	*	*
		Legend	
	Indicator is:	*Indicator is:*	*Indicator:*
two stars (**)	*widely accepted by various stakeholders.*	*measurable and data is available.*	*is difficult to manipulate.*
one star (*)	*partially accepted by the stakeholders.*	*measurable but good data is not available.*	*can be manipulated but robustness can be tested.*
no star (o)	*only brought forward by one type of stakeholder.*	*not measurable.*	*is very prone to manipulation.*

Source: Author's analysis.

referring to budget allocations, academics have frequently used RD&D budget data in econometric studies of innovation activity, and public officers in LCET support programmes as well as fund managers stressed the particular role of public RD&D budgets for energy technology innovation in the interviews. At the level of individual programmes, larger public budgets are mostly, but not always, perceived as desirable. The success of commercialization support programmes, for instance, largely depended on the existence of suitable innovative enterprises. By increasing budgets and, hence, the number of supported enterprises, the risk of picking less-promising enterprises increases.

10.3.1.2 Ease of Monitoring

Comprehensive data on national energy RD&D spending of Nordic countries is reported to and published by the International Energy Agency on an annual basis (IEA 2015b). The resolution of the data is fine enough to differentiate between LCET and other energy technologies. RD&D spending data is also available at the programme level, even though it is scattered, so that it requires some data-gathering effort to obtain a systematic overview.

10.3.1.3 Robustness

The particular presentation of RD&D spending data leaves room for manipulation. Today's share of LCET RD&D in GDP is, for instance, very high in the Nordic countries compared to other industrial states (IEA 2015b). However, when comparing to historic data, it was three times higher in Sweden in the early 1980s, which has to be seen in the context of the oil crises (IEA 2015a).

RD&D spending on LCET can also be compared to overall public RD&D spending, which represents about 3–3.5 per cent of GDP in the Nordic countries (as compared to 0.03–0.11 per cent for LCET). At the time of the oil crises, energy R&D made up more than 10 per cent of overall R&D both in Europe and the Americas, a ratio that has dropped to 2 per cent and 3 per cent respectively (IEA 2015a).

Absolute RD&D spending is very low in the Nordic countries as compared to larger countries. The US loan guarantees of US$535m to solar cell producer Solyndra and of US$465m to electric car manufacturer Tesla (Rodrik 2014) exceeded the current capacity of the Nordic countries' RD&D budgets, which seem even smaller in comparison to the support that China grants to some of its renewable energy companies, for example US$9.1bn to LDK Solar, US$7.6bn to Suntech Power, and US$7bn to Yingli Solar (Sanderson and Forsythe 2013).

Moreover, RD&D financing for LCET may well be concentrated in a few lighthouse projects, as, for example, CCS funding is in Norway, which made up more than half of RD&D to LCET between 2009 and 2012 (IEA 2015b). In contrast, it was observed that there was too little public funding for early stage enterprises that have already received seed-funding but often have difficulties securing follow-up financing (Grünfeld, Iverson, and Grimsby 2011).

Finally, RD&D spending is not adjusted for the respective costs of conducting RD&D, for example the costs for employing research staff, which are significantly higher in countries like Sweden as compared to many other European countries (Jacobsson and Rickne 2004).

10.3.2 CO_2 Emissions

The most apparent indicator for assessing the environmental effectiveness of public RD&D financing of LCET in the Nordic countries is the development of CO_2 emissions from fossil fuel combustion. It is often presented in relation to GDP growth in order to account for the size of the respective economy.

Both CO_2 emissions and the emissions intensity of the economy fell in all Nordic countries but Norway between 2000 and 2014 (IEA 2015c).

10.3.2.1 Acceptance

CO_2 emissions are not widely accepted as a significant impact indicator. On the one hand, investigated policy programmes and the laws in which they are enshrined do refer to the reduction of CO_2 emissions, and also academics comprehensively discuss the role of technology push policies for reducing CO_2 emissions. On the other hand, emission reductions do not play a major role at the programme and project level. The interviews revealed that the reduction of CO_2 emissions is seen as a 'by-product' of the (economic) success of supported enterprises and not as an indicator for success in itself.

10.3.2.2 Ease of Monitoring

CO_2 emissions data is certainly measurable and available at the national level in the Nordic countries but difficult to measure at the programme level, as the lion's share of emission reduction typically does not take place in RD&D projects but indirectly through selling and deploying LCETs on domestic and international markets. Only few programmes included CO_2 emissions in their assessment, for example Enova Norway's support for 'new energy technology', which monitored energy savings and CO_2 emission reductions both in absolute terms and in relation to provided funding (Enova 2015).

10.3.2.3 Robustness

While national-level emissions data is rather robust and an established system for monitoring, reporting, and verification is in place in all Nordic countries, there is a lot of room for manoeuvre at the programme level. Either direct or induced emissions reductions may be monitored, at both the national and international level. Moreover, the choice of the baseline for evaluating reductions, and not merely monitoring them, leaves room for manipulation. Base years may vary and business as usual scenarios rest on many assumptions.

10.3.3 Patents

All Nordic countries multiplied their share of low-carbon technology patents in total patents between 1999 and 2011, reaching about 10 per cent in 2011 (OECD 2015). This suggests that within the Nordic countries LCET became a more significant area of innovation, which may be partly driven by additional public RD&D financing in this sector. This trend is not restricted to the

Nordic countries, but it is likely more pronounced than in many other countries, so that the 'relative technological advantage' of Nordic countries in LCET may well have strengthened in this period (Haščič and Migotto 2015: 30).

10.3.3.1 Acceptance

At the national level, patents are frequently used as proxies for technological progress, both by academics and government agencies. The situation is different at the programme level where patents are mainly regarded as a means to an end. Even if not seen as ends in themselves, patents and the process of protecting intellectual property rights do play a role in the RD&D support that is provided to LCET in the Nordic countries. Patents are simply not regarded as a relevant indicator for success at the programme level.

10.3.3.2 Ease of Monitoring

Patent data of LCET is available at the national level and published regularly. In contrast, patent data is not made available in a systematic way at the programme level, so that the attribution of patents to public support instruments becomes difficult. The Finnish national innovation funding agency TEKES monitors the overall number of patents registered by supported organizations but does not provide a specific breakdown for LCET (TEKES 2015). A Danish study of the green economy compares innovation activity and patenting of green enterprises to all enterprises, showing that the trading of patents and intellectual property rights plays a larger role in green enterprises than in the overall economy (Danish Energy Agency 2012: 38).

10.3.3.3 Robustness

Patents are a robust indicator. Data is available, it can be rather easily verified so that there is little room for manipulation, and patents can to some extent be attributed to RD&D projects. Still there is a risk that funding agencies account for a full patent in cases in which they provided only a minor share of the overall project budget.

10.3.4 Turnover, Exports, Jobs

Turnover in the LCET sector, its jobs and exports are frequently used indicators in the context of RD&D financing instruments. The most developed and standardized way to measure the commercial development of subsectors of the green economy is provided in national statistics about the Environmental

Goods and Services Sector (EGSS), which is defined in the statistical guidelines of Eurostat (Eurostat 2009).

10.3.4.1 *Acceptance*

Various stakeholders stress the commercial dimension of RD&D financing of LCET. There is virtually no public support programme in the Nordic countries that does not explicitly refer to economic development. The political emphasis of commercial aspects is a view that was reaffirmed in the interviews where public officers stressed the role of commercialization potential. Even in academia the focus is increasingly put on the commercialization aspect of publically funded RD&D (Jacobsson, Lindholm-Dahlstrand, and Elg 2013).

10.3.4.2 *Ease of Monitoring*

Turnover, jobs, and exports in the LCET sector are measurable and some data is available at both the national level (in statistics on the EGSS) and at programme level. However, available data is scattered and cross-country comparisons are not possible. Sweden is the only Nordic country that has collected comprehensive data on its EGSS for more than a decade, including specific data on the subsectors renewable energy, and energy savings. The Danish EGSS statistics only cover the years 2012–14, the Finnish statistics do not include the subsectors renewable energy and energy efficiency, yet, and in Norway the statistics office is preparing for the first publication of EGSS data in 2017. The lack of official data from statistics offices is partly compensated for with data from industry associations (Mellbye and Espelien 2013; Cleantech Finland 2014).

Specific programme evaluations sometimes also include the economic outcomes of RD&D support programmes. The Danish Business Innovation Fund, which financed mainly green economy enterprises in 2010–12, required, for instance, all supported enterprises to communicate five-year turnover and employment targets. These targets were summarized and followed up in a mid-term evaluation (Deloitte 2012), but no further evaluation with actual data is available, yet. This example illustrates a typical challenge of programme evaluations. Once temporary support programmes are finalized, little priority and resources are given to evaluation.

10.3.4.3 *Robustness*

As well as data availability, quality of commercial data also varies. Due to the fact that there is no standardized way to measure commercial indicators for LCET (and the whole EGSS) the data may vary between different sources.

In particular, data from grey literature tends to be less robust. One example is Norway's renewable energy sector, for which industry sources frequently report employment of 50,000, a turnover of NOK200 bn and approximately 2000 companies in 2010. This is far higher than the figures published in a more elaborated study, which found 13,700 employees, NOK85bn turnover and 860 companies in 2010 (Mellbye and Espelien 2013).

Furthermore, economic data about the LCET sector does not reflect that employment, turnover, and exports could also be generated in other sectors. The actual figures do not reflect the net effect of the respective support policies, that is its additionality, but only their gross effects. The claim that the Danish wind power sector employs more people than the Swedish automotive industry is often made in the context of job creation. This is potentially misleading as it does not say anything about the net employment effects of past wind power support policies in Denmark.

10.3.5 Return on Investment (ROI)

In the case of public equity financing instruments, ROI is an additional commercial indicator under consideration. There is no exclusive public venture capital (VC) fund for LCET in the Nordic countries, but several public VC funds have LCET companies in their portfolio. These funds typically stress that they operate like private funds and that their main objective is ROI. This supports the findings of Yang and Sollen (2013), who found strong evidence for a *de facto* profit motive in state-owned VC in the Nordics.

The track record of public VC to LCET enterprises has a rather poor image among analysts in the Nordic countries, some of whom call it a complete absence of success stories. Due to 'poor financial returns on Cleantech investments' (Murray and Cowling 2014) the Danish Growth Fund has not made any initial VC investments into cleantech since 2011, and neither has the Norwegian public VC fund Investinor. Even the performance of private VC funds that invest in cleantech is at best mixed in the Nordic countries (Wang 2015). The absence of success may have other reasons than public VC being an inappropriate support instrument, including the poor timing of investments with respect to economic cycles and long lead times in this sector, which means that there have not been many exits, yet (Murray and Cowling 2014).

10.3.5.1 Acceptance

With the exception of (state-owned) VC fund managers, little support could be gathered for taking ROI into consideration as an indicator for the effectiveness of public RD&D financing of LCET. Several stakeholders argued that the state should support those ventures that are too risky for the private sector but potentially beneficial for society. These are most likely not the ones that promise the highest returns. While it is widely accepted that profit-orientation should be the *modus operandi* for public equity funds, a general profit target is not accepted at all. In the interviews it was suggested that benefits to the state could be assessed in a different way, that is by looking at financing costs and at the indirect impact on tax revenue that is triggered by additional commercial activity.

10.3.5.2 Ease of Monitoring

The returns from public VC investments into LCET are measurable, which is straightforward after a portfolio company has been sold (exit). There are, however, large methodological challenges in estimating the current value of existing portfolios. LCETs have a long time to market so that several of the public investments in the Nordic countries could not be exited yet, which impedes the calculation of ROI. Good data for public VC investments in the Nordic countries is not available, and even less so for LCET investments in particular, since LCET investments are typically part of larger VC funds that are not specialized into energy or cleantech.

10.3.5.3 Robustness

Due to the lack of data, it is not possible to assess the actual robustness of the indicator ROI. Still, it is rather clear how the data could be manipulated and why. Fund managers have strong incentives to overestimate the current value of their portfolio, while entrepreneurs also have to portray their respective ventures as a success story in order to receive continued financing.

10.3.6 The Ratio of Public and Private RD&D Financing

Merely looking at effectiveness is not sufficient to assess the success of a policy intervention. RD&D financing instruments in the Nordic countries showed a clear attempt not only to be effective but also to both ensure the additionality of the intervention and, to a lesser extent, monitor this additionality effect. The most common indicator for the additionality of Nordic RD&D support schemes was the ratio of public and private RD&D financing, that is the consideration of whether public financing has crowded in or crowded out private financing.

10.3.6.1 Acceptance

The ratio of public and private RD&D was clearly the indicator that was used most to investigate additionality. Its role as input indicator, however, slightly reduced acceptance, as after all additional effects on the environment, technology, and economy were sought after rather than additional financial input, as such. Furthermore, the indicator is not always easy to interpret, which further reduced acceptance.

10.3.6.2 Ease of Monitoring

Both public and private RD&D financing are measurable and data is partially available, even though data on private sector RD&D spending is less comprehensive. There is no comprehensive study about private versus public sector RD&D for LCET in the Nordic countries. At the programme level the evidence from evaluations, reports, and interviews clearly suggests very high additionality of public RD&D to LCET in the Nordics. Gaps in the innovation financing cycle of cleantech were identified by various private and public investors (Finnsson 2011). Evaluations of TEKES' (Finland) financing of environmental technology (Valovirta et al. 2014) and of Innovation Norway's Environmental Technology Scheme (Espelien et al. 2014) found high degrees of additionality. In the latter case NOK1 of financing 'triggered' NOK3.6 in private investments. Moreover, in the case of Sweden, public funding seems to crowd in private capital for cleantech investments; and co-investments are particularly common in the sub-sector of energy (Yang and Sollen 2013: 59). While, at the programme level, data on private co investments in RD&D is collected and, in many cases, even has to be collected, this does not provide any information about private RD&D activity outside publicly co-financed projects.

10.3.6.3 Robustness

The ratio of public and private sector RD&D financing is a simple input indicator and as such it avoids some of the difficulties in assessing the additionality of programme outcome. However, interviewees pointed out that the interpretation of the indicator is not self-evident. Additionality of public funds is likely if the share in total RD&D financing (public and private) remains the same or even decreases. It is more difficult to interpret when the share of public RD&D increases. This could be either due to a crisis in private RD&D financing, hence pointing towards a high degree of additionality, or due to crowding out, indicating a low degree of additionality.

One example for an increased share of public financing is the development of cleantech VC in Sweden. Private VC cleantech investments in Sweden dropped from their peak at nearly 700 million SEK in 2008 to about 50 million SEK in 2014, while dedicated public VC funds increased from about 25 million SEK to 100 million SEK (Tillväxtanalys 2015). The collapse of private VC investments despite slightly increasing dedicated public VC funds provides indication for a high degree of additionality of public funds, even though they were not successful in crowding in much private funding.

Besides its ambiguity, the robustness of this indicator is further challenged by the fact that private sector RD&D financing data is largely based on self-reporting. Companies have many options for manipulating the data they report, for example increasing their budgets by inflating the staff hours they put into an RD&D project.

10.4 EFFECTS OF INDICATOR CHOICE: POTENTIAL BIASES AND THEIR POLICY IMPLICATIONS

The analysis of six common indicators in the context of RD&D financing of LCET showed that even a basic structured assessment does not result in a clear-cut indicator-based evaluation framework. Trade-offs between comprehensiveness, acceptance, ease of monitoring, and robustness are impossible to avoid.

An argument in favour of indicator-based monitoring and evaluation is that it helps to establish accountability of policy makers. If evaluations of RD&D programmes for LCET are carried out at all, they are typically based on indicators. Accepting that these indicators only represent a subset of all available indicators, moreover a subset that is faced with heavy trade-offs, it becomes clear that the mere selection of indicators can have a major impact on evaluation results. These results then feed back into the policy-making process and may trigger changes in programme design and strategic focus.

The active selection of indicators may introduce bias into indicator based evaluation. It is important to note, though, that certain biases might be justified as the specific objectives of different programmes (e.g., technological progress or economic growth) may differ. Flexibility in the computation and presentation of indicators (i.e., lack of robustness) introduces further uncertainty about the validity of assessment results. In Section 10.4.1 to 10.4.3, some potential biases and uncertainties in the evaluation of Nordic RD&D financing of LCET are discussed and possible policy implications are mentioned.

10.4.1 A Focus on Short-Term Economic Performance and ROI

In the case study, the growth of jobs, exports, and turnover, and also profitability appeared to be increasingly important indicators of the public financing of RD&D and its commercialization. At the same time private RD&D spending in the Nordic LCET sector has recently decreased and public RD&D

spending has levelled off. The interviews made clear that, in particular, public VC instruments have moved away from cleantech due to low profit expectations and long times to market. Hence, dedicated support for LCET is not likely to perform well in assessments if much attention is paid to the indicators ROI and the (short-term) development of jobs, exports, and turnover.

A bias towards these indicators largely disregards social benefits related to the development and deployment of LCET, such as resource conservation and climate change mitigation. This improves the position of other sectors in the competition for public funds. The information and communication technology (ICT) sector is, for instance, less capital-intensive and has shorter development cycles.

However, LCET-specific support and commercial success do not exclude each other in the Nordic countries. Analyses of the Danish wind energy sector and the Swedish bioenergy sector have shown that 'medium-sized countries can be within the world's leading nations in a specific field of energy technology, if appropriate supply and demand side policies support a certain technology' (Bointner 2014: 738). In order to be commercially successful, public RD&D financing of LCET likely has to be part of a more comprehensive policy mix. Accordingly, fund managers and public officers stressed in the interviews that the business plans of several supported companies could only be worked out if demand side policies were in place. Demand-side measures include feed-in-tariffs for renewable energy in Denmark and Finland, the common green certificates market of Norway and Sweden, CO_2 taxes, and deployment subsidies for various LCETs. The main policy implication of a strong focus on short-term economic performance is, hence, that (further) dedicated support to LCET is difficult to justify if there are no additional demand side policies in place.

10.4.2 Stressing the Additionality of Financing

Despite the lack of workable indicators, additionality was strongly emphasized in both interviews and reports. This was slightly surprising as there was virtually no evidence for 'crowding-out' private capital from the Nordic LCET sector. The perceived importance of additionality can be traced back to regulatory requirements stipulated in EU state aid regulation. The investigated support instruments included various institutional mechanisms to make sure that the state does not finance 'too much', including co-investment provisions, maximum aid intensities, and limited opportunities for follow-up investments.

It would be an exaggeration, though, to understand these mechanisms as a result of a bias towards additionality in evaluation. While the importance of additionality was indeed frequently stressed, actual monitoring happened, if at all, mainly for the input indicator 'ratio of public and private funding'. This supports the thesis that 'additionality can be treated ex ante as a design criterion and ex post as an area where some evidence can be collected but where full measurement may be impossible and in any case is not justified in resource terms' (Georghiou 2002: 64). It would require further discourse analysis to better understand how the frequent discussion of additionality has influenced the design of RD&D financing measures in the Nordic countries.

The main policy implication of a large emphasis on additionality indicators in the assessment of RD&D support schemes is that it may favour cautious state intervention rather than strong industrial policy push for LCET.

10.4.3 Disregarding Decarbonization

Decarbonization was a very prominent objective in the justification and communication of LCET support measures in the Nordic countries, while at the programme level climate aspects were overtrumped by innovation objectives. Accordingly, most RD&D support to LCET companies was managed by dedicated innovation agencies like TEKES, Innovation Norway, and Vinnova.

The challenge to place LCET support within a certain policy domain is well-illustrated by an evaluation of the Norwegian Environmental Technology Scheme (Innovation Norway). The hierarchy between the scheme's objectives 'environmental effect' and 'commercial potential' was not clear and the evaluators recommended 'design[ing] explicit objectives including a clear goal hierarchy as soon as possible' (Espelien et al. 2014: 6), being very outspoken that priority should be given to commercial potential. This reflects a frequently expressed view in the case study, that is that commercial success is the best strategy to assure positive environmental impact.

Moreover, previous econometric studies have shown that little direct influence of public RD&D financing on CO_2 emissions from energy can be expected (Garrone and Grilli 2010). Considering further that there are serious methodological challenges to attribute emission reductions to specific RD&D support schemes (Miedzinski et al. 2013), it was not surprising that the indicator 'CO_2 emissions' was largely disregarded in evaluations of the analysed instruments.

The potential policy implication of disregarding CO_2 emissions as an assessment indicator lies in the selection of LCETs that are worth supporting. There is a risk that the mitigation potential of a technology becomes secondary concern in the selection of support-worthy RD&D projects and enterprises.

10.5 CONCLUDING REMARKS

The main objective of this study was to assess the performance of indicator-based evaluation in the context of public RD&D financing of LCET. The

Nordic countries provided an interesting case to study the choice of indicators in policy evaluation, their acceptance, the ease of monitoring them, and their robustness. The analysis clearly showed that a structured assessment of indicators can help to point up the trade-offs and limitations that are inherent in indicator-based evaluation. Selecting indicators can introduce bias. The discussion of LCET RD&D financing in the Nordic countries illustrated how a focus on short-term economic performance may hinder (further) dedicated support to LCET, how stressing the additionality aspect of public financing may lead to rather cautious state intervention, and how the partial neglect of CO_2 emissions in evaluation may shift the focus away from the abatement potential of supported technologies.

If such biases happen to correspond with the policy objectives behind the respective instruments and programmes, they can be justified. If, on the other hand, the ambition is to act according to the targets of the 2015 Paris Agreement, more dedicated support to LCET with substantial abatement potential and bold state interventions are needed.

REFERENCES

Anderson, K. (2015). 'Duality in Climate Science'. *Nature Geoscience*, 8(12): 898–900.

Bergek, A., S. Jacobsson, B. Carlsson, S. Lindmark, and A. Rickne (2008). 'Analyzing the Functional Dynamics of Technological Innovation Systems: A Scheme of Analysis'. *Research Policy*, 37(3): 407–29.

Bointner, R. (2014). 'Innovation in the Energy Sector: Lessons Learnt from R&D Expenditures and Patents in Selected IEA Countries'. *Energy Policy*, 73(October): 733–47.

Bozeman, B. and J. Melkers (eds) (1993). *Evaluating R&D Impacts: Methods and Practice*. Boston, MA: Springer.

Carley, S., A. Brown, and S. Lawrence (2012). 'Economic Development and Energy From Fad to a Sustainable Discipline?'. *Economic Development Quarterly*, 26(2): 111–23.

Carley, S., S. Lawrence, A. Brown, A. Nourafshan, and E. Benami (2011). 'Energy-Based Economic Development'. *Renewable and Sustainable Energy Reviews*, 15(1): 282–95.

Carlsson, B., S. Jacobsson, M. Holmén, and A. Rickne (2002). 'Innovation Systems: Analytical and Methodological Issues'. *Research Policy*, 31(2): 233–45.

Clarke, L., et al. (2014). 'Assessing Transformation Pathways'. In O. Edenhofer, et al. (eds), *Climate Change 2014: Mitigation of Climate Change. Contribution of Working Group III to the Fifth Assessment Report of the Intergovernmental Panel on Climate Change*. Cambridge and New York: Cambridge University Press, pp. 443–510.

Cleantech Finland (2014). 'Cleantech Industry in Finland 2014'. Available at: <http://www.slideshare.net/cleantechfinland/cleantech-industry-in-finland-2014> (accessed 18 December 2015).

Danish Energy Agency (2012). 'Green Production in Denmark–and Its Significance for the Danish Economy'. Danish Energy Agency, Copenhagen.

Deloitte (2012). 'Mid Term Review of the Business Innovation Fund'. Danish Business Authority, Copenhagen.

Dual Citizen LLC (2016). *2016 Global Green Economy Index (GGEI)*. Available at: <http://dualcitizeninc.com/GGEI-2016.pdf > (accessed 3 November 2016).

Edenhofer, O., et al. (2014). 'Technical Summary'. In O. Edenhofer et al. (eds), *Climate Change 2014: Mitigation of Climate Change. Contribution of Working Group III to the Fifth Assessment Report of the Intergovernmental Panel on Climate Change*. Cambridge and New York: Cambridge University Press, pp. 33–107.

Enova (2015). 'Results and Activities 2014'. Enova SF, Trondheim.

Espelien, A., L. A. Grünfeld, K. Wifstad, M. Gulbrandsen, E. Kverneggen, and A. Liverud (2014). 'Veien Fra FoU Til Marked for Miljøteknologi. Evaluering Av Miljøteknologiordningen'. 10/2014. MENON Business Economics, Oslo.

European Commission (2005). 'Impact Assessment Guidelines'. SEC2005 (791), European Commission, Brussels.

European Commission (2015). 'Eco Innovation Observatory'. Available at: <http://database.eco-innovation.eu/> (accessed 15 November 2015).

Eurostat (2009). *Environmental Goods and Services Sector* (handbook). Brussels: European Commission.

Finnsson, P. T. (2011). 'Bridging the Investment Gap in Nordic Cleantech'. Available at: <http://nordicinnovation.org/sv/nyheter/586/> (accessed 12 December 2013).

Fischer, F. (1995). *Evaluating Public Policy*. Belmont: Wadsworth Group.

Gallagher, K. S., J. P. Holdren, and A. D. Sagar (2006). 'Energy-Technology Innovation'. *Annual Review of Environment and Resources*, 31(1): 193–237.

Garrone, P. and L. Grilli (2010). 'Is There a Relationship between Public Expenditures in Energy R&D and Carbon Emissions per GDP? An Empirical Investigation'. *Energy Policy*, 38(10): 5600–13.

Georghiou, L. (2002). 'Impact and Additionality of Innovation Policy'. *Innovation Science and Technology IWT Observatory*, 40: 57–65.

GGKP (Green Growth Knowledge Platform) (2013). *Moving Towards a Common Approach on Green Growth Indicators*. GGKP, Geneva.

Ghosh, S. and R. Nanda (2010). 'Venture Capital Investment in the Clean Energy Sector'. Working Paper 11-020. Harvard Business School, Cambridge, MA.

Griliches, Z. (1992). 'The Search for R&D Spillovers'. *The Scandinavian Journal of Economics*, 94: 29–47.

Grünfeld, L. A., L. M. Iverson, and G. Grimsby (2011). 'The Need for Government Supported Capital Measures in the Market for Early Stage Risk Capital in Norway'. MENON Business Economics, Oslo.

Grupp, H. (2000). 'Indicator-Assisted Evaluation of R&D Programmes: Possibilities, State of the Art and Case Studies'. *Research Evaluation*, 9(2): 87–99.

Guedes V. S., J. Martin, D. Wilkinson, and J. Newcombe (2001). 'Reporting on Environmental Measures: Are We Being Effective'. EEA, Copenhagen.

Gupta S. et al. (2014). 'Cross-Cutting Investment and Finance Issues'. In O. Edenhofer, et al. (eds), *Climate Change 2014: Mitigation of Climate Change. Contribution of Working Group III to the Fifth Assessment Report of the Intergovernmental Panel on Climate Change* Cambridge and New York: Cambridge University Press, pp. 1207–46.

Haščič, I. and M. Migotto (2015). 'Measuring Environmental Innovation Using Patent Data'. OECD Environment Working Papers, OECD, Paris.

IEA (International Energy Agency) (2015a). 'Key Trends in IEA Public Energy Technology Research, Development and Demonstration (RD&D) Budgets'. IEA, Paris.

IEA (2015b). 'RD&D Indicators'. IEA, Paris.

IEA (2015c). 'CO2 Emissions from Fuel Combustion 2015'. IEA, Paris.

Jacobsson, S., Å. Lindholm-Dahlstrand, and L. Elg (2013). 'Is the Commercialization of European Academic R&D Weak? A Critical Assessment of a Dominant Belief and Associated Policy Responses'. *Research Policy*, 42(4): 874–85.

Jacobsson, S. and A. Rickne (2004). 'How Large Is the Swedish "Academic" Sector Really? A Critical Analysis of the Use of Science and Technology Indicators', *Research Policy*, 33(9): 1355–72.

Jaffe, A. B., R. G. Newell, and R. N. Stavins (2005). 'A Tale of Two Market Failures: Technology and Environmental Policy'. *Ecological Economics*, 54(2/3): 164–74.

Kleinknecht, A., K. Van Montfort, and E. Brouwer (2002). 'The Non-Trivial Choice between Innovation Indicators'. *Economics of Innovation and New Technology*, 11(2): 109–21.

Kolstad C. et al. (2014). 'Social, Economic and Ethical Concepts and Methods'. In O. Edenhofer et al. (eds), *Climate Change 2014: Mitigation of Climate Change. Contribution of Working Group III to the Fifth Assessment Report of the Intergovernmental Panel on Climate Change* Cambridge and New York: Cambridge University Press, pp. 207–82.

Mellbye, C. and A. Espelien (2013). 'The Norwegian Cleantech Industry—Statistics 2011'. 01/2013, MENON Business Economics, Oslo.

Mickwitz, P. (2003). 'A Framework for Evaluating Environmental Policy Instruments'. *Evaluation*, 9(4): 425–36.

Miedzinski, M. et al. (2013). *A Short Guide to Assessing Environmental Impacts of Research and Innovation Policy*. Brussels: European Commission.

Murray, G. and M. Cowling (2014). 'Evaluation of the Danish Growth Fund: Evaluation of Activities, 2010–2012'. Damvad, Copenhagen.

Neij, L. and K. Åstrand (2006). 'Outcome Indicators for the Evaluation of Energy Policy Instruments and Technical Change'. *Energy Policy*, 34(17): 2662–76.

OECD (Organization for Economic Co-operation and Development) (2015). 'Patents by Main Technology and by International Patent Classification (IPC)'. OECD, Paris.

Rodrik, D. (2014). 'Green Industrial Policy'. *Oxford Review of Economic Policy*, 30(3): 469–91.

Sanderson, H. and M. Forsythe (2013). *China's Superbank: Debt, Oil and Influence—How China Development Bank Is Rewriting the Rules of Finance*. First Edition. Singapore: Bloomberg Press.

Schoenecker, T. and L. Swanson (2002). 'Indicators of Firm Technological Capability: Validity and Performance Implications'. *IEEE Transactions on Engineering Management*, 49(1): 36–44.

Scriven, M. (1991). *Evaluation Thesaurus*. Fourth Edition. Newbury Park: Sage Publications.

Somanathan E. et al. (2014). 'National and Sub-National Policies and Institutions'. In O. Edenhofer et al. (eds), *Climate Change 2014: Mitigation of Climate Change.*

Contribution of Working Group III to the Fifth Assessment Report of the Intergovernmental Panel on Climate Change. Cambridge and New York: Cambridge University Press, pp. 1141–1205.

Sonnenschein, J. (2016). 'Understanding Indicator Choice for the Assessment of Research, Development, and Demonstration Financing of Low-Carbon Energy Technologies'. WIDER Working Paper No. 48, UNU-WIDER, Helsinki.

Sonnenschein, J. and S. Saraf (2013). 'Public Cleantech Financing in Denmark, Finland and Norway'. IIIEE Report 2013:01, Lund University, Lund.

Spangenberg, J. H. (2004). 'Reconciling Sustainability and Growth: Criteria, Indicators, Policies'. *Sustainable Development*, 12(2): 74–86.

Stavins, R. N. (2014). 'What Are the Benefits and Costs of EPA's Proposed CO2 Regulation?' Available at: <http://www.robertstavinsblog.org/2014/06/19/what-are-the-benefits-and-costs-of-epas-proposed-co2-regulation/> (accessed 12 December 2015).

Stosic, B., R. Milutinovic, N. Zakic, and N. Zivkovic (2016). 'Selected Indicators for Evaluation of Eco-Innovation Projects'. *Innovation: The European Journal of Social Science Research*, 29(2): 1–15.

TEKES (2015). 'The Impact of Tekes and Innovation Activities 2015'. Teke, Helsinki.

Tillväxtanalys (2015). 'Riskkapitalstatistik 2014: Venture Capital—Investeringar I Svenska Portföljbolag'. 06 Statistikserien, Tillväxtanalys, Östersund.

Valovirta, V., J. Lehenkari, O. Lehtoranta, T. Loikkanen, H. Bodewes, B. Mostert, and S. Zegel (2014). 'The Impact of Tekes Activities on Wellbeing and Environment'. Review 308/2014, Tekes, Helsinki.

Vedung, E. (2000). *Public Policy and Program Evaluation*. New Bruswick, NJ: Transaction Publishers.

Wang, J. (2015). 'Cleantech Investments in Denmark, Finland, Norway, Sweden: Are They Performing Well?' Thesis, Lund University.

Wilson, C., A. Grubler, K. S. Gallagher, and G. F. Nemet (2012). 'Marginalization of End-Use Technologies in Energy Innovation for Climate Protection'. *Nature Climate Change*, 2(11): 780–8.

WWF and Cleantech Group (2014). The Global Cleantech Innovation Index 2014. Available at: <http://info.cleantech.com/CleantechIndex2014.html> (accessed 12 December 2015).

Yang, Y. and I. Sollen (2013). 'Investments in the Swedish Cleantech Sector. Case Study of Swedish Public and Private Venture Capital Investments in Cleantech', Thesis, Umeå School of Business and Economics.

11

An Enquiry into the Political Economy of the Global Clean Energy Transition Policies and Nigeria's Federal and State Governments' Fiscal Policies

David Onyinyechi Agu and Evelyn Nwamaka Ogbeide-Osaretin

11.1 INTRODUCTION

There is currently a common problem facing all economies of the globe, developed and developing economies alike, which can only be solved by the collective efforts of all nations. In the views of the European Renewable Energy Council (EREC) (2004), climate change is the major challenge to sustainable development worldwide, manifesting in various forms and affecting various aspects of each of the economies. The effects of climate change propelled the United Nations (UN) to convene a summit that discussed the implications of the issue of climate change and how best to tackle the problem. The resolution of the summit is today referred to as the Kyoto Protocol to the UN Framework Convention on Climate Change (UN 1998).

It is true that the protocol stipulates some specific objectives targeted at reducing the prevalence of climate change within a set amount of years. However, it is also clear that not all those objectives can be achieved within the set time without profound commitments of the national governments that make up the UN. For instance, EREC (2004) recognizes that one of the key tasks faced by all the national governments in mitigating climate change is a profound transformation of the current energy system over the next few decades, replacing fossil fuels with renewable energies, and dramatically increasing energy efficiency.

However, practical experience shows that the actual level of commitment to the implementation of the Kyoto Protocol may vary significantly from the required level of commitment. Some such deviations are seen from the point of economies that depend more on fossil fuel for energy and revenue. To such economies moving away from the use of fossil fuel means reduction in their current revenues and that may come with overbearing consequences on the national governments.

As a member state of the UN that has signed the Kyoto Protocol, Nigeria depends significantly on revenues from fossil fuel. Statistical figures from Central Bank of Nigeria (CBN) (2012) shows that the contribution of oil revenue to Nigeria's total federally collected revenue ranged from 60–70 per cent annually in the early 1980s to an annual average of 75.5 per cent in the 1990s. However, the contribution reached an all-time peak of 88.6 per cent of total revenue in 2006 and has maintained a prominent place since then. Even the figure presented by CBN (2013) does not mean much deviation from the status quo. According to that figure, about 30 per cent of total federally-collected revenue came from non-oil sources, implying that oil revenue still contributed to about 70 per cent of total federal government's revenue in 2013.

Given the level of dependence of Nigerian governments on oil revenue, it is important to enquire into the willingness and implications of their implementing the Kyoto Protocol. Eleri, Onuvae, and Ugwu (2013) observed that the federal and state governments in Nigeria have outlined several policies and programmes in response to the achievement of clean energy policies. However, the scholars also observed that there is a general lack of capacity or political will to implement the outlined policies and programmes. Some of the policy documents like the Renewable Energy Master Plan and the National Energy Policy are yet to receive legislative attention or presidential endorsement. Although the National Policy on Climate Change and Response Strategy (NPCC-RS) was approved in 2015, implementation has yet to take place.

Many developed and developing countries are fast transitioning to clean energy. For example, while countries in North America reduced their imports of refined petroleum products from 3.083 million barrels per day in 2005 to 1.678 million barrels per day as of 2012 (that is, 45.57 per cent reduction), Nigeria rather increased her imports of refined petroleum products from154,324 barrels per day in 2005 to 180,921 barrels per day in 2012 (that is, 17.23 per cent increase). On the other hand, countries in North America increased their consumption of biofuels from 276.675 thousand barrels per day in the same 2005 to 941.86 thousand barrels per day in 2012 (US EIA 2015). Therefore, the decline in petroleum imports may not necessarily imply reduction in total energy needs of the countries in North America, but a function of a shift from fossil fuels to biofuels due to investment in clean energy sources. The same trend is observed in energy imports of other developed regions of the world. Hence, the sharp glut in global market oil prices of 2014 until now is attributed to a shift in energy demand from fossil fuel to clean energy sources.

On this basis, it is not out of place to wonder what could be the factors behind the lack of political will needed to implement the energy policies and programmes outlined by the federal and state governments in Nigeria. It is therefore pertinent to ask some critical questions and investigate the situation properly:

- First, how prepared are Nigerian governments to handle the fiscal shocks that will be associated with the implementation of clean energy transition policies?
- Second, what is the implication of deciding not to implement the clean energy transition policies especially when other national governments are fast implementing policies?

Therefore, this study provides evidence-based answers to these questions, which entails critically reviewing the various policies and programme documents, and the extent of implementation so far. The chapter further answers other important questions about the preparedness or readiness of Nigerian federal and state governments to handle the fiscal shocks that may be associated with clean energy transition. In addition, the study provides answer to the questions of the implications of the preparedness or otherwise of Nigerian governments, especially as many national economies are fast investing in research and development (R&D) targeted at hastened transition of global economies to clean energy so as to mitigate the effects of climate change.

As such, the success of the implementation of the Kyoto Protocol depends critically on the readiness and willingness of national governments. Without such readiness and willingness, the convention will be mere paper work that is not translated to any sustainable development of the global environment. Therefore, this chapter helps to reveal the extent of readiness of the Nigerian government, with the aim of showing Nigerian governments the necessary steps to take in implementing their clean energy policies without having adverse effects on their fiscal policies.

11.2 LITERATURE REVIEW

This section is devoted to reviewing the various literature arguments that exist in the area of clean energy transition on one hand, and global/national political economy on the other. The section starts with the theoretical arguments before proceeding to the empirical ones.

11.2.1 Conceptual Issues

11.2.1.1 Concept of Climate Change

Climate is the average weather in terms of its mean and variability over a certain period of time and a given area (IPCC 2001). According to the IPCC (2007) Fourth Assessment Report, 'climate change is a change in the state of the climate that can be identified from changes in the mean and/or the variability of its properties persisting for a long period usually for decades or longer'. Adejuwon (2006) defined climate change as 'observed changes in climate caused directly or indirectly by human activities, changing the composition of the global atmosphere, as well as natural climate variability observed over a given time period'. Climate change can be caused by human and natural factors. The activities of the natural factors include change in solar radiation, while the human factors include agricultural activities that lead to over use of land, high levels of deforestation, industrial and technological activities leading to shifts from organic fuel to high use of fossil fuels, land and air traffic, and so on.

Climate change has been differentiated from climate variability or fluctuation. Climate fluctuation or variability is defined as 'the variations in the mean, standard deviations, the occurrence of extremes, and so on, of the climate on all spatial and brief periods but beyond weather events' (Umar and Ibrahim 2011). Umar and Ibrahim (2011) went further to note that climate change and variability may be internal variability as a result of natural internal processes within the climate system or external variability, variations in natural or anthropogenic external forces. The major differences between climate change and fluctuation are mainly the time, the degree of variability, and the impact of variability. For all these factors, climate change is sterner.

11.2.1.2 Climate Change Mitigation

Climate change mitigations are actions put in place to reduce the intensity of radioactive emissions in order to reduce the potential effects of global warming through the control of sources of climate change. It involves the reductions of the concentrations of greenhouse gases (GHGs), either by reducing their sources or by increasing their sinks. This is, however, different from global warming adaptation, which is simply tolerating the effects of global warming/ climate change. According to the IPCC (2001), climate change mitigation is defined as attempt by humans to reduce the anthropogenic force of human behaviour on the climate system using such measures as reducing GHG emissions as well as their relative sinks. Similarly, IPCC (2007) defines adaptation as the ability for a system to adjust to climate change to moderate damage and cope with some of the results. Mitigation of climate change can be accomplished using low carbon sources, such as renewable energy, the use of energy efficient technology, or the protection of natural environment such as forests (Nordensvärd and Urban 2011). The concept of climate change mitigation is closely related to climate change adaptation.

Adaptation seeks to reduce the impacts of climate change, make the negative effect moderate, and exploit beneficial opportunities (Farauta et al. 2011). Adaptation is a proactive measure while mitigation is a corrective measure.

11.2.2 Theories of Climate Change

The theories of climate change are basically concerned with the major causes of the change in climate. These include the anthropogenic global warming (AGW), Bio-Thermostat, human forces besides GHGs, Planetary Motion theory, Ocean Current, and Solar Variability theory.

Anthropogenic global warming is the most common of the theories. This theory holds human emission of GHGs (carbon dioxide (CO_2), methane, and nitrous oxide) through activities such as burning wood and fossil fuels, and cutting down or burning forests as the main causes of climate change. This is accomplished through the mechanism called, enhanced greenhouse effect. IPCC (2007) maintains that water vapour is identified as the major greenhouse gas, responsible for about 36–90 per cent of the greenhouse effect, while CO_2 accounts for about 26 per cent. While it is believed that other external factors such as variation in solar radiation can lead to climate change, they argued that the effect could not account for rising temperatures. Though the forces of man-made gases are small, yet the cumulative positive feedback they exert leads to great climate change (IPCC 2007).

The bio-thermostat theory states that the negative effect of biological and chemical actions offset the effects of rising carbon through the mechanism of global bio-thermostat. Carbon enhances the productivity of plants, thus the more there is, the better the plants will grow. The theory thus concludes that neither carbon nor the biological processes are harmful to earth.

Another theory posits that the highest human influence on climate is the transformation of earth through deforestation, urban formation from population growth, as well as coastal developments. The IPCC estimated that about one-quarter to one-third of anthropogenic CO_2 emissions are due to deforestation. Following this theory, the ocean current theory posits that changes in global temperature are highly attributed to the slowdown of the ocean's thermohaline circulation (THC), while the planetary motion theory, first published by Milankovitch in 1941, states that climate change is mainly attributed to the natural gravitational and magnetic oscillations of the solar system. These oscillations alter the solar system and influence the earth to cause a change in climate. Similarly, solar variability theory states that solar variations account for most or all global warming. The effect of the variation in the sun affects the earth through solar wind on cosmic rays, which affects ocean surface temperatures and wind patterns.

11.2.3 Global Political and Policy Issues on Climate Change Mitigation

11.2.3.1 Global Political Economy of Climate Change Mitigation

The efficacy of climate change mitigation requires a global agreement. Actions need to be taken by different nations, such that the level of commitment of a nation will be at least proportional to its level of development and the level of emissions. There is resistance to a comprehensive global approach and global political acceptance has been seen as a major obstacle. In addition, low-income countries are faced with financial constraints given the high cost of clean energy, high-income countries are faced with political and consumers' unwillingness to pay the high price for clean energy and many developing countries are against any level of commitment that will affect their ability to grow. Moreover, despite agreements by nations to reduce emissions, the high cost of clean energy left nations uninspired by the Kyoto Protocol's targets.

Bailey and Preston (2014) opined that economic condition, resource endowment, and structure of a country determine management policies. These significantly affect the political acceptance of the country for low-carbon emission while the size and strength of the economy determine the resources available for the development of low-carbon energy as well as the political will to do so. Nations that are endowed with fossil energy will not be willing to carry out actions on low-carbon development because of the revenue accruing to the government. For instance in Russia, 28 per cent of total government revenue comes from fossil energy (Bailey and Preston 2014), while in Nigeria it accounts for about 70 per cent of federal government's revenue and about 90 per cent of foreign earnings (CBN 2013). Russia, as the fourth world largest GHG emitter, only ratified the Kyoto Protocol in November 2004, and placed a legal limit to gas flaring at 5 per cent from 2012 with implementation under way and estimated to be delayed by two to three years (Korppoo and Vantansever 2012).

Politicians will always pursue policies that they expect to gain or maintain support from political constituencies, thus they acquire more power by doing what the people will like rather than pursue climate change mitigation that will impose additional costs through carbon tax and loss of jobs for those in the carbon sector. Bailey and Preston (2014) further noted that while high-income countries are focusing on jobs, real incomes, and deficits, middle-income countries are focusing on economic development, inequality, and expansion of services, and low-income countries are concerned with poverty reduction, health, and education. The global awareness of climate change is also very low. The lack of information and knowledge about climate change has led to reluctance in the acceptance of the reality in most of the developing countries, Nigeria included. People are undereducated on carbon-pricing, carbon tax, or other tools for the reduction of emission rates. Hence, climate change mitigation only has little global political support.

Thus apart from the problems associated with global frameworks, the domestic political economy in the various countries matters much, since many countries only depend on the domestic support they can muster from their citizens to implement such a global treaty. This is especially so whenever a current administration is seeking re-election. An administration may accept irrational policies that may be contrary to global clean energy policies just to please the citizens and win their votes. Therefore, any discussion on global clean energy transition policies must also consider the national political economy angle in the subject matter (Löschel, Sturm, and Vogt 2010).

11.2.3.2 Conflicts between National Policies and Global Policies

Mitigating climate change has not witnessed much commitment as compared to the expectation from Kyoto Protocol in 1997 of reduction in emissions to about 30 per cent below what would have occurred under business as usual. In 1997, the Kyoto Protocol was signed with legal commitment given to developed countries to reduce carbon emission by 5 per cent below 1990 levels over a five-year period. This was followed by Copenhagen negotiations in 2009. German Watch (2011) noted that since the Copenhagen negotiations in 2009, many countries have not been able to deliver on their commitments to the climate change campaign as compared to the early years of the campaign. Mitigating climate change has been identified as a public good with its complex externality problem. As a public good, it is characterized with the free-rider problem hence each country will be expecting the other to shoulder the responsibility while they concentrate on their own national policies. Auerswald, Konrad, and Thum (2011) noted that a unilateral reduction in emissions by one country reduces the uncertainty associated with emission-related damages. The commitment level of countries towards the mitigation of climate change is attributed to two factors: the category that the country falls into, whether Annex 1 countries or not (i.e., industrialized and transition economies or otherwise); and the source of GHG emissions, whether from energy use in the form of transport and other related acts or from industrial and agricultural processes, such as deforestation and industrial wastes, or a combination of both. Majority of the Annex 1 countries are working on all policies effective in the reduction of the causes of climate change. On the other hand, Non-Annex 1 countries are only looking at available policies as recommended by Kyoto, probably because they were not made to pledge their commitment given their level of development (Ekins and Speck 2011). Some countries are attempting to reduce global carbon emission, while many others are subsidizing fossil fuels to the tone of about US$500 billion in 2011 and by estimated to rise to US$544 billion in 2012. The policy makers supporting subsidies believed that it is needed for price stabilization and economic security (Clements et al. 2013 in Stepp and Nicholson 2014).

Domestic policies conflicting with global policies are evident in some countries that are major GHG emitters, major economies, and have a heavy reliance on fossil fuels due to abundant domestic resources such as the US, China, and Malaysia (Nordensvärd and Urban 2011). They have not been able to introduce a low carbon economy despite their efforts even when they consider its importance for their domestic economy and their international competitiveness. They are more concerned with policies that will make their industries more competitive, thus carrying on the global policies of mitigating emissions would be a disadvantage relative to other competitors who have done nothing. For instance in the USA, government has failed to provide strict national policies on climate change. It was observed that while the USA accepted the UN Climate Change Convention in 1992, they failed to accept the Kyoto Protocol of 1998. They have only presented a weak target of 17 per cent reduction in carbon emissions as compared to other developed countries.

Malaysia is at odds with their desire to promote climate change mitigation versus the government's national policy aimed at enhancing their level of economic growth, which encompasses exploiting natural resources, utilizing fossil fuels through palm oil production, and conserving the nation's resources by creating policies of environmental protection and development. Hiding under the coverage of Non-Annex 1 countries, Malaysia has been unwilling to commit itself to tackling climate change. Though the country has accepted both the UN climate change convention and the Kyoto Protocol since 2002, it has not accepted the Copenhagen Accord. Furthermore, the country is yet to develop any national climate change policies (Nordensvärd and Urban 2011). Ozor (2009) observed that Nigeria had no effective climate change policy, and bills that should enhance good practices for sustainable environment were yet to be implemented. This is evident in the fact that the deadline to stop gas flaring in Nigeria was moved from 2008 to 2009, and then 2011.

Dolsak (2001) opined that the existence of conflict in policies to be implemented between the public and private sector in some countries has led to conflict between national policies and global policies. This has made some countries chose to continue in their old ways of energy use, others ready to act but requesting international financial assistance, and yet others ready to spend their own resources and publicly committed to the reduction of emissions. There is thus need for national policies to be in line with international policies if the global goal of mitigating climate change will be achieved. These inconsistencies in and conflict between national policies and global policies greatly hinders effective global policy implementation and the target set at the international level for clean climate.

11.2.4 Empirical Studies

Several studies have been carried out on climate change ranging from the economic implication of its impact to the different national and global policies of mitigating climate change. These studies seem to present conflicts between national policies and global policies towards climate change. Dolsak (2001) carried out an empirical investigation on factors affecting the level of commitment of different countries on the mitigation of climate change for 91 countries, both developing and developed. This was assessed using the variance between the signature and ratification of the Framework Convention on Climate Change and its actual implementation. Applying a logistic regression model, the result of the study showed that the level of commitment of a country is more significantly affected by incentives to the national government than the national government's ability to affect global climate change through the various mitigation actions. It was also found that the economic benefits from climate change do not affect a country's commitment to the mitigation of climate change.

In a study carried out by Löschel, Sturma, and Vogt(2010) on the empirical assessment of demand for climate protection using Germany, the researchers noted that policies towards climate change have political dimension and national policies on it have to be politically accepted. The study employed the willing to pay (WTP) approach for climate protection to examine willingness of the individual to climate protection. The result of the study found that the willingness to pay was very low amounting to about €12 per tonne of CO_2.

Speck (2010) analysed the debate between climate change policies, politics, and the media in Australia. In a pilot study using interviews from some leaders, it was found that climate change mitigation action has been very slow in Australia because of the media's information about uncertainty in the climate change science, weak leadership in the country, as well as the unpleasant cost of policies towards climate change mitigation from their political view even when such policies were found effective at the national and international levels.

Ayinde et al. (2010) analysed the impact of climate change on agricultural productivity in Nigeria, and examined the linkage between agricultural productivity and climate change parameters. Employing time series data for the period 1975–2005, a descriptive statistics and granger causality analysis were carried out on the data for the variables. The result showed that the climatic parameter (changes in rainfall) positively affected agricultural production while temperature was found relatively constant and does not affect agricultural output.

Koblowsky and Speranza (2010) analysed the institutional challenges to developing an effective climate change policy in Nigeria. Reviewing the roles of existing and planned institutional and legal frameworks in promoting or hindering policy implementation, the study used primary data collected between autumn 2009 and spring 2010, and found that there had been lack of agreement between political initiatives and institutions. The study also found a weak implementation of environmental laws and directives, and concluded that a policy framework on climate change is still non-existent for Nigeria as a result of divergent interest among the parties.

Furthermore, Akuru, Okoro, and Chikuni(2013) carried out a study on the impact of renewable energy deployment on climate change in Nigeria, and reviewed energy sources used in Nigeria as well as their impact on climate change. The study found that Nigeria is over-dependent on fossil-generated energy, which has had adverse effects on the environment, and thus recommended the integration of renewable energy into Nigeria's energy use. Amobi and Onyishi (2015) looked at the public policy perspective of governance and climate change in Nigeria, pointing out the problem of climate change within the governance scope. They showed that there is a direct relationship between the characters of the state and governance system on one hand and Nigeria's response to climate change on the other.

11.3 FINDINGS

There are many issues with the political economy of clean energy transition of Nigeria's governments—national and sub-national alike. Some of the issues have to do with fiscal policy implication of such transition policies. First, we can assume that the country is very willing and ready to transition. The readiness and willingness come with heavy fiscal policy implications. Secondly, we can also assume that the country is not very willing and may not be ready to transition to clean energy. These also come with heavy fiscal policy implications. We therefore present the current efforts and situation, observing some possible factors that explain the current rate of transition to clean energy in Nigeria, so as to be able to present the fiscal policy implications of the current efforts.

11.3.1 Nigeria's Fiscal Policy Stance and Current Efforts towards Clean Energy Transition

In order to transition from the current fossil energy regime to a more environmentally friendly energy regime, governments ought to take some specific steps towards clean energy transition, such as the fiscal policies. The role of fiscal policy in determining the political economy direction of any government cannot be overemphasized. Therefore, it is pertinent to ask two very important questions: What is the current fiscal policy stance of Nigerian governments and how do their fiscal policies translate to clean energy transition efforts? It is suitable to open the discussion with the observed skew in the two broad categories of expenditures among Nigerian governments—recurrent expenditures (mainly, the personnel that will run government institutions) and capital expenditures (relevant activities of the institutions that will deliver the policy mandates of the governments).

Information from the *CBN Statistical Bulletin* (2014) reveals that the proportion of capital expenditures in total expenditures of the Federal Government of Nigeria consistently declined in the first democratic era of 1981–3. However, the proportion was still better than what was obtained in subsequent military administrations of 1984–5 and 1985–93. On the contrary, the military administration of 1993–8 raised the proportion of capital expenditures on annual basis, but the tempo was not sustained in the subsequent military regime of 1998–9. Notably, since the return of democracy in 1999, the proportion of capital expenditures in total expenditures has decreased and remained very low. Particularly in 2014, the proportion of capital expenditures reduced to as low as 17 per cent of total expenditures of the federal government.

Establishing a link between the issues raised here and the main point of discussion, it is important to emphasize that clean energy transition policy is capital intensive. Good policy coordination will observe that the heavy financing needed for clean energy transition may not necessarily entail increase in total expenditures of the government, but a refocus on certain priority items that have to take greater proportion of the expenditures. Therefore, greater proportion of recurrent expenditures as observed in the beginning of this section may not guarantee the needed financing for research and development in Nigeria's clean energy transition programme. The cost of governance takes a very large chunk of the recurrent expenditures in Nigeria (BOF 2014a). Clean energy transition programmes of Nigeria must necessarily affect the size of government, thereby reducing current trend in the cost of governance, effectively investing in clean energy sources, and reducing the reliance of the country on fossil fuels as both sources of energy and revenues.

On the contrary, the fiscal policy position of state governments in Nigeria contrasts with that of Nigeria's federal government within the study period. In 1999, there was a wide gap between the share of capital expenditures and the share of recurrent expenditures in total expenditures of State governments. The gap between the two classifications of expenditures continued declining until 2012, when the proportion of capital expenditures in total expenditures exceeded the proportion of recurrent expenditures. This was not the case with federal government expenditure profile. State governments in the democratic era of 1999–2014 seem to commit more proportion of funds to capital expenditures than the federal government did within the same period. The proportion of Federal Government's capital expenditures in total expenditures maintained a downward trend, while the proportion of State governments' capital expenditures in total expenditures maintained an upward trend. Thus, we have established the existence of greater proportion of recurrent expenditures than that of capital expenditures in total expenditures and it is not clear how this trend of expenditures can affect the overall goal of clean energy transition policies of the governments in Nigeria.

To take appropriate steps towards clean energy transition, the federal government of Nigeria has set up several agencies and commissions charged with the responsibilities of researching alternative energy sources. One such agency is the Energy Commission of Nigeria (ECN) with the mandate of coming up with alternative sources of energy. However, one of the major challenges of the federal government of Nigeria is the duplication of institutions with similar responsibilities. For example, the law that established the ECN provides that the commission shall consist of fossil fuel department, nuclear energy department, solar energy department, and any other energy department that the government may determine from time to time. This implies that atomic energy may as well be determined to be relevant and therefore be made a department of the energy commission of Nigeria. However, instead of following this law, another commission was established with the mandate of developing and promoting nuclear technology—Nigeria Atomic Energy Commission (NAEC). With such duplication of institutions and agencies, it leaves the federal government of Nigeria with the option of committing insufficient funds to each of the agencies leaving the agencies with little to no output.

Statistical facts gathered from the federal government budgets of 2010–14 reveal that greater proportion of the annual budgets of Energy Commission of Nigeria goes to recurrent than to capital. In 2014, about 75 per cent of the Commission's total budget went to recurrent expenditures leaving only about 25 per cent to capital (BOF 2010, 2011, 2012, 2013, 2014a). This means that the main responsibilities of the commission may not be achieved, since the proportion that goes to capital expenditures will be too insignificant to invest in renewable energy sources or renewable energy research. Thus, it becomes necessary to ask if the whole idea of establishing such a commission is for employment creation or for deliberate efforts to transition to clean energy. If the goals of the enabling laws of ECN are to be actualized, then it is necessary to restructure the expenditure pattern of the commission in order to allow for efficiency. It is also possible that one of the factors contributing to the nature of expenditure is the existence of a parallel institution. Supposing the money allocated to the two different institutions is channelled to only one, the ratio might probably change, thereby freeing up some more money for capital projects of the single commission.

In addition, actual expenditures widen the gap between capital and recurrent expenditures. For instance, Energy Commission of Nigeria as a government agency is under the Federal Ministry of Science and Technology. The 2014 budget implementation report of the Budget Office of the Federation (BOF) (2014b) reveals that only 48.19 per cent of all the budgeted capital expenditures of the Federal Ministry of Science and Technology and its agencies were released and cash-backed. Given the nature, manner, and timing of the releases and cash-backing, only 93 per cent of the released funds were utilized. This implies that as at 2014, only 45.21 per cent of budgeted capital expenditures of the Federal Ministry of Science and Technology and its agencies were effectively implemented. The same report shows that almost 100 per cent of the budgeted recurrent expenditures of most ministries, departments, and agencies of the government were released.

The implication is that the gap between recurrent expenditures and capital expenditures of the energy commissions is so high that the commission may not fully discharge its statutory obligations. Based on the issues raised in this section, we may say that Nigeria's public institutions established for renewable energy may not have been effective due to lack of funds. This could have been alleviated by streamlining the institutions based on their statutory functions and responsibilities.

11.3.2 Possible Factors of the Observed Efforts towards Clean Energy Transition in Nigeria

The focal point of all the issues presented in Section 11.3.1 is that the institutional framework set up by the government in order to promote clean energy transition in Nigeria has not been strengthened enough to carry out their statutory obligations. Therefore, it is pertinent to find out if the inability of the government to strengthen these institutions is connected to the governments' dependence on oil revenue.

Sachs and Warner (2001) observed that it has been empirically proven that countries with abundant natural resources tend to perform poorly in terms of growth. Available statistical figures from CBN (2014) reveal that Nigeria has been reliant on revenue from fossil fuel. The portion of oil revenue in total federally collected revenues oscillated between 62 and 89 per cent during 1999–2014 fiscal years, and it reached its peak in 2006 at 89 per cent. Sachs and Warner (2001) explained the reason for such poor performance as crowding-out effect of natural resources—otherwise referred to as natural resource curse hypothesis.

The crowding-out effect manifests in two ways in Nigeria as: (1) absence of accountability and efficiency of government, and (2) neglect of the productive sector. In Section 11.3.1, we showed the proportions of federal and state governments' expenditures that go into recurrent expenditures, especially administration costs (i.e., executive, legislative, and judicial arms). John (2011) shows that in countries where natural resources account for more than 80 per cent of total government revenue, governments are likely to neglect the productive sector since the government does not really depend on them for taxes. As such, the private sector operators that are rarely taxed will not have any moral right to hold the government accountable to the people. This means that there is a connection between neglect of the productive sector and absence of accountability in government.

The situation of over-dependence on oil revenue also prevails among the sub-national governments in Nigeria just as with the federal government. Available statistical facts from CBN (2014) also reveals that the proportion of revenue from federation account in the total revenues of all the states in Nigeria reached a peak level of 70.5 per cent in 2001 and the lowest level of 37.6 per cent in 2009, but remained high at an average of 58.21 per cent throughout the period of 1999–2014.

Although governments in Nigeria (national and sub-national alike) depend heavily on oil revenues, there has been little effort on the part of the governments to invest in alternative energy sources. The shift in global emphasis from fossil fuel to renewable energy due to climate change effect should have sent a serious warning to the governments in Nigeria to diversify their energy sources rather than setting up several institutions with none of them coming up with any clear output on alternative energy sources.

One important observation is that Nigeria's federal government of Nigeria did not take the issue of transition to clean energy seriously until the second half of 2014, when the global oil price moved in a downward trend. It also dawned on the sub-national governments in Nigeria that fossil fuel may soon be abundant in Nigeria, yet not demanded in the global market given the current trend in research into alternative energy sources. This reawakening made many sub-national governments start emphasizing internally generated revenues against the previous reliance on federation accounts. The federal government has therefore started emphasizing non-oil revenue. This has spurred the governments to involve private sector operators in their decision-making process through consultations. The current cooperation between private sector operators and the government will probably produce greater commitment to the implementation of the renewable energy master plan than is currently experienced in the country. This means that if the price of oil in global market had continued rising, Nigerian governments would have continued treating the issue of clean energy transition with reluctance.

11.3.3 Fiscal Policy Implications of Nigeria's Delayed Transition While Other Economies Transition to Clean Energy

Other nations of the globe are fast transitioning to renewable energy sources, and this holds great opportunities and threats for the fiscal policy regimes of

Nigerian governments. It is therefore important to determine the fiscal policy implications of the current rate of transition policies in Nigeria while other national governments are fast shifting to clean energy regime.

Global oil price remained low during 1999–2003 (between US$20 and US$30 per barrel). However, the price started increasing on annual basis from 2004 until 2007 (when it exceeded US$90 per barrel). As an erratic market, the price was affected by the global economic/financial crisis of 2008–9 by forcing a sharp decline in oil price in 2008 from about US$90 per barrel in 2007 to about US$40 per barrel in 2008. Nevertheless, the global oil price picked up again from 2009 and continued on an increasing trend up to 2013. The period of 2008–11 seemed to be a period of incubation of concerted efforts towards renewable energy sources. The period of 2011–13 seemed to be the period of investment in efficient renewable energy sources of which some may take time to yield results, while others started yielding results in very short time. This could explain the relatively stable price of oil during 2011–13. This also implies that once the global investments in alternative energy sources start yielding expected returns, then the price of oil is bound to fall as witnessed in second half of 2014.

Available statistical figures from Index Mundi (n.d.) reveal recent prices of crude oil in the global oil market. It seems the global investments in alternative/renewable energy sources have started yielding returns. The data show that global oil price has been on the decline with highest rate of decline recorded between June 2014 and January 2015, when global oil price declined from about US$112 per barrel to about US$48 per barrel—a decline of about 57.14 per cent within six months.

The fiscal policy implications of such severe decline in global oil prices are profound for Nigerian governments. First, with a decline of 57.14 per cent in the price of the major source of revenue for Nigerian governments, it therefore means that about 40 per cent of federal government of Nigeria's revenue has been eroded by a fall in global oil prices (taking the average of 70 per cent of total revenue discussed earlier). Secondly, this decline implies that there is a shift in demand from fossil fuel to alternative energy sources. This means that, as long as Nigeria does not invest in alternative energy sources, the country will soon move from being an energy exporting country to an energy importing country. This scenario will pose great threat to the fiscal survival of the governments in Nigeria, especially in view of severe decline in the revenue base.

Nigeria is a net exporter of energy—especially crude oil, though the country imports refined petroleum products. Most of the country's imports (including energy imports) are funded with proceeds from oil exports. However, US EIA (2015) shows that many developed countries are shifting their demand for fossil fuels to demand for renewable energy. The implication is that renewable energy will soon replace fossil fuels globally. Therefore, should Nigeria not invest heavily in clean energy, it will soon move from net energy exporter country to net energy importer country, which has severe implications on the fiscal policy stance of the country at all levels of government due to the relevance of oil revenues as discussed in Section 11.3.2.

As a net exporter country, Nigeria has not been able to fund clean energy programmes and projects appropriately. It therefore seems difficult (if not impossible) for Nigeria to finance clean energy research and development as a net energy importer country. This is especially true for a country that depends heavily on energy (fossil fuel) as a source of government revenues.

11.4 CONCLUSION

From the findings presented in Section 11.3, we can conclude that Nigerian governments' future fiscal outlooks could be considered threatened except there is a major departure from the current expenditure pattern. We also conclude that recent occurrences in the global scene have forced global oil prices down with possibility of further decline in oil prices in the near future. This means that Nigerian governments' current disposure in handling the fiscal shock associated with fluctuations in the global oil market may no longer be adequate. As a way of absorbing the fiscal shocks that may be associated with fluctuations in the global oil market, the Federal Government of Nigeria sets the Oil Price benchmark below the minimum projected oil price for the year during the annual budgeting. Whatever is sold in excess of the benchmark price is deposited in the Excess Crude Account. However, the account has been mismanaged in recent times, leading to depletion of the savings.

We recommend that, in order to survive the fiscal policy shocks associated with the current global clean energy transition policies, the Nigerian governments should not ignore the following:

- Since oil revenues collected by the federal government are shared among the various tiers of government, investment in energy sources should not be restricted to only the federal government. This means that instead of having only the Energy Commission of Nigeria and Nigerian Atomic Energy Commission, funded by the Federal Government of Nigeria, the state governments should be free to have their various energy centres where research into clean energy is funded and their outputs form part of the policies of the sub-national governments.
- The current trend of expenditure among the various tiers of government in Nigeria, in favour of recurrent expenditures should be revised. Emphasis should be placed on capital expenditures with the target of equipping the energy institutions and agencies to come up with home-grown

alternative energy sources. In addition, the institutions and agencies that have overlapping responsibilities should be streamlined so as to ensure efficiency.

- Effective and efficient tax regime should be adopted to increase the volume and proportion of non-oil revenues in the face of dwindling oil revenue in order to meet up with the demand of the required investment for clean energy transition.

ACKNOWLEDGEMENTS

We are grateful to all the participants at workshop of the PECE project held at NREL, Golden, Colorado; especially to researchers who participated in country case study group, for their helpful comments. We also appreciate Channing Arndt's efforts in critiquing an early draft of this study, the administrative assistance received from Anne Ruohonen throughout the project, and equally appreciate the research assistance received from Chukwunenye Ferguson Emekaraonye, Emmanuel Ikechukwu Dick, and Uchenna Alexander Nnamani.

REFERENCES

Adejuwon, J. O. (2006). 'Food Crop Production in Nigeria. II. Potential Effects of Climate Change'. *Climate Research*, 32: 229–45.

Akuru, U. B., O. I. Okoro, and E. Chikuni (2013). 'Impact of Renewable Energy Deployment on Climate Change in Nigeria'. Paper presented at AFRICON, 9–12 September, Pointe-Aux-Piments, Mauritius. Available at: <http://www.erc.uct.ac.za/jesa/Volume26/26-3jesa-akuru-okoro-chikuni.pdf> (accessed 26 January 2016).

Amobi, D. and T. Onyishi (2015). 'Governance and Climate Change in Nigeria: A Public Policy Perspective'. *Journal of Policy and Development Studies*, 9(2). Available at: <http://www.arabianjbmr.com/pdfs/JPDS_VOL_9_2/17.pdf> (accessed 26 January 2016).

Auerswald, H., K. A. Konrad, and M. Thum (2011). 'Adaptation, Mitigation and Risk-Taking in Climate Policy'. Working Paper 3320. Munich: CESifo.

Ayinde, O. E., O. O. Ajewole, I. Ogunlade, and M. O. Adewumi (2010). 'Empirical Analysis of Agricultural Production and Climate Change: A Case Study of Nigeria'. *Journal of Sustainable Development in Africa*, 12(6): 275–83.

Bailey, R. and F. Preston (2014). 'Stuck in Transition: Managing the Political Economy of Low-Carbon Development'. Energy, Environment and Resources Briefing Paper 14/01. London: The Royal Institute of International Affairs (Chatham House). Available at: <https://www.chathamhouse.org/sites/files/chathamhouse/home/chatham/public_html/sites/default/files/20140200LowCarbonBaileyPreston.pdf> (accessed 14 August 2015).

BOF (Budget Office of the Federation) (2010). 'FGN 2010 Approved Budget'. Abuja: BOF Publications.

BOF (Budget Office of the Federation) (2011). 'FGN 2011 Approved Budget'. Abuja: BOF Publications.

BOF (Budget Office of the Federation) (2012). 'FGN 2012 Approved Budget'. Abuja: BOF Publications.

BOF (Budget Office of the Federation) (2013). 'FGN 2013 Approved Budget'. Abuja: BOF Publications.

BOF (Budget Office of the Federation) (2014a). 'FGN Approved budget'. Abuja: BOF Publications.

BOF (Budget Office of the Federation) (2014b). 'Quarterly Budget Implementation Report—Fourth Quarter 2014'. Abuja: BOF Publications.

CBN (Central Bank of Nigeria) (2012). *Statistical Bulletin*. Abuja: CBN Publications.

CBN (Central Bank of Nigeria) (2013). *Annual Report and Statement of Accounts*. Abuja: CBN Publications.

CBN (Central Bank of Nigeria) (2014). *Statistical Bulletin*. Abuja: CBN Publications.

Dolsak, N. (2001). 'Mitigating Global Climate Change: Why Are Some Countries More Committed than Others?' *Policy Studies Journal*, 29(3): 414–36.

Ekins, P. and S. Speck (2011). 'The Fiscal Implications of Climate Change and Its Policy Responses'. Technical Report prepared for UCL Energy Institute. London: University College London. Available at: <http://www.mca4climate.info/_assets/files/ClimatePolicy_FiscalSustainability_Final_Report(1).pdf> (accessed 21 August 2015).

Eleri, E., P. Onuvae, and O. Ugwu (2013). *Low-Carbon Energy Development in Nigeria: Challenges and Opportunities*. London: International Institute for Environment and Development (IIED) Publications. Available at: <http://pubs.iied.org/pdfs/G03555.pdf> (accessed 30 April 2015).

EREC (European Renewable Energy Council) (2004). 'Renewable Energy–A Key Solution to Climate Change'. Online Report of the Council. Available at: <http://www.erec.org/fileadmin/erec_docs/Documents/Publications/ClimateChangeBriefing.pdf> (accessed 10 March 2015).

Farauta, B. K., C. L. Egbule, Y. L. Idrisa, and V. C. Agu (2011). 'Farmers' Perceptions of Climate Change and Adaptation Strategies in Northern Nigeria: An Empirical Assessment'. Research Paper 15. Nairobi: African Technology Policy Studies Network. Available at: <http://www.atpsnet.org/Files/rps15.pdf> (accessed 27 July 2015).

German Watch (2011). 'Climate Change Performance Index 2010'. Available at: <http://www.germanwatch.org/klima/ccpi2010.pdf> (accessed 27 July 2015).

Index Mundi (n.d.). 'Crude Oil (Petroleum): Dated Brent, US$ per Barrel'. Available at: <http://www.indexmundi.com/commodities/?commodity=crude-oil-brent&months=360> (accessed 30 April 2011 and 22 August 2015).

IPCC (Intergovernmental Panel on Climate Change) (2001). *Climate Change 2001: Synthesis Report*. Cambridge: Cambridge University Press.

IPCC (Intergovernmental Panel on Climate Change) (2007). *Climate Change 2007: The Physical Science Basis*. Cambridge: Cambridge University Press.

John, J. D. (2011). 'Is there Really a Resource Curse? A Critical Survey of Theory and Evidence'. *Global Governance*, 17: 167–84.

Koblowsky, P. and C. I. Speranza (2010). 'Institutional Challenges to Developing a Nigerian Climate Policy'. Paper presented at the Berlin Conference on the Human Dimensions of Global Environmental Change: Social Dimensions of Environmental Change and Governance, 20 September. Available at: <http://www.diss.fu-berlin.de/docs/servlets/MCRFileNodeServlet/FUDOCS_derivate_000000001408/Koblowsky-Institutional_challenges_to_developing_a_Nigerian_climate_policy-394.pdf> (accessed 26 January 2016).

Korppoo, A. and A. Vatansever (2012). 'A Climate Vision for Russian: From Rhetoric to Action'. Policy Outlook. Washington, DC: Carnegie Endowment for International Peace.

Löschel, A., B. Sturma, and C. Vogt (2010). 'The Demand for Climate Protection–An Empirical Assessment for Germany'. Discussion Paper 10-068. Germany: Centre for European Economic Research. Available at: <http://ftp.zew.de/pub/zew-docs/dp/dp10068.pdf> (accessed 27 July 2015).

Nordensvärd, J. and F. Urban (2011). 'The Ambiguous Role of Corporations in Climate Change Mitigation: An Explorative Appraisal of Corporations in China, Malaysia and the US'. Working Paper 373. Brighton: Institute of Development Studies.

Ozor, N. (2009). 'Implications of Climate Change for National Development—the Way Forward'. Enugu Forum Policy Paper 10. Enugu: African Institute for Applied Economics.

Sachs, J. D. and A. M. Warner (2001). 'Natural Resources and Economic Development: The Curse of Natural Resources'. *European Economic Review*, 45: 827–38.

Speck, D. L. (2010). 'A Hot Topic? Climate Change Mitigation Policies, Politics, and the Media in Australia'. *Human Ecology Review*, 17(2): 125–34.

Stepp, M. and M. Nicholson (2014). 'Beyond 2015: An Innovation-Based Framework for Global Climate Policy'. A Technical Report to Centre for Clean Energy Innovation. Available at: <http://www2.itif.org/2014-beyond-2015-innovation-framework-global-climate-policy.pdf> (accessed 15 September 2015).

Umar, H. S. and H. Y. Ibrahim (2011). 'Mitigating Climate Change through Organic Agriculture: A Case Study of Farmers' Participation in Organic Farming Practices in Nasarawa State, Nigeria'. *Journal of Life and Physical Science*, 4(1):44–52.

UN (United Nations) (1998). *Kyoto Protocol to the United Nations Framework Convention on Climate Change*. Available at: <http://unfccc.int/resource/docs/convkp/kpeng.pdf> (accessed 30 April 2015).

US EIA (United States Energy Information Administration) (2015). International Energy Statistics. Available at: <https://www.eia.gov/cfapps/ipdbproject/iedindex3.cfm?tid=79&pid=79&aid=1&cid=regions&syid=2000&eyid=2012&unit=TBPD> (accessed 30 December 2015).

Part IV

Actors and Interests

Whether actors and interests are operating within a democracy, as in India, or in a one-party state, as in China, stakeholders will confront varying incentives and, depending on the nature of the incentives, will both drive and erect barriers to a clean energy transition. Isoaho, Goritz, and Schulz consider these issues for these two most populous economies in the world. Davidson, Kahrl, and Karplus maintain the focus on interests and actors under differing institutional constructs while concentrating on wind power, with a particular focus on China. They find that, despite an array of institutional and industry characteristics that are unique to China, China still encounters a series of political economy factors that are common across countries who have made significant wind power investments. Rennkamp and Bhuyan turn the attention to South Africa and the large, potentially predominant, role of interests and actors, as opposed to technical merits, in the South African nuclear power programme. Finally, Heshmati and Abolhosseini take a different tack focusing attention on Europe's interests with respect to energy security and the role that renewables might play in an enhanced energy supply security framework.

12

Governing Clean Energy Transitions in China and India

Karoliina Isoaho, Alexandra Goritz, and Nicolai Schulz

12.1 INTRODUCTION

China and India face a tremendous dilemma in addressing their developmental goals. Both need to respond to demands in poverty reduction, energy access, and urbanization, while reconsidering their development pathways that have been highly coupled to fossil fuel use. Renewable energy technologies (RETs) offer a solution for this dilemma. Decoupling economic growth from unsustainable resource consumption through the development and deployment of RETs would enable the transition to a clean energy economy, helping to reduce China and India's greenhouse gas (GHG) emissions while allowing societal and economic development (Altenburg and Pegels 2012). The rapid innovation and fall of prices in the RET sector indicates that a shift to RETs is becoming economically and technologically more feasible (World Energy Council 2013). Yet, broad consensus exists round the view that a full transition to RETs requires radical and 'deep structural' changes in the energy system (Geels and Schot 2007). This chapter analyses the key drivers and barriers for the promotion of RETs in the electric power systems of China and India. It does so by looking at how a governing coalition's ability and willingness to promote RETs is shaped by its power and cohesiveness, societal pressures, as well as a country's institutional framework, and how these factors vary between sub-national units.[1]

[1] For an extensive version of the chapter including detailed overviews of national RET governance structures and policy, see Isoaho, Goritz, and Schulz (2016).

12.2 THEORETICAL UNDERPINNINGS

A clean energy transition is characterized by a shift from a fossil fuel energy regime to a cleaner one.[2] We consider a clean energy transition occurring if the share of renewable energy in the power mix is growing faster than those of other energy sources. The vast majority of transition literature explains the processes of such shifts by analysing socio-technical and techno-economic means through which a transition could occur (Geels and Schot 2007). Recently, however, scholars are increasingly stressing the importance of political economy factors and geography-based analyses (Bridge et al. 2013; Meadowcroft 2011). While being attentive to the geographical context in which transition occurs, we align particularly with the more political economy-oriented literature.

In this study, we argue that a governing coalition's willingness and ability to implement successful RET policies is a function of three interconnected political economy factors. First and foremost, rulers are interested in staying in power. It is when this interest is seriously threatened by a specific policy problem that governing coalitions will be willing to solve it. Translating this to RET policy, a change towards a clean energy transition would be likely to occur if the governing coalition faces pressure from (potentially) powerful groups in society that are negatively affected by the current energy regime or that profit from promotion of RETs. This being said, relevant societal pressures could also be less RET-specific, for example about providing broad access to electricity. When this can be feasibly addressed using RETs (e.g., because of geographical advantages or technological advances), then they are likely to be promoted as well. Importantly, however, the political survival of the governing coalition will also depend on powerful groups interested in keeping the fossil fuel dominated status quo (Tsebelis 2002; Moe 2010). Only when the societal pressures demanding the departure from the status quo are more threatening than those demanding its preservation—or if these two pressures can be somehow reconciled—are governing coalitions willing to actively promote RETs.

Second, how power is distributed within and outside the governing coalition shapes its ability to develop and implement policy effectively (Khan 2011). Generally, once decision-makers in a governing coalition have decided to implement a certain policy, they will be more able to do so the less fragmented the coalition is and the less external opposition it faces. This is because weak and fragmented governing coalitions are more likely to have to

[2] Energy efficiency, nuclear, and carbon capture and storage (CCS) technologies play a significant part in the debates around clean energy transition, but we perceive RETs as more radical. We define RETs as 'modern technologies based on solar, wind, biomass, geothermal, and small hydropower' but focus primarily on solar and wind energy (Martinot et al. 2002: 310).

allow policy (or rent) capture in order to hold together factions within the governing coalition and to coopt other social organizations in society as a survival strategy (Migdal 1988). Thus, while certain distributions of power can strengthen a government's policy implementation capacity, others can act as a barrier.

Finally, certain institutional setups—such as constitutionally defined structures—are hard to change but strongly shape policy-making and implementation. Whether the central government can push sub-national governments to implement RETs, for example, is highly dependent on how federal 'rules of the game' structure the polity (North 1990). Some institutions, however, are easier to change, for example the allocation of budgets and responsibilities to different ministries. Section 12.3 analyses how these factors matter for China and India.

12.3 CHINA

12.3.1 Drivers

China's energy mix is dominated by fossil fuels. In 2014, coal accounted for 66 per cent of total energy consumption, followed by oil (18 per cent), hydroelectricity (8 per cent), natural gas (6 per cent), and nuclear power (1 per cent). Non-hydropower renewable energy only accounted for 2 per cent of the energy mix (BP 2015).[3] Yet, this rather bleak picture is contrasted with recent developments to push renewable energy. For the first time, China's new renewable power capacity surpassed new fossil and nuclear capacity in 2013, and the same was achieved in the following year (REN21 2015). In 2014, China also emerged as the world leader in clean energy investment with US 83.3 billion invested (FS-UNEP 2015). Strikingly, China's GHG emissions stagnated for the first time in a decade (IEA 2015a). First results from 2015 estimate that coal consumption fell by 5 per cent (Greenpeace 2014).

12.3.1.1 Shift towards Promoting RET Policy

These developments indicate that the Chinese government has been actively moving towards promoting RETs. A clear shift took place in 2005 with the establishment of the renewable energy law (Chen 2012). It developed pioneering measures ranging from targets in installed capacity to direct financial

[3] China counts large- and medium-scale hydropower projects (i.e., dams with an individual capacity of >50MW) as renewable energy. We do not consider these as renewable energy because of their negative effects on sustainability.

support policies. Since then, China has developed comprehensive renewable energy policies and included RET targets in its Five-Year Plans (FYPs), the government's key strategic social and economic development initiatives. The twelfth FYP (2011–15) set the objective to reorient China's economic growth towards sustainability, and also clearly identified new energy (i.e., hydro, nuclear, solar, and wind) as one of the key 'emerging strategic industries' to replace old pillar industries such as coal (Wang 2014).

This active promotion of RET policy has translated into concrete achievements. China is estimated to be on track meeting its 2020 RET targets from the twelfth FYP (CAT 2015a). The target of installing 30 GW in wind capacity has already been achieved, with installed capacity amounting to 89–91 GW in 2013 (Schoen 2013). In 2013, China had installed 12 GW of solar photovoltaic (PV) projects, which largely surpassed the official goal for solar power of 1.8 GW by 2020 (FS-UNEP 2015). Chinese authorities have estimated that the total cumulative installed wind capacity reached 145 GW and solar 38 GW by the end of 2015 (Hu 2015).

Of course, RET development has also been incentivized by decreasing costs, demand from the world market—especially in the case of solar PV development—and economic slowdown hitting the fossil-fuel-intensive industry (Spratt et al. 2014). In China, both small and large hydropower projects are the most competitive RETs, followed by biomass, wind power, and solar PV (IRENA 2014). Owing to the abundance of coal and rather low costs required to install conventional power plants, the RET industry still requires some support to compete with fossil fuel technologies (Liebreich 2015).

12.3.1.2 Societal Pressures: Pollution

China's ambition to promote RET policy can be partly explained by changes in societal pressure and elite priority resulting from increasing concern over air pollution. In 2012 <1 per cent of the 500 largest Chinese cities met the World Health Organization's air quality standards (Zhang and Crooks 2012). Recent studies estimate that pollution is causing the death of 1.6 million people per year (Rohde and Muller 2015). As a reaction to this environmental degradation, Chinese citizens have been increasingly voicing their complaints in letters to the government and through large-scale protests—something very unusual in China's repressive context (Tong and Lei 2014).

Most coal reserves and the majority of currently operating plants are located in the north and north-east of China, which is why they also have the most polluted cities (Cornot-Gandolphe 2014). In these provinces, the need to respond to high energy and electricity demands while simultaneously improving air quality is particularly acute. Serving as an indicator of increasing societal pressures, the cities of Beijing, Hebei, Tianjin, Shandong, Chongqing, and Shaanxi have pledged to reduce their coal use by 2017 and have set absolute coal consumption targets in order to address air quality issues (Greenpeace 2014). The leadership now acknowledges that, if not adequately addressed, societal pressures regarding emission reduction could have destabilizing effects on the political regime. Yet, regional differences in societal pressures have become clear recently: whilst senior authorities vowed to shut down coal plants in the Inner Mongolia region—the hub of coal production—because of social unrest over pollution (*The Japanese Times* 2015), they announced to build more power plants in inland provinces that are more underdeveloped.

Concern over pollution is also visible in China's cleaner growth discourse at international level, most recently in the United States–China Joint Agreement and the Intended Nationally Determined Contributions (INDCs) for the Paris Climate Summit (COP21). As part of the former, China announced new policies to tackle fossil fuel use, such as a cap on coal consumption at 4.2 billion tonnes until 2020, and a nationwide emissions-trading scheme to reduce the price gap between coal and clean energy sources (White House 2015). In the INDCs China added, inter alia, a target to make carbon dioxide (CO_2) emissions peak by 2030 or earlier (CAT 2015a). During COP21, China released a national policy recommendation to further reduce coal-fired generation at the same time as Beijing city government issued a red pollution alert, the most serious level as per air quality index, for the first time ever (Phillips 2015). These changes suggest a shifting elite preference from purely high economic growth towards incorporating environmental concerns in the governing coalition's strategy to stay in power.

12.3.1.3 Increasing Institutional Capacity

Bureaucratic and institutional changes have also been a key driver for China's RET policy development. In general, the State Council (i.e., the central government) sets the broad directions for policy development, whereas the ministries are responsible for policy drafting and formulation (Chen 2012). However, considerable changes to recentralize energy management at the top level occurred in the early 2000s, in parallel to the period when the problem of air pollution started to get more serious.

The National Leading Group on Climate Change Energy Conservation and Emissions Reduction (NLGACCERCER) was established in 2007 to act as an advisory and coordinating body in energy-related areas where several administrative bodies under the State Council have purview over clean energy policies (Zhang et al. 2013). Moreover, in 2008, the National Development and Reform Commission (NDRC)—the highest rank ministry in charge of China's macroeconomic and social development—was granted more power in the fields of energy, climate, and carbon reduction policy. In a first step, the National Energy Administration (NEA) was established under the NDRC to function as the agency responsible for RET formulation and implementation at the national level. The NLGACCERCER's climate change department was placed under the NDRC as well, thus assigning the NDRC a central role in climate policy formulation in addition to addressing energy questions.

Finally, the National Energy Commission (NEC) was created in 2012 to further streamline energy policy decision-making. The NEC functions as a 'super ministry', and is in charge of drafting the national energy development strategy (Bao and Gordon 2013). Locating clean energy and climate policy within the most influential agencies at the top level can be seen as an attempt to avoid the struggles and buy-ins that often emerge when clean energy policy drafting is dealt with within weaker ministries.

12.3.2 Barriers

Despite these positive policy and institutional developments, the implementation of central policies, especially in the case of energy, remains highly complex. This is, in part, visible in the regulatory, institutional, and political failures that China has faced in RET development.

12.3.2.1 Insufficient Grid Capacity

A major barrier for power generation from RET is the insufficient grid infrastructure, which is not able to absorb a large share of their power capacity (IRENA-GWEC 2013). The NDRC has aimed to prioritize renewable energy over coal by proposing green power dispatches and pilot trading platforms that allow energy generators and big end-users to negotiate prices (Chu 2015; White House 2015). To better integrate RETs in the current system, and help take power from distant western regions to major cities in the east, many smart grid lines are also under construction in line with the 12t FYP goal of developing an ultra-high-voltage power transmission (UVH) grid across the country (Mathews and Tan 2015). In 2015, however, an estimated 15 per cent of wind and 9 per cent of solar output were still curtailed due to grid constraints. As a result, Chinese regulators suspended the approval of new wind projects in 2016 in the most wind resource-intensive regions (Song and Hong 2016). Hence, much of the RET energy that should be replacing fossil fuel power is still going to waste in the current system.

12.3.2.2 Vested Interests

Political and institutional barriers also help explain why moving towards high penetration of RET in the energy power system has not been advancing as fast as it could have. One of the key political barriers is that China's energy strategy is still in many ways guided by the requirements of agencies that promote conventional energy: actors with strong vested interests in the power sector have significant political influence at the top levels of the CPC (Communist Party of China). Although it is difficult to accurately analyse the internal composition of the CPC owing to its secrecy, there is strong evidence that these actors have built powerful factions within the party (Xiaofei 2013).

The power sector is particularly prone to political corruption (Moses 2014). A case in point is Li Peng, former Chinese premier from 1988 to 1998, and his family, who are considered figureheads of the oil and coal factions of the CPC (Hornby 2014). Li Peng ran China's energy monopoly throughout the 1990s, staffing the top management positions with his relatives, running what Bezlova (2002) calls a 'family fiefdom'. While the monopoly was broken up into five power generation firms, Li's children, Li Xiaopeng and Li Xaolin, became the heads of two of them. In 2012, Li Xiaopeng was then promoted to governor of the coal-rich Shanxi region (Hornby 2014). Liu Zhenya, president of China's largest power SOE (state-owned enterprise) State Grid Corporation, is another example of the overlap of vested interest in the electric power sector and political power in the CPC. Liu, who has been openly opposing the central government's plans to break up the firm he heads (Zhu and Lague 2012), is also an alternate member of CPC's central committee, one of China's top ruling bodies.

This formal and informal amalgamation of political and business power implies that strong incumbent SOEs have the means to challenge and influence the top authorities in cases where their interests conflict (Bergsager and Korppoo 2013). Today, the large power generation companies and two grid companies are major players in conventional and renewable energy power generation (Dai 2015). The government's ability to make SOEs responsible for RET growth has been one of the main enabling factors for the impressive development. Yet, the power of SOEs has also blocked 'potentially system-disruptive' developments: the grid companies have successfully resisted the State Council's mandates to make the transmission system more favourable for RETs (Moe 2015). Although President Xi Jinping has directly targeted and weakened the vested interest groups that oppose energy reforms by launching anti-corruption campaigns in the power sector (Hornby 2014), we argue that the vested interest of certain SOEs and political elites, as well as their associated fear of losing power remain an important political barrier for RET transition in China.

12.3.2.3 Administrative Hierarchy

The institutional reforms brought about many new central level agencies with purview and interest over energy questions. These bodies—including within the NDRC—are often competing for authorship (Zhang et al. 2013). At the local level, in turn, implementation has become more decentralized as of 2014 (Zhao 2014). As a result, responsibilities during RET implementation are not always clearly allocated, increasing the risk of both interest bargaining and issues in accountability (Kostka 2014). Local SOEs have a reputation of getting away with violating environmental standards as local governments have limited capacities to enforce compliance, and international oversight is not permitted (Curtin 2015). This is particularly difficult in areas where the companies are major contributors to the local economic development. Nonetheless, as the top officials of local SOEs are evaluated in the cadre system, provincial leaders have more control over them than over managers of private sector firms (Harrison and Kostka 2014). This is important when considering that the largely private-owned small and medium enterprise sector is concentrated in very energy-intensive economic sectors.

12.3.2.4 Central vs Local Level

Local authorities are responsible for policy implementation, and they also have considerable autonomy regarding the economic development of their areas and rights over land use (Qi and Wu 2013). This autonomy and preference over economic development risks outweighing central RET policy within local jurisdictions (Dai 2015). In 2015, state agencies granted more power to local authorities through the power sector reform initiatives: all new coal power plants and RET projects can now officially be approved by provincial governments. Contrary to the objective of this state measure, there is evidence that local officials have taken advantage of their new powers to boost economic development. In the Shanxi and Inner Mongolia provinces, for example, many coal-fired power plants that were previously discarded by top authorities because of their environmental impact have now been re-opened by local leaders (Lingyu 2016). Furthermore, the CPC has sought to give environmental issues more weight in the local cadre evaluation as a result of increasing societal pressures. Yet, these efforts have met with resistance from local level leaders (Kostka 2014). Given that the average term length is between three to four years, local leaders tend to adopt short-term policy solutions that contradict central level initiatives and may not be sustainably viable in the long term (Kostka 2014). Hence, to achieve stronger RET policy enforcement, central authorities have to more effectively address imbalances within its institutional structures.

12.4 INDIA

12.4.1 Drivers

In India, numerous mechanisms are in place to support RET development and deployment, such as grants to develop technologies, tax incentives, and generation-based incentives (Hogg and O'Regan 2010). Wind has been receiving support since the early 1980s, and support for solar took off since 2008. As of November 2015, India had a total installed capacity of grid-interconnected power of 38.3 GW from renewable energy sources, with wind and solar accounting for 24.8 GW and 4.7 GW respectively (MNRE 2016).

12.4.1.1 Ambition

The Indian government wants to command a total renewable energy generation capacity of 175 GW by 2022: 100 GW from solar power, 60 GW from wind power, 10 GW from SHP, and 5 GW from biomass-based power projects (GoI 2015). The costs are immense, with only the solar goal expected to cost between US$100–113 billion (Ghosh 2015). Although India is one of the largest investors in renewable energy, with US$7.4 billion in 2014 (FS-UNEP 2015), this is far from what is needed. At RE-invest, the investors' conference for renewable energy held in February 2015, the Modi-led government secured pledges from national and international companies to deploy 266 GW of renewable energy in the next five years, and financial institutions committed to finance RE projects amounting to 78 GW (Bhaskar 2015; Ghosh 2015). At COP21, Prime Minister Narendra Modi launched the International Solar Alliance where 120 states committed to promote solar energy and mobilize >US$1 trillion of investments by 2030 to install 1 TW of solar power (Ananthakrishnan 2015). With India expected to be the second largest solar market by 2030 (IEA 2015b), this alliance is a crucial step towards this goal.

Importantly, India's ambitions with regard to RET promotion have reached a new level with the election of Modi as prime minister in 2014. Although the former prime pinister, Manmohan Singh, had launched the solar mission with a goal of installing 20 GW by 2020, Modi has increased this target fivefold. He is also encouraging foreign investment especially in the solar sector, and wants to attract US$100 billion to the sector until 2022 (Parkes 2015). He has already secured a US$1 billion deal with the US Export–Import Bank to facilitate shipping equipment from the United States (ibid.), and a US$2.25 billion deal with the German government for solar and other RETs (*Reuters* 2015).

12.4.1.2 Pollution, Energy Access, and Regional Variation

Unlike in China, protests against air pollution are rare in India. Although 13 of the 20 most polluted cities worldwide are located in India (Chauhan 2015), environmental standards seem to be perceived as barriers to economic growth and job creation. For example, despite living in one of the world's most polluted cities, the population of Vapi (Gujarat) protested against higher environmental standards (Barry and Bagri 2014). This is mainly explained by the city's dependence on its large and highly-polluting pharmaceutical and chemical industries. Although extreme, the example of Vapi appears representative of the priorities of large parts of the Indian population. Thus, in contrast to China, health issues related to environmental degradation (although existing) do not seem to be a central concern for the majority of society, or, at least, clearly less important than economic needs.

Providing broad and stable access to electricity to its fast-growing population is, however, a key priority to both India and China (Ghosh and Ganesan 2015). Understanding regional variation of political and geographical circumstances in India is crucial to explain the growth in RETs in the last years. Installed solar power capacity is largely concentrated in the north-west of the Indian sub-continent, particularly in the three states of Gujarat, Rajasthan, and Madhya Pradesh, being home to 60 per cent of the country's solar capacity (Figure 12.1). Part of the reason is geographical. While the north-west of India has particularly high solar radiation, eastern states are less rich

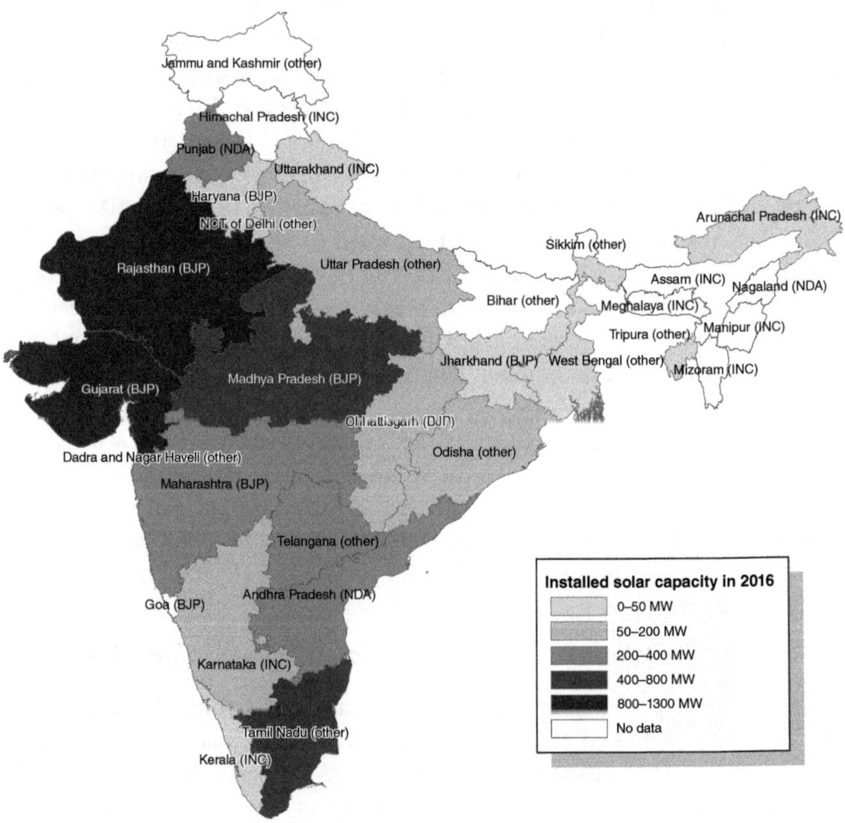

Figure 12.1 Installed solar power capacity and state governing parties in India.

Source: Authors' depictions, based on data from MNRE (2016b). Names of state governing parties are in parentheses.

in this regard, yet have considerable coal deposits. Thus, while energy demand can be met more easily with solar energy in the north-west, coal still tends to have the upper-hand in most remaining states in regards to economic feasibility.

This 'geographical feasibility' is often accompanied by a 'political feasibility'. To clarify this, it is important to understand India's federal division of responsibilities with regard to energy policy. Whereas the federal level is partly responsible for policy-making, the state level is responsible for both policy-making and implementation. Together with the general independence of Indian states, this implies that politics at the state level are particularly important with regard to policy implementation. Figure 12.1 shows that the states with large installed solar capacities are dominated by the Hindu nationalist Bharatiya Janata Party (BJP) and its coalition governments, the National Democratic Alliance (NDA). As we argued in Section 12.2, the cohesiveness and power of a governing coalition are key determinants of its capacity to implement policy. While we certainly do not claim that the BJP is a cohesive and powerful party in all of the states it governs (similarly to the INC—Indian National Congress), there is strong evidence that this was the case for the north-western states, and particularly in Gujarat, where the current Indian prime minister, Narendra Modi, was chief minister for 12 years. Roy (2013) describes how Modi managed to create a strong support base cutting across class and caste—usually *the* political cleavages in India. The characteristics of Modi's government are similar to authoritarian regimes of East Asia: cohesive, basically without alternative and strong opposition, fostering strong state–business relationships, and high governance and policy implementation capacities (Roy 2013). Basically, Modi transformed Gujarat into an economic and solar powerhouse, where national, international, and specifically solar businesses invested heavily. In 2014, Gujarat was home to 40 per cent of India's solar capacity (Pearson and Chakraborty 2014) and additionally became the nation's third largest wind energy producer in 2015 (Parkes 2015). To a large part, this massive expansion of RETs was the reason that the Modi government successfully provided stable energy access for all its citizens. Thus, the case of Gujarat underpins the interplay of conducive geographic and political factors. The combination of high solar and political capacity made RETs a feasible tool to address the key social demand of energy access.

This regional success is closely related to India's recent massive push for solar power on a national level. When Modi became prime minister in 2014, Gujarat's success story strengthened his and voters' confidence that this is replicable on a national level. Additionally, Modi can use the international finance and national RET programmes to support RET expansion in BJP-ruled states with high solar potential, thereby consolidating his party's power, and simultaneously contributing positively to international climate negotiations.

Finally, decreasing costs of implementation are another factor influencing Modi's renewable energy ambition. Prices for solar and wind have been falling drastically over recent years. The costs for onshore wind are nearly at grid-parity and solar is catching up rapidly in India (Liebreich 2015). Moreover, ambitious policy announcements increase the confidence of global players in the sector, which leads to further a price decrease.

12.4.2 Barriers

While future energy scenarios for India predict that renewable and nuclear energy will play a more important role, fossil fuels are expected to remain the main energy sources with an estimate of 56 per cent in power generation capacity for 2030 (CAT 2015b). In the current Indian power mix, renewable energies represent only 13 per cent, whereas fossil fuels still have a share of 70 per cent (MoP 2015). Coal has a particular importance. It is the most used and cheapest energy source in the country and accordingly accounts for over 60 per cent of the power mix. Importantly, former and current governments have pushed the expansion of coal. The annual production shall be increased from a current level of 600 million metric tonnes to 1.5 billion metric tonnes in 2020 (EIA 2015), which translates into opening a new mine every month (Rose 2015). Different projections for India suggest that coal use is set to increase between two and a half and three times compared with current levels (Dubash et al. 2015).

Vested interests in coal have arguably played a crucial role in this. An illustrative example of the crony capitalism surrounding coal in India is the so-called 'Coalgate' scandal. The former Singh government is accused of having allocated coal blocks inefficiently and under market value and having used a subjective and opaque system instead of competitive bidding (*The Hindu* 2015). According to estimations, this scandal has cost the country US$31 billion (Mathiesen 2014). Moreover, coal-related interest groups are particularly powerful politically in the coal-rich regions in India's eastern states.

However, India introduced a coal tax for imported and national coal in 2010, which was doubled in March 2015 to US$3.2. The revenues from the tax are going into the National Clean Energy Fund, which had generated an estimated US$2.7 billion by the end of the 2014/15 fiscal year. Whether the money, however, is spent towards its actual goal—RET research and innovation—is questioned (Ghosh 2015). Critics argue that the fund is used to balance the books of several ministries (Krithika and Mahajan 2014). This contributes to the already difficult RET finance environment in India (Chaudhary et al. 2014), which makes RET projects more expensive than in other countries. Subsequently, we look at further factors explaining this adherence to coal and other barriers to RET expansion in India.

12.4.2.1 Discoms

One of the greatest barriers to energy transition in India are state-government-owned electricity distribution companies, the so-called discoms. Of 29, 21 are deeply indebted, with over US$2.5 billion of debt in 2014 (Pearson and Chakraborty 2014). This has great implications for renewable energies. First, discoms are unable to meet their renewable purchase obligations (Krithika and Mahajan 2014). They prefer to feed in coal-generated power to the grid, as this has been considerably cheaper in the past. Feeding in RETs, they fear, will further increase their debt. Second, RET investors avoid investing in states with highly indebted discoms, as they fear that their generated electricity will not be bought and that discoms might default on them. Third, the discoms are too indebted to invest in major grid improvements, which not only lead to more losses and more debt, but also make the RET-generated power less viable. All this creates a highly adverse environment for RET promotion.

Theft, a badly implemented subsidy scheme for the rural and poor, and technical losses are main reasons for this immense indebtedness. In India, 20 per cent of electricity is 'lost' during transmission and distribution, double the world average (Aniti 2015). Electricity theft, that is 'non-technical' losses amounts to US$16.2 billion per year (Northeast Group 2015). Well-off farmers who do not fear prosecution apparently account for most of the thievery.[4] Interestingly, theft seems to increase before elections (Golden and Min 2012). Thus, in many Indian states, governing coalitions appear so vulnerable, they need to allow thievery of electricity, especially during elections. Mismanaged subsidies are a related reason for indebted discoms. Originally, these subsidies were introduced exclusively for farmers (Antholis 2014). However, many non-farmers, businessmen, and wealthier citizens free-ride on this subsidized power, as in many states the government appears to allow this practice or to be incapable of stopping it. The only way discoms seem to be able to reduce the speed of increasing debt is to cut power regularly.

Modi-governed Gujarat is one of the rare exceptions in this regard. It provides its citizens with 24-hour access to electricity (Pearson and Chakraborty 2014). Electricity for agriculture was split from the rest, and farmers received subsidized tariffs only for a few hours per day. Other paying customers received uninterrupted supply at normal rates. Officials checked who merits subsidized power. Similar strategies in other states were met with extreme and violent responses (Katakey and Singh 2014). Consequently, only Gujarat's (and Punjab's) discoms received A+ ratings from the Ministry of Power for their performance in 2013–14 (*The Economic Times* 2015).

[4] Personal communication with Detlef Sprinz (25 January 2016).

In contrast to other state governments, Modi's governing coalition was strong enough to execute such undertakings, which were extremely unpopular in the short term but very effective in the long term.

12.4.2.2 Land Rights

Another enormous barrier for the fast deployment of RETs is land acquisition (Ghosh 2015). Land in India is very scarce but essential for large-scale solar projects and wind farms. The process of land acquisition for private companies, however, is expensive (four times the market price in rural areas) and extremely complicated (70–80 per cent of the affected families have to agree and social impact assessments have to be conducted). Thus, a normal process would last 58 months (Kazmin 2015). This is largely the legacy of a law introduced in 2013 by the Congress party, in order to strengthen the rights of land owners. Before 2013, land owners felt mistreated by the state, because it could expropriate them and sell it to businesses, if they would use the land for 'public good'. As part of his drive to attract more investment, Modi tried to change this law and enable authorities to take land without social impact assessments and without farmers' consent, although buyers would have had to pay above the market price as before. His proposal met strong resistance from the Congress and other parties with big agrarian constituencies. As these parties still hold the majority in the Indian upper house, the Rajya Sabha, they were able to repel Modi's land reform. This not only is perceived by many as the biggest failure of the current Modi government but also illustrates how difficult it is to pass laws in the Indian polity (Kazmin 2015).

12.5 DISCUSSION AND CONCLUSION

A successful transition to RET-based energy systems requires radical changes in the current energy regimes. Although governments play a key role in pushing such change, their ability and willingness to do so depends particularly on a range of political economy factors, which also interplay with geographical and technological ones. Our goal was to analyse how these factors drive and impede clean energy transition with RETs in China and India.

In China, societal pressures due to increasing air pollution has been one key factor pushing the CPC to close coal plants and turn to environment-friendly power production in order to secure its political survival. This has been accompanied by an increasing institutional and political capacity for clean energy policy-making at the central level. Nevertheless, the state appears to be constrained by powerful political vested interest in the power sector on the one hand and imbalances within institutional structures on the other. Importantly, there are arrangements where provincial level actors or incumbent SOEs have the capacity to override top-level incentives when interests conflict, which has often been problematic for RET development.

In contrast to China, environment-related societal pressures have not been the key driver for RET promotion in India. Rather, a number of Indian states—above all Modi-governed Gujarat—profited from a combination of beneficial geographic and political circumstances, which allowed them to address their populations' growing demand for electricity through the promotion of RETs. The recent upswing in support of RETs at the national level is closely associated with the fact that leaders of the RET-feasible states are also leading the national government. Yet, severe barriers for a clean energy transition pertain. India plans to increase coal production massively, which can be seen as a symptom of the vested interests in the sector. Moreover, Indian discoms are in very bad shape financially because they are often used as a political tool to strengthen the governing coalition's political survival. Finally, the problems of land acquisition in India present a major barrier for renewable energy projects, which often need large tracts of land.

Our analysis generates three broad implications. First, it helps unveil how politics and the balance of power between contending interest groups drives or hinders system change. We saw that vested interests in incumbent electric power systems can inhibit the promotion and deployment of RETs in electricity. Second, the comparison shows that the drivers of change can be significantly different in distinct contexts. In both countries, changes in the overarching macro-structures were key to driving RET promotion. These dynamics, however, were very different. In China, unrest from increasing levels of pollution has effectively *coerced* the governing coalition to promote a cleaner electric power system. In India, however, we found that it has been much less a change in attitudes or beliefs but rather in financial and political feasibility that has *allowed* or *incentivized* certain state governing coalitions to promote RETs as engines for energy access. Third, in China principal–agent problems between the central and provincial levels are a key barrier to change, whereas in India the causal drivers and barriers to RET promotion can be found largely at the sub-national state level. It is thus key to observe how politics depends on and interplays with regional contexts.

Concluding, this study highlights two key areas for further research. First, we see the need for more in-depth analyses of the dynamics surrounding energy transition in sub-national entities. Second, it would be interesting to further investigate why massive pollution in India has not created the same societal pressures (in the form of protests) as in China. Specifically, whether this is mainly a reflection of lower economic development levels or rather of more complex political economy factors.

REFERENCES

Altenburg, T. and A. Pegels (2012). 'Sustainability-Oriented Innovation Systems: Managing the Green Transformation'. *Innovation and Development*, 2(1): 5–22.

Ananthakrishnan, G. (2015). 'Modi Launches International Solar Alliance'. *The Hindu*, 1 December.

Aniti, L. (2015). 'India Aims to Reduce High Electricity Transmission and Distribution System Losses'. US Energy Information Agency (EIA), 22 October.

Antholis, W. J. (2014). 'Narendra Modi's Power Obsession: Indian Energy Reform', 28 July. The Brookings Institution, Washington, D.C.

Bao, Y. and H. Gordon (2013). 'The Evolution of China's Energy Institutions: Centralization versus Decentralization'. *China Institute Occasional Paper Series*, 1 (1): 1–20. China Institute, University of Alberta, Edmonton.

Barry, E. and N. T. Bagri (2014). 'Narendra Modi, Favoring Growth in India, Pares Back Environmental Rules'. *The New York Times*, Asia Pacific, 4 December.

Bergsager, H. and A. Korppoo (2013). *China's State-Owned Enterprises as Climate Policy Actors: The Power and Steel Sectors.* Copenhagen: Nordic Council of Ministers

Bezlova, A. (2002). 'China Corruption Probes Signal Power Plays'. *Asia Times Online*, 1 November.

Bhaskar, U. (2015). 'Renewable Energy: India's Sunrise Sector'. *LiveMint*, 13 March.

BP (British Petroleum) (2015). 'BP Energy Outlook to 2035'.

Bridge, G., S. Bouzarovski, M. Bradshaw, and N. Eyre (2013). 'Geographies of Energy Transition: Space, Place and the Low-Carbon Economy'. *Energy Policy*, 53(February): 331–40.

CAT (Climate Action Tracker) (2015a). 'China'.

CAT (2015b). 'India'.

Chaudhary, A., A. Narain, C. Krishna, and A. Sagar (2014). 'Who Shapes Climate Action in India? Insights from the Wind and Solar Energy Sectors'. *IDS Evidence Report 56: Rising Powers in International Development.* Brighton: Institute of Development Studies.

Chauhan, C. (2015). '13 Out of World's Top 20 Polluted Cities in India, Only Three in China'. *Hindustan Times*, New Delhi, 5 June.

Chen, G. (2012). *China's Climate Policy.* London and New York: Routledge.

Chu, J. (2015). 'RE100: China's Fast Track to a Renewable Future'. *China Analysis 2015.* Beijing: The Climate Group.

Cornot-Gandolphe, S. (2014). 'China's Coal Market: Can Beijing Tame "King Coal"?' OIES Paper CL 1, Oxford Institute of Energy Studies, Oxford.

Curtin, J. (2015). 'Climate Hero or Villain? An Analysis of China's Plans for Low Carbon Transition'. Institute of International and European Affairs, Dublin.

Dai, Y. (2015). 'Who Drives Climate-Relevant Policy Implementation in China?' *IDS Evidence Report 134: Rising Powers in International Development.* Brighton: Institute of Development Studies/Tsinghua University.

Dubash, N. K., R. Khosla, N. D. Rao, and K. R. Sharma (2015). 'Informing India's Energy and Climate Debate: Policy Lessons from Modelling Studies'. Climate Initiative, Research Report, April, Centre for Policy Research, New Delhi.

EIA (Energy Information Agency) (2015). 'India's Coal Industry in Flux as Government Sets Ambitious Coal Production Targets'. US Energy Information Agency, 25 August.

FS-UNEP (Frankfurt School-United Nations Environment Programme) (2015). 'Global Trends in Renewable Energy Investment 2015', edited by A. McCrone, U. Moslener, E. Usher, C. Grüning, and V. Sonntag-O'Brien. Frankfurt School of Finance & Management–UNEP Centre and Bloomberg New Energy Finance, Frankfurt.

Geels, F. W. and J. Schot (2007). 'Typology of Sociotechnical Transition Pathways'. *Research Policy*, 36(3): 399–417.

Ghosh, A. (2015). 'The Big Push for Renewable Energy in India: What Will Drive It?' *Bulletin of the Atomic Scientists*, 71(4): 31–42.

Ghosh, A. and K. Ganesan (2015). 'Policy: Rethink India's Energy Strategy'. *Nature*, 521(7551): 156–7.

GoI (Government of India) (2015). 'General Budget 2015–16: Highlights and Summary'. Press Information Bureau, Government of India, New Delhi.

Golden, M. and B. Min (2012). 'Theft and Loss of Electricity in an Indian State'. IGC Working Paper, February. International Growth Centre, London.

Greenpeace (2014). 'The End of China's Coal Boom: 6 Facts You Should Know'. Greenpeace, East Asia.

Harrison, T. and G. Kostka (2014). 'Balancing Priorities, Aligning Interests: Developing Mitigation Capacity in China and India'. *Comparative Political Studies*, 47(3): 450–80.

Hogg, K. and R. O'Regan (2010). 'Renewable Energy Support Mechanisms: An Overview'. In M. Bonass and M. Rudd (eds), *Renewables: A Practical Handbook*. New York: PriceWaterhouseCoopers, pp. 31–48.

Hornby, L. (2014). 'China's Anticorruption Drive Focuses on Energy Industry'. *Financial Times*, 28 May.

Hu, W. (2015). 'China Solar Power Capacity Reaches 37.95 GW by End-Sep'. *China Coal Resource*, 21 October.

IEA (International Energy Agency) (2015a). 'Energy and Climate Change: World Energy Outlook Special Report'. International Energy Agency, Paris.

IEA (2015b). 'India Energy Outlook: World Energy Outlook Special Report'. International Energy Agency, Paris.

IRENA (International Renewable Energy Agency) (2014). 'Renewable Power Generation Costs in 2014'. International Renewable Energy Agency, Abu Dhabi.

IRENA-GWEC (International Renewable Energy Agency/Global Wind Energy Council) (2013). China Wind Policy.

Isoaho, K., A. Goritz, and N. Schulz (2016). 'Governing Clean Energy Transitions in China and India: A Comparative Political Economy Analysis'. WIDER Working Paper No 2016/28. UNU-WIDER, Helsinki.

Katakey, R. and R. K. Singh (2014). 'India Fights to Keep the Lights On'. *Bloomberg Business*, 5 May.

Kazmin, A. (2015). 'India: Land in Demand'. *The Financial Times*, 7 July.

Khan, M. H. (2011). 'The Political Settlement and Its Evolution in Bangladesh'. Unpublished Monograph, School of Oriental and African Studies, University of London.

Kostka, G. (2014). 'Barriers to the Implementation of Environmental Policies at the Local Level in China'. Policy Research Working Paper WPS7016.,World Bank, Washington, D.C.

Krithika, P. R. and S. Mahajan (2014). 'Governance of Renewable Energy in India: Issues and Challenges'. Background Paper, TERI-NFA Working Paper Series 14, The Energy and Resources Institute, New Delhi.

Liebreich, M. (2015). 'Global Trends in Clean Energy Investment'. Paper presented at the Bloomberg EMEA Summit, 12 October, Bloomberg New Energy Finance, London.

Lingyu, K. (2016). 'Local Govt's Use Newfound Power to Approve Dirty Coal-Fired Power Plants'. *Caixin Online*, 15 January.

Martinot, E., A. Chaurey, D. Lew, J. R. Moreira, and N. Wamukonya (2002). 'Renewable Energy Markets in Developing Countries'. *Annual Review of Energy and the Environment*, 27(1): 309–48.

Mathews, J. A. and H. Tan (2015). *China's Renewable Energy Revolution*. Basingstoke: Palgrave Macmillan.

Mathiesen, K. (2014). 'Coalgate: India Urges Supreme Court Not to Close Coal Mines'. *The Guardian*, 1 September.

Meadowcroft, J. (2011). 'Engaging with the Politics of Sustainability Transitions'. *Environmental Innovation and Societal Transitions*, 1(13): 70–5.

Migdal, J. S. (1988). *Strong Societies and Weak States: State–Society Relations and State Capabilities in the Third World.* Princeton, NJ: Princeton University Press.

MNRE (Ministry of New and Renewable Energy) (2016). 'Programme/Schemewise Physical Progress in 2015–16 (Up to the Month of January, 2016)'. Ministry of New and Renewable Energy, New Delhi.

Moe, E. (2010). 'Energy, Industry and Politics: Energy, Vested Interests, and Long-Term Economic Growth and Development'. *Energy*, 35(4): 1730–40.

Moe, E. (2015). *Renewable Energy Transformation or Fossil Fuel Backlash: Vested Interests in the Political Economy*. Basingstoke: Palgrave Macmillan.

MoP (Ministry of Power) (2015). 'Total Installed Capacity (as on 31.12.2015)'. Ministry of Power, New Delhi.

Moses, R. L. (2014). 'Power Struggles: Seeing China's Anti-Graft Drive in a Different Light'. *The Wall Street Journal*, 27 June.

North, D. C. (1990). *Institutions, Institutional Change and Economic Performance*. Cambridge: Cambridge University Press.

Northeast Group (2015). 'India Smart Grid: Market Forecast (2015–2025)'.

Parkes, R. (2015). 'India: Realising Its Potential'. *Renewable Energy Focus*, 14 September.

Pearson, N. O. and D. Chakraborty (2014). 'Modi Signals Solar Revolution for Power Market: Corporate India'. *Bloomberg Business*, 13 March.

Phillips, T. (2015). 'Beijing Issues First Pollution Red Alert as Smog Engulfs Capital'. *The Guardian*, 7 December.

Qi, Y. and T. Wu (2013). 'The Politics of Climate Change in China'. *WIREs Climate Change*, 4(4): 301–13.

REN21 (2015). 'Renewables 2015 Global Status Report'. REN21 Secretariat, Paris.

Reuters (2015). 'Germany Offers India $2.25 Billion for Solar, Clean Energy', 5 October.

Rohde, R. A. and R. A. Muller (2015). 'Air Pollution in China: Mapping of Concentrations and Sources'. *PLoS ONE*, 10(8): 1–14, Article e0135749.

Rose, D. (2015). 'Why India Is Captured by Carbon'. *The Guardian*, 27 May.

Roy, P. (2013). 'The Political Economy of Growth under Clientelism: An Analysis of Gujarat, Tamil Nadu and Pakistan'. PhD Thesis, Department of Economics, SOAS, University of London.

Schoen, L. (2013). 'Why Is China Taking Action on Clean Energy and Climate Change?' *China FAQs Issue Brief*, May, World Resources Institute, Washington, D.C.

Song, R. and M. Hong (2016). 'China's 1-2-3 Punch to Tackle Wasted Renewable Energy', 27 April, World Resource Institute, Washington, D.C.

Spratt, S., W. Dong, C. Krishna, A. Sagar, and Y. Qi (2014). 'What Drives Wind and Solar Energy Investment in India and China?' *IDS Evidence Report 87: Policy Anticipation, Response and Evaluation*. Brighton: Institute of Development Studies.

The Economic Times (2015). 'Gujarat's 4 Discoms Get A+ Rating', 10 August.

The Hindu (2015). 'What's the Coal Scam About?' 12 March, Chennai.

The Japanese Times (2015). 'China Plants to Shut Down after Violent Protest', 7 April.

The White House (2015). 'U.S.–China Joint Presidential Statement on Climate Change'. Office of the Press Secretary, 25 September.

Tong, Y. and S. Lei (2014). *Social Protest in Contemporary China, 2003–2010. Transitional Pains and Regime Legitimacy*. London and New York: Routledge.

Tsebelis, G. (2002). *Veto Players: How Political Institutions Work*. Princeton, NJ: Princeton University Press.

Wang, Y. (2014). 'A Review of Renewable Energy Legislation and Policies in China'. In E. Moe and P. Midford (eds), *The Political Economy of Renewable Energy and Energy Security: Common Challenges and National Responses in Japan, China and Northern Europe*. Basingstoke: Palgrave Macmillan, pp. 197–220.

World Energy Council (2013). 'World Energy Perspective: Cost of Energy Technologies'. World Energy Council, London.

Xiaofei, L. (2013). 'Chinese Political Transition Makes Change Possible in Energy Approach'. *Oil and Gas Journal*, 111(2): 50–2.

Zhang, Q. and R. Crooks (2012). 'Toward an Environmentally Sustainable Future: Country Environmental Analysis of The People's Republic of China'. Asian Development Bank, Philippines.

Zhang, S., P. Andrews-Speed, X. Zhao, and Y. He (2013). 'Interactions Between Renewable Energy'. *Energy Policy*, 62(November): 342–53.

Zhao, S. (ed.) (2014). *China's Search for Energy Security: Domestic Sources and International Implications*. London and New York: Routledge.

Zhu, C. and D. Lague (2012). 'In China, a Power Struggle of a Different Order'. *New York Times*, 17 October.

13

Towards a Political Economy Framework for Wind Power

Does China Break the Mould?

Michael R. Davidson, Fredrich Kahrl, and Valerie J. Karplus

13.1 INTRODUCTION

Wind power is considered one of the most cost-effective options for reducing carbon emissions from the global electricity system. However, achieving higher penetrations of wind energy presents a number of unique development and integration challenges. Most of the existing literature has focused on general solutions to technical engineering and economic problems, as well as the design of dedicated policy support. There has been far less focus on the political economy of wind development, which involves the features of countries, regions, and systems that shape incentives for developing and integrating wind power. Given that this broader context mediates the effectiveness of dedicated technical and policy approaches to promote wind, this is an important gap.

Drawing on global experience, this chapter develops an analytical framework for understanding the spectrum of political economy conflicts that arise when introducing and scaling wind power within an electricity system. We apply this framework to China, a country that has very different electricity sector institutions from those found in most other countries. We argue that China's wind development and integration challenges can be understood through a general political economy framework, and show how high levels of wind energy curtailment in China are an expected result of clashes among actors and interests.

13.2 BACKGROUND: WHY DEVELOP A POLITICAL ECONOMY FRAMEWORK FOR WIND

The physical properties of wind energy—its variability, forecast uncertainty, and location relative to demand centres—create technical challenges for existing systems. Electricity systems have historically been designed to accommodate generation over which operators had greater certainty and more control. Accommodating wind requires revisiting established planning and operational procedures, challenging prevailing political and economic authorities under institutional arrangements.

Similar to other large-scale technological changes, efforts to increase the share of wind energy on electric grids must confront existing political and economic institutions. In the case of wind, such political economic conflicts are numerous. Although they have been examined in a handful of contexts (Fischlein et al. 2010; Kahrl and Wang 2014; Krishna et al. 2015; Lehmann et al. 2012), analysis of the political impediments to wind development and integration tends to focus more on public acceptance than on institutional design (Haggett 2008). The redistribution of economic rents that ensues can lead to a reshaping of political influence (Jacobsson and Johnson 2000), motivate incumbents to minimize adverse impacts through political channels (Stigler 1971), and at its outset prompt resistance that slows the pace of institutional and technological change (Mahoney and Thelen 2010). Technically efficient wind integration strategies, such as new market designs or enlarging balancing authority areas, may be slowed, altered, or dropped altogether when they challenge established practices, norms, and interests.

Wind power development interacts with ongoing transformation efforts in electricity sectors worldwide. Since the 1980s, a number of countries have restructured their electricity industries, transitioning from regulated, vertically integrated natural monopolies to unbundled ownership structures with competition in the generation and, in some cases, retail segments of the industry. Each jurisdiction has its own unique pathway defined by prior institutional context and proximate justifications for reform, which affects the degree to which market competition can be facilitated. Costs, economic transfers, and economic behaviour associated with wind development and integration occur, and must be understood, within these unique institutional contexts.

Developing countries share some similarities in approaches to restructuring. Typically, countries with a rapidly expanding 'green-field' electricity system and expected high demand growth will emphasize attracting capital over efficiency gains that result from competition. Public ownership is more prevalent and may be retained even following unbundling. Providing electricity services at prices affordable to low-income populations complicates liberalizing retail tariffs and may hide inefficient cross-subsidization. Weak or resource-limited government institutions for administration, information collection, and verification can hinder cost-effective regulatory design and implementation (Jamasb et al. 2005; Williams and Ghanadan 2006).

One might expect China to be different from other emerging economies because of its unique institutional history, which before market-oriented reforms in 1978 consisted of a planned economy layered on top of a largely federalized system of governance established over centuries of dynastic rule. But is China really different? The extent to which China's experience with wind energy in its electricity system reflects a more universal set of political economy challenges is ultimately an empirical question.

13.3 POLITICAL ECONOMY FRAMEWORK FOR WIND POWER DEVELOPMENT AND INTEGRATION

We develop a framework to probe institutional bottlenecks affecting the development of wind electricity. We distinguish between political institutions (the degree of centralization and regulatory philosophy) and economic institutions (industry structures and approaches to price formation). Our analysis sheds light on how alternative configurations of political and economic institutions, through their influences on actors and interests, lead to different decisions—and thus outcomes—for wind investment and integration (see Figure 13.1).

Our framework is rooted in a broader literature that evaluates the impact of political and economic institutions on the rate and direction of transition within large-scale fixed infrastructure systems with long lifetimes (Markard 2011). We adopt the Northian definition of institutions as 'humanly devised constraints' that shape interaction and can be both formal and informal (North 1991: 97). We develop our framework with examples from a wide

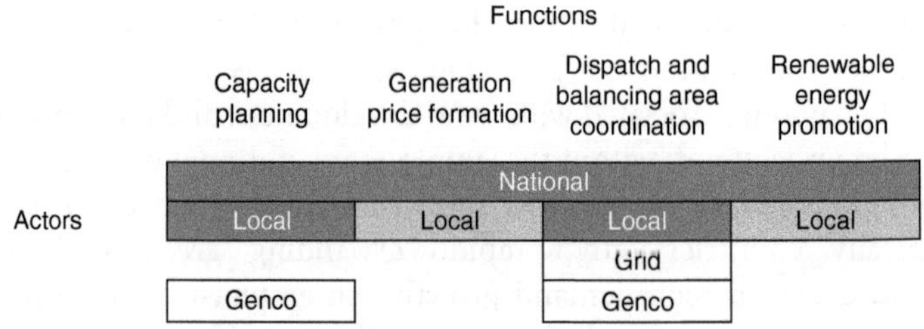

Figure 13.1. Economic and political actors in key wind sector functions in China.

Note: National, national government agencies; Local, local government agencies; Grid, grid companies; Genco, generation companies. Within government, dark shading indicates primary role, light indicates an oversight role.

Source: Authors' compilation.

range of contexts, mainly in the US and Europe. We then ask whether this framework holds in the case of China, drawing on rich experience in wind development witnessed in recent years.

13.3.1 Political Institutions

Political institutions comprise the first dimension of the framework. Here, we define political institutions as governmental roles housed within political bodies, and the vertical and horizontal relationships that connect them. Broadly, these roles can be divided into policy and regulatory roles. Policy roles include treatment of state and non-state entities, long-term plans for electricity reforms, environmental and energy policy design, and in some cases pricing, which has implications for the design of renewable energy incentives. Regulatory roles typically involve implementing the policy regime in a fair and efficient manner, for instance by policing abuses of market power, determining costs, and overseeing pricing, planning, dispatch, and other electricity sector functions.

Countries vary in size, network structures, and resource endowments, which affect the viability of creating markets and influencing government priorities (Jamasb 2006). In general, countries with well-developed electricity systems have different goals from those still in early stages of development: the former may be aiming to optimize efficiency and provide greater choice to market participants, whereas the latter are typically trying to attract private capital to an over-burdened publicly funded system. The institutions and ideological basis for creating complex markets are also more developed in the former, whereas the latter may not share the basic regulatory premise of valuing reduced government intervention (Williams and Ghanadan 2006).

13.3.1.1 Governance of Power Systems: Dimensions of Diversity

We consider four distinct dimensions in the governance of power systems, with systems lying on a spectrum between extremes within each dimension (Table 13.1).

First, countries and supranational bodies differ in the extent to which policy and regulatory functions are distinct and separate. We refer to this as 'horizontal separation'. Canonical power system designs emphasize the importance of separation to ensure system operation is free from interference by the regulated economic actors and the political actors that set the rules (Joskow 2008). The argument for separation extends to ensuring that regulatory bodies have sufficient authority to compel changes in the sector. In the US, this separation is pronounced, with the Federal Energy Regulatory Commission (FERC) at the federal level and public utility commissions at the state level

Table 13.1. Four dimensions of governance that affect power system outcomes

Governance dimension	Description
Horizontal separation	To what extent are policy and regulatory functions distinct and separate?
Vertical separation	To what extent are policy and regulatory functions concentrated at the central government level or decentralized to local governments?
Ownership	To what extent is ownership public or private?
Economic planning	To what extent are economic planning and investment planning centralized in government agencies or decentralized to market participants?

Source: Authors' compilation.

responsible for coordinating regulation, and Congress, state legislatures, and executive agencies charged with formulation of policy.

Second, countries and supranational bodies differ in the extent to which policy and regulatory functions of the power sector are concentrated at the central level, or vested with subordinate levels of government, as in a federal system. We refer to this as 'vertical separation'. In many cases, functions are spread across different levels of government, and may come into conflict. For instance, in the US regional wholesale markets are overseen by the FERC, but infrastructure siting decisions and retail electricity prices within those markets are overseen by state regulatory commissions.

Third, countries differ in the extent to which the government directly controls power generation, transmission, and distribution assets through direct ownership or majority or minority controlling stakes. Developing countries, in general, tend to maintain higher government ownership of assets, particularly if direct ownership is deemed central to 'developmental state' priorities (Johnson 1995). Developed countries vary in government ownership of assets. In France, the dominant electricity provider, EDF, is a government majority-owned utility. In the US, most electricity is provided by investor-owned utilities, although the federal government continues to own a significant amount of generation capacity and publicly-owned municipal utilities continue to be important providers.

Fourth, countries differ in their historical and current relationship between the government and the economy more broadly. Some countries still rely on elements of central planning, whereas other countries have a long history of regulated markets. The relative reliance of governments on markets versus planning—either in the present, or historically—is often reflected in governance of the power sector. For example, economies such as China, India, and the former Soviet Union used to be planned economies, and despite adopting capitalist structures, elements of central planning still persist in their power sectors to this day.

13.3.2 Economic Institutions

The second key dimension of this framework is economic institutions, which encompasses the structure of power markets, the relationship between producers and consumers, and the institutions that allocate costs and shape economic behaviour. These institutions vary significantly across electricity sectors, as competition has been introduced to different extents in different parts of the electricity industry in different locations.

13.3.2.1 Industry Structure: Traditional and 'Standard' Restructuring Models

Delivering electricity requires the coordination of five main activities: generation, transmission, system operation, distribution, and retail. In most countries, the first large-scale electricity companies were vertically integrated utilities (VIUs) that owned and controlled all five aspects under a single roof. Governments typically opted for public ownership and operation of VIUs, or created private franchises that obligated utilities to serve all customers at cost-based prices in exchange for a guaranteed rate of return on invested capital, a model known as 'cost-of-service'.

As in other network industries with cost-based tariffs, electricity regulators face significant information asymmetries with respect to which costs the utility should be allowed to pass on to customers. In response to this and a wide array of other factors, in the 1990s a number of countries began to introduce competition in parts of the sector. Based on three decades of reforms, there now exists a 'standard liberalization prescription' that specifies which and in what order certain activities should be made competitive, the appropriate methods for regulating activities that remain monopolistic, and the necessary institutions to ensure well-functioning markets (Joskow 2008: 11–13). In practice, owing to differing motivations for restructuring as well as varieties of institutions, countries have rarely implemented the textbook liberalization approach.

These arrangements differ in their requirements on regulatory institutions. The 'standard' restructured model has the largest diversity and complexity of actors, whereas more vertical arrangements have fewer regulated entities. Across all industry structures, regulators must develop sufficient expertise to evaluate the prudency of investments by network companies. The creation of markets brings additional regulatory complications, as the need for specialized knowledge to validate some costs gives way to the need to recognize and quantify the exercise of market power. In countries with a large public sector and less experience with competition regulation—the case for many developing and former centrally planned economies—this may be even more challenging.

Different generation price formation regimes are typically, but not always, associated with different industry structures. Generation price formation falls into three generic categories:

- Cost-of-service: prudent costs approved by regulator and included in rates.
- Benchmark: price based on the (usually average) cost of a benchmark technology, possibly determined through 'yardstick' competition.
- Organized markets: e.g., energy-only market, energy prices determined through bilateral contracts and short-term wholesale market.

Higher penetrations of wind generation generally increase system-wide costs of accommodating wind variability and uncertainty, often referred to as 'integration' costs. These costs are compensated to different extents and in different ways under different pricing mechanisms.

Even in regions with competitive wholesale markets, investment in renewable energy is typically driven by incentives that exist outside of the market. The two most common forms of incentive are: (i) feed-in tariffs (FITs) or generation-based tax credits, where renewable generators are paid a fixed price per unit generation (kilowatt-hour) delivered to load-serving entities (LSEs); and (ii) renewable energy quotas, such as renewable portfolio standards (RPS), which require LSEs to procure a certain share of their sales from renewable energy. Hybrids or combinations of these price and quantity mechanisms are common.

13.3.3 Actors and Interests and Wind Energy

The political and economic institutions discussed in Section 13.3.2 engage a set of actors and their interests, which shape power sector decisions related to wind energy. In the end, all actors—which include generators, transmission operators, and distribution companies, dispatch authorities, regulators, and policy makers—play distinct roles in setting the agenda and determining rules that govern the power sector, and are impacted to varying degrees by the system-level outcomes of these decisions.

13.3.3.1 Political Institutions

Depending on their arrangements, political institutions may simultaneously enable and constrain wind: for instance, if policy sets targets for expanding the share of renewable electricity, but also constrains dispatch decisions to time scales not amenable to efficiently integrate wind, conflicts can (and do) emerge. A beneficial political arrangement in one setting may lead to poor wind integration outcomes in another due to reorientation of interests.

First, the horizontal integration of policy and regulatory functions can have clear benefits for wind in countries that have achieved political consensus on the benefits of renewable energies. In this case, regulatory and market institutions may be more easily revisited or altered through administrative measures to reflect how policy incentives play out in the actual operation of the power system and the settlement of its costs. However, if closely entwined policy and regulatory functions are vulnerable to capture by powerful incumbent interests (including pre-existing fossil generators), the incentives for wind integration will be weaker. Even if interest politics plays a smaller role, more frequent interference with the regulatory system may lead to economically suboptimal outcomes and harm long-term development potential of the sector.

Second, vertical integration (centralization) of decision-making authority can ensure that new capacity is optimally located to reflect resource quality and generation needs, whereas greater autonomy for decision makers at subordinate levels could (though not necessarily) lead to suboptimal allocations based on local political conditions. Likewise, increasing geographic scope of transmission and system operation decisions can be favourable to wind integration. By contrast, in the US, transmission siting authority is vested at the state level, even within larger multi-state balancing authorities (MIT 2011).

Third, a high degree of state ownership in the power system can enable or constrain wind generation, hinging on the extent to which state-owned enterprises (SOEs) act as agents subordinate to the state, or conversely, the extent to which SOEs effectively capture regulatory and policy functions. If policy priorities at the top shift in favour of wind, state-owned wind developers are direct beneficiaries.

Characteristics of a nation affect the ability of its economy to generate or adapt to new technologies and practices (Porter and Stern 2001). Instead of claiming that one institutional form is universally superior for wind energy integration and therefore should be grafted onto another with an expectation of similar performance impact, we submit that the more important task involves understanding how institutions shape outcomes and how potential interventions interact with legacy structures in ways that create momentum towards desirable outcomes. For example, as will be discussed later in this chapter, there are aspects of how legacies of planned economic systems result in rigid quota setting (on generation within and trade across provinces in China, for example) that are not compatible with the short-term flexibility required to efficiently integrate wind power.

13.3.3.2 Economic Institutions

Traditional VIUs and restructured markets create different incentives for wind energy expansion. For VIUs, the profit objective for deploying wind will include network expansion and operation costs, and coordinated network and generation expansion has the potential to reduce overall social costs (GE Energy 2010). However, VIUs may not have sufficient incentives to control costs, leading to inefficient investments.

First, how to achieve coordination across functions in support of wind can be a major challenge, depending on market structure. For competitive generation markets, not only are network and generation expansion separated, so are many other economic activities essential to integrate wind, such as price formation at different timescales and ancillary services. Wind may have reduced revenue streams because of an inability to participate in forward energy and balancing markets, and because many markets were not designed to accommodate particularities of wind, there may not be a sufficient variety of market products to incentivize integration.

Second, who determines the terms of access of wind generators to the network? VIUs, which integrate transmission and generation functions, should have an incentive to connect wind quickly. However, in restructured markets, connection is a critical step in wind farm development, and associated costs (including network enhancements) are particularly contentious. Calculating appropriate costs and the degree of socialization will depend on whose calculation it is, whether an ISO, a transmission company, or an integrated network utility. Connection delays can also result in disproportionate hardship on farm owners as a result of cash flow issues, as almost all costs are concentrated in upfront capital.

Third, once connected, wind integration depends on dispatch rules of the system operator. ISOs and VIUs alike will generally try to minimize short-run operational costs, benefitting wind with near-zero marginal cost. Owners of transmission networks that also do system operation may have incentives to dispatch generators connected to transmission lines with favourable tariffs. These incentives in operation also depend on policy and regulations for curtailment, including circumstances in which it is allowed and compensation, if any.

13.4 CHINA CASE STUDY

We apply this framework of interactions of political and economic institutions to help shed light on wind integration outcomes in China, the world's largest energy consumer. China has the world's highest installed capacity of wind energy, but also faces the most severe wind integration situation: curtailment rates—forced spillage of available wind electricity by the grid operator typically for economic or grid stability reasons—have been in the double-digits for at least five years, reaching 40 per cent in some regions during 2015 (NEA 2016). Delay in grid connection to wind farms is another important barrier to smooth expansion of wind energy, with wind installations lacking appropriate connection surpassing 16 GW in 2015 (GWEC 2016). By applying our framework, we can begin to gain a qualitative sense of whether, and how, political economy challenges explain wind development and integration outcomes—or conversely, whether technocratic fixes in the form of capacity targets, price support, transmission build-out, and wind dispatch requirements will be sufficient or sufficiently accepted by the affected parties—to catalyse a low-carbon electricity transition in China.

Many of China's market-based electricity sector liberalization efforts during the early 2000s have been abandoned or significantly altered from their ideal prescription owing to historical legacies and institutional priorities, neither of which are independent. These echo challenges in other country and regional settings, and hence provide a valuable case of the varied political economy impacts of wind energy transitions across the developing world. Figure 13.1 shows the actors in China that participate in decisions related to electricity system functions (capacity planning, generation price formation, dispatch and balancing area coordination, and renewable energy promotion policies) most relevant for the development and grid integration of wind, as a function of industry structures (national, local).

13.4.1 Planning and Project Approval of Wind Farms

Planning—deciding on future generation capacity and transmission needs—and project approval are critical determinants of wind development and grid integration outcomes. Systems vary in the degree to which planning functions are carried out by technical or political bodies. Project approval, also generally by a government or affiliated office, is required before initiating construction of new capacity and transmission projects.

China is perhaps most distinct in the extent of its government involvement in the planning and project approval processes as well as in industry decision-making through state control of firms engaged in all stages of wind farm construction and operation. This reflects, in part, the institutional legacy of China's planned economy, which has persisted longer in electric power than in many other sectors. In practice, it means that generation capacity and transmission planning is largely driven by the supply side, in a top-down manner, targeting a specific installed base without explicit incentives to optimize around system operation.

Planning for wind capacity expansion largely occurs at the central level but siting and integration have been complicated by conflicts with local institutions. The NDRC (National Development and Reform Commission) (together with the NEA (National Energy Administration), after it was created in 2008) sets the national wind capacity target through medium- and long-term industry development plans (Ling and Cai 2012). The national target is then allocated to provinces. Provinces can also volunteer to host large wind bases, as happened in Gansu in 2006–7, and which later obtained NDRC approval, benefitting from central support for long-distance transmission (Davidson et al. 2017). Provincial as well as national officials have incentives to target capacity expansion as it adds to investment, boosting gross domestic product. Capacity expansion also creates local jobs and demand for the output of one of China's strategic renewable energy industries, which was just emerging in the late 1990s and early 2000s. In China, minimizing cost is only one consideration in planning decisions, which reflect many other factors including local economic goals, industrial policy, technological feasibility, and profit-sharing arrangements among local stakeholders.

Historically, wind projects have included government contract projects and concession projects. Before 2003, government contract projects dominated, in which the government directly awarded project development rights to one consortium. After 2003, the concession model was introduced, in which the NDRC selected favourable resource locations for projects and allowed potential developers to bid through a tender process (Han et al. 2009).

Although the concession system enabled rapid development of wind capacity, several features undermined its effectiveness. First, projects were initially selected on a least-cost basis, prompting bidders to offer unrealistically low prices that later undermined quality (Han et al. 2009). Second, pressure to bid at low cost was exacerbated by targets on the largest generation companies to expand renewables to 5 per cent of their total capacity (not generation) (Liu and Kokko 2010). In 2009, bidding with an electricity price was replaced with region-specific benchmark pricing for wind projects based on resources. By this time, there was already evidence that some capacity was of exceptionally poor quality, with turbines producing far less than rated output, requiring more downtime for maintenance, or even collapsing (Han et al. 2009).

Capacity thresholds that determine the level of government at which authorization could be granted also created conflicts between stakeholder interests and wind integration, leading to a situation in which generation expansion rapidly outpaced transmission buildout and exacerbated curtailment. In the early years of wind development in China, all new wind projects required central approval (Han et al. 2009). Development accelerated significantly when approval authority for wind farms sized <50 MW was granted to provincial authorities; indeed, a large number of wind farms built during this period have a capacity of 49.5 MW, as provincial government approval was often the preferred, often faster option to launch wind farm construction.[1] Many wind farm developers at this time were SOEs with access to low cost financing and were rewarded principally according to capacity constructed.[2] However, at this size, the grid company was even more reluctant to connect capacity, especially capacity in remote areas that required significant additional transmission (Yang et al. 2012).

The conflicts related to the planning and project approval process in China are not entirely unique. In other markets, such as the US, it is easier to build new capacity than to site new transmission. In terms of this mismatch, a high degree of federalism plays a similar role in both the US and China. Crossing state (provincial) lines with new transmission requires additional co-ordination and approvals in both countries; however, in China grid expansion arguably faces less resistance from citizens and groups concerned about aesthetic or environmental impacts. Instead, in China resistance arises because the interests and constraints of grid authorities are fundamentally different from those of wind farm developers and their local government champions.

13.4.2 Generator Cost Recovery

The transition to higher wind penetrations often creates economic conflicts between wind and thermal generators, tied to cost recovery. The extent of these conflicts depends on a number of factors, described in Sections 13.4.2.1–13.4.2.4, that are common to both China and other country contexts.

13.4.2.1 Support Mechanisms and Dispatch

In China, wind energy development is incentivized through FITs. FITs fix the price, but not the quantity, of wind power. If the FIT price is sufficiently high, renewable energy developers may rapidly expand installed wind capacity, reducing operating hours, market prices, and revenues for thermal generators. In countries with economic dispatch, this physical and economic displacement of thermal generation occurs primarily through the dispatch merit order, as wind has very low marginal costs.

In China, dispatch order is determined administratively rather than according to marginal cost, and the operating hour impact of higher wind penetrations on thermal generators is, to some extent, negotiated. In most provinces, operating hours for each generating unit are determined through an annual

[1] Cities also competed by offering favourable arrangements for obtaining land for siting wind farms (Liu and Kokko 2010).

[2] At the end of 2008, 90 per cent of China's wind developers were SOEs (Liu and Kokko 2010).

planning process, and system operators (grid companies) dispatch units to meet targets set through this process.[3] In a small number of provinces, dispatch is based on a preset order, with non-dispatchable renewable energy receiving dispatch priority.[4]

Nationwide, China has had a 'mandatory procurement' (*quan'e shougou*) policy for renewable energy since 2005. Wind curtailment rates, however, have been much higher than those seen in other countries with similar levels of wind penetration (Kahrl and Wang 2014). In April 2015, the NDRC issued new rules requiring local planning departments to prioritize renewable generation in annual plans, as part of a broader reform package (NDRC and NEA 2015).

The ongoing nature of dispatch reforms to promote higher utilization of renewable energy in China reflects a conflict between: (i) renewable energy generators and the political establishment, which are keen to promote renewable energy and reduce renewable energy curtailment, and (ii) thermal, and particularly coal, generators, which are keen to limit reductions in their operating hours.

13.4.2.2 Fixed-Cost Recovery

Lower operating hours have a significant impact on coal generators' ability to recover their fixed costs in China, because of the energy-only benchmark approach to setting their wholesale tariffs. Under this approach, all coal generators receive the same price for each megawatt-hour of output, with the price benchmarked against the levellized cost of a supercritical coal unit. This benchmark tariff requires an estimated number of fully loaded operating hours, which for coal units in China is typically around 5000 hours, to convert fixed costs (in yuan per megawatt-year) to a variable price (yuan per megawatt-hour).[5] As the number of operating hours falls, the wholesale price that generators require to recover their fixed costs increases nonlinearly (Figure 13.2).

13.4.2.3 Operating Cost Recovery

At higher wind penetrations, thermal generators are generally required to change their operating practices. This includes (i) maintaining higher reserve

[3] For more detail on this planning process and how it intersects with system operations, see Kahrl and Wang (2014).

[4] This policy is known as 'energy efficient dispatch' (*jieneng diaodu*). For political economic reasons, energy efficient dispatch has proved difficult to extend to other provinces (see Kahrl et al. 2013).

[5] More specifically, the levellized fixed cost (*LFC*, in yuan per megawatt-hour) is calculated as $LFC = AFC/AOH$, where *AFC* is the annual fixed cost (in yuan per kilowatt-year) and *AOH* is the annual operating hours (in hours per year).

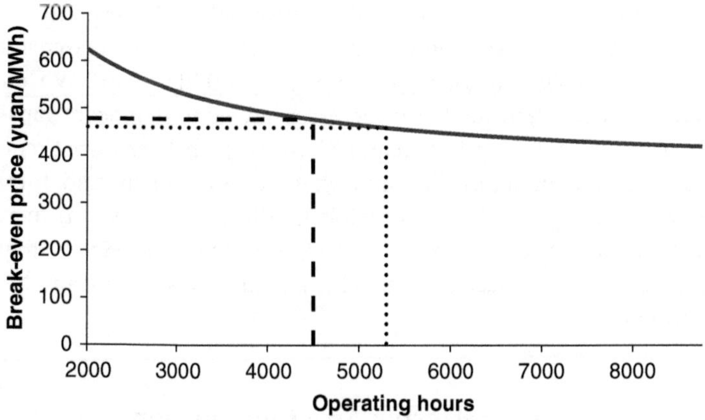

Figure 13.2. Break-even price for a supercritical coal unit in China as a function of operating hours.

Note: This example assumes a capacity cost of 530 yuan/kW-year and an energy cost of 0.36 yuan/kWh.

Source: Authors' calculation based on E3 (2015).

levels, or operating further below rated capacity, to account for the higher uncertainty in wind availability; (ii) more frequent, faster, and deeper changes in output ('ramps') to respond to changes in wind availability; and (iii) more frequent start-ups and shutdowns to respond to the uncertainty and variability in wind output. These kinds of changes in operating practices increase operating costs for coal and other dispatchable generators. Collectively, these additional costs tend to be very small as a portion of system costs, but more palpable as a share of generator profits. In many countries, they are recovered either on a regulated cost basis or through energy and ancillary services markets. In China, however, thermal generators are still not directly compensated for a significant portion of their additional cost of accommodating wind generation.

13.4.2.4 Cost Premium Recovery

China's 2005 Renewable Energy Law created a national surcharge to pay for the higher cost of renewable energy. This surcharge, initially set at 0.001 yuan/kWh, is collected in each province through a 'renewable price surcharge' (*kezaisheng nengyuan fujia*). Grid companies collect these funds separately and use them to pay premiums to renewable energy generators within their own province. Where revenue collection exceeds payment obligations, the funds are collected centrally and redistributed to provinces where payment obligations exceed revenues.

The drawback to this approach is that, if renewable generation grows faster than total demand and/or if coal prices fall, the pool of revenues to pay premiums to renewable generators will be insufficient. In response to rapid growth in renewable energy, government agencies have increased the surcharge twice, to 0.008 yuan/kWh in 2011 and to 0.015 yuan/kWh in 2013. However, the lag between surcharge increases has led to significant gaps—10.7 billion yuan by the end of 2011 (Yu and Xiao 2014)—between what renewable generators are owed under the FIT, and what they are actually paid. Increasing the renewable energy surcharge is politically difficult because: (i) increasing the surcharge raises electricity prices that are already perceived to be high, and (ii) the surcharge is collected nationally and involves significant transfers among provinces.

13.4.3 Balancing Area Coordination

A properly functioning electricity system needs to instantaneously balance supply and demand within a small tolerance. Meeting uncertainties in demand and supply while respecting various system security constraints traditionally requires centralized system operation of dispatching plants (i.e., specifying production quantity). The geographic purview of a system operator is known as a 'balancing area'.

Coordinating neighbouring balancing areas has important benefits for integrating high penetrations of wind and solar energy: aggregating geographically distant resources tends to reduce resource variability; aggregating conventional energy sources increases total system flexibility; and access to more balancing options reduces integration costs such as reserves (GE Energy 2010). As a result of the grid operation institutions in China—including significant vertical separation of operations and planning, and complex horizontal overlapping authorities—the benefits of the large transmission network are not fully realized for wind integration.

13.4.3.1 Structure of China's Grid Operations

Electricity in China, following 2002 reforms, is served primarily by two large central state-owned grid companies, State Grid Corporation of China and China Southern Power Grid Company, and one local grid company, the Inner Mongolia Grid Company. State Grid is further organized into five grid regions, each consisting of roughly five provinces. Within State Grid and Southern Power Grid, direct subsidiary relationships of provincial grid companies within regional grids create nominal lines of authority.

Electric power operations in China involve a range of vertical and horizontal linkages among grid and government institutions. Power plants are for the most part dispatched by the provincial grid company enhancing vertical separation, although there is large heterogeneity across regions. Larger facilities and those serving grid-balancing functions may be directly dispatched by the regional or national grids. Quota-setting and heuristic dispatch ordering take place mostly at the provincial level. Regional grids help coordinate inter-provincial connections whereas the national grid helps coordinate inter-regional connections. These two coordination processes are key to the functioning of the system, and increasingly important for integrating large quantities of variable wind energy.[6] Central policies aim to increase inter-regional exchanges of electricity to exploit remote resources of wind, solar, coal, and hydropower (NEA 2014).

The annual generation planning process ensures that provinces can meet demand with supply and that generators will receive sufficient quotas to maintain profitability. Wind and other renewable energy may be incorporated at this stage by removing a portion of its expected generation from the total available quota. Transmission contracts for exchanges between balancing areas are negotiated in tandem with this process. The institutions involved in annual generation planning and their respective goals are varied (see Table 13.2).

After the heavily negotiated annual plans are finalized, the grid company's goal is to ensure these targets are met by allocating to shorter time periods and adjusting for intra-annual changes in supply or demand. They may be censured by government regulators if they deviate too much (SERC 2011). At the same time, they are faced with the possibly conflicting policy for mandatory procurement of wind. Short-term balancing operations *within* balancing areas are thus heavily constrained. Short-term adjustments *between* balancing areas are even more difficult because quotas are not easily convertible between regions, and the rigid transmission contracting process is difficult to renegotiate (Davidson et al. 2017).

These inflexibilities have led to increasingly high rates of wind curtailment across all major wind regions of China. In the northeast, where a large fraction of coal generators are must-run combined heat and power, Jilin and neighbouring Liaoning experienced 32 per cent and 10 per cent curtailment, respectively, in 2015 (NEA 2016). In the northwest, where a significant fraction of coal plants are directly dispatched by the regional operatorwind curtailment in Gansu reached the country's largest of 39 per cent in 2015 (NEA 2016). In this case, without reforming generation planning—addressing horizontal overlapping authorities—vertical integration has had limited benefits.

Within this rigid planning framework, policy makers in China have piloted various mechanisms—both market and administrative—to increase renewable energy dispatch, primarily focusing on the provincial grid. Energy-efficient dispatch, established in 2009, reorients renewable energy and high-efficiency coal to the top of annual plans, but does not compare plants in different

[6] For a more detailed treatment of institutional coordination issues, see Kahrl and Wang (2014).

Table 13.2. Actors and interests in annual generation planning in China

Actors	Interests
Grid companies	Increase efficiency of delivery (i.e., reduce losses) Utilize transmission lines with energy-based compensation (typically, ultra-high voltage)
Coal-fired power companies	Lobby for higher quotas
Wind companies	Lobby to reduce planned quantities of conventional generation
Provincial governments	Lower local electricity price Promote local generation over imports
National government	Minimize frictions among provinces Conserve resources nationally for energy security and environmental goals

Source: Authors' compilation.

provinces and does not completely do away with the basic quota system to ensure profitability (Davidson et al. 2017). Reforms announced in early 2015 highlight reducing the total amount in the plan (hence available for quota) as well as prioritizing renewable energy in cross-region transfers (NDRC and NEA 2015).

Vertical integration through regional power exchanges has been piloted and disbanded multiple times as many institutions sought to stall reform. According to one account, provincial governments and generation companies were united in opposition: provincial governments did not want to give up autonomy over planning decisions through a regional market, and risk-averse generation companies had already grown accustomed to the guaranteed revenue streams under the quota system (Wen 2014).

13.5 CONCLUSION

Energy system transitions, by introducing or replacing one technology or practice with another, inevitably create winners and losers. To better assess the landscape of political economy obstacles to energy system transition, in this chapter we developed an analytical framework to understand the political economy of wind energy—a high potential source of zero-carbon dioxide electricity.

At the generator level, the political economy impacts of wind power are in part driven by wind's physical characteristics. Wind displaces conventional dispatchable resources (e.g., coal, natural gas generation) because of its low marginal costs, but requires dispatchable resources for balancing because of its limited predictability and variability. This may reduce capacity utilization of conventional dispatchable generators and force them to operate in new conditions, creating costs that may or may not be remunerated under existing market or regulatory rules.

Yet, as we showed, the political economy impacts of wind extend beyond transfers among generators. Greater investment certainty for wind generators is underwritten by electricity customers, and governments may transfer more of that risk onto specific classes of customers (e.g., residential or industrial customers). Regional dispatch can reduce the operating challenges of wind power, but integrating local electricity systems into more regionally coordinated dispatch creates economic transfers between higher- and lower-cost regions. The losers in these political economy conflicts will often resist policies that support wind, renewable energy, and energy transition.

By applying our framework to China, a country with electricity sector institutions that are very different from those in most other countries, we demonstrated how wind development and integration challenges in China can also be understood within a more general political economy framework. On the basis of our analysis, we surmise that vertical separation (i.e., degree of federalism) plays a very important role in explaining wind integration outcomes in China, perhaps as important as in the US and Europe. We also find evidence that underneath the veneer of stronger horizontal integration in China the disparate interests of actors can lead to poor coordination across functions such as generation and transmission planning, or generation planning and dispatch, with consequences for wind development. Fleshing out on-the-ground implications of vertical and horizontal separation for wind integration is an important topic for future empirical work.

China is, in many ways, an extreme case because of the severity of its wind energy curtailment problem. As such, China presents a cautionary tale of the perils of not proactively identifying and addressing potential political economy conflicts. We argue that, although the technical challenges of renewable integration may have reasonably straightforward solutions, addressing political economy challenges by their nature must be built into longer-term political and economic strategy. In developing policies to facilitate low-carbon energy transitions, governments should ensure that they simultaneously acknowledge and address potential political economy conflicts.

REFERENCES

Bird, L., J. Cochran, and X. Wang (2014). 'Wind and Solar Energy Curtailment: Experience and Practices in the United States'. NREL Technical Report NREL/TP-6A20-60983, National Renewable Energy Laboratory (NREL), Golden, CO. Available at: <http://www.nrel.gov/docs/fy14osti/60983.pdf> (accessed 28 March 2016).

Davidson, M. R., C.-T. Li, and V. J. Karplus (2017). 'Grid Operations and Renewable Energy Integration in China'. Mimeo (based on interviews conducted in China's Northwest and Inner Mongolia), MIT Joint Program on the Science and Policy of Global Change, Cambridge, MA.
E3 (2015). 'Generation Cost Model for China'. Energy + Environmental Economics (E3), San Francisco, CA. Available at: <https://www.ethree.com/documents/generation_cost_model_for_china/generation_cost_model_for_china.xlsx> (accessed 28 March 2016).
Fischlein, M., J. Larson, D. M. Hall, R. Chaudhry, T. Rai Peterson, J. C. Stephens, and E. J. Wilson (2010). 'Policy Stakeholders and Deployment of Wind Power in the Sub-national Context: A Comparison of Four US States'. *Energy Policy*, 38(8): 4429–39.
GE Energy (2010). 'Western Wind and Solar Integration Study'. NREL Technical Report NREL/SR-550-47434, National Renewable Energy Laboratory (NREL), Golden, CO. Available at: <http://www.nrel.gov/docs/fy10osti/47434.pdf> (accessed 28 March 2016).
GWEC (Global Wind Energy Council) (2016). 'Global Wind Statistics 2015'. GWEC, Brussels. Available at: <http://www.gwec.net/wp-content/uploads/vip/GWEC-PRstats-2015_LR.pdf> (accessed 28 March 2016).
Haggett, C. (2008). 'Over the Sea and Far Away? A Consideration of the Planning, Politics and Public Perception of Offshore Wind Farms'. *Journal of Environmental Policy & Planning*, 10(3): 289–306.
Han, J., A. P. J. Mol, Y. Lu, and L. Zhang (2009). 'Onshore Wind Power Development in China: Challenges behind a Successful Story'. *Energy Policy*, 37(8): 2941–51.
Hogan, W. W. (2002). 'Electricity Market Restructuring: Reforms of Reforms'. *Journal of Regulatory Economics*, 21(1): 103–32.
Jacobsson, S. and A. Johnson (2000). 'The Diffusion of Renewable Energy Technology: An Analytical Framework and Key Issues for Research'. *Energy Policy*, 28(9): 625–40.
Jamasb, T. (2006). 'Between the State and Market: Electricity Sector Reform in Developing Countries'. *Utilities Policy*, 14(1): 14–30.
Jamasb, T., R. Mota, D. M. Newbery, and M. G. Pollitt (2005). 'Electricity Sector Reform in Developing Countries: A Survey of Empirical Evidence on Determinants and Performance'. Policy Research Working Paper WPS3549, World Bank, Washington, D. C. Available at: <http://www-wds.worldbank.org/external/default/WDSContentServer/WDSP/IB/2005/03/30/000012009_20050330110431/Rendered/PDF/wps3549.pdf> (accessed 28 March 2016).
Joskow, P. L. (2008). 'Lessons Learned from Electricity Market Liberalization'. *The Energy Journal*, 29(2): 9–42.
Kahrl, F. and X. Wang (2014). 'Integrating Renewables into Power Systems in China: A Technical Primer—Power System Operations'. Regulatory Assistance Project, Beijing.
Kahrl, F., J. H. Williams, and J. Hu (2013). 'The Political Economy of Electricity Dispatch Reform in China'. *Energy Policy*, 53: 361–9.
Krishna, C., A. D. Sagar, and S. Spratt (2015). 'The Political Economy of Low-Carbon Investments: Insights from the Wind and Solar Power Sectors in India'. Institute of Development Studies, Sussex.

Lehmann, P., F. Creutzig, M.-H. Ehlers, N. Friedrichsen, C. Heuson, L. Hirth, and R. Pietzcker (2012). 'Carbon Lock-Out: Advancing Renewable Energy Policy in Europe'. *Energies*, 5(2): 323–54.

Ling, Y. and X. Cai (2012). 'Exploitation and Utilization of the Wind Power and Its Perspective in China'. *Renewable and Sustainable Energy Reviews*, 16(4): 2111–17.

Liu, Y. and A. Kokko (2010). 'Wind Power in China: Policy and Development Challenges'. *Energy Policy*, 38(10): 5520–9.

Mahoney, J. and K. Thelen (2010). 'A Theory of Gradual Institutional Change '. In J. Mahoney and K. Thelen (eds), *Explaining Institutional Change: Ambiguity, Agency, and Power*. Cambridge: Cambridge University Press.

Markard, J. (2011). 'Transformation of Infrastructures: Sector Characteristics and Implications for Fundamental Change'. *Journal of Infrastructure Systems*, 17(3): 107–17.

MIT (Massachusetts Institute of Technology) (2011). 'The Future of the Electric Grid: An Interdisciplinary Study'. MIT Energy Initiative, Cambridge, MA.

NDRC and NEA (National Development and Reform Commission and National Energy Administration) (2015). 'Guiding Opinion Regarding Improving Electricity System Operational Adjustments for Increased and Complete Clean Energy Generation'. Development and Reform Notice 518, 20 March, NDRC/NEA, Beijing. Available at: <http://www.sdpc.gov.cn/zcfb/zcfbtz/201503/t20150323_668203.html> (accessed 28 March 2016).

NEA (2014). 'Energy Sector Strengthening Air Pollution Prevention Plan'. NEA, Beijing. Available at: <http://www.nea.gov.cn/2014-05/16/c_133338463.htm> (accessed 28 March 2016).

NEA (2016). 'Wind Industry Development Statistics 2015'. NEA, Beijing. Available at: <http://www.nea.gov.cn/2016-02/04/c_135073627.htm> (accessed 28 March 2016).

North, D. C. (1991). 'Institutions'. *Journal of Economic Perspectives*, 5(1): 97–112.

Porter, M. E. and S. Stern (2001). 'Innovation: Location Matters'. *MIT Sloan Management Review*, 42(4). Available at: <http://sloanreview.mit.edu/article/innovation-location-matters/>(accessed 28 March 2016).

Scrase, I. and A. Smith (2009). 'The (Non-)Politics of Managing Low Carbon Socio-technical Transitions'. *Environmental Politics*, 18(5): 707–26.

SERC (State Electricity Regulatory Commission) (2011). 'National Electricity Exchange and Market Operations Report'. SERC, Beijing.

Stigler, G. J. (1971). 'The Theory of Economic Regulation'. *The Bell Journal of Economics and Management Science*, 2(1): 3–21.

Ventosa, M., P. Linares, and I. J. Pérez-Arriaga (2013). 'Power System Economics'. In I. J. Pérez-Arriaga (ed.), *Regulation of the Power Sector*. London: Springer.

Viétor, B., T. Hoppe, and J. Clancy (2015). 'Decentralised Combined Heat and Power in the German Ruhr Valley: Assessment of Factors Blocking Uptake and Integration'. *Energy, Sustainability and Society*, 5(1): 5.

Wen, H. (2014). 'East China Electric Market Experiment Beginning to End'. *Southern China Energy Observer*, 27 January. Available at: <http://hvdc.chinapower.com.cn/news/1036/10367311.asp> (accessed 28 March 2016).

Williams, J. and R. Ghanadan (2006). 'Electricity Reform in Developing and Transition Countries: A Reappraisal'. *Energy*, 31(6/7): 815–44.

Yang, M., D. Patiño Echeverri, and F. Yang (2012). 'Wind Power Generation in China: Understanding the Mismatch between Capacity and Generation'. *Renewable Energy*, 41: 145–51.

Yu, H. and Q. Xiao (2014). 'Old Debts for the Renewable Energy Premium Are Hard to Collect'. *China Energy News*, 17 November. Available at: <http://paper.people.com.cn/zgnyb/html/2014-11/17/content_1500391.htm> (accessed 4 April 2016).

14

The Social Shaping of Nuclear Energy Technology in South Africa

Britta Rennkamp and Radhika Bhuyan

14.1 INTRODUCTION

Why is the South African government pursuing a nuclear energy programme, despite abundant and accessible fossil and renewable energy resources? The South African electricity sector has historically been mainly coal-fuelled. One nuclear power plant has contributed five per cent of electricity since the 1980s. Renewable energy contributes another five per cent.

We argue that the government has chosen to procure an additional 9.6 GW of nuclear capacity, because of a significant coalition of nuclear energy supporters. The role of coalitions in shaping policy outcomes has long been established. The constructivist literature on science and technology policy explains why decision-makers sometimes choose less practical and less cost-efficient technologies over others. The concept of social shaping of technology suggests that the interplay of social, political, economic, and cultural factors in a society shape the design and implementation of a technology. The literature identifies political factors leading to the prioritizing of nuclear technologies over other alternatives. We apply this theoretical perspective to the recent nuclear programme in South Africa. A discourse network analysis helps to establish the political arena of nuclear energy and to identify the coalitions supporting and opposing the nuclear energy programme, as well as the arguments that motivate their positioning (Leifeld 2012). The analysis uses data from 350 media articles, relevant policy documents, and background interviews.

14.2 SOCIAL SHAPING OF NUCLEAR ENERGY TECHNOLOGIES AND THE ROLE OF DISCOURSE COALITIONS

Existing research from the sociology, political science, and nuclear scholarship identifies various motivations for governments to pursue nuclear programmes, despite other options. Sociologists framed theories of social shaping of technology (SST), social construction of technology (SCOT), and actor network theory (ANT) in the 1980s and 1990s. These approaches dismiss the idea of 'technological determinism', which assumes technological choices are purely economically or technically determined. Constructivist theories suggest that organizational, political, economic, and cultural factors 'pattern the design and implementation of technology' (Williams and Edge 1996). The reasons for acceptance or rejection of a technology emerge from the societies themselves (Bijker 1995; MacKenzie 1998). 'Relevant social groups' shape technological trajectories according to their interests and interpretations (Bijker 1995: 269).

Constructivists argue that choices are inherent to a technology trajectory. The logic of choice between technological trajectories makes technology a negotiable matter. Choices may be irreversible and lead towards long-term 'lock-in' situations (Bijker 1995; Williams and Edge 1996). The concept of choice relates directly to the politics that emerge from prioritizing one technology over another. Technologies are never neutral. Choices trigger controversy among their supporters and opponents (Williams and Edge 1996; Latour 2005). Complex social, political, and cultural dynamics are inherent to technological knowledge production and technology choices (Bijker 1995; Williams and Edge 1996; Latour 2005).

Scholars have identified factors that explain the choice of energy technologies, which may not be economically or technically advantageous. The analysis of the social construction of large technological systems demonstrates how technology choices correspond to the political and economic structure of a nation. The analysis of the United Kingdom's (UK) and German electricity sectors suggests that the political system shapes the governance structure of the power sector. Centralized governments produce large centralized electricity systems. Decentralized systems favour smaller distribution and generation infrastructure (Hughes 2011).

Autocratic elements within a political regime were critical in the adoption of large and centralized nuclear energy technologies (Winner 1986, 2000; Temples 1980). Countries that adopted nuclear technology show strong connections between nuclear power and nationhood (Jasanoff and Kim 2009). Nationhood is an idea built through public discourses and debates about national identity and a nation's path of future development. National public discourses often emerge from energy technology and its politics, as energy counts as a basic building block of industrialization, technological progress, socio-economic development, and consequential economic power. Governments often procure energy technologies because of prestige, rather than economic necessity (Hughes 2011).

A comparative analysis of six nuclear nations identifies six drivers that sustain commercial nuclear power programmes (Sovacool and Valentine 2012: 250):

(1) national security and secrecy
(2) technocratic ideology
(3) economic interventionism
(4) centrally coordinated energy stakeholder network
(5) subordination of opposition to political authority and
(6) social peripheralization.

The characteristics of countries that are following a nuclear path include closed political systems that minimize opposition, transparency, and accountability; economies with a history of central planning and government intervention; as well as strong national commitment to technological progress (Sovacool and Valentine 2012).

Public discourse reflects the debates about national identity, a nation's path of future development, and its significance in the global arena. The way a nation exploits its energy resources is an essential component of this process (Bouzarovski and Bassin 2011). Nuclear technology programmes are state interventions that create winners and losers, as any other public policy. Distributional conflicts motivate actors to shape coalitions in support or opposition of these policies that represent their beliefs, ideas, and interests (Sabatier 1988; Hajer 1995).

Discourse coalitions differ from traditional political coalitions or alliances, because there is a linguistic basis for political coordination between various parties. 'Story-lines, not interests, form the basis of the coalition, whereby story-lines potentially change the previous understanding of what the actors' interests are' (Hajer 1995: 66). This chapter presents an analysis of the discourse coalitions that emerge in support and opposition to the nuclear programme in South Africa.

14.3 NUCLEAR POWER POLITICS IN SOUTH AFRICA

South Africa may soon count as both old and new nuclear power. Nuclear power was a substantial commitment of the apartheid government, which ruled the country on behalf of the white minority between 1948 and 1994.

Under apartheid rule, engineers built nuclear capacity in nuclear fuel production, a weapons programme, uranium enrichment, and research since the 1950s. The programme came with a high cost, as large parts of the electricity produced from the reactor went into fuel and uranium production (Marquard 2006). The remainders of this programme are Africa's only commercial nuclear plant in Koeberg outside Cape Town, a research reactor in Pelindaba near Pretoria, a state-owned nuclear cooperation, a regulator, and a coalition of nuclear engineers.

The African National Congress (ANC) actively opposed nuclear power in its role of a liberation movement and during its early days as a ruling party (ANC 1994). This position changed gradually, as the scientist and technicians involved in the pro-nuclear coalition continued pushing for a continuation of nuclear energy.

The pebble-bed modular reactor (PBMR) was used in the ANC's first nuclear programme. The programme aimed to commercialize a German licensed small reactor design, which did not succeed in a research and development (R&D) programme in Germany and lost public funding in 1998 (BMU 2006). The main problem with the pebble bed reactor was that it created very high temperatures at its core, which could not safely transform into process heat. The temperatures inside the reactor could only be estimated, as the movement of the pebbles inside the reactor destroyed all measuring devices (Moormann 2008). South African and German scientists convinced the ANC government to fund the technology from 1993 until 2010. The PBMR received more than ZAR10 billion of public funding, accounting for one-third of the Department of Trade and Industry's R&D budget (Creamer 2010; Hogan 2010).

The PMBR programme originally envisioned producing electricity in smaller reactors than pricy pressurized water reactor (PWR) plants like Koeberg. Yet, it became part of a wider 'nuclear policy' published in 2008, which aimed to revive the manufacturing processes including uranium enrichment and nuclear fuel production. The policy framework aimed to build new PWR plants by 2015 and create more than 10,000 new jobs in the industry (DME 2008).

The release of the nuclear policy coincided with the world's financial crisis and severe shortages in electricity supply. International rating companies evaluated South Africa's utility Eskom's financial prospects negatively, which did not correspond with the government's ambition to acquire new nuclear power plants, 'owned and operated' through Eskom (DME 2008).

In 2011, the ANC restated its plans for the nuclear build programme three days after the accidents in Fukushima. The nuclear build programme suggests building six nuclear plants to add 9.6 GW of capacity to the current national capacity of about 40 GW. The programme refers to the country's electricity plan, the integrated resource plan (IRP), that outlines South Africa's electricity mix 2030. The IRP was the first integrated, participative plan for the future of the electricity sector, which grounds in the energy white paper (DME 1998).[1] The plan proposed two scenarios with 9.6 GW of nuclear power (DoE 2011). Initially, the plan was set out as a flexible planning instrument, subject to updates every two years as economic and technical parameters may change. The original IRP was grounded in higher assumptions of economic growth than in reality, which translated into a higher demand (RSA 2010; DoE 2011). The original IRP stated clearly that growth assumptions have to be revised in the future (DoE 2011). The IRP update report (DoE 2013) revised the lower electricity demand and lower growth assumptions. The report established several scenarios taking alternative electricity generating technologies and prices into account. It recommended delaying or abandoning the nuclear built programme if the electricity demand does not unfold accordingly; solar, hydro, and gas options become available; or the cost of procuring nuclear energy exceeds a certain cost.[2] The IRP update report did not undergo a public consultation process and was not presented for parliamentary approval. The original IRP remains the main policy basis for the government's plans to expand nuclear power generation capacity.

14.3.1 Discourse Coalitions in Support and Opposition of the Nuclear Build Programme

The government's plans to expand nuclear energy triggered a polarizing public debate between two coalitions in support and opposition of the programme. Fifty-five organizations are publicly involved in the debate on the nuclear programme—Figure 14.1 shows these actors as circles. The organizations range from business (including industry, consulting firms, trade unions, and business associations) to governmental departments, state agencies, state-owned enterprises, and civil society organizations, which include non-governmental organizations (NGOs) and academic institutions.

Actors share a connection if they put forth the same argument to support or oppose the nuclear programme. These arguments appear as squares. The coalition in support of nuclear power shares connections in light lines. The actors in the opposing coalition connect through dark lines. Arguments

[1] The Energy White Paper (DME 1998: 58) ruled out that new nuclear capacity should be added before 2007. The paper qualifies that 'whether new nuclear capacity will be an option at that point or beyond will depend largely on the environmental and economic merits of other energy sources relative to nuclear and its potential and public acceptability, construction lead-times and load characteristics'.

[2] The IRP update report (DoE 2013) suggests delaying the decision to procure nuclear energy 'if the demand is lower than 270 TWh, if hydropower, [...] rooftop photovoltaic or shale gas options succeed. The programme should be cancelled if the procurement price exceeds US$6,500 per kWh.'

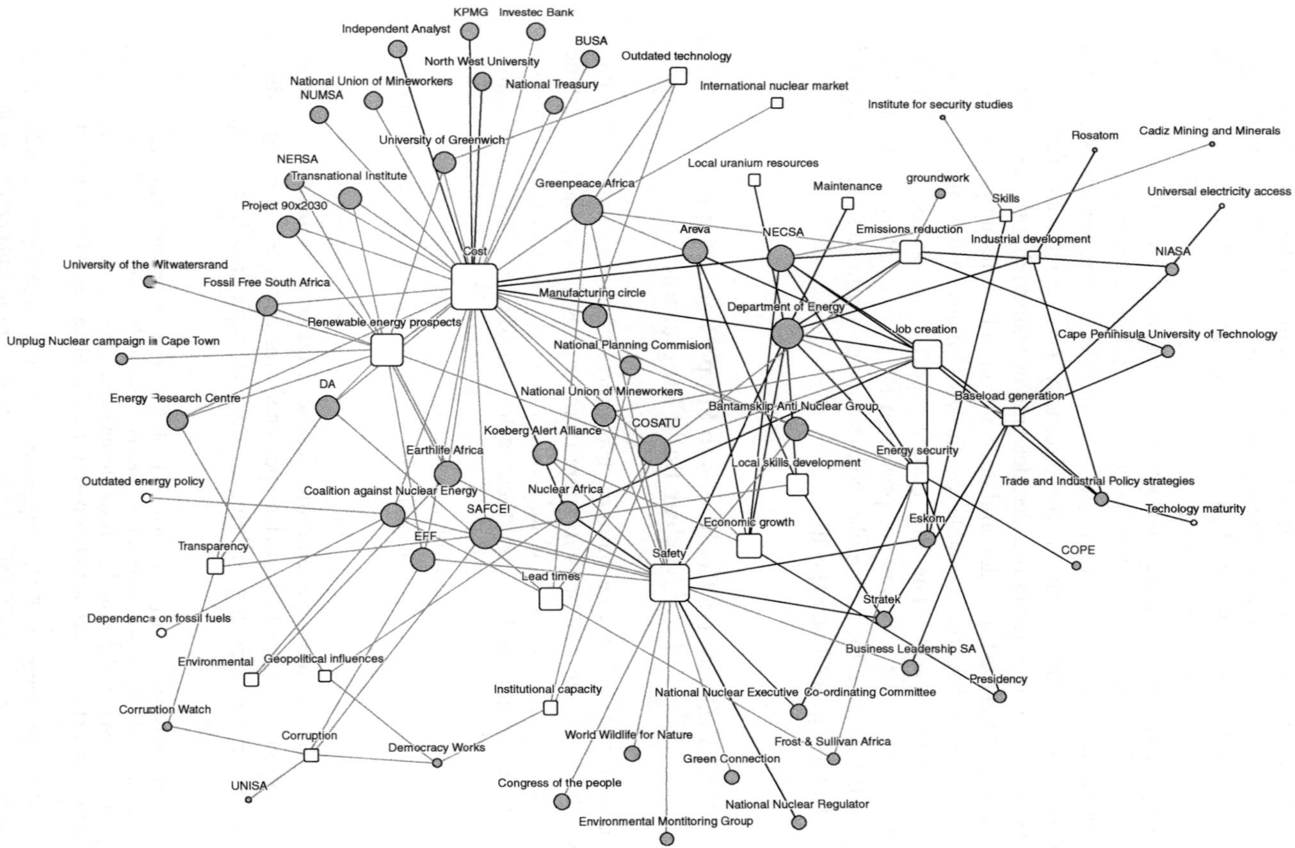

Figure 14.1. Discourse coalitions in South Africa's nuclear power programme.

Source: Authors' compilation.

that share both lines are the most contested arguments. Actors and arguments that share many connections are more relevant to the discourse. The discourse network emerges from a dataset of 350 articles, which made 64 statements in favour or against the nuclear programme. Arguments and actors were coded and their discursive relationships appear in the network.

The actors put 24 arguments forth in support or opposition of the nuclear energy programme. Cost and safety emerge as the main contested issues in the current nuclear debate. Impacts on emissions reductions, economic growth, job creation, and skills development are also contested in the public debate. The four most frequent arguments, calculated in Eigenvector centrality, are cost, safety, renewable energy prospects, and job creation.

14.3.1.1 The Coalition of Supporters

The coalition of supporters consists mainly of government departments, agencies, and state-owned enterprises, private business, as well as a few outspoken individuals in universities and consulting companies. The Department of Energy (DoE) is the central actor in this coalition, followed by the Nuclear Energy Cooperation South Africa (NECSA). Further state actors in support of the programme are the utility Eskom, the Presidency, the Nuclear Regulator, and the Nuclear Energy Committee (NNEECC). The DoE has become very active in advancing the nuclear programme. The Department's main arguments for nuclear power include cost, baseload generation, energy security, emissions reductions, job creation, industrial, and skills development.

South Africa's state-owned nuclear cooperation, NECSA, supports the programme arguing for benefits in job creation, industrial development, energy security, and skills development. The organization employs about 100 of the country's nuclear engineers and scientists. NECSA has high stakes in the nuclear debate, as the organization is supposed to play a significant role in the future programme. The ownership of the new fleet is not yet clearly defined. Eskom owns and operates Koeberg, but there is uncertainty whether the utility will be in the position to manage the new plants as well or if NECSA will step in. NECSA denied media reporting that the organization is under financial constraints and unable to pay salaries (Joubert 2015). The annual report disclosing finances for 2015 is still outstanding.

The NNEECC was first established in 2008 and headed by the former President Kgalema (Creamer 2012). The committee consisted of a group of ministers supplemented with a committee of technical experts, which were exchanged to political appointments. Motlanthe did major preparations and networking efforts with international vendors to identify financing for the nuclear programme. In 2013, Jacob Zuma took over the committee's leadership (City Press 2013; Donelly and Faull 2013). This move reflects the president's growing interest in the nuclear programme. His main arguments for nuclear power are economic growth and its contribution to energy security of the country in the long term. Cost has been its only publicly stated constraint to the programme (Zuma 2016). In 2014, the committee changed its purpose to overview the entire electricity sector. The committee continues supporting the nuclear programme arguing that nuclear energy will contribute to energy security and that the technology is safe. The National Nuclear Regulator echoes safety affirmations.

Business representation falls into three categories. International vendors, namely Rosatom of Russia and Areva of France are in the first category. Rosatom argues that the main benefit will be industrial development, as the company plans to involve the South African nuclear industry in a 'global value chain' through high rates of locally produced components of nuclear technology. Areva argues for job creation, contributions to economic growth, and skills development. The second category is domestic business representation. The Nuclear Atomic Industry Association (NIASA) and Business Leadership SA share the arguments for local skills development, economic development, and baseload generation. The third category consists of consulting companies and academic institutions. The consulting firms, Stratek and Nuclear Africa, as well as the universities of the North West and the Cape Peninsula are institutional homes to a handful of individuals who are very actively engaged in supporting nuclear energy in the public debate. Their arguments concentrate mainly on the cost of the nuclear programme and its contributions to emissions reductions, job creation, baseload, and economic growth.

14.3.1.2 The Coalition of Opponents

The coalition of opponents to the nuclear programme connects some government departments and agencies, business associations and trade unions, political parties, academic institutions, and many NGOs. The opposing coalition is larger in terms of numbers of actors and arguments. This coalition does not show a central actor pushing against the programme. It is a wide range of 40 actors with a majority of civil society organizations. The main arguments against the nuclear programme are the cost, safety of nuclear technology, and alternative solutions that include an expansion of the renewable energy programme. Numerous local and international NGOs share these views, motivated through environmental conservation concerns or place-based concerns about the prospects of having nuclear power plants built in their proximity.

The main business actors in this coalition are business associations and trade unions. The main mine worker unions (NUM, NUMSA, and COSATU) argue against the programme. The unions express concern about harmful impacts of an expensive nuclear programme on economic growth and job creation (COSATU 2015). The trade unions historically protect labour rights in the mining sectors. COSATU argues that the baseload generation argument for nuclear energy does not hold, as coal-fired plants can provide sufficient baseload to supplement renewable energy. COSATU also expresses its concerns about the possibility of human error and natural disasters, which led to the accidents in Ukraine, the United States (US), and Japan. They highlight the lack of safe waste storage and a policy framework to store waste in a secure manner. Business Unity South Africa (BUSA), South Africa's largest business association, shares the unions' concern about the cost of the programme. The association supports the recommendations of the national planning commission to delay the decision on an 'extremely expensive technology option', as electricity demand has declined (Paton 2014). A few governmental actors form part of the opposing coalition to the nuclear programme. The National Treasury has kept its eye on the cost of the nuclear programme, which is its principal argument that may lead to halting the programme.

When President Jacob Zuma first took office, he assembled a planning commission to develop a vision for the country's development path until 2030. The first draft did not touch on the nuclear programme, but the final draft recommended to delay the decision and to conduct environmental and economic analyses to understand the feasibility of the programme first (NPC 2011). The Energy Commissioner argued against the nuclear programme stating that it would not contribute to solving the current electricity crisis and that it is more expensive than other options. He argued that Eskom does not have the institutional capacity to operate a fleet of six power plants, which would leave the operation to foreign companies. The National Energy Regulator (NERSA) argued against the nuclear programme, because of its high cost and opportunity cost of an investment in nuclear power, which would be better placed in renewable energy, which allows for power generation from independent power producers.

The opposition party, Democratic Alliance (DA), argued against nuclear power using similar arguments. Long lead times of building nuclear power plants will not solve the current supply shortages. The ANC should rather look into affordable renewable energy options than invest in an expensive nuclear programme. The Economic Freedom Fighters (EFF) argued that the cost for the programme is too high, the technology is not safe, and emissions reductions can be achieved through expanding the renewable energy programme (Ndlozi 2015).

14.3.2 Political Discourse Shaping South Africa's Nuclear Technology Decision

Sections 14.3.2.1–14.3.2.8 contextualize the main arguments and actors in the discourse networks within the framework of analysis of political factors that favour nuclear programmes, according to Sovacool and Valentine (2012).

14.3.2.1 Cost

Cost is the most controversial issue in the current debate on the nuclear build programme. Both coalitions refer to the cost argument. Thirty-nine statements from the opposition question the affordability of the programme, while the supporters affirm the affordability of the programme. The polarizing debate on costs emerges from the variety of cost estimates available publicly and the lack of cost assessment for the South African programme.

The costs per kilowatt of new build nuclear power vary from different sources and high or low cost scenarios between US$1,500 and US$8,000 (Caetano and Rennkamp 2014; Thomas 2010). Taking the lowest and the highest price estimates for the nuclear build programme of 9.6 GW, prices range between US$14.4 billion and US$76.6 billion. The National Treasury collects annual revenue of roughly ZAR780 billion/US$46 billion.

Delayed scenarios that increase the initial cost estimates are likely in nuclear power plant building, because every plant is innovative. Any changes that engineers need to make in the construction require approval from the national regulator to ensure the safety of the plant. These additional bureaucratic processes often cause delays. Recent nuclear plants built in Flamanville, France, and Olkiluoto, Finland, cost more than twice the originally estimated budget and delayed construction times were over ten years (WNN 2015).

Large public infrastructure works bear additional risk factors. Strikes of highly unionized workers may cause delays. Scarce civil engineering and road infrastructure may cause unforeseen logistical challenges. Both risk factors have caused delays during the construction of the coal-fired plants in Medupi and Kusile (SAPA 2015; Steyn 2015a).

14.3.2.2 Opposition to Political Authority

The contested price tag of the nuclear build programme caused major debates within the government and the energy stakeholder network. Several ANC internal battles emerged from this issue. The controversial dismissal of former finance minister, Nhlanhla Nene, had its roots in his critical eye on the build programme, among other issues (Brummer 2015; Paton 2015). Old and new finance minister, Pravin Gordhan, committed to funding only projects that fall within the budget (England 2015). Jacob Zuma stated in his state of the nation speech in February 2016, 'we will only procure nuclear on a scale and pace that our country can afford' (Zuma 2016). The energy minister stated repeatedly that nuclear energy 'will be affordable' and that it had calculated the cost of the build programme, but cannot release any of the cost estimates (Gqirana 2015).

The National Planning Commission commissioned a study on the cost implication of a nuclear programme, which revised the assumption of IRP (ERC 2013). The update of the IRP reported similar dynamics and suggested that the nuclear decision 'can possibly be delayed' (DoE 2013: 8). The update process of the IRP suggests that there are controversies within the department, as department internal analysts suggested delaying the nuclear decision. These controversies have not been reflected in the public discourse.

The debates on the cost of the programme closely relate to the power battles internal to the ANC and the opposition to political authority, as Sovacool and Valentine (2012) put it. In a single party system with nearly two-thirds majority of the vote, internal opposition creates substantial checks and balances. Traditional opposition work from alternative political parties is compromised, as they are unlikely to win a national election and threaten the power position of the ruling elite.

14.3.2.3 Economic Interventionism

The debate on cost also relates to the concept of 'economic interventionism'. The nuclear programme would be South Africa's largest public expenditure in the history of the country. The public procurement process will have to correspond to the constitutional rules. These require that public expenditure corresponds to five principles: 1) value for money; 2) open and effective competition; 3) ethics and fair dealing; 4) accountability and reporting; and 5) equity. Public procurement procedures in principle need to correspond to each of these pillars (RSA 1996, n.d.). The rules correspond to the government's attempts for Black Economic Empowerment (BEE), which targets (previously) disadvantaged groups of individuals, small businesses, and local producers to benefit from economic development. Inequalities in ownership of South Africa's asset base continue to be a major structural problem, which these rules aim to correct. The procurement rules specify more details on the relationship between price and social development criteria (localization, community development) in specific programmes. A 'public protector' oversees these processes (RSA 2014). A government gazette supports the procurement of the nuclear programme (DoE 2015a).

Another typical characteristic of economic interventionism is the attempt to advance job creation and industrial development through large publicly procured infrastructure projects. The analysis of discourse coalitions showed that the supporting coalition argues for industrial development and job creation as positive outcomes of the nuclear programme. 'Localization' is a very specific objective in South African economic policy-making. The idea of introducing local content requirements into public procurement rules is common in developing countries. The aim is to minimize the amount of imported technology and to produce as many components as possible locally. South Africa's Department of Science and Technology developed a 'localization strategy' that aims to mainstream the approach of localizing technological components across the economy (DST 2008). Localization always comes with the trade-off between potential job creation locally, a higher technology price, and delays caused through technological learning.

Nuclear energy technology also falls under the localization strategy (DST 2008). The DoE aims to achieve the following objectives through localization, based on its nuclear energy policy: 1) lead the supply of nuclear products and services; 2) create jobs; 3) become self-sufficient in all aspects of the nuclear fuel cycle; and 4) contribute to energy security and economic growth (DoE 2013). Uranium enrichment can be mined as a by-product of gold and copper mining. A small industry of three companies contributes to 1 per cent of the world's uranium supply. This is a relatively small share, given that South Africa's accessible uranium reserves add up to 4.8 per cent of the world's known resources (Van Wyk 2013).

The representatives of the Russian nuclear industry nurtured the South African government's aspiration to revive nuclear manufacturing processes locally and to create jobs in the sector. Rosatom spokesperson, Segey Novikos, sees potential for 60 per cent localization of the nuclear plants in South Africa through Rosatom (SABC 2013). Former energy minister, Ben Martins, pointed out that localization, job creation, and skills development were the main 'tenets' to 'create a better life for our people' (Campbell 2013). The dreams of reactivating the nuclear value chain with a local enrichment and nuclear fuel programme are 'unrealistic' with outdated technologies and no access to an enrichment plant, according to a representative of the Nuclear Energy Industry Association. Localization of the construction increases the cost of the programme, but the actual numbers are unknown.[3] Areva criticized the South African localization plans as being unrealistic, as only 10 per cent of local companies were qualified to contribute local content, which makes 50–80 per cent targets difficult to implement (Steyn 2015b).

14.3.2.4 Job Creation and Skills

Job creation is the third contested impact of the nuclear build programme. The coalition of supporters uses this argument frequently in support of the programme. In light of a national unemployment crisis with unemployment as high as 25 per cent and youth unemployment at 50 per cent, the national government's discourse centres on prioritizing job creation (RSA 2010). The programme can create similar employment to Koeberg in each new power plants, which adds up to roughly 10,000 permanent jobs in the operation. Job creation in construction, design, and manufacturing are uncertain.

The debate on job creation questions who the beneficiaries would actually be. Former energy minister, Dipuo Peters, saw job creation potential along the

[3] Interview, South African Nuclear Industry Association (2013).

value chain including the uranium-mining sector. The controversial Gupta family acquired a uranium mine, but required additional funding to operate it (Shamase 2011; Gosam 2016). Greenpeace argues that job creation in the nuclear programme is unlikely to make a difference, as the technology is highly specialized and import-driven. COSATU argues along similar lines, nuclear programme will have minimal effects on South Africa's unemployment crisis. Most likely, this will benefit specialists in the nuclear establishment and foreigners. The estimates for job creation vary from 27,000 (Eskom) to 70,000 (DoE) on up to 400,000 (Areva) (Biyase 2010; Felix 2013).

The debate on skills and their development intertwines with the debate on job creation. NECSA appears in the opposing coalition once, because of the statement that there are not enough skills in the nuclear sector, despite their training efforts. Others doubt whether the skills development in the past still suffices to manage a large scale build programme in the future (Wild 2013). NGOs argue that the country lacks the nuclear engineers to operate the fleet safely (Abbas 2015). Eskom denies any skills shortages (Wild 2013).

14.3.2.5 Safety

Safety concerns are the second main contested argument in the discourse network. The coalition of opponents mainly uses this argument along three lines. The first argument refers to the risks of nuclear accidents. Greenpeace argues that there is a lack of regulatory capacity to deal with nuclear accidents at the scale of Fukushima (Macleod 2011; Donelly 2015). The bulk of the NGOs, political opposition parties, and trade unions share this general concern. The second argument refers to the risk from radiation under normal operation. The third argument refers to waste and its storage. Nuclear waste can never be stored safely and can jeopardize the health of the generations to come according to Earthlife and the Coalition against Nuclear Energy in South Africa (Bega 2011).

The main argument of the nuclear supporters is that Koeberg has been operated safely for 30 years. Eskom, Nuclear Africa, Kgalema Motlanthe, the DoE, and the National Nuclear Regulator share this point (Bega 2011). Safety and environmental concerns of the nuclear programme motivated numerous local and international NGOs to engage publicly in the discourse on nuclear energy. Earthlife and Greenpeace produced studies to inform the debate (Earthlife 2007; Adam et al. 2011). Almost half of the actors in the opposing coalition are NGOs.

There are a number of local organizations that tackle the immediate consequences of nuclear power plants near major urban spaces. Yet, civil society engagement remains largely at the small scale of individual research, publications, and media interviews.

The factor of 'social peripheralization' implies that NGOs are only marginally influential in the political discourse (Sovacool and Valentine 2012). NGOs are a substantial component of the coalition of opponents. Overall, 16 active NGOs represent a population of 55 million people and their focus is mainly on knowledge production, media outreach, and local initiatives. There are no visible nationwide campaigns.

14.3.2.6 Energy Security, Renewable Energy, and 'Baseload'

A third debate centres on energy security and the feasibility of nuclear technologies over other alternatives. The third most frequent argument in the opposing coalition relates to the feasibility of nuclear versus renewable and other technologies. Actors argue that the prospects of renewable energy technologies make the nuclear programme unnecessary. Their short construction lead times can help solving the current supply shortages. Further arguments relate to higher cost-effectiveness and safety, as well as the fact that South Africa's abundant solar and wind resources that have not been fully exploited yet.

The coalition of supporters counters arguments for renewable energy with the need for 'baseload generation' (Pressly 2013). This argument refers to baseload power needed to address the intermittency of solar and wind technologies during the night and in case of no wind. The actors supporting nuclear energy argue mainly that nuclear power provides basepower, which renewable energy does not. NIASA, NECSA, CPUT, and the DoE continue this line of argument that the country needs to reduce emissions. Minister Joemat-Pettersson summed up this view asking the question, 'How are we going to reduce our carbon footprint and increase our baseload if we are not going to do nuclear energy?' (Gqirana 2015).

The main debates on the technical issues on energy security, as well as baseload and alternative supply technologies reflect the structure of the power sector and the energy network. Increasing the influx of renewable energy would favour independent power producer and change the current structure of the power sector towards decentralization. A power sector reform towards decentralization would create power losses in the current beneficiaries of the sector, which are mainly the state-owned companies operating coal and nuclear plants, Eskom, and to some extent NECSA. Attempts to restructure Eskom have failed as the process of the Independent System Markets Operator (ISMO) illustrate. The bill was dropped after five years of consultation.

In sum, the debate on baseload versus renewable energy technology reflects the polarizing positions on the current and future structure of the power system and its stakeholder network.

14.3.2.7 Secrecy, Transparency, and Corruption

The political opposition parties, as well as NGOs, argue against the nuclear programme, because of the lack of transparency in the policy process and risk for corruption. Secrecy surrounding the relationship with Russia, the cost of the programme and controversial appointments as Minister of Energy fuelled these arguments. The World Wildlife Fund (WWF) argues that the nuclear programme can be compared to the arms deal in the way the cabinet runs the process secretly ignoring the Promotion of Access to Information Act (PAIA) (Fakir and Pienaar 2016). The NNEECC's 'proceedings and documents are classified under the Minimum Information Security Standard Act (MISS Act) as TOP SECRET'. The Committee reports to the cabinet. No agenda or minutes can be shared with the public (Zuma 2015).

The DoE has denied any allegations of secrecy arguing that the documents need to be kept classified in order not to compromise the integrity of the process or confuse the public (DoE 2015b; Gqirana 2015; Zulu 2015; Le Cordeur 2016). The arguments about corruption and lack of transparency from the opposition to the nuclear programme emerge from the unknown cost of the programme, the lack of transparency in the procurement process, and the relations with Russia.

14.3.2.8 International Prestige and Geopolitical Dimensions

The veil of secrecy on cost and procurement reaches into the geopolitical dimensions of the nuclear energy programme. The nuclear technologies will require technology transfer from international vendor companies. These companies are typically closely linked to their national governments and make presidential diplomacy an essential component of international nuclear procurement. The South African government entertains close relationships to the world's major nuclear powers. Most of the world's installed nuclear capacity sits in the US, France, Japan, and Russia (IAEA 2015). An invitation to join the BRICS economics club catalysed the government's attempts to expand relations with other emerging economies. The BRICS are one of the few clubs that mostly favour nuclear power. In this group, Russia has been the country that the South African government has least established relations with. The nuclear programme offered an opportunity to fill the gap. Russia has been revitalizing its nuclear programmes under Putin's rule. Zuma had made the nuclear programme his presidential priority. He used the engagement with the BRICS nations to explore partnerships for implementation. According to media reports, Putin and Zuma negotiated the terms of the nuclear cooperation during the BRICS summits and several bilateral visits between 2010 and 2015 (Hunter and Faull 2014; Gosam 2016).

The DoE worked towards signing Memorandi of Understanding (MoU) with several countries, which are eligible to supply nuclear technology to South Africa. These MoU were very different in nature, length, and detail with the MoU with Russia the most lengthy and detailed (DoE 2015b). The opposition to nuclear power suspected that Russian suppliers had already been chosen as preferred partners without following public procurement rules. The DoE kept the MoU with Russia secret and did not reveal the details until the *Mail and Guardian* published a translation from the Russian version of the same MoU. Russia's nuclear provider, Rosatom, had published it on its website (RSA/RF 2014).

14.4 CONCLUSION

The analysis has shown the main controversies and coalitions in favour of and in opposition to the South African nuclear build programme. The structure of the coalitions and the main arguments in the debate can explain why the South African government chose to procure nuclear power over other alternative options. The nuclear build programme shapes two polarized coalitions. The opposing coalition is larger and consists mainly of civil society organizations concerned with environmental issues and good governance. The most powerful actor is the National Treasury, which is the main political counterweight to supporters in the DoE, NECSA, and the Presidency. The coalition of supporters is much smaller and consists mainly of government actors, business, and a few experts in academia and consulting firms. The supporting coalition may be smaller, but has higher stakes in the programme, as well as immediate financial and employment benefits in case of success.

The arguments in the discourse network relate to the six political conditions that help to explain why they should favour nuclear programmes. First, the debate on the cost of the nuclear programme reveals the main political battles within the ANC. The subordination of opposition to political authority is particularly striking in the single party system. Most checks and balances occur within the ruling party, while the formal opposition poses no immediate electoral threat. The National Treasury is therefore the main political counterweight to the nuclear programme. Second, the debate on cost also relates to the favouring factor of economic interventionism, as the state plays a strong role in picking winners through public procurement and setting rules for localization and job creation for relatively specialized nuclear workers who have very different profiles from the millions of unemployed in South Africa. Third, the debate on safety reflects an active civil society engagement in knowledge production, but without major political awareness campaigns and public outreach to ANC voters. Civil society engagement is not peripheral, but it has not mobilized those who can make a difference with their votes. Fourth, the debate on energy security, baseload, renewables, and emissions reductions reflects the clashes in the energy stakeholder network in a centrally organized political system between those who want to preserve the status quo and those who argue for reform. The defenders of the status quo aim to preserve existing privileges, while supporters argue for decentralization and uptake of new technology options. Fifth, security concerns are secondary in today's nuclear debate. International prestige mattered more in the past than today. Secrecy is a major characteristic of the current nuclear governance style and the ANC has preserved the culture of secrecy of the past in managing the nuclear build programme. The lack of transparency in the procurement, classification of documentation on any environmental and cost assessments, and hidden details on the nature of international relations fuel doubts on the legitimacy of the programme. Sixth, geopolitical influences play a significant role in the implementation of the programme. Secrecy conferred on the agreements with Russian agencies and institutional counterparts has fuelled suspicions of corruption. The ANC's political culture of secrecy reflects elements of an autocratic rather than a democratic, open, and transparent regime.

The analysis revealed how the political conditions that favour nuclear programme play out in South Africa. Economic factors such as cost, job creation, the health of the financial sector, and state-owned entities turned out to be significant, which the original framework did not account for. Security concerns become insignificant, as the economic factors turn out to be the main determinant of success or failure of the programme. Benefits of job creation, financial prospects for constrained public entities, and uranium mining prospects for government-related businesses are tangible gains for the coalition of supporters, which suggests that South Africa's nuclear programme is not primarily about generating electricity.

REFERENCES

Abbas, G. (2015). 'Interfaith Group Says No to Nuclear Energy Deals'. *Cape Argus*, 2 July.

Adam, F., D. Fig, D. Gilbert, M. Kantey, F. Musana, N. Schulz, M. Steele, R. Teule, and S. Thomas (2011). 'The True Cost of Nuclear Power in South Africa'. Greenpeace Africa, Johannesburg.

ANC (1994). 'The Nuclear Debate: Proceedings of the Conference on Nuclear Policy for a Democratic South Africa'. 11–13 February, Environmental Monitoring Group and ANC Science and Technology Group, Cape Town.

Bega, S. (2011). 'Cabinet's Approval of Nuclear Plans Ignores Tragedy'. *The Star*, 19 March.

Bijker, W. E. (1995). *Of Bicycles, Bakelites and Bulbs. Towards a Theory of Socio-Technical Change*. Cambridge, MA: MIT Press.

Biyase, L. (2010). 'Areva Predicts a New Nuclear Plant'. *The Star*, 19 February.

BMU (2006). 'Gemeinsames Übereinkommen über die Sicherheit der Behandlung abgebrannter Brennelement und über die Sicher heit der Behandling radioaktiver Abfälle, Bericht der Bundesrepublick Deutschland für die zweite Überprüfstagung'. BMU, Berlin.

Bouzarovski, S. and M. Bassin (2011). 'Energy and Identity: Imagining Russia as a Hydrocarbon Superpower'. *Annals of the Association of American Geographers*, 101(4): 783–94.

Brummer, S. (2015). 'Nuclear Price Tag Set Nene against Jacob Zuma'. *Mail & Guardian*, 18 December.

Caetano, T. and B. Rennkamp (2014). 'Nuclear Power: Structural Change, Lock-Ins and Trade-Offs'. ERC Research Paper prepared for the IAEA, IAEA, Vienna.

Campbell, K. (2013). 'SA Government Wants Country to be Global Nuclear Competitor'. *Engineering News*, 25 November.

City Press (2013). 'Government Denies Sidelining Motlanthe'. *News24*, 26 July.

COSATU (2015). 'COSATU Statement on the Introduction of Nuclear Energy', COSATU Press Statement, 31 July.

Creamer, T. (2010). 'PBMR Pullback a "Painful" Decision, Hogan Admits'. *Engineering News*, 15 April.

Creamer, T. (2012). 'Motlanthe-Led Nuclear Committee Meets'. *Engineering News*, 6 August.

DME (Department of Minerals and Energy) (1998). 'White Paper on the Energy Policy of the Republic of South Africa'. DME, Pretoria.

DME (2008). 'Nuclear Energy Policy for the Republic of South Africa'. DME, Pretoria.

DoE (Department of Energy) (2011). 'Integrated Resource Plan for Electricity'. *Government Gazette*.DoE, Tshwane.

DoE (2013). 'Integrated Resource Plan for Electricity (IRP) 2010–2030'. Update Report, DoE, Tshwane.

DoE (2015a). 'Nuclear Programme: Determination under Section 31(1) of the Electricity Regulation Act 4 of 2006'. *Government Gazette*, 606, DoE, Pretoria.

DoE (2015b). 'Presentation of International Cooperation Agreements to the Portfolio Committee on Energy'. 1 September, DoE, Cape Town.

Donelly, L. (2015). 'SA's Nuclear Safety Is Reassessed'. *Mail & Guardian*, 1 August.

Donelly, L. and L. Faull (2013). 'Zuma Slips into Nuclear Driver's Seat'. *Mail & Guardian*, 26 July.

DST (Department of Science and Technology) (2008). 'Technology Localisation Plan: A Framework for Engagement in National Technology Localisation Actions'. DST, Tshwane.

Earthlife (2007). 'The Following Submission Provides Ten Reasons that Earthlife Africa Cape Town Is Opposed to Nuclear Energy'. Earthlife Africa, Cape Town.

England, A. (2015). 'Pravin Gordhan Pledges to Keep a Grip on South Africa's Spending'. *Financial Times*, 20 December.

ERC (Energy Research Center) (2013). 'Towards a New Power Plan'. Research Report Prepared for the National Planning Commission, ERC, Cape Town.

Fakir, S. and G. Pienaar (2016). 'Government's Nuclear Plans—Will It Be the Arms Deal 2.0?' *Daily Maverick*, 21 January.

Felix, J. (2013). 'Nuclear Plants "Still a Top Priority" for SA'. *The Mercury*, 26 June.

Gosam, L. (2016). 'Zuma, the Gupta's and the Russians'. *Rand Daily Mail*, 19 February.

Gqirana, T. (2015). 'Nuclear Build Will Be Affordable'. *Mail & Guardian*, 1 September.

Hajer, M. A. (1995). *The Politics of Environmental Discourse: Ecological Modernization and the Policy Process*. Oxford: Oxford University Press.

Hogan, B. (2010). 'Address by the Minister of Public Enterprises, Barbara Hogan, to the National Assembly, on the Pebble Bed Modular Reactor'. Available at: <http://www.dpe.gov.za/newsroom/Pages/ADDRESS-BY-THE-MINISTER-OF-PUBLIC-ENTERPRISES,-BARBARA-HOGAN,-TO-THE-NATIONAL-ASSEMBLY,-ON-THE-PEBBLE-BED-MODULAR-REACTOR.aspx> (accessed 22 February 2016).

Hughes, T. (2011). 'The Evolution of Large Technological Systems'. In W. E. Bijker, T. P. Hughes, and T. Pinch (eds), *The Social Construction of Technological Systems: New Directions in the Sociology and History of Technology*. Cambridge, MA: MIT Press, pp. 45–76.

Hunter, Q. and L. Faull (2014). 'Jacob Zuma's Secret Nuke "Stitch-Up"'. *Mail & Guardian*, 26 September.

IAEA (International Atomic Energy Agency) (2015). 'Nuclear Power Reactors in the World'. IAEA, Vienna.

Jasanoff, S. and S. H. Kim (2009). 'Containing the Atom: Sociotechnical Imaginaries and Nuclear Power in the United States and South Korea'. *Minerva*, 47(2): 119–46.

Joubert, J. J. (2015). 'Nuclear Body Faces Financial Meltdown'. *Sunday Times News*, 8 March.

Latour, B. (2005). *Reassembling the Social. An Introduction into Actor-Network Theory*. Oxford: Oxford University Press.

Le Cordeur, M. (2016). 'Joemat-Pettersson Wants Nuclear under IPP Office'. *Fin24*, 23 February.

Leifeld, P. (2012). 'Discourse Network Analyzer'. Eawag, Dübendorf.

MacKenzie, D. (1998). *Knowing Machines: Essays on Technical Change*. Cambridge, MA: MIT Press.

Macleod, F. (2011). 'Avoid Disaster on Our Shores'. *Mail & Guardian*, 18 March.

Marquard, A. (2006). 'The Origins and Development of Energy Policy in South Africa'. PhD Thesis, University of Cape Town, Cape Town.

Moormann, R. (2008). 'Safety Re-Evaluation of the AVR Pebble Bed Reactor Operation and Its Consequences for Future HTR Concepts'. Available at: <http://www.iaea.org/inis/collection/NCLCollectionStore/_Public/39/099/39099096.pdf> (accessed 3 March 2016).

Ndlozi, M. Q. (2015). 'EFF Urges the South African Government to Strongly Consider Germany's Advice on Nuclear Energy'. Economic Freedom Fighters, Johannesburg.

NPC (National Planning Commission) (2011). 'National Development Plan 2030'. NPC, Pretoria.

Paton, C. (2014). 'Nuclear Is Not Power Priority, Says Business'. *Business Day LIVE*, 12 November.

Paton, C. (2015). 'Nuclear Must Be Affordable, Says Nene'. *Business Day LIVE*, 24 August.

Pressly, D. (2013). 'Full Steam Ahead for Nuclear Expansion'. *IOL Business News*, 15 May.

RSA (Republic of South Africa) (1996). 'Constitution of the Republic of South Africa, No. 108 of 1996'. RSA, Pretoria.

RSA (2010). 'The New Growth Path: The Framework'. Department of Economic Development, RSA, Pretoria.

RSA (2014). 'Public Protector Thuli Madonsela on Public Procurement System'. RSA, Pretoria.

RSA (n.d.). 'Guidelines for Public Procurement. The National Treasury, Pretoria:.

RSA/RF (Russian Federation) (2014). 'Agreement between the Government of the Russian Federation and the Government of the Republic of South Africa on Strategic Partnership and Cooperation in the Fields of Nuclear Power and Industry'. English translation available at: <http://www.ee.co.za/wp-content/uploads/2015/02/amaBhungane-English-translation-of-Russian-nuclear-cooperation-agreement.pdf> (accessed 20 November 2015).

Sabatier, P. A. (1988). 'An Advocacy Coalition Framework of Policy Change and the Role of Policy-Oriented Learning Therein'. *Policy Sciences*, 21: 129–68.

SABC (South African Broadcasting Corporation) (2013). 'Interview with Sergey Novikov, Rosatom Spokesperson'. Available at: <www.youtube.com/watch?v=3agXfl2sXi8> (accessed 20 March 2016).

SAPA (South African Press Association) (2015). 'Eskom Contractors Fire More Than 1,000 Workers'. *Mail & Guardian*, 27 March.

Shamase, N. (2011). 'Row over "Unpaid" Bills at Gupta Family's Mine'. *Mail & Guardian*, 9 September.

Sovacool, B. K. and S. V. Valentine (2012). *The National Politics of Nuclear Power: Economics, Security, and Governance*. New York: Taylor & Francis.

Steyn, L. (2015a). 'Sinking into Eskom's Black Hole'. *Mail & Guardian*, 2 May.

Steyn, L. (2015b). 'SA's Nuclear Deal with Russia Is Far from Done'. *Mail & Guardian*, 19 February.

Temples, J. R. (1980). 'The Politics of Nuclear Power: A Subgovernment in Transition'. *Political Science Quarterly*, 95(2): 239–60.

Thomas, S. (2010). 'The Economics of Nuclear Power: An Update'. Boell Stiftung, Heinrich Berlin.

Van Wyk, J. (2013). 'South Africa's Nuclear Future'. Working Paper, South African Institute of International Affairs, Johannesburg.

Wild, S. (2013). 'SA Need Skilled Nuclear Scientists'. *Mail & Guardian*, 18 October.

Williams, R. and D. Edge (1996). 'The Social Shaping of Technology'. *Research Policy*, 25: 865–99.

Winner, L. (1986). *The Whale and the Reactor. A Search for Limits in an Age of High Technology*. First Edition. Chicago, IL: Chicago University Press.

Winner, L. (2000). 'Do Artifacts Have Politics?' In D. MacKenzie and J. Wajcman (eds), *The Social Shaping of Technology*. Philadelphia, PA: Open University Press, pp. 521–44.

WNN (World Nuclear News) (2015). 'Claims Updated in Olkiluoto 3 Delay Arbitration'. *World Nuclear News*, 3 August.

Zulu, T. (2015). 'Progress with the Nuclear New Build Programme'. Press Statement by the Director-General of the Department of Energy, Mr Thabane Zulu. Available at: <http://cdn.bdlive.co.za/images/pdf/nuclear.pdf> (accessed 7 March 2016).

Zuma, J. (2015). 'President Jacob Zuma's Written Reply to Questions in the National Assembly'. Available at: <http://www.thepresidency.gov.za/pebble.asp?relid=19365&t=79> (accessed 1 March 2016).

Zuma, J. (2016). 'State of the Nation Address 2016'. Available at: <http://www.thepresidency.gov.za/pebble.asp?relid=21584> (accessed 1 March 2016).

15

European Energy Security

Challenges and Green Opportunities

Almas Heshmati and Shahrouz Abolhosseini

15.1 INTRODUCTION

Energy security can be considered as both an international and a national security issue that may lead to cross-national tensions. Europe is facing some challenges and also has possible opportunities to avoid or to overcome these challenges. The crisis in Ukraine, with its importance as an energy corridor for Europe, is a threat to energy security at a time when political tensions have increased between Ukraine and Russia. Iran could be considered as an alternative source of energy for the European market, but there is a longstanding dispute between Iran and the West involving economic sanctions. Iran is not only the holder of the largest natural gas reserves and the fourth largest crude oil reserves in the world (BP 2014), but it is also located in a geopolitical area.

The European Union (EU) is determined to diversify its energy supply in order to reduce its dependency on Russian supplies. In this regard, renewable energy technologies are an alternative for power generation and subsequent dependency reduction. A sizable investment has been made in the EU for developing renewable energies. The advanced technologies of Europe enable these countries to produce renewable energy more cost-effectively and the power generated by these sources continues to increase. However, profitability of renewable energies depends on the prices in the depletable energies market. Thus, developing the renewable energy market is very sensitive to price fluctuations in primary energy sources. In this regard, integrating distributed renewable energy sources and smart grids within local marketplaces for trading renewable energy in small units can be a promising combination for enhancing renewable energy deployment. Considering Europe's intention to organize a single harmonized system, the member countries are required to apply suitable mechanisms to support renewable energy enhancement.

In this research, we review the relevant literature on the current state and effectiveness of developing renewable energy on energy security. The review uses primary energy import sources, possible alternatives, and considers how energy security in Europe is affected by the sources. For this purpose, the Herfindahl-Hirschman Index (HHI) is calculated to measure the level of energy security for selected countries. We also examine EU energy policy, try to analyse the reasons why Europe should adopt a new energy policy direction, and we suggest alternative solutions for enhanced energy supply security. The aim is to suggest suitable solutions for energy security in Europe through energy supply diversification, including alternative energy corridors, to reduce dependency on Russian supplies and enhance the power generated by renewable energy sources in order to meet energy targets based on the EU 2020 strategy including greenhouse gas (GHG) emission reduction,[1] renewable energy deployment, and energy efficiency enhancement. The information used for the chapter is derived mainly from official reports and is based on secondary data sources.

The rest of this chapter is organized as follows: Section 15.2 discusses the effects of renewable energy development on energy security. Section 15.3 deals with the challenges in the Middle East and the balance of energy demand and supply in Europe. This is followed by a discussion of import sources and possible alternatives for Europe in Section 15.4. Section 15.5 presents a policy framework aimed at enhancing energy security in Europe. Section 15.6 provides a conclusion.

15.2 THE EFFECTS OF RENEWABLE ENERGY DEVELOPMENT ON ENERGY SECURITY

Ongoing concerns about energy security have been raised since the Arab oil embargo in 1973 when oil importing countries in Europe and elsewhere were faced with the effects of high oil prices and limited energy supplies. Although concerns about energy security started waning with low oil prices in the 1980s, oil prices increased again over the next decades. Considering the importance of energy for economic development and growth, industrialized countries focused on alternative policies to enhance alternative energy sources such as nuclear power and renewable energy. Over recent years, climate change and environmental protection have been at the core of energy policies together with energy security. Nuclear power and renewable energy sources have been

[1] Europe 2020 is a ten-year strategy proposed by the European Commission in March 2010 for advancement of the economy of the EU. It aims for smart, sustainable, inclusive growth with greater coordination of national and European policy (European Commission 2010).

suggested as alternative sources since the 1970s. However, since the Fukushima Daiichi nuclear disaster in Japan in 2011, energy policies in relation to nuclear power plants have been reshaped around the world.

Deese (1979) discusses the economic, political, and security aspects of energy. Energy security is achieved when stable sources of energy are available at affordable prices. It is possible to attain high energy security through internal (domestic) or external (imported) energy supplies. Russia has been the main source, providing 41 per cent of total European imports of energy in 2014 (BP 2015a). Germany has been importing 45 per cent of its natural gas demand from Russia. Considering the disputes that have occurred between Russia, Ukraine, and Western powers over recent years regarding gas transmission and political interventions, European countries are looking for sustainable sources of energy to diversify their energy supply sources. Policy makers have therefore been trying to enhance the deployment of renewable energy sources through different incentive policies such as feed-in-tariffs and renewable portfolio standard carbon tax.

15.2.1 Natural Gas and the Interplay with Renewables and the Environment

The literature about natural gas and the interplay with renewables, particularly related to temporal transitions, energy/environmental policies, and energy security, has been developing rapidly. Creutzig et al. (2014) view the energy transition in Europe as catching two European birds (mitigating climate change and energy security crisis) with one renewable stone. The governance of natural gas transit in Europe is investigated by Bouzarovski, Bradshaw, and Wochnik (2015). They develop a theoretical framework to explore the regulatory practices and spatial features of this unexplored infrastructural realm. This chapter reveals emerging new socio-technical assemblage and institutional orders other than the traditional organizational arrangements.

The interaction of risks associated with natural gas and renewable resources for electricity is analysed by Esposito, Krupp, and Carley (2015). They identify risks relating to development, construction of power plants and transmission systems, planning, costs, and policies decisions, and show how the risks can potentially be offset. Lee et al. (2012), in the case of the United States (US), find natural gas and renewable energy technologies enjoy many complementarities arising from their similarities, but find that their dissimilarities provide the biggest opportunities for mutually beneficial collaboration. Mediavilla et al. (2013) refer to the physical limits and temporal conditions in the transition from an oil economy towards renewable energies. The temporal dynamics of energy transitions is studied by Sovacool (2016). The focus is on the speed at which transition can take place and causal complexity underlying the conceptions and definitions of energy transitions.

Fischhendler and Nathan (2014), in their study of energy security in Israel, find the concept of energy security open to manipulation and various interpretations by inter-ministerial committees. The security concept differs from environmental acceptability and interdependency, supply reliability, and geopolitical benefits. Verdeil et al. (2015) also discuss the challenge of the transition to development of urban natural gas in the Mediterranean Metropolis. The issue of accurately forecasting the availability of natural gas to ensure sustainable energy policy is emphasized by Darda, Guseo, and Mortarino (2015). The use of alternative reserve estimates for South Asian natural gas shows where in the region the reserves allow for building a sustainable power reserve system for natural gas to meet increasing energy demand in the medium term. Ghezelbash et al. (2015) emphasize the use of high net present values of selected systems in the assessment of performance of natural gas expansion. Kahrl et al. (2013) analyse the changes necessary to increase the share of natural gas in China's electricity mix.

15.2.2 Renewable Energy Development

As mentioned in Section 15.2.1, renewable energy sources have been developed to enhance energy security and emission reduction. European countries are at the forefront when it comes to improving efficiencies and using renewable energy. Also, tax policies have been applied by these countries in order to reduce liquid fossil fuel consumption. It is expected that liquid fuel consumption in the Organisation of Economic Co-operation and Development (OECD) countries will be 14.0 million b/d (barrels per day) in 2040 which is 0.8 million b/d lower than the 2010 level (EIA 2014). The importance of developing renewable energy has been highlighted by climate change and energy security issues through the excessive consumption of fossil fuels, political instabilities in the Middle East, and uncertainty around supply disruptions due to political disputes in Ukraine. Germany is considered to be the forerunner when it comes to generating renewable energy, because it was able to raise the share of renewable energy sources in its fuel-mix to more than 10 per cent in 2014.

Renewable energy has been developing rapidly in recent years. Reductions in costs due to economies of scale and use of advanced technologies have made it possible for countries to generate renewable energy more efficiently and cost-effectively. Table 15.1 compares power generation costs and cost outlooks based on different sources of energy over the period 2000–15 and 2015–50, respectively.

Table 15.1. Power generation costs (2000–15) and cost outlook (2015–50) based on sources of energy, US$ per megawatt hour

	2000	2010	2014	2015	2000–15 (%)
Power generation costs: Gas (CCGT) US	62	58	52	47	−24
Hydro	52	70	67	66	25
Geothermal	48	84	77	72	50
Gas (CCGT) Europe	47	80	80	76	62
Coal US	57	81	80	80	40
Wind onshore	76	116	82	86	14
Gas (OCGT) US	117	110	96	91	−23
Coal (Europe)	77	109	92	92	19
Nuclear	74	96	100	102	37
Gas (OCGT) Europe	94	142	139	134	42
Solar PV	509	333	145	135	−74
Biomass	109	144	140	139	28
Coal w/CCS US	84	135	146	156	86
Coal w/CCS Europe	107	152	154	164	53
Wind offshore	NA	205	180	184	NA
Solar CSP	207	238	226	200	−4
Wave-tidal	NA	280	296	302	NA
	2015	**2020**	**2030**	**2050**	**2015–50 (%)**
Power generation cost outlook: Solar PV	135	117	89	51	−62
Geothermal	72	71	69	66	−8
Hydro	66	66	66	66	1
Wind onshore	86	83	79	70	−18
Solar CSP	200	178	141	88	−56
Gas (CCGT) US	47	61	70	92	94
Nuclear	102	101	98	92	−9
Gas (CCGT) Europe	76	85	92	102	34
Coal US	80	92	101	114	43
Wind offshore	184	172	151	116	−37
Coal Europe	92	112	122	130	42
Biomass	139	138	137	133	−5
Coal w/CCS US	156	157	150	137	−13
Coal w/CCS Europe	164	175	167	149	−9
Wave-tidal	302	274	226	154	−49
Gas (OCGT) US	91	111	124	159	76
Gas (OCGT) Europe	134	147	157	174	31

Notes: CCGT=combined cycle gas turbine; OCGT=open-cycle gas turbine; CCS=carbon capture and storage; CSP=concentrated solar power; PV=photovoltaic; NA=not available; US=United States.
Source: Adapted from EI (2015).

As can be seen in Table 15.1, Solar PV accounted for the largest reductions in costs during the last decade. Renewable technologies may not be compatible with conventional fuels due to unit costs, but they could be feasible if we were to consider associated externalities such as carbon emissions and social effect.

Also, economies of scale are crucial for reducing unit costs. Bohi and Toman (1993) studied energy security by looking at externalities and policies. Table 15.1 shows that Solar PV is expected to be the least expensive power-generation technology by 2050. The unit cost for this technology reduced by 74 per cent during 2000–15 and it is forecast that this will reduce further by another 62 per cent over 2015–50. Therefore, Solar PV could be considered as an alternative source of energy in decent sunlight conditions, for example, in Central Europe.

Currently, nuclear power is being considered as an alternative to fossil fuels in European countries, but these countries are also coming under pressure to stop the operation of existing nuclear power plants. There has been a de facto moratorium on the construction of new nuclear power plants, with phase-outs announced by Sweden, Germany, Switzerland, Spain, Belgium, and the Netherlands (Asif and Muneer 2007). Additionally, high primary energy prices combined with energy support policies have been the main driver in enhancing renewable energy sources. Many OECD countries have enacted national policies to support sustainable development through clean technologies. These policies deal with a wide range of objectives such as energy security, market competition, and environmental protection. Economic feasibility is at the core of renewable energy development. Without this essential parameter, renewable energy technologies will not be able to compete with conventional fossil fuels. Table 15.2 shows the growth rates of wind and solar energy consumption as the main renewable energy sources, based on region, over the period 2010–14.

Based on the figures presented in Table 15.2, we see a rapidly increasing trend (as high as a 3–4 digit growth rate) during 2010–14 for wind and solar energy consumption. In parallel with non-oil and gas producing countries, even countries located in the Middle East are determined to develop solar energy for environmental reasons. Among individual countries, the US had

Table 15.2. Wind and solar energy consumption in 2010–14, in terawatt hour (TWh)

	Wind power			Solar power		
	2010	2014	Change, %	2010	2014	Change, %
North America	105.5	202.1	91.6	1.3	19.3	1384.6
South and Central America	3.5	17.7	405.7	a	1.1	>2000.0
Europe & Eurasia	152.6	261.6	71.4	23.2	99.7	329.7
Middle East	0.2	0.3	50.0	0.1	1.1	1000.0
Africa	2.2	6.1	177.3	0.3	2.1	600.0
Asia-Pacific	79.4	218.3	174.9	6.4	62.6	878.1
World	343.4	706.2	105.6	31.4	185.9	492.0

Note: 'a' indicates less than 0.05 level of significance.

Source: Authors' calculations based on BP Statistical Review of World Energy (BP 2015a).

the highest consumption of wind energy (183.6 TWh), which was more than the total consumption in the top five consumer countries in Europe (171.1 TWh); the US also accounted for 26 per cent of the total global consumption of wind power (BP 2015a). Europe had a lower growth rate than North America, but it ranked first in relation to total wind and solar energy consumption. In other words, developing renewable energy sources has already reached a high level in Europe.

15.3 MIDDLE EAST CHALLENGES AND ENERGY BALANCE IN EUROPE

The Middle East is considered to be the main source of crude oil supply in the world. Almost 35 per cent of the total crude oil traded on the world market is supplied by Middle Eastern exporters. Russia and the Middle East are major exporters of natural gas and crude oil to Europe. Considering the potential for terrorist attacks and political instability in these regions, Europe requires alternative sources of energy supply. Other factors such as climate change and crude oil prices make it even more important to switch from fossil fuels to renewable energy sources. The Yom Kippur War (1973), the Iranian Revolution (1979), the Iran–Iraq War (1980), the First Persian Gulf War (1991), the Second Persian Gulf War (2003), the Arab Spring in Egypt, the Libyan and Syrian Civil Wars (2011), and the Yemen Civil War (2014), together with the Saudi Arabian invasion of Yemen (2015) have all contributed to continued political instability in the Middle East. Owen (2004) discusses oil supply insecurity: control versus damage costs. Sen and Babali (2006) focus on problems of and solutions to security concerns for oil supply in the Middle East.

In the future, natural gas will play a critical role in the global energy market due to its advantages relating to prices, availability, cleanness, and security. Currently, European countries greatly depend on Russia as a supplier of natural gas. After the Fukushima disaster, nuclear power has lost its priority in the energy portfolio of EU-28 countries and alternative clean energy resources are being seriously considered. However, the generation capacity of renewable energy sources is not enough to cover the gap created by shutting down nuclear power plants completely. If this trend aggregates with climbing demand in the natural gas sector, Europe will depend more on foreign energy resources and major producers.

In the past, the EU built an integrated single energy market to be used as an asset in relations with neighbouring supplying countries. Therefore, when we talk about supply diversification, not only different sources of energy but also different exporters and transit routes should be taken into account for the

EU. Europe therefore needs to design an energy policy based on regional integration in order to use it as leverage for making long-term and stable partnerships with major suppliers. This new policy should create an ability to develop concrete mechanisms to deal with emergency situations arising from changing attitudes on the supply side. Vivoda (2009) asks whether diversification of oil import sources and energy security is a key strategy or an elusive objective. Nuttall and Manz (2008) discuss a new energy security paradigm for the twenty-first century.

Considering this perspective, Europe needs to try and have a close relationship with the Caucasus region to improve supply security and to establish an international partnership. The Caucasus has started an ongoing trend of capturing larger shares of the EU energy market, while export capacity is affected by increasing domestic demand. Furthermore, there are some constraints on exports because of limited investments, lack of transport infrastructure, lack of technology, and only a few available transit routes. Europe depends on Russian energy sources, and Qatar's exports cannot decrease Europe's energy vulnerability. Russia's position as a dominant supplier in the European energy market along with the EU's limited internal resources is forcing the EU to consider other suppliers in the Persian Gulf region. In this regard, Iran could serve as a candidate for changing the EU's energy portfolio.

15.3.1 Energy Policy in Europe

The location of existing oil reserves does not align with population concentration and energy use structures around the world. As an example, consumption in Asia-Pacific, Europe, and North America accounts for almost 77 per cent of the total world consumption, while they control only 19 per cent of the global oil reserves (BP 2015a). At the same time, the former Soviet Union, Middle East, South America, and Africa consume 26 per cent while they control 81 per cent of the world's oil reserves.

A majority of energy suppliers are located in politically unstable regions, such as the Middle East, Latin America, and Africa, which poses high potential risks due to political instability (in Iraq, Syria, Venezuela, and Nigeria). There was a longstanding dispute between Iran as a key supplier in the Middle East and Western powers, but this has been resolved recently. The changed relations could offer great potential for an alternative solution for the EU's supply diversification plans. Therefore, EU countries need to use a specific energy policy framework or energy diplomacy in order to take advantage of competition among suppliers including Iran.

Energy security can also be improved by replacing more vulnerable supplies with stable sources of supplies. More than 40 per cent of Europe's imported natural gas through pipelines comes from Russia (BP 2015a). The gas supply dependency of Ukraine and Belarus on Russia is 71 per cent and 100 per cent respectively. Considering that Ukraine is the main transit route for Russia to export natural gas to Europe, the EU, as a third party, could suffer because of any supply disruptions caused by a Ukraine–Russia dispute (gas pricing, transit fee, or other political issues affecting the two countries' relations).

'Optimal policy can be achieved by pricing both energy security and greenhouse gas abatement and pursuing each technology to the point where its additional cost is equal to the marginal benefits achieved in both dimensions' (Brown and Huntington 2008). There are three key targets under the EU 2020 growth strategy that are supposed to be met by the year 2020. First, a 20 per cent cut in greenhouse gas emissions from the 1990 level. Second, 20 per cent of energy to come from renewable energy sources. Third, a 20 per cent improvement in energy efficiency. The effects of renewable energy technologies, energy efficiencies, and market regulation on carbon emission reduction are very important for this purpose. Different energy policies may interact with each other. These interactive effects should be accounted for in policy-making.

Indicators that are considered for supply diversification and energy security include import dependency, fuel-mix, and stocks of critical fuels (Bhattacharyya 2011). The percentage of dependency on fuel imports shows the potential of the risk. This dependency may not be the same for all fuel types. Some countries are self-sufficient in producing one fuel type but they have to import another one. A ratio of the fuel-mix can be used by countries in order to diversify sources of energy supply. The EU has developed renewable energy sources in recent years and the region has also changed the major source of motor fuels from gasoline to diesel.

15.3.2 The Monopolized Energy Market and the Necessity of Including New Suppliers

Undoubtedly, Russia is the largest supplier of natural gas for EU countries. Russia has the second largest natural gas proven reserves in the world and enjoys a monopoly in the European energy market, but we cannot ignore the other players in the market. Some countries such as Azerbaijan, Algeria, Nigeria, and Qatar have specific shares in gas supply to Europe. But in the future we must observe some variations in these supplies which may cause certain problems for EU countries. A similar situation may apply in the case of crude oil. Natural gas reserves have not been concentrated as much as crude oil, but Russia together with the Caspian Sea and the Middle Eastern regions is the owner of about half of the total natural gas reserves in the world.

Major consumers, including the EU, will be more dependent on the same regions for importing oil and gas. We should keep in mind that there is rapid growth in energy consumption in developing countries such as China and India; they are looking for sources of energy in these regions too. So finding a new source of energy and a sustainable supplier is very important not only for Europe, but also for major emerging economies like China and India. In other words, the Middle East is a kind of battlefield between Western and Eastern powers to win a greater share of oil and gas resources to obtain long-term influence over the sector. Therefore, the security of oil and gas supply to the EU in the long term depends on having access to producing areas.

15.4 IMPORT SOURCES AND POSSIBLE ALTERNATIVES FOR EUROPE

Comprising more than 40 per cent of the world's crude oil and natural gas proven reserves, the Middle East is the key region for energy supply. The free flow of oil to world markets from the Persian Gulf region forms a vital part of major security issues. Cooperation between Iran–Europe, Iran–US, and Iran–Persian Gulf states could show the Middle East to be a stable region that could be important for both Europe and the US. DeRosa and Hufbauer (2008) investigate the consequences of the normalization of economic relations for Iran's economy and the US. Katzman (2012) discusses US concerns about and policy responses in relation to Iran. Europe would have access to an alternative source of energy and the US would save a large amount of military expenses if this happened. Political stability in the Middle East would allow for the US's estimated military expenses in the Persian Gulf to be reduced (Delucchi and Murphy 2008).

Table 15.3 presents a security index for some countries over 1980–2012 in the EU, including Germany, Spain, France, Italy, the Netherlands, Sweden, and the UK. Net energy imports have been used as an energy security index in this table. Net energy imports are calculated as a percentage of energy use (production is excluded).

The HHI can be used to measure the level of energy security. However, this index has a disadvantage as it does not take political risks into consideration. The UK, Germany, Italy, and Spain had HHI above 2000, indicating that these countries were highly concentrated for power generation (Bhattacharyya 2011). We calculated the HHI of fuel-mix concentration for the selected countries for the period 2000–14. Figure 15.1 shows the index values. Consistent with Bhattacharyya's study, our calculations show clearly that these countries rely heavily on fossil fuels for power generation.

With the exception of France, we see a declining trend in the fuel-mix concentration since 2009. This shows the effect of the 2008–9 crisis on energy consumption. Policy makers tried to support alternative energy sources and

Table 15.3. Energy security index for some countries in the European Union

	1980	1990	2000	2010	2012
Germany	48.03	46.98	59.79	59.78	59.88
Denmark	95.02	41.03	−48.84	−20.68	−16.94
Spain	76.70	61.60	74.10	73.15	74.19
France	72.57	50.06	48.12	48.17	47.11
Italy	84.79	82.73	83.58	82.52	79.40
The Netherlands	−11.58	7.83	21.38	16.31	17.26
Sweden	60.16	37.11	35.82	34.72	28.53
Europe Union	44.88	42.21	43.90	51.25	

Source: World Bank (2015).

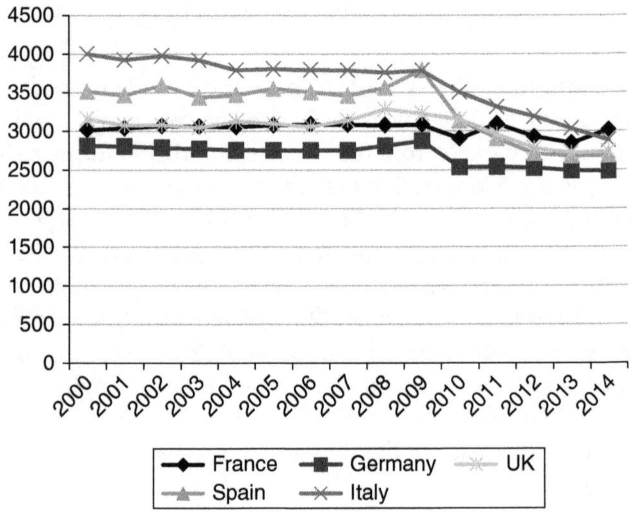

Figure 15.1. Fuel-mix concentration level using HHI (2000–14).
Source: Authors' calculations.

enhance energy efficiency which led to decreasing levels of concentration. We find that Germany's energy policy was more stable than that of other sample countries. Also, Germany had the lowest HHI among the five selected countries, while Italy and Spain showed significant improvements in HHI measurements after 2009. However, the level of concentration in the countries chosen indicates that supportive policies should be the focus of policy makers for diversifying energy sources. The diversity in the fuel-mix of primary energy consumption in some European countries is shown in Table 15.4, which gives the level of fuel-mix concentration of energy consumption in seven European countries for 2010 and 2014.

Unlike Scandinavian countries, which rely mainly on hydropower, EU-28 will be relying on natural gas resources rather than on clean energy in the

Table 15.4. Fuel-mix of primary energy consumption in seven European countries (%)

	Year	Oil	Natural gas	Coal	Nuclear	Hydro	Renewable
France	2010	33.04	16.72	4.79	38.39	5.67	1.35
	2014	32.38	13.60	3.79	41.52	5.98	2.74
Germany	2010	36.03	22.91	23.94	9.95	1.35	5.82
	2014	35.85	20.51	24.89	7.07	1.48	10.19
Italy	2010	42.50	39.71	7.97	–	6.51	3.26
	2014	38.01	34.32	9.07	–	8.66	9.94
The Netherlands	2010	49.75	39.16	7.89	0.09	–	2.20
	2014	48.83	35.64	11.10	1.11	–	3.21
Poland	2010	27.45	13.47	56.37	–	0.84	1.98
	2014	24.87	15.36	55.28	–	0.52	4.08
Spain	2010	49.77	20.71	5.54	9.29	6.41	8.28
	2014	44.74	17.82	9.02	9.77	6.69	12.03
UK	2010	35.25	40.41	14.92	6.74	0.38	2.34
	2014	36.88	31.93	15.70	7.66	0.69	7.03

Source: Authors' calculations, based on BP Statistical Review of World Energy (BP 2015a).

future. This implies that these countries need to develop relations with the main suppliers and transit countries. Although part of Europe's energy demand could be covered by domestic suppliers in the east and north of the continent, the EU imported around 78 per cent of crude oil consumption for 2014 (BP 2015a). Despite Europe being at the forefront of renewable energy deployment in recent years, the table shows that fossil fuels still remain the main source of primary energy consumption in the region.

15.4.1 Import Sources for Europe

Currently, Russia is considered to be the main source of energy for Europe. Pipelines are not only a means of transportation for natural gas and crude oil transmission, they also play a critical role related to geopolitics and energy security. The Russia–Ukraine dispute in 2009 over pricing, when Russia cut off gas supply to Ukraine and allowed increased flows to South Eastern Europe and to some parts of Central and Western Europe, is the best example of pipeline politics leading to an energy crisis in Europe. The importance of pipeline transmissions is also relevant in Central Asia and the Caucasus. Europe is looking for supplies for reducing its dependency on Russia.

In 2014, more than 30 per cent of energy trade movement through pipelines in Europe was supplied by Russia. Europe imported 365.70 bcm (billion cubic metres) of natural gas (320.8 pipeline, 44.9 Liquefied Natural Gas, LNG) in 2014 (BP 2015a). A major part of the LNG import demand in Belgium, France,

Italy, Spain, and UK is supplied by Qatar and Algeria. The other suppliers (outside Europe) are Nigeria and Trinidad and Tobago. About 3 per cent of European LNG consumption is covered by other European suppliers.

Given current tensions between Russia, Ukraine, and Turkey, there is increasing concern about potential disruptions to the security of energy supply to EU consumers in the long term; pipeline connections running from the East to the West should be given special priority in order to mitigate these (Lise, Hobbs, and Van Oostvoorn 2008). Iran could be considered as a potential source in this regard. Security and the diversification required for EU energy supply highlight the significance of an energy corridor bridging the large natural gas reserves of the Caspian region, including those in Iran, with the EU market (Mavrakis, Thomaidis, and Ntroukas 2006). The EU has failed to create coherent energy security and energy foreign policy since the Russian–Ukrainian gas conflict in January 2006. Supply diversification through the aforementioned countries and implementing an energy policy of decreasing their overall gas demand will enable EU countries to reduce Russia's gas exports (Umbach 2010).

15.4.2 Renewable Energy Technologies as Alternative Sources

At individual country level, Germany has the highest rate of economic growth together with the highest level of installed capacity of renewable energy sources (Moutinho, Moreira, and Silva 2015). Natural gas is able to bridge the transition period required for renewable energy technologies to facilitate larger energy deployment in order to make it feasible from an economic point of view. It is forecast that natural gas consumption in Europe will be 650 billion cubic metres of natural gas (bcma) in 2020 and 780 bcma in 2030, while conventional gas production will decline to 230 bcma in 2020 and 140 bcma in 2030 (Weijermars et al. 2011).

Some scholars believe that transition progress for renewable energy sources such as wind and solar will happen after 2020; even the growth rate of consumption will increase rapidly during the next decade. Also, renewable energy markets are not formed easily due to cost disadvantages and subsidized fossil fuels (Jacobsson and Bergek 2004). Some countries such as Indonesia took advantage of falling oil prices during 2014–15 to reduce subsidies paid for fossil fuels, but still a large amount of subsidies are paid by oil-rich countries. Because of the negative and irreversible externalities associated with conventional energy production, what is required is enhancing renewable energy supply technologies.

Economic policies can be used as incentives for enhancing production and use of renewable energy sources. Also, charging taxes on emission generation or fossil fuel consumption can be used as a supportive policy to promote the deployment of renewable energy. There are three types of supportive mechanisms that are widely used by states to promote renewable energy technologies: feed-in-tariffs (FIT), tax incentives, and renewable portfolio standard (RPS). Both FIT and RPS mechanisms have been applied by the EU to develop renewable energy technologies and it therefore has experience with both of these mechanisms. FIT policy has led to the rapid expansion of power generated by renewable energy sources and it has been employed more than the RPS mechanism (Rickerson and Grace 2007).

A comparison of these policies indicates that FIT is an appropriate policy for developing renewable energy sources when a low level of risk for investors is required. However, the RPS mechanism works well when the government wants to use a market view policy. Europe intends to organize a single harmonized FIT system though this is impossible because of different policies across the countries in the EU. The RPS system has not been implemented in Europe because the FIT system has been used by most EU countries. Hence, FIT policies are suitable for encouraging development of renewable energy sources, while the RPS mechanism should be applied to renewable energy sources promoted to a certain level (Abolhosseini and Heshmati 2014).

Considering the outlook for the renewable energy market, what is required is a marketplace where small volumes of power generated can be traded. If such a marketplace does not exist then enhancing renewable energy will be limited to individual households to cover their own demands. Power generation by renewable energy sources can also benefit from integration of technologies. Integrating distributing and renewable energy sources and smart grids within local marketplaces for trading renewable energy in small units can be a promising combination for developing renewable energy sources across the EU (Heshmati and Abolhosseini 2014).

The 2020 strategy sets three critical targets for renewable energy enhancement to meet by 2020. This includes GHGs reduction, renewable energy deployment, and energy efficiency improvement. These targets could be achieved by developing renewable energy sources, technological change, and market regulation on carbon emissions. An estimated model (Heshmati, Abolhosseini, and Altmann 2015) shows that the role of governmental policy-making is more important than economic growth.

15.5 A NEW EUROPEAN POLICY FRAMEWORK

Energy security has moved up the EU's priority list. A high import dependency is not considered a problem in itself but it becomes an issue when supply is interrupted. In spite of Europe's struggle for energy saving and improving the share of renewable energy in its energy basket, the region remains highly dependent on imports, especially of natural gas. The EU's collaboration has to extend to the Persian Gulf, East Mediterranean, and the Caucasus area because Europe plays a critical role in the supply side of energy markets. Limited natural gas resources in North Africa may lead to an increased focus on Azerbaijan, Turkmenistan, Iraq, Qatar, and Iran as substitute sources. Due to existing infrastructure problems, limited investments in gas fields, and underdeveloped transit routes in these countries, Europe has to share the costs and benefits with the owners of energy resources.

A common energy policy applied by EU member countries enables them to have a well-interconnected market in order to avoid possible supply disruptions. These countries can create a competitive market for suppliers through international collaborations with all market players. There are also economies of scale in building necessary infrastructures. Considering certain targets for reducing carbon emissions which have been defined by different scenarios, it is crucial for the EU to choose an optimal and proper direction for enhancing energy efficiency. The share prices of major manufacturers of solar panels in China and wind turbines in Denmark have decreased due to a sharp decrease in crude oil prices. This suggests that clean energy policies are influenced more by economic conditions than by environmental concerns.

In this research we aimed to define a policy framework for the EU's energy security. Considering the small share of power currently generated by renewable energy sources, it is necessary for EU countries to design an applicable green pathway in order to achieve energy supply security (Heshmati 2014). In order to improve energy efficiency and power generation by renewable energy sources, high-level commitment, investment resources, and efficient management are required for the development and implementation of policies and programmes (Gellings 2009). There are some barriers that should be removed in order to facilitate the market creation process.

The Middle East and the Caucasus are two main strategic regions for the supply of natural gas to Europe. For renewable energy to be enhanced, a marketplace for green energy needs to be designed in order to promote market liquidity. Developing renewable energy has relied on public support and economic incentive programmes, but this has been affected by the economic crisis. A proper marketplace for trading the power generated by distributing renewable energy sources installed by households, combined with support policies for enhancing energy efficiency will lead to promoting market liquidity (see Heshmati and Abolhosseini 2014; Heshmati, Abolhosseini, and Altmann 2015).

A multidimensional policy approach is required for achieving sustainable energy security in Europe. Important dimensions of this policy include improved security, lowered dependency, increased share of clean and renewable energy, diversified energy sources, and common energy policies. As such, it is necessary to create a balanced security situation incorporating all potential market partners through international cooperation. For this to happen, a proper policy design, effective policy-making for internal efficiency, and diplomacy in external energy are required so as to be able to take advantage of international cooperation.

15.6 SUMMARY AND CONCLUSION

The free flow of oil to world markets from the primary energy sources is a vital part of energy safety issues. Energy security can be considered as both international and national security issues that may lead to cross-national tensions. Europe is faced with some challenges and also has possible opportunities to overcome these challenges. The Ukrainian crisis and its importance as an energy corridor for Europe is a threat to energy security when political tensions increase between Ukraine and Russia. Iran could have been considered as an alternative source of energy supply for Europe, but there has been a longstanding disagreement between Iran and the West involving economic sanctions.

The nuclear deal between Iran and Western powers in July 2015 (effective on 16 January 2016) may facilitate possible changes that could build a bridge between Iran's natural gas sources and major consumers in Europe. Iran is not only the holder of the largest natural gas reserves and the fourth largest crude oil reserves in the world, but it is also located in a geopolitical area. The EU is diversifying its energy supply in order to reduce its dependency on Russia. In this regard, renewable energy technologies are an alternative means of power generation. The EU has made sizable investments and the use of advanced technologies enables the countries to produce renewable energy more cost-effectively. The power generated by these sources also continues to increase. This is consistent with sustainable development goals.

REFERENCES

Abolhosseini, S. and A. Heshmati (2014). 'The Main Support Mechanisms to Finance Renewable Energy Development'. *Renewable and Sustainable Energy Reviews*, 40(December): 876–85.

Abolhosseini, S., A. Heshmati, and J. Altmann (2015). 'A Review of Renewable Energy Supply and Energy Efficiency Technologies'. IZA Discussion Paper, Institute for the Study of Labor, Bonn.

Asif, M. and T. Muneer (2007). 'Energy Supply, Its Demand and Security Issues for Developed and Emerging Economies'. *Renewable and Sustainable Energy Reviews*, 11(7): 1388–413.

Bhattacharyya, S. C. (2011). *Energy Economics: Concepts, Issues, Markets and Governance.* London Springer-Verlag.

Bohi, D. R. and M. A. Toman (1993). 'Energy Security: Externalities and Policies'. *Energy Policy*, 21(11): 1093–109.

Bouzarovski, S., M. Bradshaw, and A. Wochnik (2015). 'Making Territory through Infrastructure: The Governance of Natural Gas Transit in Europe'. *Geoforum*, 64(August): 217–28.

BP (Bristish Petroleum) (2014). 'BP Statistical Review of World Energy, June 2014'. Available at: <http://www.bp.com/content/dam/bp-country/de_de/PDFs/brochures/BP-statistical-review-of-world-energy-2014-full-report.pdf>(accessed 8 April 2015).

BP (2015a). 'BP Statistical Review of World Energy, June 2015'. Available at: <https://www.bp.com/content/dam/bp/pdf/energy-economics/statistical-review-2015/bp-statistical-review-of-world-energy-2015-full-report.pdf> (accessed 9 July 2015).

Brown, S. and H. G. Huntington (2008). 'Energy Security and Climate Change Protection: Complementarity or Tradeoff?' *Energy Policy*, 36(9): 3510–13.

Creutzig, F., J. C. Goldschmidt, P. Lehmann, E. Schmid, F. von Blucher, C. Breyer, B. Fernandez, M. Jakob, B. Knopf, S. Lohrey, T. Susca, and K. Wiegandt (2014). 'Catching Two European Birds with One Renewable Stone: Mitigating Climate Change and Eurozone Crisis by Energy Transition'. *Renewable and Sustainable Energy Review*, 38 (October): 1015–28.

Darda, M. A., R. Guseo, and C. Mortarino (2015). 'Nonlinear Production Path and an Alternative Reserves Estimate for South Asian Natural Gas'. *Renewable and Sustainable Energy Review*, 47(July): 654–64.

Deese, D. A. (1979). 'Energy: Economics, Politics, and Security'. *International Security*, 4(3): 140–53.

Delucchi, M. A. and J. J. Murphy (2008). 'US Military Expenditures to Protect the Use of Persian Gulf Oil for Motor Vehicles'. *Energy Policy*, 36(6): 2253–64.

DeRosa, D. A. and G. C. Hufbauer (2008). 'Normalization of Economic Relations: Consequences for Iran's Economy and the United States'. Report prepared for the NFTC, National Foreign Trade Council, Washington, DC.

EIA (Energy Information Administration) (2014). 'International Energy Outlook 2014'. US Energy Information Administration, Washington, DC.

EI (Energy Intelligence) (2015). 'The Energy Cost Report'. *EI New Energy*, IV(26) (Accessed through Personal Communication, 21 July 2015).

Esposito, D., J. Krupp, and S. Carley (2015). 'Interaction of Risks Associated with Natural Gas- and Renewable-Based Electricity'. *Electricity Journal*, 28(8): 69–84.

European Commission (2010). 'Europe 2020: A European Strategy for Smart, Sustainable and Inclusive Growth'. European Commission, Brussels.

Fischhendler, I. and D. Nathan (2014). 'In the Name of Energy Security: The Struggle over Exploration of Israeli Natural Gas'. *Energy Policy*, 70(July): 152–62.

Gellings, C. W. (2009). *The Smart Grid: Enabling Energy Efficiency and Demand Response.* Liburn, GA: Fairmont Press.

Ghezelbash, R., M. Harzaneh-Gord, H. Behi, M. Sadi, and H. S. Khorramabady (2015). 'Performance Assessment of a Natural Gas Expansion Plant Integrated with a Vertical Ground-Coupled Heat Pump'. *Energy*, 93(2): 2503–17.

Heshmati, A. (2014). 'An Empirical Survey of the Ramifications of a Green Economy'. IZA Discussion Paper 8078, Institute for the Study of Labor, Bonn.

Heshmati, A. and S. Abolhosseini (2014). 'Market Design for Trading Commoditized Renewable Energy'. IZA Discussion Paper 8375, Institute for the Study of Labor, Bonn.

Heshmati, A., S. Abolhosseini, and J. Altmann (2015). *The Development of Renewable Energy Sources and Its Significance for the Environment.* Singapore: Springer.

Jacobsson, S. and A. Bergek (2004). 'Transforming the Energy Sector: The Evolution of Technological Systems in Renewable Energy Technology'. *Industrial and Corporate Change*, 13(5): 815–49.

Kahrl, F., J. Hu, H. Kwok, and J. H. Williams (2013). 'Strategies for Expanding Natural Gas-Fired Electricity Generation in China: Economics and Policy'. *Energy Strategy Reviews*, 2(2): 182–9.

Katzman, K. (2012). 'Iran: U.S. Concerns and Policy Responses'. Congressional Research Service, Washington, DC.

Lee, A., O. Zinaman, J. Logan, M. Bazilian, D. Arent, and R. L. Newmark (2012). 'Interactions, Complementarities and Tensions at the Nexus of Natural Gas and Renewable Energy'. *The Electricity Journal*, 25(10): 38–48.

Lise, W., B. F. Hobbs, and F. Van Oostvoorn (2008). 'Natural Gas Corridors between the EU and Its Main Suppliers: Simulation Results with the Dynamic GASTALE Model'. *Energy Policy*, 36(6): 1890–906.

Mavrakis, D., F. Thomaidis, and I. Ntroukas (2006). 'An Assessment of the Natural Gas Supply Potential of the South Energy Corridor from the Caspian Region to the EU'. *Energy Policy*, 34(13): 1671–80.

Mediavilla, M., C. de Castro, I. Capellan, L. J. Miguel, I. Arto, and F. Frechoso (2013). 'The Transition towards Renewable Energies: Physical Limits and Temporal Conditions'. *Energy Policy*, 52(January): 297–311.

Moutinho, V., A. C. Moreira, and P. M. Silva (2015). 'The Driving Forces of Change in Energy-Related CO_2 Emissions in Eastern, Western, Northern and Southern Europe: The LMDI Approach to Decomposition Analysis'. *Renewable and Sustainable Energy Reviews*, 50(October): 1485–99.

Nuttall, W. J. and D. L. Manz (2008). 'A New Energy Security Paradigm for the Twenty-First Century'. *Technological Forecasting and Social Change*, 75(8): 1247–59.

Owen, A. D. (2004). 'Oil Supply Insecurity: Control versus Damage Costs'. *Energy Policy*, 32(16): 1879–82.

Rickerson W. and R. C. Grace (2007). 'The Debate over Fixed Price Incentives for Renewable Electricity in Europe and the United States: Fallout and Future Directions'. Heinrich Böll Foundation, Washington, DC.

Sen, S. and T. Babali (2006). 'Security Concerns in the Middle East for Oil Supply: Problems and Solutions'. *Energy Policy*, 35(3): 1517–24.

Sovacool, B. K. (2016). 'How Long Will It Take? Conceptualizing the Temporal Dynamics of Energy Transitions'. *Energy Research & Social Science*, 13(March): 202–15.

Umbach, F. (2010). 'Global Energy Security and the Implications for the EU'. *Energy Policy*, 38(3): 1229–40.

Verdeil, E., E. Arik, H. Bolzon, and J. Markoum (2015). 'Governing the Transition to Natural Gas in Mediterranean Metropolis: The Case of Cairo, Istanbul and Sfax (Tunisia)'. *Energy Policy*, 78(March): 235–45.

Vivoda, V. (2009). 'Diversification of Oil Import Sources and Energy Security: A Key Strategy or an Elusive Objective'. *Energy Policy*, 37(11): 4615–23.

Weijermars, R., G. Drijkoningen, T. J. Heimovaara, E. S. J. Rudolph, G. J. Weltje, and K. H. A. A. Wolf (2011). 'Unconventional Gas Research Initiative for Clean Energy Transition in Europe'. *Journal of Natural Gas Science and Engineering*, 3(2): 402–12.

World Bank (2015). 'Energy Imports, Net (% of Energy Use)'. Available at: <http://data.worldbank.org/indicator/EG.IMP.CONS.ZS> (Accessed on 29 June 2015).

Part V

Incumbency

Clean energy transitions entail sustained periods of collision between incumbents and new entrants. The tensions between new entrants and incumbents are not purely economic, but range across political, regulatory, and cultural landscapes as well. In this part, authors explore these dynamics through case studies from three different perspectives.

Looking at recent regulatory disputes in Wisconsin, USA, Astoria explores how a seemingly straightforward conflict over solar electricity rates in fact reveals how deeply the regulatory framework has been shaped by the biophysical characteristics of fossil fuels. The case illuminates the crucial role of legal and institutional reform in achieving clean energy transitions. Shifting focus to the wires that constitute electricity grids, Eid, Hakvoort, and de Jong examine how the seemingly neutral term 'smart grid' manifests itself in deeply different ways in China, the EU, and USA. Drawing out the widely varied industrial organization of each setting, the authors discuss how real-world implementation of smart grids encodes deeper signals of social priorities, economic interests, and political economic constraints. Finally, shifting to one of the largest remaining redoubts of incumbency, Asmelash explores the challenges of—and pathways towards—a binding multilateral agreement to phase out fossil fuel subsidies. Building upon a wide-ranging discussion of institutional architecture and international legal precedent, the chapter focuses attention on what role should be played by key multilateral institutions such as the World Trade Organization (WTO).

16

Incumbency and the Legal Configuration of Hydrocarbon Infrastructure

Ross Astoria

16.1 INTRODUCTION

Through its intended nationally determined contribution (INDC), the United States (US) has committed to reducing its greenhouse gas emissions by between 26 and 28 per cent in 2025 relative to a 2005 baseline. The US commitment is founded solely on regulations promulgated by administrative agencies under existing statutory authority, including the Clean Air Act, the Energy Policy Act, and the Energy Independence and Security Act (UNFCCC 2016). These regulations include fuel-economy standards for various classes of vehicles, energy conservation standards for appliances and some commercial buildings, and, most prominently, carbon dioxide emission standards for new and existing electrical generating units (EGUs). The regulations for new EGUs are known as New Source Performance Standards (EPA 2015b) and those for existing EGUs, as the Clean Power Plan (EPA 2015a). The Environmental Protection Agency (EPA) promulgated both under the authority of the Clean Air Act.

In accordance with the Clean Air Act's 'cooperative federalism', the Clean Power Plan decentralizes the US's mitigation effort by setting emission targets for the individual states. The individual states are then to design a state implementation plan (SIP) which meets those emission targets. At present, then, whether the US successfully eliminates greenhouse gas emissions from its electrical power sector and transitions to clean energy or whether it backslides on its mitigation commitment is heavily dependent upon state policy and politics.

Greenhouse gas abatement touches almost every aspect of policy, and in this chapter, I use a recent utility rate case from the state of Wisconsin to illustrate some of the political and policy difficulties states will confront during the transition to renewables. In this rate case, the state's largest capital owned utility, WE Energies, sought tariff rates which severely undermined the value of distributed photovoltaic (PV) energy production to the homeowner and thus the viability of the distributed photovoltaic industry in Wisconsin. The Public Service Commission of Wisconsin (PSCW) approved the rates and advocates of the renewable energy industry sued, winning a lower court decision reversing a portion of the tariff. The case docket and ensuing litigation strongly suggest that the incumbent utility used its economic power to undermine the formation of solar industry in Wisconsin. That is, the rate case is an instance of monopolistic rent-seeking, the typical response to which is to propose disciplining the badly behaving monopoly through either more regulation or more competition (forced by regulation).

This economic view of the conflict between the incumbent utility and the solar industry, however, is insufficient to identifying the contradictions between hydrocarbon infrastructure and renewable energy infrastructure. First, the economic view falls short in describing how the law organizes the relevant markets and other institutions of exchange. Second, the economic view fails to take into consideration the biophysical aspects of hydrocarbon generation and renewable energy generation. Hence, in the first half of this chapter, I sketch the salient aspects of the WE Energies rate case which make it appear an instance of monopolistic rent-seeking. In the second half of the chapter, I analyse how the biophysical aspects of hydrocarbon generation have been scribed into the law of electrical power utilities. The biophysical point of view, *inter alia*, incorporates the first and second laws of thermodynamics into its analysis of socioeconomics. The concept of energy returned on investment (EROI) is a corollary of the second law of thermodynamics, and allows us to identify three contradictions between hydrocarbon infrastructure and renewable infrastructure. Using examples from the WE Energies rate case, we can see how the law, legal institutions, and the legally organized markets for electricity are configured around and support hydrocarbon infrastructure.

As a general matter, then, for the US to transition to renewables and eliminate greenhouse gas emissions, law, legal institutions, and the markets which they organize must be reconfigured to support renewable energy infrastructure.

16.2 STRATEGIC RENT-SEEKING IN WISCONSIN

Wisconsin is situated on the west coast of Lake Michigan and the port city of Milwaukee is its largest. The early economy of Wisconsin revolved around agriculture and the extraction of its northern pine forests, the timber from which was transported either to Milwaukee or, more likely, to Chicago, from where it was exchanged for the grains from the fertile Midwestern plains. The area between Chicago and Milwaukee became a leading developer of tractor equipment. Along with timber, the farm equipment was shipped west to the farmers who used it to build up their homesteads and crop yields. This foundation, along with a railroad system and access to navigation over the Great Lakes, led eastern Wisconsin to develop into a classic Keynesian-Fordist economy—centralized manufacturing of raw materials with unionized labour and a fairly robust welfare state (Cronon 1992; Hurst 1964). The west of the state was and is agricultural. In line with the general trend in the US, during the 1980s, the Keynesian-Fordist economy in eastern Wisconsin began to deteriorate. A few silos of Keynesian-Fordist manufacturing remain in the vicinity of Milwaukee, and a remnant of both the timber industry and the Keynesian-Fordist economy is found in the paper and pulp mills of northern Wisconsin.

The law has organized the electrical power utilities in Wisconsin under three different forms: capital-owned utilities, municipally owned utilities, and rural electrical cooperatives. WE Energies is a capital-owned electrical power utility (officially, 'The Wisconsin Electrical Power Corporation' or WEPCO doing business as WE Energies). In July 2015, WE Energies' holding company, formerly called Wisconsin Energy, completed the US$9.1 billion purchase of Integry Energy Group, at which point it became the 'WEC Energy Group' (WEC) (Content 2015).

According to Energy Information Agency figures from 2012, WEC Energy Group's electrical utilities held load territory covering 35 million megawatt hours (MWh) of loads per year, about 25 million of which are WE Energies' (EIA 2012). For comparison, the next largest investor-owned electrical utility, Wisconsin Power & Light, held load territory encompassing 10 million MWh of annual loads (EIA 2012).

On 31 January 2014, WE Energies filed a request with the PSCW to open a rate docket for the 2015 test year. The request revised rates extensively and included new tariffs for distributed generation (PSCW Case No. 5-UR-107). The new WE Energies tariff decrease the value of distributed PV to its owner in five ways. First, like all residential customers, distributed generation owners would pay an increased customer-related facilities charge of US$0.5202/day rather than the previous US$0.30/day (PSCW 2014a: 35). WE Energies concurrently reduced the volumetric charge from 13.9 cents to 13.4 cents (PSCW 2014h: appendix B). Second, distributed generation owners would pay a facilities charge for the installation of a second meter (required to meter imports and exports separately) (PSCW 2014a: 56). The PSCW partially approved this facility charge (PSCW 2014h: 70–1). Third, the crediting of exports was altered from annual to seasonal, reducing the ability of solar generation to spread its surplus generation over a longer period of time. Fourth, owners of distributed generation will receive the wholesale price (rather than the retail price) for their exports, a difference of about 8 cents/

KWh (PSCW 2014b). Fifth, owners of distributed generation would pay a new customer-related demand charge, which, according to WE Energies 'will recover distribution costs that are not recovered by the facilities charge of the underlying rate' (PSCW 2014a: 56–7). WE Energies also stated that the demand charge is supposed to 'recover the costs of stand-by generation' (2014a: 57). The demand charge for solar would have been US$3.794/kW/month (PSCW 2014c).

RENEW Wisconsin is a non-profit organization which advocates for the renewable energy industry in Wisconsin. They intervened before the PSCW and along with the Alliance for Solar Choice eventually filed suit against the demand charge (but not the other features of the tariff). RENEW's policy director, Michael Vickerman, offered testimony on the economic impacts of the new tariffs to owners of PV. Excluding the generally applicable increase in the facilities charge, Vickerman calculated that for a 5 kW PV system installed in 2013 whose total annual output is equivalent to half of the resident's annual loads, the new tariff reduces the value to the owner by 35 per cent, from US $834 per year to US$542 per year. If the PV system is sized to cover 95 per cent of loads, it becomes more likely that production will exceed loads and that the new annual netting procedure and new export price would be applicable more frequently. Under this scenario, Vickerman calculated that the value of the PV to the owner drops 47 per cent, from US$834 per year to US$389 per year (PSCW 2014d: 20–2).

In other states, such as Minnesota, Maine, New York, and California, stakeholders are having a robust discussion on how to value distributed energy resources in a manner which facilitates the transition to clean energy and expands economic opportunity, while also preserving capital's fiscal position. Because the utility made no effort to engage the renewable energy stakeholders, because the PSCW voted against the recommendation of staff, and because the one (dated) study on the value of solar concluded it to be beneficial to WE Energies, this rate case seems a classic example of the sort of strategic rent-seeking which canonical economic thought instructs us to expect from monopolies (Kaiser 2014; PSCW 2014e: 27–32). In any case, the judge presiding over the legal challenge to the demand charge thought so. During the oral hearing, he commented on the evidence WE Energies presented on behalf of the demand charge:

> [t]he strong impression that this Court has when we look at the evidence that was presented is…these are stories, but they're not empirical, and they happen to be stories spun by a company that is facing competition from the people who are now going to be paying these higher rates. So if these were our standards in most cases, let's hear the competitor tell us what his competitor is doing but without evidence, but without empirical evidence, we would expect it to be skewed
>
> (*Alliance for Solar Choice v. Public Service Comm. of Wisconsin* 2015: 60).

The judge seems to be indicating that he believes the monopoly utility to being using the economic power of its incumbency to hinder the emergence of economic competition. Incumbents protecting their economic position by raising obstacles to competition from renewable energy will no doubt hinder the transition to renewable generation. However, focusing on economic incumbency conceals the degree to which presently existing infrastructure and the laws which configure that infrastructure presuppose and entrench hydrocarbon generation. The incumbent, then, is not so much any particular firm, but hydrocarbons *per se*, a subject to which we now turn.

16.3 THE BIOPHYSICAL APPROACH AND HYDROCARBON'S LEGAL INFRASTRUCTURE

The biophysical approach incorporates the first and second law of thermodynamics into the analysis of socioeconomics. Here, we are interested in the second law of thermodynamics and its relationship to the legal regime which has arisen around and supports hydrocarbon infrastructure. The second law of thermodynamics states, roughly put, that without exogenous inputs of energy, a closed system become less organized, more random. One way to measure the exogenous inputs of energy into the socioeconomic system is energy returned on investment, or EROI. EROI is 'the ratio of energy returned from an energy-gathering activity compared to the energy invested in that process' (Hall and Klitgaard 2012: 310). It is derived from the ecologist's practice of analysing the transportation of energy through trophic food webs.

EROI for all fossil fuels declines over the history of the industry. In 1930, petroleum and gas had EROIs as high as 100:1, and perhaps higher. In 1970, the EROI of petroleum and gas was down to 30:1, and the EROI from contemporary tar sands extraction is as low as 2–4:1. At best, corn-base ethanol has an EROI of 3:1. In 1930, coal had an EROI of 80:1, which had declined to 30:1 by 1970. Hydropower retains a high EROI of about 100:1, while wind turbines have an EROI of about 18:1. Solar PV presently has an EROI of about 7:1, and some analyses show as high as 15:1. The EROIs for wind and solar, however, do not include the embodied energy of whatever devices, such as storage, might be needed for a renewable grid (Hall, Lambert, and Balogh 2014).

Prieto and Hall estimate that an EROI of 3:1 is required at the well-head merely to extract, refine, and transport the petroleum to the place of use. EROIs at the well-head must be of the order of 8:1 if one wishes the workers at the well to be able to support a family and as high as 12:1 if one wishes that family to have access to health care (Prieto and Hall 2013). In general, energy-gathering activities which have high EROI ratios allow for the satisfaction of a greater number of life activities above mere necessity. Declining EROI ratios make societal activities which are high on the hierarchy of 'energy needs' (such as art or flying to Las Vegas) increasing difficult to achieve while also satisfying the needs of other members of society.

Scholars still dispute the appropriate boundaries for determining EROI for particular energy-gathering activities and in many cases the necessary data is unavailable or incomplete. However, neither the orders of magnitude nor the general trend in EROI values are contentious. The fossil fuels extracted during the twentieth century represented extraordinary high EROIs and show a continual decline. The best first principle from ecology suggests that hydrocarbon EROIs can be expected to decline further, and can be intuitively observed in quests for 'extreme energy' such as Arctic petroleum drilling endeavours.

On the other hand, since renewable generation is in its infancy, their EROIs are expected to increase, although this is by no means guaranteed. The EROIs on both wind and solar PV, for instance, already exceed that of tar sands. Whether renewables will be able to obtain ratios high enough to support their own expansion remains open, especially at the rate required to avoid disastrous global warming. Less likely still is that renewable generation will obtain EROIs high enough to satisfy the consumption norm which organized the US's productive capacities during the previous century of high EROI hydrocarbons.

From EROI we may introduce the notion of hydrocarbon infrastructure, which is infrastructure which is possible and useful only when high EROI hydrocarbons are available. The electrical power utility in the US is a paradigmatic example of hydrocarbon infrastructure. The electrical power industry, however, is not only an organization of technology and human capacity. It is also a legal apparatus, which responds to and is configured by the biophysical aspects of hydrocarbons. As such, the electrical power utility has two critical relationships to the depletable reservoir of high EROI hydrocarbons. First, it is the apparatus through which that depletable store is converted from a use-value to an exchange-value. Second, it is parasitic upon that same depletable store.

First, the primary use-value of hydrocarbons is found in their chemical bonds. The bonds represent both a store of energy (heat) and an arrangement of atoms from which a substance may be made. For the hydrocarbons' use-value to be realized as an exchange-value, the use-value must be made social by bringing it to the market, as the metaphor goes. At the market, other members of society may access the hydrocarbon's use-value by exchanging it for some other use-value. The exchange is not, of course, completed by bartering use-values, but through the exchange of the money commodity, that is, currency. In the case of the electrical power utility, the use-value of the hydrocarbon (heat) is converted to a commodity (the kWh) by processing it through an extensive number of technological devices. With a monthly bill, the 'customer' then exchanges that kWh for currency. The currency is thereby brought under the legal control of the utility, which then dispenses that currency to different social interests, that is, to capital in the form of dividends and coupon payments, to labour in the form of wages, to politicians in the form of campaign contributions, and so forth.

To further illustrate the utility's power over the dispensation of the use-value of high EROI hydrocarbons, consider the cost-of-service study (COSS) which WE Energies used in the recent rate case. Electrical power utilities serve a variety of customer types, typically residential, commercial, and industrial. The utility incurs costs in delivering electrical power to these different classes, but because all customer classes are reliant upon the same grid, a method is needed to assign those costs to the different customer classes. Once the costs are assigned, the rate paid by the different customer types can be calculated. A COSS assigns costs across customer types and a great number of methodologies are available.

The Citizens' Utility Board (CUB) submitted testimony opposed to WE Energies' requested rates. One objection was that in previous rate cases WE Energies had used several COSS methodologies and then assigned costs and rates by comparing and weighting the different outcomes of the different methodologies. In this case, however, the CUB noted that WE Energies used only a single COSS methodology known to assign costs disproportionately to residential rate payers as compared to commercial and industrial rate payers (PSCW 2014f: 11). In disproportionately assigning costs of service to residential customers, the CUB argued that WE Energies was concurrently assigning to the residential rate class a disproportionate responsibility for paying those costs as compared to the commercial and industrial rate classes. Although concealed as a technical accounting procedure, the COSS is a political instrument for assigning the benefits and costs of hydrocarbon's use-value among different social interests.

Second, the electrical power utility could not exist or perpetuate itself in the absence of the reservoir of high EROI hydrocarbons and is, therefore, parasitic upon it. The various devices which transform the chemical bonds of the hydrocarbon into heat, steam, mechanical motion, and then an electrical current represent very large magnitudes of embodied energy. Those physical devices include not only the steam turbines, electromagnets, transformers, and conductors but also the steel rails over which the coal is transported to the plant. Without access to the high EROI reserve of fossil fuel, the devices which convert it into a current could not themselves have been brought into being.

In sum, the electrical power utility is a legal apparatus which is both parasitic upon a depletable reserve of high EROI hydrocarbons and realizes the use-value of those hydrocarbons, channelling their benefits and costs to various segments of society.

16.4 THREE CONTRADICTIONS AND THE CONFIGURING LEGAL APPARATUS

Renewable energy infrastructure and hydrocarbon infrastructures both involve a myriad of technological devices. Here, however, I focus on the biophysical characteristics of coal-fired and distributed PV generation. These two types of generation were central to the rate case and also show distinctly the contradiction between the legal regime which has arisen around hydrocarbon infrastructure and that which must be constructed around renewable infrastructure. The incumbent, then, is not simply any particular utility which exercises its entrenched economic power to exclude competition. Rather, the law entrenches hydrocarbon infrastructure itself and high EROI coal is itself the incumbent. Other types of generation, such as wind and nuclear, will have a mix of biophysical characteristics which overlap those of solar and coal. This overlap means they will play an important part of the transition away from hydrocarbon generation and to renewable generation, but that role is not discussed here.

16.4.1 Efficiency Logics and Territory

The efficiency logics of coal-fired and distributed PV generation are different, which manifests in different legal relationships with territory.

Coal's (formerly) high EROI means that it can be combusted at a perpetual high heat, and thus produce a perpetual high voltage. To match this perpetual high voltage as required by Ohm's law, hydrocarbon generation needs loads that are temporally disbursed. In this way, the turbine can be turning as near to capacity as possible. Hence, utilities developed loads by, for instance, giving away appliances and through construction projects. The amusement park, for instance, was a creature of the utilities meant to provide the turbine with an off-peak load (Nye 1992: 122–32). These temporally dispersed loads are also physically separate from each other, so to bring them under their dominion, the utility must acquire and retain control of the territory in which those loads are located. In the US this was accomplished through the law and, during the formative years of the industry, the utilities convinced states to transfer to them certain powers of sovereignty, including exclusive franchises over territory, eminent domain, limited negligence liability, and access to premises (Hempling 2013: 15–34 (territory), 57–8 (eminent domain), 61–3 (liability); Wisc. Statutes 1983).

This impulse to control the development of a territory is so strong that the law has made some attempt to restrain it. Utility law in Wisconsin prohibits public utilities and nonutility affiliates from engaging in real estate practices, residential real estate development, property management, or residential or commercial construction (Wisc. Statutes 1999a). However, this constrain is almost immediately undermined in the next section of law, which permits public utilities or nonutility affiliates to consult or make financial arraignments with third parties engaged in residential or commercial construction (Wisc. Statutes 1999b). Hence, WE Energies' holding company, WEC, owns Wispark, a 'full-service real estate development company' which develops 'real estate projects that support and complement key programs of Wisconsin Energy Corporation (WEC)' (Wispark Webpage 2016).

In contrast, distributed PV is intermittent and produces low voltages. The perception of intermittency is conditioned by a century's access to hydrocarbon infrastructure, but nevertheless counsels confining loads to the time and wattage of generation. Since voltage deteriorates linearly over a conductor, low voltages do not travel far. Hence, the efficiency logic of distributed PV is to bring loads both physically and temporarily proximate to the generation and to minimize or eliminate all other loads. This is almost the exact opposite of the efficiency logic of hydrocarbon generation. Hence, distributed PV does not require the control of territory or the gifting of toasters.

However, distributed PV does require legal protection against shading as well as the authority to enter so as to remove that shading, or, at the least, the upper hand in negotiating over the prevention and removal of shading. In 1982 the Supreme Court of Wisconsin considered whether a homeowner had a cause of action against a neighbour whose trees shaded the homeowner's solar thermal water heater. Building off of the common law's Doctrine of Ancient Lights, the Supreme Court of Wisconsin found that the homeowner could bring a nuisance suit over the shading and the Wisconsin legislature then codified the ruling (*Prah v. Maretti* 1982). Likewise, to receive a payment from the state's solar incentive programme, a homeowner must show the installed PV system will have '10 percent or less obstacle shading' (Focus on Energy 2016).

16.4.2 Decommodification and Jurisdiction

Under hydrocarbon infrastructure, the utility is the 'producer' of a commodity, the kWh, and the homeowner is the 'consumer' of that commodity. The utility produces the kWh to maximize its exchange-value and not for its use-value, which is incidental to the utility.

In contrast, for distributed PV, the 'consumer' and the 'producer' are the same. The distinction, then, no longer provides a felicitous description of socioeconomic relations. For distributed PV, the kWh is not brought to the market as a means of realizing its exchange-value by swapping it for the money commodity. The electric current is generated for its use-value, and its exchange-value is incidental to the homeowner. Distributed PV, then, represents the decommodification of a critical aspect of human well-being (power, heat, and light) which the utilities had formerly commodified (Nye 1992: 234).

However, the legally organized electrical power 'markets' are configured around electrical power as a commodity which utilities generate for its exchange-value. The law of jurisdiction, as scribed in the US Constitution itself, configures both the retail markets and the wholesale markets which have commodified electrical power. Early in the development of the electrical power industry in the US, a detente was reached between municipal ownership and capital ownership of utilities. The law acknowledged capital-owned utilities as monopolies (supposedly 'natural') in exchange for having a state public utility commission regulate rates so as to be 'just and reasonable'. Wisconsin was one of the first states to legislate this arrangement in 1907 (Nord 1975).

The power industry's construction of transmission across state lines disrupted the detente. In the early 1920s, Narragansett Electric Light Company, located in Rhode Island, entered into a contract with the Attleboro Electric Steam Company, located in Massachusetts, to supply electrical current. As required by law, the Narragansett Company then filed the rate schedule contained in the contract with the Rhode Island Public Utilities Commission (PUC), which accepted it. Several years later, the Narragansett attempted to revise the contract through a proceeding before the Rhode Island PUC, claiming that because of increased generating costs, it was suffering an operating loss. The Rhode Island PUC agreed and revised the rates accordingly. The Attleboro Electric Steam Company filed suit against the Rhode Island PUC for its approval of the new rates, and the parties appealed the case to the Supreme Court of the US. The Supreme Court found that the current crossing the state boundary separating Rhode Island from Massachusetts was interstate commerce. The Supreme Court then ruled that the commerce clause of Article I, Section 8, prohibited the Rhode Island PUC from exercising jurisdiction over the transaction between Narragansett and Attleboro. It further concluded that jurisdiction over interstate transactions of electrical power was reserved to the federal government (*Public Utilities Commission of Rhode Island et al. v. Attleboro Steam & Electric Co.* 1927). In response, the Congress passed the Federal Power Act which assigned to the Federal Power Commission jurisdiction over 'the transmission of electric energy in interstate commerce and to the sale of electric energy at wholesale in interstate commerce' (Federal Power Act 1935).

Hence, out of the Supreme Court's interpretation of the commerce clause, there emerged a regulatory framework in which the federal government, through the Federal Power Commission (now the Federal Energy Regulatory Commission (FERC)) regulates the interstate wholesale electrical power industry while states regulate the intra-state retail side of that industry.

Accordingly, during the restructuring period of the mid-1990s, the FERC used a sequence of orders to coax the regional wholesale markets into being and now supervises them to ensure that rates are 'just and reasonable' (see FERC Orders 888, 889, 890 (FERC 1996a, 1996b, 2007)). Those wholesale markets entrenched existing hydrocarbon infrastructure, including both generating assets and the high-voltage transmission which that generation requires.

Recent litigation over FERC orders requiring the organized wholesale markets to allow the participation of demand response in the wholesale markets illustrates the entrenchment. Voltage and loads must be perpetually and instantaneously matched. As initially configured, the wholesale markets achieved that matching by increasing generation rather than by reducing load. In those markets, load-serving entities (LSEs, i.e., distribution utilities) submit their projected daily demand to the market, usually a day ahead and for durations of an hour, although each regional wholesale market configures its own market and there is variation among them. Generators then submit their generation bid for those same time increments. All trading takes places in dollars per MWh. Generator bids are stacked in order of price, and the next least cost bid required to meet the demand is assigned the clearing price. The clearing price is adjusted for reliability, congestion, and line losses at particular nodes in the grid, producing the localized marginal price, or LMP. The LMP is then paid by all LSEs and received by all dispatched generators.

Into this configuration of the wholesale markets, FERC Order 719 (FERC 2008) required market organizers to allow demand-side management (DSM) to participate in the bidding. FERC Order 745 (FERC 2011) further established the mechanism for compensating DSM. According to a 2009 FERC study (FERC 2009: x), DSM could reduce peak load by up to 150 GW nationally, something many generators were not interested in, and the Electrical Power Supply Association (EPSA) brought a suit. EPSA claimed that Orders 719 and 745 exceeded FERC's jurisdictional authority because DSM fell on the retail side of the jurisdictional divide. The appellate court agreed: '[d]emand response—simply put—is part of the retail market. It involves *retail* customers, their decision whether to purchase *at retail*, and the levels of *retail* electricity consumption' (*EPSA v. FERC* 2014, emphasis in the original).

The Supreme Court reversed the appellate court on the grounds that DSM 'directly affects wholesale rates' and that FERC has not 'regulated retail sales' (*FERC v. EPSA* 2016). Nonetheless, esoteric matters of legal jurisdiction continue to configure the organized markets around pre-existing hydrocarbon infrastructure. As legally organized, generation of MWh transmitted over high-voltage conductors is the reference position for the organized wholesale markets. Other techniques for managing electrical power management have no jurisdictional home. Consider, for instance, that under the appellate court's rationale in *EPSA v. FERC*, the exported generation from distributed PV would likely be considered a retail commodity just like DSM, but that under

WE Energies' distributed generation tariff the exports from distributed PV are compensated at the wholesale rate. The Supreme Court's ruling in *FERC v. EPSA* disturbs the jurisdiction configuration of hydrocarbon infrastructure, but does not alter it. Especially given the mechanism by which DSM providers are to be compensated (from the LSEs in which the DSM is located), *FERC v. EPSA* likely marks an important aspect of the US's transition to renewables. It does not by itself, however, reconfigure those markets around renewables generally or distributed PV specifically. The commodity exchanged in the wholesale markets, the MWh, remains for distributed PV the wrong order of magnitude.

16.4.3 Guaranteeing and Disciplining Capital

The devices which compose the electrical power utility, such as the turbine, represent large magnitudes of embodied energy. The initial financial capital required to construct those devices is therefore also large and requires long payback periods over which it is subject to many risks. To assuage these risks and facilitate capital formation, states guarantee capital's return on investment (ROI). Indeed, in *Missouri ex rel. Southwestern Bell Telephone Co. v. Public Service Commission* 1923) the Supreme Court interpreted the Fifth Amendment's Taking Clause to require state public service commissions to approve rates large enough to preserve financial capital's ROI and thereby the utility's fiscal position. This guarantee of capital is illustrated by WE Energies' rate case. In its final ruling, the PSCW began its decision by settling on the appropriate return on equity (ROE) to be received by the shareholders. In this instance, PSCW staff facilitated negotiations between WE Energies, the CUB, the Wisconsin Industrial Energy Group, and the Wisconsin Paper Council. They agreed to reduce WE Energies' proposed ROE from 10.4 per cent to 10.2 per cent. From this prioritization of ROE, WE Energies' revenue requirements were then calculated, and from that how much the utility was to collect from ratepayers (PSCW 2014g: 3, 2014h: 17–20). The presently existing legal configuration disciplines the formation of capital towards hydrocarbon infrastructure.

In contrast, each individual distributed PV requires a small fiscal outlay and the payback period is comparatively short. As compared to hydrocarbon generation, then, risks to capital are greatly reduced. Nonetheless, relative to the pocketbook of the PV owner, the capital outlay for distributed PV might yet be substantial. Further, transitioning to a renewable energy infrastructure will require capital outlays of at least the same order of magnitude as a business-as-usual scenario over the coming decades. Some state-recognized financing mechanism, then, is needed to facilitate and discipline capital formation towards investments in renewables.

A third-party installer arrangement is one such mechanism. Under such an arrangement, the installer finances and installs the distributed PV system, and the installer and the homeowner then share the economic benefit of the homeowner's reduced power costs. However, under this arrangement the installer appears to be selling kWh to the homeowner and ambiguity exists as to whether, as a matter of law, the third-party installer should be classified as a utility subject to regulation by the public service commission. This legal ambiguity exists in Wisconsin, and appears to be preventing the formation of a third-party installer industry. In its rate case, WE Energies solicited a prohibition on third-party installers by seeking a requirement that distributed generators own their own equipment (PSCW 2014a: 60). The PSCW rejected WE Energies' proposal, but without resolving the ambiguity about the legal status of third-party installers (PSCW 2014h: 89).

16.5 CONCLUSION: RENTS, LAW, AND MAKING MARKETS FOR RENEWABLES

From the merely economic point of view, WE Energies' rate case is an instance of incumbent rent-seeking. However, when we view the electrical power utility as a paradigmatic instance of hydrocarbon infrastructure composed of technological devices, human capacities, and a supporting legal apparatus, we discover more substantive contradictions between distributed PV and coal-fired generation.

The utility's legal apparatus is configured around the biophysical characteristics of high EROI hydrocarbons, coal in particular. Hydrocarbon generation tends towards a centralized legal configuration which exercises sovereign functions over an exclusive territory, which is the 'producer' selling a commodity to a 'consumer', and which has the state secure capital's fiscal position over long periods of time. Distributed PV generation, in contrast, implies a tendency towards physically and temporally confining loads to the proximity of generation, represents the decommodification of an important part of life (heat, light, and power), negates the distinction between 'producer' and 'consumer', and requires the state to support a different type of capital formation and discipline.

These biophysical differences, of course, are both tendencies and extreme limits. They are not 'natural' and their existence and perpetuation requires constant effort. The utility's monopoly over territory, the commodification of the kWh, and the state's preservation of capital's fiscal position were all earned by shrewd manoeuvring during the industry's formative years. The utilities, for instance, used differential rates between residential and industrial customers to undercut industries' self-generation, and installed snake lines to strategically service otherwise unprofitable communities so as to undercut the political impetus for municipal ownership in these communities (Nye 1992: 316). During the initial years, it was unclear as to what, exactly, the utility was selling and whether electrical power ought to be classified as labour, capital, a raw material, or a service (Nye 1992: 234). Hence, the legal apparatus now aggregated around hydrocarbon infrastructure is not merely a formal recognition of facts on the ground, but participated and continues to participate in bringing those facts into existence. In this regard, as in so many others, the law is performative, calling into being institutions and modes of living. To complete the transition to renewables, the law and legal institutions must similarly be configured around and bring into existence renewable generation and infrastructure.

The task of transitioning to renewables in the US is not merely one of pricing carbon or finding the right level of subsidization of renewables. While these benefit the transition, they continue to operate under the conceit that greenhouse gas emissions are a 'market failure' which needs 'correcting'. Markets and the law which constitutes them are always and everywhere configured around certain assets and to the benefit of particular interests. To transition to renewables, then, law and the markets the law structures do not so much need to be 'corrected' as configured around renewables generation in the first place.

REFERENCES

Alliance for Solar Choice v. Public Service Comm. of Wisconsin (2015). 'Case No. 15CV153', Oral Ruling Transcript dated 30 October 2015 (on file with author).

Content, T. (2015). 'Wisconsin Energy, Integrys Deal Powered by Natural Gas'. *JSOnline*, 27 June. Available at: <http://www.jsonline.com/business/wisconsin-energy-integrys-deal-powered-by-natural-gas-b99524619z1-310407601.html> (accessed 29 January 2016).

Cronon, W. (1992). *Nature's Metropolis: Chicago and the Great West*. New York: W.W. Norton.

EIA (Energy Information Agency) (2012). 'Top Five Retailers of Electricity, with End Use Sectors, 2012', Table 3 in Wisconsin Electricity Profile 2012. Available at: <http://www.eia.gov/electricity/state/archive/2012/wisconsin/>(accessed 29 January 2016).

EPA (Environmental Protection Agency) (2015a). 'Carbon Pollution Emission Guidelines for Existing Stationary Sources: Electric Utility Generating Units', 80 Fed. Reg. 64,661 (23 October 2015) (to be codified at 40 C.F.R. Part 60) (referred to as the Clean Power Plan or CPP).

EPA (2015b). 'Standards of Performance for Greenhouse Gas Emissions from New, Modified, and Reconstructed Stationary Sources: Electric Utility Generating Units', 80 Fed. Reg. 64509 (23 October 2015) (to be codified at 40 C.F.R. Part 60, 70, 71, and 98) (referred to as New Source Performance Standards).

EPSA v. FERC (2014). 753 F.3d 216, 223 (D.C. Cir. 2014).

Federal Power Act (1935). § 201 (codified at 16 U.S.C. §824(b)(1)).

FERC (1996a). 'Order 888, Promoting Wholesale Competition through Open Access Non-Discriminatory Transmission Service by Public Utilities, Recovery of Stranded Costs by Public Utilities and Transmitting Utilities', 61 Fed. Reg. 21,540 (codified at 18 C.F.R. pts. 35 and 37).

FERC (1996b). 'Order 889, Open Access Same-Time Information System (formerly Real-Time Information Networks) and Standards of Conduct', 61 Fed. Reg. 21,736 (codified at 18 C.F.R. pt 37).

FERC (2007). 'Order 890, Preventing Undue Discrimination and Preference in Transmission Service', 72 Fed. Reg. 12266 (codified at 18 C.F.R. pts.35 and 37).

FERC (2008). 'Order 719, Wholesale Competition in Regions with Organized Electric Markets', 73 Fed. Reg. 64,100 (codified at 18 C.F.R. §35.28(g)(1)).

FERC (2009). 'A National Assessment of Demand Response Potential'. June. Available at: <http://www.ferc.gov/industries/electric/indus-act/demand-response/dem-res-adv-metering.asp> (accessed 29 January 2016).

FERC (2011). Order 745, 'Demand Response Compensation in Organized Wholesale Energy Markets', 76 Fed. Reg. 16657 (codified at18 C.F.R. §35.28(g)(1)(v)).

FERC v. EPSA (2016). 136 S. Ct. 760, 733, 577 U.S. __, 14.

Focus on Energy (2016). 'Renewable Energy: Solar Electric (PV) Incentive Application'. Available at: <https://focusonenergy.com/sites/default/files/FOE_SolarPV_Application_R8_20160211_Fillable.pdf> (accessed 21 April 2016).

Hall, C. A. S. and K. A. Klitgaard (2012). *Energy and the Wealth of Nations: Understanding the Biophysical Economy*. New York: Springer.

Hall, C. A. S., J. G. Lambert, and S. B. Balogh (2014). 'EROI of Different Fuels and the Implications for Society'. *Energy Policy*, 64: 141–52.

Hempling, S. (2013). *Regulating Public Utility Performance: The Law of Market Structure, Pricing and Jurisdiction*. Chicago, IL: American Bar Association.

Hurst, J. W. (1964). *Law and Economic Growth: The Legal History of the Lumber Industry in Wisconsin, 1836–1915*. Cambridge, MA: Belknap Press of Harvard University Press.

Kaiser, L. (2014). 'WE Energies Funded Study Say Solar Actually Helps—Not Harms—Ratepayers'. *The Sheppard Express*, 8 October. Available at: <http://shepherdexpress.com/article-24145-we-energies-funded-study-says-solar-actually-helps%E2%80%94not-harms%E2%80%94ratepayers.html> (accessed 31 January 2016).

Missouri ex rel. Southwestern Bell Telephone Co. v. Public Service Commission (1923). (262 U.S. 276).

Nord, D. (1975). 'The Experts versus the Experts: Conflicting Philosophies of Municipal Utility Regulation in the Progressive Era'. *Wisconsin Magazine of History*, 58(3): 219–36.

Nye, D. E. (1992). *Electrifying America: Social Meanings of a New Technology*. Cambridge, MA: MIT Press.

Prah v. Maretti (1982). 108 Wis.2d 223.

Prieto, P. A. and C. A. S. Hall (2013). *Spain's Photovoltaic Revolution: The Energy Return on Investment*. New York: Springer.

PSCW (Public Service Commission of Wisconsin) (2014a). 'Docket 5-UR-107, Direct-WEPCO-WB-Rogers' (WE Energies Direct Testimony). All PSCW materials are available at: <http://psc.wi.gov/apps40/dockets/content/detail.aspx?dockt_id=5-UR-107 (accessed 29 January 2016).

PSCW (2014b). 'Docket 5-UR-107, Ex. WEPCO/WG-Rogers-11 Schedule 7' (WE Energies exhibit schedule 7).

PSCW (2014c). 'Docket 5-UR-107, Ex.-WEPCO/WB-Rogers-11 Schedule 8' (WE Energies exhibit schedule 8).

PSCW (2014d). 'Docket 5-UR-107, Direct-RENEW-Vickerman' (RENEW Direct Testimony).

PSCW (2014e). 'Docket 5-UR-107, Direct-PSC-Singelary' (PSC Singelary Direct Testimony).

PSCW (2014f). 'Docket 5-UR-107, Direct-CUB-Wallach' (CUB Direct Testimony).

PSCW (2014g). 'Docket 5-UR-107, Direct-PSC-Spanjar' (PSC Spanjar Direct Testimony).

PSCW (2014h). 'Docket 5-UR-107, PSC Final Order in Case No. 5-UR-107' (PSCW Final Order).

Public Utilities Commission of Rhode Island et al. v. Attleboro Steam & Electric Co. (1927). 27 U.S. 83.

UNFCCC (United Nations Framework Convention on Climate Change) (2016). INDC website. Available at: <http://www4.unfccc.int/submissions/indc/Submission%20Pages/submissions.aspx)> (accessed 29 January 2016).

Wisc. Statutes (1983). §196.171 (allowing entrance for purposes of examination and criminalizing the prevention or hindrance of such entrance).

Wisc. Statutes (1999a). §196.796(2)(a)–(d) (prohibiting utility ownership of commercial or residential construction corporations).

Wisc. Statutes (1999b). §196.796(3)(4)–(5) (allowing utilities to consult and finance commercial or residential construction corporations).

Wispark Webpage (2016). Available at <http://www.wispark.com/> (accessed 25 May 2016).

17

Global Trends in the Political Economy of Smart Grids

Cherrelle Eid, Rudi Hakvoort, and Martin de Jong

17.1 INTRODUCTION

The global transition towards sustainable, secure, and affordable electricity supply is driving changes in the consumption, production, and transportation of electricity. In the past decade there has been a dash for 'smart' in power systems, resulting in a consensus that a 'smart grid' will pave the way to decarbonization, reliability, and efficiency in the electricity sector. On the basis of the terminology alone, smart grids have received a lot of support in government policy for the theoretical benefits implementation might bring.

Smart grids can be defined as electricity networks that enable two-way communication and power exchange between electricity consumers and producers, utilizing information and communication technology (ICT) to manage demand, and ensure safe and secure electricity distribution (DOE 2006; Hall and Foxon 2014). A smart grid can support the reliability of the grid with the penetration of distributed generation and electric vehicles (EVs), and can deliver possibilities for real-time management of electricity demand, production, and storage.

However, the smart grid concept is both broad and vague, allowing actors to adopt a strategic position, but not to systematically favour the emergence of a shared vision for a smart grid (Tricoire 2015). Within the subjectivity of the term (nobody would support a 'dumb grid'), the assumption exists that trade-offs between sustainability, affordability, and security of supply would be reduced with the transition towards a smart grid. It can be observed that the political economic context influences the motives for investing in smart grids. Nonetheless, not each investment necessarily contributes to the sustainability and affordability objectives from a greater social perspective. The conflicts between different policy objectives and the interests of the actors involved present an interesting point for research. In this chapter, the authors clarify which (set of) goal(s) are emphasized by smart grids in the United States (US), Europe, and China according to industry structure, regulatory context, and the power of energy policy.

17.2 THE HISTORY OF 'SMART GRID' TERMINOLOGY

The functionalities of the smart grid are not recently discovered concepts. Schweppe, Richard, and Kirtley described these functions in a report called 'Homeostatic Control: The Utility/Customer Marketplace for Electric Power' (Schweppe, Richard, and Kirtley 1981). In this report, Schweppe, Richard, and Kirtley referred to homeostatic control as a way of maintaining internal equilibrium between electricity supply and demand with the use of economic signalling and information and communication technology.

However, the term 'smart grid' itself was not used until 2005. In that year a report from the Institute of Electrical and Electronics Engineers (IEEE) was published, named 'Toward a Smart Grid: Power Delivery for the 21st Century' (Amin and Wollenberg 2005). In this report the electric grid was likened to a F15 aircraft with 'self-healing' possibilities in case of emergency, similar to the homeostatic description by Schweppe. In this colourful metaphor, the F15 aircraft is able to continue flying even after losing one wing due to fault detection and automation. The use of detection and automation was suggested to improve transmission grid operations.

17.2.1 Defining Smart Grids

Technically speaking, it is not straightforward to define whether a grid is 'smart' or not 'smart'. Most systems, at least at the high voltage levels, have technologies in place in order to sustain reliability of supply with supervisory control and data acquisition systems (SCADA). However, distribution grids have traditionally been managed in a passive manner and therefore smart grids generally refer to new developments on the distribution side. Aspects of smart grids that can represent such developments are: (i) the installation of physical 'smart devices', and (ii) the (real-time) operational management of those devices.

The smart meter is frequently seen as a prerequisite for smart grids. This digital meter measure consumption data in short time intervals of typically 15 minutes. This consumption data can be communicated to different actors like consumers themselves who can then adjust their consumption levels. Furthermore, this data can be of interest to the utility, the distribution service operator (DSO), retailer, and/or aggregator for billing purposes and evaluation of provided demand response. Usage data along with price signals are the key components of the 'homeostatic feedback' Schweppe envisioned and are central to any smart grid.

In addition to the smart meter, there are other devices that enable insight for the consumer and provide automated feedback on signals through, for example, in-home displays and in-home automation. *Distributed energy resources* (DER) are different types of units that enable local production, alternative consumption, and/or storage. Examples of distributed generation (DG) units are solar photovoltaics (PV) and combined heat and power (CHP). *Battery storage* can provide important value with the increase of self-consumption from electricity generation, reduction of peak consumption, reduction of system-wide generation costs, losses, and network congestions. *Electric vehicles* (EVs) can also be part of a smart grid. EVs, for example, can act as a storage unit for generated electricity.

17.2.2 Real-Time Management and Control

The distributed energy resources do not automatically operate efficiently after installation. The installation of such devices becomes profitable when combined with contracts for variable pricing, direct control, and automation (Faruqui et al. 2010; Aghaei and Alizadeh 2013; Geelen et al. 2013). The interactions of such devices can be called demand side flexibility or demand response. Demand response refers to ability of the demand to 'respond' to triggers like prices or direct control. There are multiple purposes for which the operation of smart devices can be optimized: economic, environmental, and network purposes (Conchado et al. 2011). The demand response provision is seen as one of the major added values of the smart grid (Faruqui, Harris, and Hledik 2010).

17.3 ACTOR PERSPECTIVES ON SMART GRIDS

In addition to technical changes, smart grid innovations can require institutional changes in market design, actor roles, and responsibilities. Different actors involved in the electricity sector might have diverse interests in smart grid developments. Those different actors are, for example, the (public) service utilities, DSO, retailers, aggregators, and electricity customers. Smart grid assets relate to different technical functionalities, which can provide benefits and costs for (some of the) actors involved. Investment in smart grid assets simultaneously influences targets related to affordability, sustainability, and reliability (JRC and DOE 2012). For example, distribution automation has many benefits for reliability due to the automated/self-healing ability of the network, but can also reduce expenses for network expansion (affordability). At the same time, with large DG penetration, distribution automation can help reach sustainability targets. Alternatively, depending on the way in which the dynamic tariff takes account of policy goals, smart metering and smart appliances can reduce costs for electricity usage, and help support reliability and sustainability objectives. Therefore, due to its multiple functionality, a smart grid can present a different 'toolset' for each of the actors involved.

The actors themselves operate within the context of the industry structure. Depending on the structure applied, more or less room is given for competition between actors. Figure 17.1 presents examples of the most common industry structures. In the US the integrated utility structure prevails (Brooks 2015). In Europe, retail competition is applied due to the European Commission's laws regarding functional and legal unbundling of network operators (Newbery 2002; CEER 2013). Unbundling refers to splitting network operation from supply or production activities in order to allow non-discriminatory grid access to all market parties. Switzerland, however, is not part of the European Union (EU) and currently has a wholesale competition model. In China, a single buyer model is applied, where there are two state-owned utilities that deliver electricity in their service area. Due to this diversity, the actors involved are motivated by different interests on smart grid investments. This issue will be further described in Section 17.4.1.

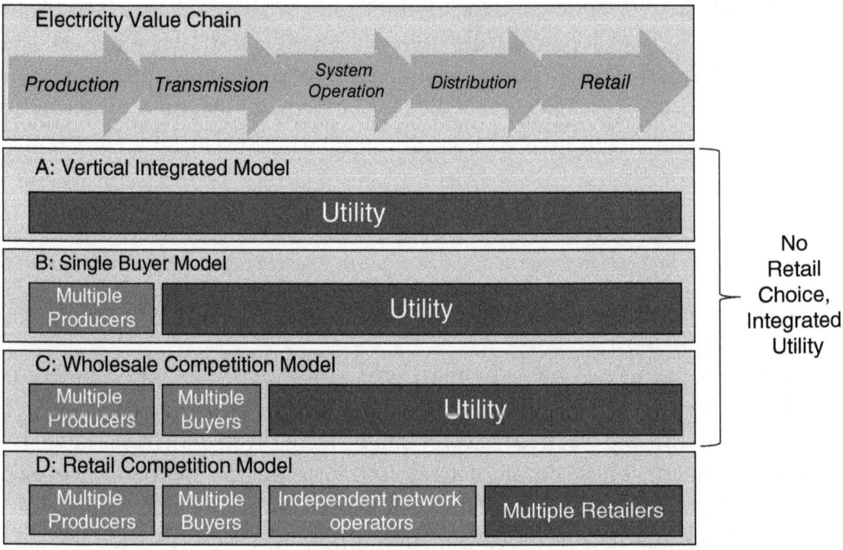

Figure 17.1. Possible industry structures due to sector liberalization.

Source: Authors' illustration adapted from Batlle and Ocaña (2013).

17.3.1 The Integrated Utility

A strictly vertically integrated utility owns and manages the entire chain of electricity production, transport, and retail. When the production units are not owned by the public utility (which is more common) the utility is either a single buyer of electricity (with one-sided competition), or there is a wholesale competition model in place where different buyers and sellers compete for the lowest price at any moment in time. In this chapter we refer to an integrated utility when there is a single retailer present in the area of service, but there is still competition possible in the wholesale market. The integrated utility can be seen in options A–C in Figure 17.1.

The aptitude of smart grid investments depends on the regulatory scheme applied in the sector. For example, under cost of service regulation, capital expenses (CAPEX) in smart grids could be more easily approved if the regulator deems such investment as prudent, especially if investments would help to solve imminent grid congestion or supply constraints. Furthermore, the integration of IT and smart grid devices might reduce the need for grid reinforcements or production investments by optimizing the integration of distributed generation. However, many smart grid investments not only involve CAPEX, but also increase operational expenses (OPEX), for example for the procurement of flexibility in real-time operations.

On the other hand, under incentive regulation, as applied elsewhere for integrated distribution companies, incentives might exist to reduce OPEX, for which smart grids might or might not help. Policy makers could support smart grid investments (CAPEX and OPEX) by allowing them to remain outside the regulatory benchmark.

Another issue with smart grid developments and the position of integrated utilities is that due to the monopoly position of the utility, the value, and hence, the price of demand response (of flexible consumption) is not competitively set. Consequently, investments may be hampered by a lack of proper economic incentives discouraging cost-efficient innovation.

17.3.2 The Distribution Service Operators

In order to give electricity consumers retail choice, it is necessary to unbundle the distribution activity from electricity retail and supply as has taken place in Europe. Generally, the DSO's main task is to keep electricity reliability levels above thresholds by installing enough network capacity and maintaining the grid. Furthermore, the DSO is responsible for providing free third-party access to consumers and producers. Through the established regulatory scheme for the DSO, for example cost of service or incentive regulation, the DSO can recover its incurred costs.

Certain investments in smart grid control devices and metering improve the DSO's ability to get insight into developments and can therefore decrease operational and investment expenses. A DSO can benefit with smart metering from a reduction in metering costs. Furthermore, the procurement of flexibility can delay the need for investments in the network.

In Europe, most DSOs are subject to incentive regulation, which means that their expenses should reduce with an efficiency factor each year. However, the procurement of flexibility through smart grid solutions can increase the operational expenses in time. This can counteract the tendency of the DSO to embark on this route. In several European countries, there is therefore a debate about whether smart grid investments should be left outside the regulatory benchmark.

With respect to unbundling of generation and supply versus transmission and distribution, there is an important difference between the level of unbundling. When the DSO is administratively or legally unbundled (i.e., separated from production and supply while remaining under the same holding company), the holding could maintain (financial) links between the network and generation company. In this case, the smart grid investments by the network company might implicitly benefit other companies in the same holding. On the other hand, with ownership unbundling, the DSO and the generation company are different firms which are strictly separated with respect to the ownership of the assets. In this case, the allocation of smart grid benefits would provide clear benefits that could be transparently allocated to the actors providing the added value.

The Council of European Energy Regulators (CEER) emphasized that the DSOs should provide a level playing field for other actors in the electricity supply chain (CEER 2014). In a response by the association of European Distribution System Operators (EDSO), the DSOs accentuated that all actions that influence grid operations should be carefully assessed (for example installation of new EV charging stations and DG units). Furthermore, they emphasized that if the required regulation for a new activity is of such size that it becomes closely monitored by the regulator, then it should probably be directly done by the DSO itself as an already regulated entity (CEER 2014; EDSO 2015).

17.3.3 Retailers

Unlike the vertical integrated utility, in a retail competition model retailers are competing for their share of electricity consumers. For retailers, smart metering with insight into real-time consumption could provide more insight into consumption load curves and price elasticity and consequently could result in efficient trading for the supply of electricity. Smart metering and real-time data transfer could enable the possibility of tailored contracts for direct control of devices and real-time pricing, and furthermore could support the provision of incentives for new flexibility services (Hakvoort and Koliou 2014).

Besides supply of traditional electricity retail services, the smart grid could open up new business opportunities for retailers, for example with real-time trading of electric flexibility services on balancing markets, ancillary services, or congestion markets (Eid et al. 2016). This role could also be fulfilled by the aggregator, which specifically focuses on enabling, management, and trading of aggregated flexibility as presented in Section 17.3.4.

17.3.4 New Entities: Aggregators and Energy Service Companies

Due to integration of real-time data management and control, business models could arise for new actors in the electricity supply value chain. First of all, the service of electricity supply could be offered by traditional retailers (as presented in Section 17.3.3). However, this could also be provided by aggregators or energy service companies (ESCOs). ESCOs exist, for example, in the UK and might combine offers for a range of supplied services like electricity, heating, cooling, and gas supply for a certain urban district (Hannon and Bolton 2015). In addition to the provision of multiple services to the customers, the customer could also trade services to the system with contracts for flexibility through an aggregator (Eid et al. 2016). This aggregator is different to the traditional retailer due to the fact that it specifically focuses on the trading of flexibility services on markets.

17.3.5 Consumers

Different to the traditional passive role that residential electricity consumers normally have within the electricity supply chain, the smart grid could open up possibilities for active engagement through real-time insight in consumption data, price changes, local production, and self-consumption of electricity. Both in the design phase of smart grid projects and in the operational phase, this customer engagement is possible. For example, smart grid projects in the Netherlands and Germany show the active engagement of consumers in the design phase of the smart grid with involvement regarding the installed technologies and the way in which flexibility is managed. From previous experience, it can be seen that privacy should be addressed carefully already in the design stage of the smart grid project to support consumer engagement (McDaniel and Smith 2009; Cuijpers and Koops 2012).

With the application of dynamic pricing and the installation of in-home displays and energy management systems, consumers can have more control of their consumption and might actively participate in reducing their energy costs and their impact on emissions.

17.4 POLICY PERSPECTIVES ON SMART GRIDS

Often, policy makers promise very positive prospects with smart grids through enabling real-time interactions that enhance sustainability, affordability, and reliability for the electricity sector. Section 17.4.1 presents the main sources for political and economic tensions with the investments for smart grids. Later, from Section 17.4.2, different developments of smart grids in the US, Europe, and China are presented.

17.4.1 Sources for Socio-Political Tensions

Depending on industry structure, regulatory context, and push from energy policy, a certain (set of) goal(s) could be emphasized with the smart grid. The following subsections highlight those factors, after which developments with the application of smart grids are described in the US, Europe, and China. Table 17.1 provides an overview of the most important aspects of the diverse policy perspectives on smart grids described in Sections 17.4.2, 17.4.3, and 17.4.4.

17.4.1.1 The Impact of Industry Structures

As discussed in Section 17.3, the industry structure provides important insight with regard to the type of actors involved in the electricity supply value chain.

Table 17.1. Summary of policy perspectives on smart grids

	United States	Europe	China
Industry structure	Mostly vertically integrated or wholesale competition	Retail competition	Vertically integrated
Regulatory model	Cost of service/rate of return regulation	Incentive regulation for DSO	Rate of return regulation
Energy policy	Bottom-up	Hybrid	Top-down
Initial smart grid interests	Reliability and recovery of investments for utilities	Affordability and sustainability	Supply surge of electricity demand in reliable and sustainable manner
Smart grid developments	Smart metering applied in many places, but no greater smart grid vision. End user left passive in many cases	Smart metering roll-out only fully completed in Sweden and Italy. Remaining problems are the role of the DSO	Large-scale projects, including smart metering, micro grids, and EV pilot projects are deployed on larger scale, where direct control is applied by state grid company

Source: Authors' illustration.

In the US the integrated utility model prevails, with around 70 per cent of electricity sold coming from integrated utilities. Almost half of these are public utilities (municipality, cooperatives, and others) and the rest are private (Brooks 2015). The customers in those places are bound by the local (public) service utility for the electricity services.

The integrated utility could directly utilize the insight into consumer data for both network and (local) supply optimization. For integrated utilities, however, the possibility exists that due to their monopoly position, the price of demand response (of flexible demand) is not competitively set. Therefore, such investments might lead to too high benefits for the utility. This raises questions about the strategic behaviour of utilities in preventing society from adopting cost-efficient innovation. This is also of importance for China, due to the fact that all electricity for residential consumers comes from one of two state-owned companies.

Alternatively in Europe, the retail competition model prevails with the unbundled DSO. Depending on the type of unbundling, legal or ownership unbundling has effects on the possible benefits that the DSO could transfer to the retailer associated company. When the DSO is only administratively or legally unbundled (i.e., separated from production and supply while remaining under the same holding company), the holding could maintain (financial) links between the network and generation company. In this case the smart grid investments by the network company might implicitly benefit other companies in the same holding, creating a competitive advantage towards other retailers. On the other hand, under full ownership unbundling, the DSO and the generation company are separate firms that are strictly separated with respect to the ownership of the assets. The allocation of smart grid benefits would provide clear benefits which could be transparently allocated to the actors providing added value.

17.4.1.2 The Impact of the Regulatory Model

As discussed in Section 17.3, the industry structure influences which (type of) actors are involved in the electricity supply chain. However, the business model for the network operators and other monopolistic entities depends heavily on the applicable regulation. Due to the fact that the electricity network is monopolistic by nature, electricity transport remains a regulated utility.

There are different ways in which regulators can settle the remuneration for regulated companies. Incentive-based regulations motivate utilities to reduce OPEX and/or CAPEX in line with an efficiency factor. Alternatively, with rate of return or cost of service regulation, CAPEX can be more easily recovered if the regulator deems such investment as prudent. If such smart grid investments require CAPEX (for example, the installation of smart meters), this can be recovered if the regulator agrees. However, many smart grid investments not only involve CAPEX but also OPEX, for example with the procurement of flexibility in real-time operations of the network to delay network expansions. A strict cost of service regulation on CAPEX could hamper a holistic smart grid vision, due to the fact that smart grid assets do not operate 'smartly' without additional long-term operational expenses for procurement, management, and remuneration of electric flexibility that is activated from end-users.

On the other hand, under incentive regulation as applied elsewhere for integrated distribution companies, incentives might exist to reduce OPEX, for which smart grids might or might not help. Policy makers could support smart grid investments by allowing them (CAPEX and OPEX) to remain outside the regulatory benchmark.

17.4.1.3 The Impact of Energy Policy

In order to motivate the sector to contribute to sustainability objectives, policy makers could provide in appropriate financial instruments to motivate investments in smart grids. Policy could be defined strictly top-down, bottom-up, or in a hybrid model (both top-down and bottom-up) and could indicate whether lower policy levels have less or more freedom and power in defining their own strategies.

In some places, certain aspects of the smart grid, for example the smart meter, could be legally enforced by law (top-down), or this could be left to the interests of the utilities involved and the consumers. In some places in Europe (Italy and Sweden), the DSO has been obliged to install smart meters for all consumers. In the EU in general a hybrid approach can be observed due to the settlement of (top-down) binding targets (for example the 2020 objectives for sustainability). The principle of subsidiarity, however, implies that member states are free to develop their own energy strategies and implement them in the most appropriate way (bottom-up).

In the US, where federal funding is available for smart metering (the American Recovery and Reinvestment Act of 2009), utilities still have the freedom to decide whether to use this possibility or not, presenting a bottom-up approach for actual implementation. By contrast, in China, a top-down approach is applied, with the state grid company depending entirely on the policy directions given for the roll-out of smart grids.

17.4.2 The United States' 'Smart': Reliability of Supply

In the US, policy interest in the smart grid arose due to frequent electricity interruptions in 2005, acting as a driver for innovation in the electricity sector (Lin, Yang, and Shyua 2013). Later, in December 2007, the concept of the smart grid was established in US legislation, where in the Energy Independence and Security Act of 2007, the smart grid was named as a main pillar for reform (Kaplan et al. 2009). As described in this policy report, this modernized grid would have a range of features. It would be able to 'self-heal', to motivate the participation of the electricity consumer, to provide power of a quality suitable for twenty-first century needs, to resists attacks, to accommodate all generation and storage options, to enable markets, to optimize assets, and to operate economically efficiently. The American Recovery and Reinvestment Act of 2009 (ARRA), commonly referred to as the Stimulus or the Recovery Act, provided US$4.5 billion of funding to modernize the electricity power grid.

Thirty of the largest utilities in the US have fully deployed smart meters to their customers. The states of California and Texas are leading with the penetration of smart meters. Some utilities have included the possibility of alternative pricing programmes or even including a larger smart grid vision with in-home energy management systems and displays in New York and Wisconsin (Edison Foundation 2014). Unlike in Europe, the US electricity sector is mostly structured by the integrated utility model and most residential consumers are contracted with their municipality utility. Apart from in Texas (DEFG 2015), retail choice is not common in the USA. Therefore the penetration of smart metering is carried out in most states through centralized roll-outs, probably due to the high rate of integrated utilities which have a monopoly position to do so. 'However, several states allow customers to opt-out of smart meters. For those customers typically an initial fee and a monthly opt-out fee are required. The number of customers who have requested to opt-out of their smart meter is relatively low' (Edison Foundation 2014: 1).

Alongside the emphasis on reliability in many smart meter roll-outs, a large range of projects involve experimentation with different smart grid technologies. For example, the University of Delaware's Vehicle to Grid (V2G) project presents an interesting business case for trading flexibility services to PJM, the local Transmission Service Operator (Kempton et al. 2009). Furthermore, the US presents a range of micro grid projects, in which electricity, heat, and gas supply can be locally managed (DOE 2014). Micro grids are grids that incorporate IT to coordinate local production with consumption and eventually enable operations in both disconnected and connected mode. The main motives for those projects are reliability and energy independence especially in cases of severe weather events (Bower et al. 2014).

Most utilities, however, continue with smart metering investments, due to the CAPEX nature of those investments and the possibility of receiving a return on investment, but leave out the further smart grid vision with a participative end-user. There is no interest in this due to the reduced income for utilities given the nature of current regulation in many places (cost of service or rate of return regulation).

However, recent energy policy topics are not related to smart meter deployment, but rather to how to deal with the penetration of distributed energy resources. Net metering results in overall reduction of energy sales and recovering the costs of fixed assets by utilities is being jeopardized (Eid et al. 2014). Therefore, utilities are currently focusing on the addition of or increase in fixed charges to make up for such costs.

17.4.3 Europe's 'Smart': Affordability and Sustainability in the Liberalized Sector

The smart grid concept has been fixed in European policy since publication of the European Commission's report 'European Smart Girds Technology Platform: Visions and Strategy for Europe's Electricity Networks of the Future' (European Commission 2006). In contrast to the US, the European system has not faced significant reliability problems. The main drivers for transition were described as the need for new directions to cope with secure and sustainable electricity supplies in the future. Europe's ambitious sustainability objectives do favour a need for new solutions to meet the 2020 objectives of: reducing electricity demand by 20 per cent, increasing renewable share to 20 per cent, and decreasing CO_2 emissions by 20 per cent compared to 1990. The European Commission set a target for 80 per cent of European households to be equipped with a smart meter by 2020, if the roll-out of smart meters is assessed positively (European Union 2009).

Regarding the regulatory context, in Europe the DSO is regulated by incentive-based regulation, meaning that costs for OPEX and/or CAPEX should be reduced in time. Since, in many places in Europe, parts of the network soon need replacement, DSOs are interested in options to delay the need for network investments. However, the economic rationale is totally dependent on the regulatory scheme that is in place for recovering costs for both CAPEX and OPEX related expenses for smart grids. Research shows that 'unless the DSO controls electric vehicle charging within an active system management approach, the DSO would have to heavily invest into low- and medium-voltage lines to compensate for local peak demand resulting from EVs. This example from EVs clearly demonstrates the trade-off between CAPEX and OPEX and resulting potentials to avoid unnecessary costs for DSOs' (Ruester et al. 2014: 232).

However, OPEX expenses will rise for DSO with smart grid investments and procurement of flexibility for the DSO to replace network expansions. This is not the investment rationale of the DSO and therefore is not supported by European regulation. A different situation for the UK exists due to the fact that the regulatory Office of Gas and Electricity Markets (OFGEM) adjusted regulation for the DSO to take into account other objectives like smart metering (Ruester et al. 2014).

As the DSO is a meant to be a market facilitator in a retail competition model, providing non-discriminatory third-party access to the grid, current legislation limits the DSO to procuring flexibility rather than investing in grid reinforcements. However, as stated by Lunde, Røpke, and Heiskanen, 'dynamic tariffs that vary in both time and place might discriminate customers by increasing price in a geographic area specifically with capacity problems, and not in neighbouring areas without capacity problems' (Lunde, Røpke, and Heiskanen 2015).

In Europe, so far, only Italy and Sweden have completed a full smart meter roll-out and in both cases the 'degree of smartness' has tended mostly to focus on remote meter reading (KEMA 2012a). The motive for installing smart meters in Sweden was the legal requirement to provide monthly invoices based on actual meter readings from 2009 onwards (Bartusch et al. 2011). The roll-out of smart meters would reduce electricity metering costs for the DSOs (Capgemini 2008; KEMA 2012a). However, currently the Swedish DSOs provide time-of-use tariff options in order to shift consumption from peak hours to off-peak hours (Bartusch et al. 2011).

In Italy, the electricity producer Enel initiated smart-meter roll-out in 2000, initially to reduce non-technical losses (KEMA 2012b). More recently, ENEL has also set out a path to move from the roll-out of smart meters to a demand response market platform. Furthermore, the Netherlands has been presenting different interesting cases of smart grid pilot projects since 2012.[1] The projects were subsidized by the Dutch government and in order to expand their experimental scope, these projects have been allowed to function outside the Dutch regulatory context. From those projects it is visible that the unbundling of the DSO from the traditional supply chain might lead to different hurdles for coordination between network capacity limitations and demand response programmes. Therefore, the European DSOs are possibly moving towards new roles in order to attain policy objectives that are of common interest (EvolvDSO 2014).

17.4.4 China's 'Smart': Dealing with a Surge in Electricity Demand

Chinese energy policy has focused on growth of the electricity sector in order to keep up with the surge in electricity demand. At the same time, policy efforts have focused on sustainability to reduce emissions and the negative

[1] See an overview of those pilot projects online at: <http://www.netbeheernederland.nl/smartgrids/> (accessed 17 October 2016).

health effects from the high share of fossil-fuel-powered production in China (RAP 2013; Brunekreeft et al. 2015). In this context, the Chinese government has acknowledged the importance of smart grids in its twelfth Five-Year Plan for National Economic and Social Development. The report explicitly sets up the goal of accelerating smart grid developments. In line with this policy focus power companies in China, especially the grid operators and academic institutions, are actively promoting their views on smart grids and developing, testing, and deploying smart grid technologies (Brunekreeft et al. 2015).

Chinese transmission and distribution are under the monopoly of one of the two state-owned enterprises. The National Development Reform Commission sets the price by which the monopolies can sell their electricity by means of rate of return regulation. By the end of 2015, more than 30 million smart meters had been installed in China. This was completed through unified bidding processes set up by the two state grid enterprises.[2] Furthermore, 342 of the 352 planned smart grid projects had been completed; including charging and battery swap stations for EVs, smart substations, 20 smart grid demonstration project and smart distribution networks in 68 cities.

Furthermore, the ambitious renewable targets in China have stimulated investments in wind and solar power. The grid is required to be upgraded rapidly in order to handle electricity flows coming from such intermittent production units. China is currently the number one installer of wind power capacity, and number two, after the US, in wind power production. Reduced capacity factors have been attributed to high amounts of forced curtailment, which reached as high as 50 per cent in some regions in 2012 (Davidson 2013). The inflexible planning processes that gave preference to incumbent generators, combined with the volatile nature of production from renewable sources, had disadvantaged wind production (Davidson 2013; Paulson Institute 2015).

Furthermore, most wind, solar, and coal electricity generation is located in the north and far west of China and most electricity consumption is located in the east. Due to this distance between supply and demand, large transmission lines are planned and under construction to connect generation units with consumption locations. Recently, high levels of investment have been made in the transmission network for ultra-high voltage (UHV) lines. With China being the only country to deploy UHV technology on a large scale, its investment efforts support the international position of Chinese UHV technologies in gaining global market share (Paulson Institute 2015).

In Chinese terms, the smart grid involves a broad portfolio of information and communication technologies. This also includes modern grid technologies such as UHV transmission grids or heat-resistant wires. Consequently,

[2] For more information on the smart meter installation and smart grid projects, see: <http://www.reportlinker.com/p03837879-summary/China-Smart-Electric-Meter-Industry-Report.html> and <http://www.geidco.org/html/qqnycoen/col2015100801/column_2015100801_1.html>.

smart grids in China focus on all sections of the power system, with a specific focus on the integration of RES (Yu, Yang, and Chen 2012; Brunekreeft et al. 2015). The two state-owned Chinese network companies have invested very highly in transmission network upgrades. Due to their monopolistic nature and their close connection with policy, objectives and directives can be directly implemented in a top-down manner.

In addition to the high-voltage transmission perspective of smart systems in China, there are different policy objectives for 'smart grids' at the distribution level. Currently, different pilot projects in distribution networks are operational, with one of the most important projects being EV pilot projects. China's four ministries jointly launched the nationwide pilot programme in 2009 which aimed to roll out ten new pilot cities each year.

Out of all these cities, Shenzhen had the most ambitious plan, that is to deploy 9,000 EVs, followed by Beijing and Shanghai with 5,000 and 4,157 EVs respectively. Shenzhen, known as the pilot project city of China, has one of the largest EV fleets in the world. Shenzhen will be the first city to successfully complete attempts to liberalize sectors (Li et al. 2015). Due to the overcrowded nature of some cities, the government has set maximum quotas on car ownership, but there are no restrictions for EV owners, making EVs an attractive option for potential car owners. However, there remains a problem with regard to sufficient access to charging stations. Furthermore, the National Energy Administration is planning 30 micro grid demonstrations as outlined in the renewable energy development plan.

17.5 CONCLUSIONS AND POLICY RECOMMENDATIONS

The authors of this chapter have provided an overview of developments of smart grids within different policy contexts. We defined three main causes of political–economic tensions with smart grid developments in the US, Europe, and China, namely *industry structure*, *regulatory models*, and the *impact of energy policy*.

Firstly, the industry structure defines what actors are involved in the electricity supply chain, ranging from a single (state-owned) utility to a regulated network operator with multiple retailers whom compete for their share of customers. Secondly, the regulatory model impacts how the utility is motivated to invest in smart grid assets, or not, depending on the way its costs are being recovered. Thirdly, the impact of energy policy can differ depending on how energy policy is set in legislation and at what level this is done.

In the US, due to the industry structure and the type of regulation, which is cost of service or rate of return based, the utility invests in smart grid assets if the regulator approves those as prudent. In the past, reliability issues have been the major factor for the development of the initial smart grid. However, after the installation of smart meters, in most places utilities are not interested in investments for a holistic smart grid vision, simply because the regulation does not incentivize them to do so. With their focus on cost recovery, the current emphasis is on how to handle the reduction in energy sales due to the high penetration of Solar PV with net-metering practices, and the connection of other distributed energy resources.

The situation is different in Europe where the industry structure is retail competition-based with incentive regulation. Incentive regulation motivates network operators to reduce operational and/or capital expenses over time with an efficiency factor. It is not clear if investments in smart grids are being recovered where they are being restricted by the regulatory framework. In Europe, the policy interest in smart grids began in order to help reach the highly ambitious sustainability objectives. The European Energy directive sets a binding target regarding smart metering if it is assessed that the roll-out would be positive, but member states are still allowed to set up their own roadmaps to reach an overall 80 per cent target of smart metering in 2020 (European Union 2009). Consequently, in both Europe and the US a holistic view of smart grids could be hampered due to the fact that generally operational expenses with smart grids will increase, but the regulatory scheme does normally not cover those expenses. Especially in Europe, due to the unbundling of the DSO, the new role of the DSO in smart grids remains unclear.

In China, the surge of the national electricity demand has been the driving force for smart grid policies. The two electricity state-owned enterprises in China have direct links with policy and depending on the stated policy directions by the National Energy Administration, smart grid projects can be approved and implemented in a relatively rapid manner. Most smart grid projects involve electric vehicle pilots and micro grids which are directly managed by the utility. The quick approach that China has taken results in a large-scale smart grids development, including smart metering, smart distribution networks, numerous micro grids, and EV pilot projects.

The EU and the US have formulated policies related to the roll-out of smart metering. However, smart grid investments do not (yet) primarily focus on sustainability targets. A holistic smart grid vision would open up possibilities for increased bottom-up participation and better integration of DER at low voltage levels and local energy management. However, due to the traditional regulatory funding schemes for utilities and DSOs involved in those places, the operational expenses for the activation of local flexibility are currently not equally supported. Furthermore, the cooperation models between actors involved in smart grids still remain unclear in many places in Europe, particularly in relation to the role of the DSO.

It is therefore recommended that regulators allow smart grid investments to remain outside of the regulatory framework. This should not only be the case for smart grid capital expenses, but also for smart grid operational expenses in order to support smart grid developments beyond the installation of smart meters. At the same time, the regulatory institution should be aware of the possibility for excessive benefits that can result from smart grid developments and should specify new indicators for utility regulation within the smart grid context.

A further important dilemma might be related to incumbent power producers and integrated utilities which are dependent on their energy sales from large gas, coal, or nuclear production units. The aspect of stranded costs might reduce the interest for alternative 'smart' investments and developments. Therefore, the role of policy makers should be to reduce regulatory uncertainty and support developments that have a long-term sustainable effect for the energy sector as a whole.

REFERENCES

Aghaei, J. and M.-I. Alizadeh (2013). 'Demand Response in Smart Electricity Grids Equipped with Renewable Energy Sources: A Review'. *Renewable and Sustainable Energy Reviews*, 18: 64–72.

Amin, S. M. and B. F. Wollenberg (2005). 'Toward a Smart Grid: Power Delivery for the 21st Century'. *IEEE Power Energy Magazine*, 3: 34–41.

Bartusch, C., F. Wallin, M. Odlare, I. Vassileva, and L. Wester (2011). 'Introducing a Demand-Based Electricity Distribution Tariff in the Residential Sector: Demand Response and Customer Perception'. *Energy Policy*, 39(9): 5008–25.

Brooks, C. (2015). 'The Periodic Table of the Electric Utility Landscape: A Series of Visual Tools for Enhanced Policy Analysis'. *The Electricity Journal*, 28(6): 82–95.

Brunekreeft, G., T. Luhmann, T. Menz, S.-U. Müller, and P. Recknagel (2015). *Regulatory Pathways for Smart Grid Development in China*. Berlin: Springer.

Capgemini (2008). 'Demand Response: A Decisive Breakthrough for Europe' (Online). Available at <https://www.capgemini.com/resource-file-access/resource/pdf/Demand_Response__a_decisive_breakthrough_for_Europe.pdf> (accessed 10 July 2015).

CEER (2013). 'Status Review on the Transposition of Unbundling Requirements for DSOs and Closed Distribution System Operators'. Council of European Energy Regulators, Brussels.

CEER (2014). 'The Future Role of DSOs: A CEER Public Consultation Paper'. Council of European Energy Regulators, Brussels.

Conchado, A., P. Linares, O. Lago, and A. Santamaría (2011). 'How Much Should We Pay for a DR Program? An Estimation of Network and Generation System Benefits'. Working Paper, Comillas Pontifical University, Madrid. Available at: <http://www.iit.upcomillas.es/pedrol/documents/conchado13.pdf> (accessed 10 October 2015).

Cuijpers, C. and B.-J. Koops (2012). 'Smart Metering and Privacy in Europe: Lessons from the Dutch Case'. In S. Gutwirth, R. Leenes, P. de Hert, and Y. Poullet (eds), *European Data Protection: Coming of Age*. Dordrecht: Springer, pp. 269–93.

Davidson, M. (2013). 'Politics of Power in China: Institutional Bottlenecks to Reducing Wind Curtailment through Improved Transmission'. *IAEE*, Q4: 40–2.

DEFG (2015). 'Annual Baseline Assessment of Choice in Canada and the United States 2015'. Available at: <http://defgllc.com/publication/abaccus-2015-annual-baseline-assessment-of-choice-in-canada-and-the-united-states/> (accessed 18 January 2016).

DOE (US Department of Energy) (2006). 'Benefits of Demand Response in Electricity Markets and Recommendations for Achieving Them'. DOE, Washington, D.C. Available at: <http://energy.gov/sites/prod/files/oeprod/DocumentsandMedia/DOE_Benefits_of_Demand_Response_in_Electricity_Markets_and_Recommendations_for_Achieving_Them_Report_to_Congress.pdf> (accessed 15 November 2015).

DOE (2014). 'Microgrid Activities' (Online). Available at: <http://energy.gov/oe/services/technology-development/smart-grid/role-microgrids-helping-advance-nation-s-energy-syst-0> (accessed 1 December 2015).

Edison Foundation (2014). 'Utility-Scale Smart Meter Deployments: Building Block of the Evolving Power Grid'. Edison Foundation, Washington, D.C. Available at: <http://www.edisonfoundation.net/iei/Documents/IEI_SmartMeterUpdate_0914.pdf> (accessed 15 October 2015).

EDSO (European Distribution System Operators) (2015). 'European Distribution System Operators for Smart Grids'. Response to CEER Consultation on the Future Role of the DSO. EDSO, Brussels.

Eid, C., P. Codani, Y. Perez, J. Reneses, and R. Hakvoort (2016). 'Managing Electric Flexibility from Distributed Energy Resources: A Review for Incentives, Aggregation and Market Design'. Delft University of Technology Working Paper, Paper submitted and under review for publication.

Eid, C., J. Reneses Guillén, P. Frías Marín, and R. Hakvoort (2014). 'The Economic Effect of Electricity Net-Metering with Solar PV: Consequences for Network Cost Recovery, Cross Subsidies and Policy Objectives'. *Energy Policy*, 75: 244–54.

European Commission (2006). 'European Smart Girds Technology Platform: Visions and Strategy for Europe's Electricity Networks of the Future. Brussels'. Available at: <https://ec.europa.eu/research/energy/pdf/smartgrids_en.pdf> (accessed 10 May 2015).

European Union (2009). 'Directive of 2009/72/EC of the European Parliament and of the Council of 13 July 2009 Concerning Common Rules for the Internal Market in Electricity and Repealing Directive 2003/54/EC. Off. J. Eur. Union L211, L 211/55 – L 211/93'. doi:10.1126/science.202.4366.409.

EvolvDSO (2014). 'Development of Methodologies and Tools for New and Evolving DSO Roles for Efficient DRES Integration in Distribution Networks'. Newsletter Issue No. 1. Available at: <http://www.evolvdso.eu/private/documenti/WP7/Newsletter-Draft-Versions/Newsletter_evolvDSO_OCT2014.aspx> (accessed 10 April 2015).

Faruqui, A., D. Harris, and R. Hledik (2010). 'Unlocking the €53 Billion Savings from Smart Meters in the EU: How Increasing the Adoption of Dynamic Tariffs Could Make or Break the EU's Smart Grid Investment'. *Energy Policy*, 38(10): 6222–31.

Geelen, D., A. Reinders, and D. Keyson, D. (2013). 'Empowering the End-User in Smart Grids: Recommendations for the Design of Products and Services'. *Energy Policy*, 61: 151–61.

Hakvoort, R. and E. Koliou (2014). 'Energy Management and Demand Side Response'. In U.C. Sharma, R. Prosad, and S. Sivakumari (eds), *Energy Science and Technology*. New Delhi: Studium Press LLC.

Hall, S. and T. J. Foxon (2014). 'Values in the Smart Grid: The Co-Evolving Political Economy of Smart Distribution'. *Energy Policy*, 74(November): 600–9. Paper for ESEIA-IGS Conference on 'Smart and Green Transitions in Cities/Regions', 24–25 April 2014. Available at: <http://doi.org/10.1016/j.enpol.2014.08.018> (accessed 11 September 2015).

Hannon, M. J. and R. Bolton (2015). 'UK Local Authority Engagement with the Energy Service Company (ESCo) Model: Key Characteristics, Benefits, Limitations and Considerations'. *Energy Policy*, 78: 198–212.

JRC (Joint Research Centre of the European Commission) and DOE (2012). 'Assessing Smart Grid Benefits and Impacts: EU and U.S. Initiatives'. Scientific and Policy Report by the Joint Research Centre of the European Commission. Available at: <http://publications.jrc.ec.europa.eu/repository/handle/JRC73070> (accessed 25 June 2015).

Kaplan, S. M., F. Sissine, A. Abel, J. Wellinghoff, S. G. Kelly, and J. J. Hoecker (2009). *Smart Grid: Modernizing Electric Power Transmission and Distribution*. Alexandria, VA: TheCapitol.Net.

KEMA (2012a). 'Development of Best Practice Recommendations for Smart Meters Rollout in the Energy Community'. Final Report. Available at: <https://www.energy-community.org/portal/page/portal/ENC_HOME/DOCS/2506178/0633975AB8B77-B9CE053C92FA8C06338.PDF> (accessed 20 August 2015).

KEMA (2012b). 'Global Inventory and Analysis of Smart Grid Demonstration Projects'. KEMA, Arnhem.

Kempton, W., V. Udo, K. Huber, K. Komara, S. Letendre, S. Baker, D. Brunner, and N. Pearre (2009). 'A Test of Vehicle-to-Grid (V2G) for Energy Storage and Frequency Regulation in the PJM System'. Research Paper. Available at: <http://www.udel.edu/V2G/resources/test-v2g-in-pjm-jan09.pdf> (accessed 11 June 2015).

Li, Y., C. Zhan, M. de Jong, and Z. Lukszo (2015). 'Business Innovation and Government Regulation for the Promotion of Electric Vehicle Use: Lessons from Shenzhen, China'. *Journal of Cleaner Production*, 134: 371–83.

Lin, C.-C., C.-H. Yang, and J. Z. Shyua (2013). 'A Comparison of Innovation Policy in the Smart Grid Industry across the Pacific: China and the USA'. *Energy Policy*, 57: 119–32. Available at: <http://doi.org/10.1016/j.enpol.2012.12.028> (accessed 10 May 2014).

Lunde, M., I. Røpke, and E. Heiskanen (2015). 'Smart Grid: Hope or Hype?' *Energy Efficiency*, 9(2): 545–62. Available at: <http://doi.org/10.1007/s12053-015-9385-8> (accessed 10 November 2015).

McDaniel, P. and S. W. Smith (2009). 'Security and Privacy Challenges in the Smart Grid'. *IEEE*: 75–7.

Newbery, D. M. (2002). 'Problems of Liberalising the Electricity Industry'. *European Economic Review*, 46(4/5): 919–27.

Paulson Institute (2015). 'Power Play: China's Ultra-High Voltage Technology and Global Standards' (online). 9 April. Available at: <http://www.paulsoninstitute.org/think-tank/2015/04/09/power-play-chinas-ultra-high-voltage-technology-and-global-standards/> (accessed 12 October 2015).

RAP (2013). 'Recommendations for Power Sector Policy in China: Practical Solutions for Energy, Climate and Air Quality'. RAP, Beijing. Available at: <http://www.raponline.org/document/download/id/6869> (accessed 10 October 2015).

Ruester, S., S. Schwenen, C. Batlle, and I. Pérez-Arriaga (2014). 'From Distribution Networks to Smart Distribution Systems: Rethinking the Regulation of European Electricity DSOs'. *Utilities Policy*, 31: 1–9. Available at: <http://doi.org/10.1016/j.jup.2014.03.007> (accessed 10 April 2015).

Schweppe, F. C., D. Richard, and J. L. Kirtley (1981). 'Homeostatic Control: The Utility/Customer Marketplace for Electric Power'. MIT Energy Laboratory Report. Available at: <http://dspace.mit.edu/bitstream/handle/1721.1/60510/EL_TR_1981_033.pdf?sequence=1> (accessed 10 March 2014).

Tricoire, A. (2015). 'Uncertainty, Vision, and the Vitality of the Emerging Smart Grid'. *Energy Research & Social Science*, 9: 21–34.

Yu, Y., J. Yang, and B. Chen (2012). 'The Smart Grids in China—A Review'. *Energies*, 5(5): 1321–38.

18

Falling Oil Prices and Sustainable Energy Transition

Towards a Multilateral Agreement on Fossil-Fuel Subsidies

Henok Birhanu Asmelash

18.1 INTRODUCTION

Governments worldwide subsidize the consumption and production of fossil fuels to the tune of USD$0.5–USD$5.3 trillion, depending on how subsidies are defined and measured (IEA 2015a; Coady et al. 2015). The adverse economic and environmental effects of these subsidies have long been recognized, but countries have been reluctant to remove them mainly for political economy reasons (Burniaux and Chateau 2014; Granado et al. 2010; OECD 1998; Parry et al. 2014; Pitt 1985). While momentum for fossil-fuel subsidy (FFS) reform has been building over the last few years, the sharp drop in international oil prices since the second half of 2014 has intensified calls for—and efforts to bring about—the phasing out of FFSs by making the withdrawal of these subsidies less politically controversial (IEA 2015a). It is, however, uncertain how long the decline in oil prices will persist and, if past experiences are anything to go by, the prices are likely to rise again (IEA 2015a). When oil prices start rising again, the momentum for subsidy reform could dissipate and governments may find themselves under mounting pressure to (re)introduce subsidies (Coady and Shang 2015; Kojima 2009). This means that, in addition to seizing the opportunity offered by falling oil prices to reform FFSs,[1] the key challenge is to ensure the durability of the initiated reforms

[1] A number of countries including, Angola, China, India, Indonesia, Iran, and Malaysia have already initiated or accelerated their subsidy reform seizing the opportunity offered by the low oil prices (IEA 2015a).

(Coady and Shang 2015; IEA 2015a). This entails the need for ways to lock in FFS reforms, and it is primarily in this context that an international agreement on FFSs merits consideration. Such an agreement would tie the hands of governments and thereby enhances the credibility of FFS reforms. It would also help governments resist pressure from interest groups to bring back subsidies in the wake of oil price hikes. Currently, there is no binding international agreement to phase out FFSs. The Kyoto Protocol's call for progressive reduction or phasing out of subsidies in all greenhouse gas emitting sectors, and the G20's pledge to phase out or rationalize inefficient FFSs, go some way in this direction, but they fall far short of a legally binding commitment to end FFSs.

Against this backdrop, this chapter seeks to explore the challenges of and prospects and possible avenues for negotiating a binding multilateral agreement to phase out FFSs. Fossil-fuel combustion is the largest source of greenhouse gas emission, the increasing atmospheric concentration of which leads to global warming and climate change (IPCC 2014a). In recognition of the critical role that sustainable energy transition will play in combating climate change, more than 160 countries have set renewable energy targets and policies to promote the development and deployment of renewable energy sources (IRENA 2015). The subsidization of fossil fuels runs counter to these policies. By artificially lowering fossil fuel prices, FFSs encourage the wasteful consumption of carbon-intensive fuels and undermine the competitiveness of renewables (Bridle and Kitson 2014). Their elimination would not of itself bring about the transition, but it will clear one of the major obstacles in achieving sustainable energy transition. While FFS reforms take place at the national level, an international agreement will provide legal certainty and the necessary context and basis for undertaking such reforms at the national level.

This chapter is intended to provide a basic framework for the necessary discussions on strengthening the current international legal framework for phasing out FFSs. Section 18.2 will introduce the basic notion of FFSs. Section 18.3 makes the case for FFS reform mainly from a sustainable energy transition perspective. Section 18.4 examines the main barriers to FFS reform and how the collapse in oil prices can help in overcoming them. Section 18.5 reviews existing intergovernmental initiatives to phase out FFSs. Section 18.6 begins by outlining how a binding multilateral agreement helps the global effort to phase out FFSs, before assessing the key issues and challenges in reaching such an agreement. Section 18.7 sums up the discussion.

18.2 UNDERSTANDING FOSSIL-FUEL SUBSIDIES

Despite its frequent use, the term 'subsidy' is 'notoriously difficult' to define. Existing definitions range from as narrow as a direct budgetary payment by a government to a producer or consumer to as broad as any government interventions that affect prices or costs (UNEP 2008). Since an enumeration of these definitions serves little purpose, this chapter will focus on the only legally binding international definition of subsidies, which is contained in the World Trade Organization (WTO) Agreement on Subsidies and Countervailing Measures (SCM Agreement).

For the purpose of the SCM Agreement, a subsidy is deemed to exist if: (i) 'there is a financial contribution by a government or any public body within the territory of a Member or any form of income or price support in the sense of Article XVI of the GATT [General Agreement on Tariffs and Trade]', and (ii) 'a benefit is thereby conferred' (WTO 1995: Art. 1.1). A financial contribution, as spelt out in Article 1.1(a) (1) of the SCM Agreement, may take the form of: (a) a direct transfer of funds (e.g., grants, loans, and equity infusion) or potential direct transfer of funds or liabilities (e.g., loan guarantees); (b) government revenue that is otherwise due is foregone or not collected (e.g., tax credits and other fiscal incentives); (c) provision of goods or services other than general infrastructure or government purchase of goods; or (d) government payments to a funding mechanism or government entrustment or direction to a private body to carry out one of the type of functions illustrated in (a)–(c) and which would normally be vested in the government and the practice, in no real sense, differs from practices normally followed by governments. These four broad categories of financial contributions capture more than what is normally perceived as a subsidy, but, as noted by the Appellate Body in *US-Softwood Lumber*, the inclusion of an exhaustive list of financial contributions is in itself an indication that not all government measures capable of conferring benefits would necessarily constitute a subsidy within the meaning of the SCM Agreement (WTO 2004). Some of the most notable exclusions are of particular importance in the context of energy subsidies. For example, regulatory measures, including border measures such as tariffs and export restraints, are excluded from the subsidy definition of the SCM Agreement, although they may eventually confer the same benefit as the above-mentioned financial contributions.[2] Also excluded are implicit subsidies that arise from government inaction (or inadequate action), such as the non-internalization of negative externalities or the adoption of lax environmental regulations (Bigdeli 2008). A failure to internalize negative externalities through taxes or other mechanisms is likely to confer the same benefit as any other financial contribution on those who create the externalities. Indeed, it is precisely for this reason that the International Monetary Fund (IMF) incorporates negative externalities associated with the use of fossil fuels into its global FFS estimates.

[2] For example, a restriction on coal exports, either in the form of quotas or a total ban, may increase the quantity of coal in the domestic market and hence provide an advantage for consumers in terms of lower prices in the same way a government provision of coal (below market price) does. For more details, see Rubini (2009).

18.3 FOSSIL-FUEL SUBSIDIES AND ENERGY TRANSITION: THE CASE FOR REFORM

Despite the emergence of alternative energy sources, the world remains dependent on fossil fuels for more than 80 per cent of its energy (IEA 2015a). This heavy dependence on hydrocarbons, however, has become a serious cause for climate change and energy security concerns. The Intergovernmental Panel on Climate Change (IPCC) has concluded that the warming of the climate system is unequivocal (IPCC 2014a). The global average temperature has increased by about 0.85°C during the period 1880–2012 (IPCC 2014b). This is in large part due to the unprecedented increase in the atmospheric concentration of greenhouse gas. Continued greenhouse gas emissions will cause further warming and increase the likelihood of severe, pervasive, and irreversible impacts for people and ecosystems (IPCC 2014b). Mitigating these risks requires substantial and sustained reduction in greenhouse gas emissions. This is, however, unlikely to be achieved unless the world halts its unabated use of fossil fuels. Carbon dioxide emission from fossil-fuel combustion is the single largest contributor to greenhouse gas emissions (IEA 2015b). Meeting the internationally agreed goal of limiting global average temperature increase to no more than 2°C requires the vast majority of proven fossil-fuel reserves to remain buried underground (IEA 2012). The only way this could be achieved is through a massive improvement in energy efficiency and a rapid transition of the global energy system from one that relies heavily on fossil fuels to a system that depends mainly on renewable energy sources such as solar and wind (Barnosky 2015). This much-needed energy transition is already under way, but not at a pace fast enough to avoid catastrophic and irreversible consequences of climate change (IEA 2015a).

FFSs are one of the major obstacles holding back the necessary transition. There are at least three, to some extent overlapping, ways in which FFSs hamper the development of renewable energy sources (Bridle and Kitson 2014; Bridle et al. 2014). First, FFSs undermine the competitiveness of renewables by artificially lowering the cost of fossil fuels. Second, FFSs tend to divert investment away from renewables by enhancing the relative attractiveness of the fossil-fuel industry. Third, given the inherently long-term nature of energy projects, subsidy-induced fossil-fuel investments could 'lock in' unsustainable energy infrastructure for decades to come (Unruh 2000). The literature is replete with studies showing the economic and environmental gains from the removal of FFSs. One of the earliest studies on the subject found that removing FFSs would reduce global carbon dioxide emissions by 9 per cent and lead to a global welfare gain of US$33 billion (Larsen and Shah 1992). Subsequent studies have yielded largely similar results. Merrill et al. (2015) recently found that removing fossil-fuel consumption subsidies alone would result in global greenhouse gas emission reduction of up to 13 per cent by 2050.

Why, then, are countries reluctant to harvest this 'low-hanging fruit'? The answer to this question usually takes two forms. The first response is related to the concern that since FFSs are commonly justified under the guise of protecting poor households against high energy prices, their removal could restrict such households' access to energy. Studies, however, have shown that untargeted FFSs are costly and inefficient means of making energy affordable to the poor. Rather than helping the poor, such subsidies disproportionately benefit high income households (IEA 2011; del Granado, Coady, and Gillingham 2010). The second explanation is more compelling and popular; eliminating FFSs is complex and politically difficult. Experience from previous subsidy reform efforts suggests that reforming FFSs can face stiff resistance from the public and vested interests (Koplow 2014). Section 18.4 will discuss the nature of these political barriers and whether—and if so how—low oil prices can help in overcoming them.

18.4 FALLING OIL PRICES AND FOSSIL-FUEL SUBSIDIES: FUEL FOR REFORM

The price of crude oil went from a peak of about US$115 a barrel in June 2014 to below US$50 in January 2015. The decline has continued ever since and crude oil now costs less than US$30 a barrel, falling more than 70 per cent since the second half of 2014. The environmental implications of this sharp drop in oil prices are far from straightforward. On the one hand, low oil prices encourage the overconsumption of carbon-intensive fuels and reduce the incentives for energy conservation. Low oil prices could also make renewables relatively even more expensive and hence scare off much-needed investment in the renewable energy sector (Cheon and Urpelainen 2012). On the other hand, falling oil prices presents an invaluable opportunity to phase out FFSs (Coady et al. 2015; IEA 2015a; World Bank 2015). The major challenge is how to maximize the opportunities and mitigate the challenges.

There is a broad consensus in the literature that the main barriers to FFS reform are political. Subsidy reforms face strong opposition both from vested interests and from the public at large. The primary opposition to subsidy reform comes from specific interest groups that benefit from the status quo (Overland 2010). Subsidy reforms create 'winners' and 'losers', like most other policy reforms. The political economy of reforms suggests that reforms become extremely difficult to implement when the 'losers' are more powerful or better able to organize themselves than the 'winners' (Haggard and Webb 1994). This is particularly the case for FFSs. Different studies show that subsidy benefits tend to be highly concentrated in the hands of specific groups (with higher levels of energy consumption), while the costs are widely spread across the general population. Those who benefit from the status quo obviously stand to lose from the removal of subsidies. And hence they have strong incentives to lobby for the retention of subsidies. In contrast, members of the general public have much less incentive,[3] as well as less information, to lobby for subsidy reform. This lack of countervailing lobbying for subsidy reform strengthens the vested interests' chance of successfully blocking subsidy reforms. The drop in oil prices, however, reduces the incentives for such lobbying. Under the current low oil prices, the change in fuel prices after the removal of subsidies would not be as dramatic as it would have been under high oil prices. By limiting the price increase from subsidy removal, low oil prices mitigate the cost of subsidy reforms for those groups that benefit from subsidies (Benes et al. 2015).

Another source of resistance to subsidy reform comes from the general public. In theory, one may expect the poor, in whose name FFSs are usually justified but who benefits very little from subsidies, to support rather than protest against subsidy reforms. In practice, however, public resistance is often the cause of FFS reform reversal in many countries (Cheon 2015; Victor 2009). Perhaps this is down to the lack of information about the costs of FFSs (and the benefits of their reform) among the general public and the fact that unlike their long-term economic and environmental benefits, the short-term impacts of FFS reforms (e.g., increase in fuel prices and general inflation) are more visible and easier to detect for the general public. Some countries have implemented compensatory measures such as direct cash transfers to poor households to offset the increase in fuel prices from subsidy reforms. Such targeted subsidies are, however, complex and difficult to implement, especially for countries with limited institutional capacity (Fattouh and El-Katiri 2012). Here, again, low oil prices reduce the risk of public resistance to subsidy reforms and the need for compensatory measures. Low oil prices means that the removal of subsidies is unlikely to cause a significant increase in fuel prices.

18.5 INTERGOVERNMENTAL INITIATIVES TO PHASE OUT FOSSIL-FUEL SUBSIDIES

There is a growing recognition that FFSs are not only economically inefficient but also harmful for the environment. The most visible aspect of this recognition is the proliferation of calls for the phasing out FFSs and efforts to do so

[3] This is because the cost of subsidies is likely to be much smaller in per capita terms than the benefit to vested interests (Morgan 2007).

over recent years. This section will highlight and briefly discuss the efforts that have been made thus far at the intergovernmental level. Although these efforts are yet to yield the desired results, they represent a major step forward in the quest to end FFSs, and offer valuable lessons for future multilateral efforts to eliminate FFSs.

One of the first intergovernmental responses to calls for FFS reform came in the form of the Kyoto Protocol to the UNFCCC (UN Framework Convention on Climate Change). FFS reform was included, albeit implicitly, in the Protocol's non-exhaustive list of policies and measures to tackle climate change. Article 2.1(a) (c) of the Protocol states that:

> Progressive reduction or phasing out of market imperfections, *fiscal incentives, tax and duty exemptions* and *subsidies in all greenhouse-gas emitting sectors* that run counter to the objective of the Convention and application of market instruments. (UN 1998, emphasis added)

Most FFS schemes fall under this provision, but the requirement to phase out FFSs is neither comprehensive nor mandatory. First, it applies only to the so-called Annex I countries and excludes some of the leading fossil-fuel subsidizing countries. Second, none of the policies and measures set out in Article 2 are mandatory, even for the Annex I countries (Feaver, McGoldrick, and Boyd-Wells 2010). The protocol gives the signatories absolute discretion to implement any particular policy or measure, as long as they meet their emission reduction commitments.

Perhaps the most concerted effort yet to phase out FFSs at the intergovernmental level has been that of the G20. At their Pittsburgh Summit in September 2009, G20 leaders agreed:

> To *phase out and rationalize over the medium term inefficient fossil-fuel subsidies* while providing targeted support for the poorest.... (G20 Leaders 2009, Leaders' Statement, emphasis added)[4]

A similar commitment was also made by the member states of the Asia-Pacific Economic Cooperation (APEC) shortly after (APEC 2009). Together, the G20 and APEC membership, which comprises a wide range of countries with the highest FFSs, accounts for about 83 per cent of global oil consumption, making the commitment even more remarkable (Aldy 2015). However, translating this commitment into action proved to be a slow and difficult process for several reasons. The first of these concerns the scope of the commitment. In the absence of a universally agreed upon definition and given the diverse range of interests involved, defining FFSs was one of the most contentious issues during the negotiations (Lang 2011). Since they were not able to agree

[4] Available at:< https://www.treasury.gov/resource-center/international/g7-g20/Documents/pittsburgh_summit_leaders_statement_250909.pdf> (accessed 16 October 2016).

on a common definition, the resultant compromise was a commitment to phase out 'inefficient FFSs', whereby each country determines for itself what constitutes an 'inefficient' FFS. The explicit reference to 'inefficient' subsidies was meant to permit some exceptions, but it is already hard enough to define FFSs, let alone to identify the inefficient ones. This provides a way for G20 countries to define FFSs as narrowly as they wish, and thereby claim not to have any FFSs. As noted by Koplow (2012), it was by defining FFS narrowly that eight out of the twenty G20 members claimed to have no inefficient FFSs, while those that reported having inefficient FFSs reported considerably lower figures than estimates had suggested. However, given that members are entitled to adopt their own definition of 'inefficient FFSS', there is no legal grounds for holding them accountable for the discrepancy.

Other noteworthy intergovernmental forums that have joined the fight against FFSs include the G7, the UN Conference on Sustainable Development (UNCSD), and the Friends of Fossil-Fuel Subsidy Reform (FFFSR). The G7, which is composed of seven G20 members, regularly reaffirms its commitment to phase out FFSs, but has never moved beyond such rhetorical commitments. The UNCSD has the potential to expand the geographical scope of the commitment to phase out FFSs due in part to its almost universal membership. However, this potential has yet to be fully realized. The issue of FFS reform was discussed during the Rio+20 Conference in 2012, but the outcome document—General Assembly Resolution 66/288—was confined to reaffirming the G20 and APEC non-binding commitments (with additional qualifying language) to phase out *harmful* and inefficient FFSs that encourage wasteful consumption and *undermine sustainable development* (UN 2012). Finally, FFFSR is an informal grouping of eight non-G20 countries—Costa Rica, Denmark, Ethiopia, Finland, New Zealand, Norway, Sweden, and Switzerland—formed in June 2010 to advocate for FFS reforms. Since then, the group has evolved to become a vital player in the international efforts to phase out FFSs. It has recently launched a communiqué calling for the elimination of FFSs in the context of the Paris climate change conference (FFFSR 2015). The communiqué has been endorsed by a number of countries and encourages the international community to advance FFS reform through three interrelated principles: (i) increased transparency, (ii) greater ambition in the scope of reform, and (iii) the provision of targeted support for the poorest.

18.6 TOWARDS A MULTILATERAL AGREEMENT ON FOSSIL-FUEL SUBSIDIES

Despite the series of non-binding intergovernmental agreements discussed in Section 18.5, FFSs remain prevalent around the world (Bast et al. 2015).

Against this backdrop, this section examines the prospects for, and challenges of moving towards a binding multilateral agreement on FFSs. Negotiating such an agreement is obviously highly complex and time consuming. It is also politically challenging. However, the growing recognition of the need to eliminate inefficient and environmentally harmful FFSs, together with the fall in global oil prices, has created a unique opportunity not only to reform FFSs but also to lock in those reforms.

18.6.1 The Need for a Multilateral Legal Regime

It is widely agreed that climate change is a global problem that requires global policy response. The issue of FFSs is no exception. Insofar as the environmental and economic impacts of FFSs are not limited to the subsidizing country only, it is in the interest of all countries to cooperate in phasing out FFSs. It was the recognition of this fact that led to the intergovernmental efforts discussed in the Section 18.5. However, those efforts contain only voluntary commitments with no incentives (or punishments) to encourage or force the countries concerned to undertake or sustain FFS reforms. Past precedent suggests that such commitments may not translate into actual subsidy reforms, and, even when they do, the reforms tend to be vulnerable to oil price shocks, public protest, and changes of political regime. Without any mechanism that ties their hands, reluctant governments often find it easier to renege on their voluntary commitments in times of high and rising oil prices, in the face of popular opposition, or in the run-up to elections (Vagliasindi 2013). It is thus necessary to transform these fragile political commitments into legally binding obligations to ensure that countries undertake and stick to subsidy reforms.

18.6.2 Key Issues and Challenges Ahead

In this section we discuss some of the key issues and challenges in the pursuit of a multilateral framework for FFS reforms. The likelihood and success of a multilateral agreement on FFSs largely rests on the international community's ability to: (i) define what constitutes a 'fossil-fuel subsidy', (ii) create an effective mechanism for enhancing transparency, (iii) address the concerns of developing countries and oil-exporting countries, and (iv) establish enforceable commitments with implementation timelines. At the heart of all these challenges lies the issue of political will to reach a multilateral agreement on phasing out FFSs. Despite the growing consensus on the need to eliminate FFSs, countries are still reluctant to undertake legally binding commitments. The recent history of international negotiations suggests that building the necessary political will for a binding multilateral agreement on phasing out FFSs will require, among other things, a core group of 'like-minded' countries that push for action. Until the creation of the FFFSR in 2010, the global effort to phase out FFSs had no such support. The FFFSR's active participation in the fight against FFSs indicates that the group is well positioned to lead international efforts to phase out FFSs.

18.6.2.1 Defining Fossil-Fuel Subsidies

It is not possible to forge a successful multilateral legal regime for FFS reform without a clear and reasonable definition of FFSs. Past efforts to define subsidies are characterized by disagreement on how broadly subsidies should be defined, prompting the inclusion of vague language such as 'inefficient' and 'harmful' in FFS reform commitments. Defining what is and what is not a subsidy is problematic. Defining subsidies too broadly runs the risk of including virtually all public sector activity and undermining the policy space of governments. Such a definition is more likely to be contested by governments that are reluctant to undertake comprehensive subsidy reforms. Defining subsidies too narrowly runs the risk of excluding a range of government support measures to the fossil-fuel industry. It would also allow countries to circumvent their commitment by simply replacing one form of subsidy with another. Thus, the first challenge will be finding a reasonable definition of FFSs. In this regard, the WTO's subsidy definition offers a sound starting point as it 'has been tried and tested through a rigorous negotiating process and is supported by extensive legal analysis and jurisprudence' (GSI 2010). It is, however, worth noting that this definition was designed with the specific purpose of disciplining trade-distorting subsidies and has never been tested in a dispute involving FFSs.

18.6.2.2 Enhancing Transparency

Transparency is another necessary prerequisite for a successful subsidy reform both at the national and international level. It provides a clear and comprehensive picture of FFSs, which is essential to assess their costs and benefits. It also helps to garner the necessary public support for subsidy reforms by exposing the vested interests involved (Laan 2010). Moreover, accurate information about the extent and nature of subsides is crucial to get negotiations off the ground (Laan 2010; Steenblik 2010). Such information is not yet forthcoming, however, either because 'governments themselves do not have full records on the range of support measures in place in their jurisdictions', or because they 'often do not see it in their best interests to disclose subsidies' (GSI 2012).[5] The problem is exacerbated at the international level by the

[5] Shaffer, Wolfe, and Le (2015) further note that countries 'might worry about providing adverse information for a potential legal dispute'.

'absence of an international system or protocol to comprehensively assess and monitor [FFSs]' (GSI 2012). The voluntary self-reporting and peer-review mechanisms of the G20 and APEC have shown that such mechanisms are unlikely to be sufficient to improve transparency (Casier et al. 2014). It is therefore imperative to devise a mandatory notification or reporting mechanism, which facilitates the collection of data across countries through a commonly agreed reporting format and methodology.

18.6.2.3 Clear and Enforceable Commitments

One of the shortcomings of previous intergovernmental agreements to eliminate FFSs was the lack of clear-cut commitments to phase out FFSs. Apart from their non-binding nature, the commitments were vague and indefinite. It is desirable to clearly and explicitly set out each and every country's commitment to phase out FFSs. Perhaps the most effective way to do so is to ban certain FFSs and/or impose a quantitative limit on the overall level of subsidies a country may provide.[6] Given how deeply entrenched FFSs are in many countries, it is unrealistic to expect countries to eliminate their FFSs overnight (Beaton et al. 2013). A commitment to progressively eliminate FFSs is more realistic than an outright ban. But, in order to be effective, such a commitment needs to be complemented by clear and explicit country-specific benchmarks and timelines for implementation. Moreover, a strong and effective enforcement mechanism is central to the successful implementation of such commitments.

18.6.2.4 Ensuring Wide Participation

The ultimate purpose of negotiating a multilateral agreement to phase out FFSs is to reduce greenhouse gas emissions and accelerate the transition away from fossil fuels. Such a purpose is best achieved when all countries participate in the process. However, it is clear that not all countries are yet ready to get rid of their FFSs. This has to do with the level of subsidies, the policy rationales behind them, the technical and institutional capacity to undertake reform, and the political economy of each country. These factors magnify the long-standing divergence between developed and developing countries, and between energy-producing and energy-consuming countries. Energy-producing countries use subsidies to achieve a much wider objective than that of protecting poor households from high and rising fuel prices. For them, subsidies are also a means of sharing national natural resource wealth and maintaining their natural comparative advantage (Krane 2014). Although studies have shown that subsidies are inefficient in achieving such objectives, it will take time and effort to

[6] Here examples can be drawn from the SCM Agreement's outright ban on export contingent and import substitution subsidies, and the Agreement on Agriculture's (AOA) *de minimis* limit.

bring these countries on board. During the 2012 Rio+20 Conference, for example, oil-producing countries such as Qatar, Saudi Arabia, and Venezuela successfully blocked the proposal to phase out FFSs.

Developing countries present another major hurdle in reaching a multilateral agreement on FFSs. This is partly because subsidies lie at the heart of their poverty alleviation strategies and they lack the necessary technical and institutional capacity to undertake subsidy reforms (Commander 2012). Further fuelling the problem is the fact that consumption subsidies, which are relatively easy to identify and hence have become the focus of international subsidy reform efforts, are more prevalent in developing countries than in developed countries. This makes any agreement to phase out FFSs much more burdensome for developing countries, thereby reducing their willingness to undertake binding commitments. One possible means of overcoming this problem is the provision of capacity-building and technical assistance support for developing countries that join the agreement. Another way is to incorporate flexible rules for developing countries whereby they will be given longer transition periods to eliminate their FFSs.

18.6.3 The Quest for an Institutional Home: In Whose Court Is the Ball?

The issue of FFS reform has been raised in international forums ranging from the UNFCCC to the WTO. This confirms the cross-cutting nature of the issue. But, it also shows that the quest to phase out FFSs has no single institutional home at the international level. This is not necessarily a problem per se, but the process of reaching and implementing any major international agreement requires such an institutional framework (Alvarez 2005). A single international organization is needed to coordinate the fragmented international efforts, to provide a forum for negotiations towards an international agreement, and oversee the implementation of the agreement. Since all the organizations involved in the fight against FFSs are not equally suitable, it is pertinent to ask: which international organization is best suited to take on this role?

It is relatively easy to exclude some of the organizations from the outset because they lack a secretariat, which can carry out a 'leadership role' in an international treaty-making process (e.g., G20, FFFSR, UNCSD), because they have a limited membership size (e.g., APEC, IEA, and OECD), or because the issue of FFS reform does not fall within their direct mandate (e.g., Energy Charter Treaty). The same goes for organizations that are not forums for negotiations as such (e.g., UNEP, the IMF, and the World Bank). This leaves us with the UNFCCC and the WTO, and we next examine each of them in turn to see which one is better equipped to fill the institutional vacuum and provide an institutional home for negotiating and overseeing the implementation of an international agreement on FFSs.

18.6.3.1 The UNFCCC

Since environmental concerns are at the forefront of the quest to end FFSs, the international environmental regime is the most logical starting point in the search for an institutional home for global FFS governance. The UNFCCC stands out in this respect as the only legitimate and comprehensive forum for international climate negotiations. It entered into force in 1994 and has been ratified by more than 196 countries as of 2015. This almost universal membership makes it an ideal forum for multilateral FFS governance. The elimination of FFSs fits well with its ultimate objective of reducing and stabilizing atmospheric greenhouse gas emission levels so as to prevent dangerous anthropogenic climate change (UN 1992: Article 2). As discussed in Section 18.5, the Kyoto Protocol to the UNFCCC recognizes the phasing out of subsidies in greenhouse gas emitting sectors as one of the policy measures countries should take to reduce greenhouse gas emissions. What remains to be done now is to make this recognition more specific, convert it into a binding commitment, and extend its scope of application to the entire UNFCCC membership.

The draft negotiating text of the Paris Agreement had two provisions related to FFSs: the first urging 'parties to reduce international support for high-carbon investments, including international [FFSs]', and the second listing 'the phasing down of high-carbon investments and [FFSs]' as one of the climate finance options. The inclusion of such provisions in a legally binding international agreement would have been a great step forward, but neither of these provisions made it into the final text of the agreement. What is left in the final text of the Paris Agreement is Article 2(c), which obliges member state to '[make] finance flows consistent with a pathway towards low greenhouse-gas emissions and climate resilient development' (UNFCCC 2015). Like its predecessor (i.e., Tokyo Protocol), the Paris Agreement leaves the choice of specific emission reduction policies and measures to the discretion of each country.

This is due in part to the high regard with which the principle of national sovereignty is held within the international environmental regime (Lang, Wooders, and Kulovesi 2010). The main focus of international climate negotiations that have taken place under the auspices of the UNFCCC has largely been on determining countries' emission reduction commitments, but not on the specific policies and measures countries should take to meet their commitments. It is therefore too optimistic to expect the UNFCCC to mandate specific policies and measures in the first place and, second, to create a multilateral legal regime for FFS reform. This has been confirmed by the

Paris Agreement, which refrained from prescribing specific emission reduction policies and measures.

18.6.3.2 Is the Ball in the WTO's Court?

The WTO is not an environmental organization per se, but this does not mean that it has no role to play in the protection of the environment or that it cannot serve as a forum for global FFS governance. Sustainable development that protects and preserves the environment is acknowledged as one of the overarching objectives of the WTO in the first paragraph to the preamble to the Marrakesh Agreement establishing the WTO. FFSs have also been the subject of discussion in the WTO ever since concerns over the dual pricing practices of energy-exporting countries were raised in the early 1980s (Lang, Wooders, and Kulovesi 2010). The relevant question is therefore not whether the FFS issue falls under the domain of the WTO, but rather whether the WTO is best suited to address them. In the remainder of this sub-section we argue in favour of a positive answer to this question.

It is easier to work around an existing agreement than trying to negotiate a whole new one. And it goes without saying that this could be achieved best in an organization that already possess the necessary experience and expertise in dealing with the issues at stake. As noted, the WTO has the only binding multilateral agreement applicable to FFSs. The SCM Agreement was, however, drafted with a view to discipline trade distorting subsidies. Whether and to what extent it could also discipline environmentally harmful subsidies is a matter of debate. While it is not necessary to recount the entirety of the debate here, it is important to note that the potential role of the SCM Agreement in disciplining FFSs is limited to the extent that FFSs are also trade distorting. There are two types of trade-distorting subsidies under the SCM Agreement: prohibited subsidies and actionable subsidies. In order to qualify as prohibited subsidies, FFSs must be contingent upon export performance or upon the use of domestic over imported goods (SCM Agreement: Article 3). Since countries do not subsidize fossil-fuel exports or provide import substitution subsidies to fossil-fuels, FFSs hardly qualify as prohibited subsidies under the SCM Agreement. Whereas to qualify as 'actionable subsidies', FFSs must meet the specificity and adverse effects requirements set out in Articles 2 and 5 of the SCM Agreement, respectively. The only circumstance in which FFSs may qualify as actionable subsidies is when they benefit fossil-fuel producers or energy-intensive industries (Asmelash 2015). The specificity requirement entails that the subsidy has to be provided to an industry/enterprise or group of industries/enterprises (SCM Agreement: Arts 1(2) and 2). Fossil-fuel production subsidies could meet this requirement as they tend to be industry specific, but most fossil-fuel consumption subsidies are provided to all enterprises throughout the economy, and hence are de jure non-specific. They may be deemed de facto specific within the meaning of Article 2(1)(c) insofar as they only or disproportionately benefit energy-intensive industries (Howse 2010). However, establishing de facto specificity is complex and requires subsidy-specific factual analysis. Even then, there is the adverse effects requirement. Energy is a key input for energy-intensive industries, but it is not the only input. This makes the task of establishing the causal link between the subsidized imports and the adverse effect thereof problematic. As such, existing rules of the SCM Agreement appear to be inadequate to discipline FFSs. This is partly reflected in the absence of WTO challenges to FFSs. The rules need to be strengthened if they are to discipline not only trade-distorting, but also inefficient and environmentally harmful FFSs.

The WTO has already shown interest in tackling environmentally harmful subsidies in the fisheries sector. The Doha Round negotiation on fisheries subsidies was elicited by concerns about the contribution of fisheries subsidies to overfishing and overcapacity (Bigdeli 2008). Given the similarity of purpose, it is more logical to address both issues under the auspices of the same organization. The WTO also has well-established institutional machinery capable of providing technical assistance for developing countries, estimating and analysing the economic and environmental impacts of FFSs, enhancing transparency, and facilitating multilateral negotiations. Despite its limited success, the SCM Agreement has already created a notification and surveillance mechanism, which could serve as a useful basis for developing a more effective monitoring and notification mechanism for FFSs.

Another reason that makes the WTO an appropriate forum for FFS governance is the binding nature of its dispute settlement mechanism. The Kyoto Protocol has also created a unique compliance mechanism, but it is not comparable to that of the WTO, which is strengthened by the threat of withdrawal of concessions or the imposition of trade barriers.

18.7 CONCLUSION

FFSs are one of the major obstacles in achieving sustainable energy transition. Aside from imposing a heavy burden on government budgets, they encourage wasteful energy consumption and impede investment in renewable energy sources. However, despite growing reform efforts, they remain prevalent around the world. The current political commitments need to be transformed into legally binding commitments in order to accelerate FFS reforms and ensure the durability of the initiated reforms. While this will require strong political will and leadership, the combination of the recent fall in oil prices, the growing consensus on the need to reform FFSs and the formation of intergovernmental coalitions for FFS reforms (e.g., FFFSR) has created enough momentum to at least initiate the process. The main challenges will be to define what constitutes FFSs, to bridge the developed–developing country gap, and to set out enforceable obligations with implementation timelines. Moreover, the crosscutting nature of the issue meant that the global effort to phase out FFSs is fragmented across multiple organizations. The global effort to phase out FFSs lacks a single institutional home at the international level. However, as the only international organization with a proven track record of serving as a forum for negotiating and implementing binding multilateral subsidy rules, the WTO seems to be the most appropriate forum to coordinate international efforts and serve as a forum for negotiating a binding multilateral agreement on FFSs. It is up to the FFFSR to emulate what the 'Friends of Fish' did for fisheries subsidies and put FFSs on the WTO's negotiating agenda. Norway and New Zealand (on behalf of the FFFSR) have already taken the first step by raising the issue of FFS reform in the WTO's Committee on Trade and Environment, but this needs to be followed up by a formal proposal for negotiations. Although reaching a multilateral agreement has become increasingly difficult in the WTO, the agreement reached at the 2015 WTO Ministerial Conference to abolish agricultural subsidies proves that if there is a will, there is always a way.

REFERENCES

Aldy, J. E. (2015). 'Policy Surveillance in the G20 Fossil-fuel Subsidies Agreement: Lessons for Climate Policy'. Discussion Paper 2015–70. Harvard Project on Climate Agreements, Cambridge, MA.

Alvarez, J. E. (2005). *International Organizations as Law Makers*. Oxford: Oxford University Press.

APEC (Asia-Pacific Economic Cooperation) (2009). 'APEC Leaders' Declaration: Sustaining Growth, Connecting the Region'. 14–15 November, Singapore.

Asmelash, H. B. (2015). 'Energy Subsidies and WTO Dispute Settlement: Why Only Renewable Energy Subsidies are Challenged'. *Journal of International Economic Law*, 18(2): 261–85.

Bast, E., A. Doukas, S. Pickard, L. Van Der Burg, and S. Whitley (2015). 'Empty Promises: G20 Subsidies to Oil, Gas and Coal Production'. Overseas Development Institute, London.

Barnosky, A. D. (2015). 'Transforming the Global Energy System is Required to Avoid Sixth Mass Extinction'. *MRS Energy and Sustainability*, 2(10): 1–13.

Beaton, C., I. Gerasimchuk, T. Laan, K. Lang, D. Vis-Dunbar, and P. Wooders (2013). *A Guidebook to Fossil-fuel Subsidy Reform for Policy-makers in Southeast Asia*. Winnipeg: IISD.

Benes, K., A. Cheon, J. Urpelainen, and J. Yang (2015). 'Low Oil Prices: An Opportunity for Fuel Subsidy Reform'. Columbia University, New York.

Bigdeli, S. (2008). 'Will the "Friends of Climate" Emerge in the WTO? The Prospect of Applying the "Fisheries Subsidies" Model to Energy Subsidies'. *Carbon and Climate Law Review*, 2(1): 78–88.

Bridle, R. and L. Kitson (2014). 'The Impact of Fossil-fuel Subsidies on Renewable Electricity Generation'. IISD, Winnipeg.

Bridle, R., L. Kitson, and P. Wooders (2014). 'Fossil-fuel Subsidies: A Barrier to Renewable Energy in Five Middle East and North African Countries'. IISD, Winnipeg.

Burniaux, J.-M. and J. Chateau (2014). 'Greenhouse-Gases Mitigation Potential and Economic Efficiency of Phasing Out Fossil-fuel Subsidies'. *International Economics*, 140: 71–88.

Casier, L., R. Fraseer, M. Halle, and R. Wolfe (2014). 'Shining a Light on Fossil-fuel Subsidies at the WTO: How NGOs Can Contribute to WTO Notification and Surveillance'. *World Trade Review*, 13(4): 603–32.

Cheon, A. (2015). 'Instruments of Political Control: National Oil Companies, Oil Prices and Petroleum Subsidies'. *Comparative Political Studies*, 48(3): 370–402.

Cheon, A. and J. Urpelainen (2012). 'Oil Prices and Energy Technology Innovation: An Empirical Analysis'. *Global Environmental Change*, 22(2): 407–17.

Coady, D., I. Parry, L. Sears., and B. Shang (2015). 'How Large Are Fossil-fuel Subsidies?' IMF Working Paper WP/15/105. IMF, Washington, DC.

Coady, D. and B. Shang (2015). 'Energy Subsidies in Developing Countries: Treating the Disease while Symptoms Abate'. Available at: <http://www.voxeu.org/article/energy-subsidies-developing-countries> (accessed 16 October 2016).

Commander, S. (2012). 'A Guide to the Political Economy of Reforming Energy Subsidies'. IZA Policy Paper 52, Institute for the Study of Labor, Bonn.

del Granado, J. A., D. Coady, and R. Gillingham (2010). 'The Unequal Benefits of Fuel Subsidies: A Review of Evidence for Developing Countries'. IMF Working Paper WP/10/202. IMF, Washington, DC.

Fattouh, B. and L. El-Katiri (2012). 'Energy Subsidies in the Arab World'. Arab Human Development Report Research Paper Series 2002, UNDP, New York.

Feaver, D., W. McGoldrick, and V. Boyd-Wells (2010). 'Is Australia's EAP a Prohibited Export Subsidy?' *Journal of World Trade*, 44(2): 319–47.

FFFSR (Friends of Fossil Fuel Subsidy Reform) (2015). 'Fossil-fuel Subsidy Reform Communiqué', 17 April. Available at: <http://fffsr.org/communique/> (accessed 16 October 2016).

G20 Leaders (2009). 'Leaders' Statement: The Pittsburgh Summit', 24–25 September.

GSI (Global Subsidies Initiative) (2010). 'Defining Fossil-fuel Subsidies for the G20: Which Approach is Best?' Global Subsidies Initiative Policy Brief. Available at: <https://www.iisd.org/gsi/sites/default/files/pb5_defining.pdf> (accessed 16 October 2016).

GSI (2012). 'Reforming Fossil-fuel Subsidies to Reduce Waste and Limit CO2 Emissions while Protecting the Poor'. GSI, Geneva.

Haggard, S. and S. B. Webb (eds) (1994). *Voting for Reform: Democracy, Political Liberalization and Economic Adjustment*. New York: Oxford University Press.

Howse, R. (2010). 'Climate Mitigation Subsidies and the WTO Legal Framework: A Policy Analysis'. IISD, Winnipeg.

IEA (International Energy Agency) (2011). 'World Energy Outlook 2011'. IEA, Paris.

IEA (2012). 'World Energy Outlook 2012'. IEA, Paris.

IEA (2015a). 'World Energy Outlook 2015'. IEA, Paris.

IEA (2015b). 'Energy and Climate Change: World Energy Outlook Special Report'. IEA, Paris.

IPCC (2014a). *Climate Change 2014: Mitigation of Climate Change. Contribution of Working Group III to the Fifth Assessment Report of the Intergovernmental Panel on Climate Change*. Cambridge: Cambridge University Press.

IPCC (2014b). 'Climate Change 2014: Synthesis Report—Contribution of Working Groups I, II and III to the Fifth Assessment Report of the Intergovernmental Panel on Climate Change'. IPCC, Geneva.

IRENA (International Renewable Energy Agency) (2015). 'Renewable Energy Target Setting'. IRENA, Abu Dhabi.

Kojima, M. (2009). 'Government Response to Oil Price Volatility: Experience of 49 Developing Countries'. World Bank, Washington, DC.

Koplow, D. (2012). 'Phasing Out Fossil-Fuel Subsidies in the G20: A Progress Update'. Earth Track, Inc. and Oil Change International, Cambridge.

Koplow, D. (2014). 'Global Energy Subsidies: Scale, Opportunity, and Barriers to Reform'. In A. Halff, B. K. Sovacool, and J. Rozhon (eds), *Energy Poverty: Global Challenges and Local Solutions*. Oxford: Oxford University Press, pp. 316–37.

Krane, J. (2014). 'Navigating the Perils of Energy Subsidy Reform in Exporting Countries'. Baker Institute Policy Report 58. Available at: <http://bakerinstitute.org/files/8679/> (accessed 16 October 2016).

Laan, T. (2010). 'Gaining Traction: The Importance of Transparency in Accelerating the Reform of Fossil-fuel Subsidies'. GSI, Geneva.

Lang, K. (2011). 'The First Year of the G20 Commitment on Fossil-fuel Subsidies: A Commentary on Lessons Learned and the Path Forward'. IISD, Winnipeg.

Lang, K., P. Wooders, and K. Kulovesi (2010). 'Increasing the Momentum of Fossil-fuel Subsidy Reform: A Roadmap for International Cooperation'. IISD, Winnipeg.

Larsen, B. and A. Shah (1992). 'World Fossil Fuel Subsidies and Global Carbon Emissions'. Policy Research Working Papers WPS 1002, World Bank, Washington, DC.

Merrill, L., M. Harris, L. Casier, and A. M. Bassi (2015). 'Fossil-fuel Subsidies and Climate Change: Options for Policy-makers within Their Intended Nationally Determined Contributions'. IISD, Winnipeg.

Morgan, T. (2007). 'Energy Subsidies: Their Magnitude, How They Affect Energy Investment and Greenhouse-gas Emissions, and Prospects for Reform'. UNFCCC, Geneva.

OECD (Organization for Economic Co-operation and Development) (1998). 'Improving the Environment Through Reducing Subsidies—Parts I–III'. OECD, Paris.

Overland, I. (2010). 'Subsidies for Fossil-fuels and Climate Change: A Comparative Perspective'. *International Journal of Environmental Studies*, 67(3): 303–17.

Parry, I., D. Heine, E. Lis, and S. Li (2014). 'Getting Energy Prices Right: From Principle to Practice'. IMF, Washington, DC.

Pitt, M. M. (1985). 'Equity, Externalities and Energy Subsidies: The Case of Kerosene in Indonesia'. *Journal of Development Economics*, 17(3): 201–17.

Rubini, L. (2009). *The Definition of Subsidy and State Aid: WTO and EC Law in Comparative Perspective*. Oxford: Oxford University Press.

Shaffer, G., R. Wolfe, and V. Le (2015). 'Can Informal Law Discipline Subsidies?' *Journal of International Economic Law*, 18(4): 711–41.

Steenblik, R. (2010). 'Subsidies in the Traditional Energy Sector'. In P. Joost (ed.), *Global Challenges at the Intersection of Trade, Energy, and the Environment*. Geneva: The Graduate Institute, pp. 183–92.

UN (1992). 'UN Framework Convention on Climate Change'. Available at: <https://unfccc.int/resource/docs/convkp/conveng.pdf> (accessed 16 October 2016).

UN (1998). 'Kyoto Protocol to the United Nations Framework Convention on Climate Change', 10 December 1997. Available at: <http://unfccc.int/resource/docs/convkp/kpeng.pdf> (accessed 16 October 2016).

UN (2012). 'General Assembly Resolution 66/288: The Future We Want, A/RES/66/288'. United Nations, New York.

UNEP (United Nations Environment Programme) (2008). 'Reforming Energy Subsidies: Opportunities to Contribute to the Climate Change Agenda'. Available at: <http://www.unep.org/pdf/pressreleases/reforming_energy_subsidies.pdf> (accessed 16 October 2016).

UNFCCC (UN Framework Convention on Climate Change) (2015). 'Paris Agreement, FCCC/CP/2015/L.9', done at Paris, 12 December.

Unruh, G. C. (2000). 'Understanding Carbon Lock-in'. *Energy Policy*, 28(12): 817–30.

Vagliasindi, M. (2013). 'Implementing Energy Subsidy Reforms: Evidence from Developing Countries'. World Bank, Washington, DC.

Victor, D. (2009). 'Untold Billions: Fossil-fuel Subsidies, their Impacts and the Path to Reform. The Politics of Fossil-fuel Subsidies'. GSI, Geneva.

WTO (1995). 'Agreement on Subsidies and Countervailing Measures (SCM Agreement)'. Available at: <https://www.wto.org/english/docs_e/legal_e/24-scm.pdf> (accessed 16 October 2016).

WTO (2004). 'WTO Appellate Body Report, United States—Final Countervailing Duty Determination with Respect to Certain Softwood Lumber from Canada', WT/DS257/AB/R, 19 January.

World Bank (2015). 'Global Economic Prospects: The Global Economy in Transition'. World Bank, Washington, DC.

Part VI

Sector Reform

This part of the book provides four unique analyses of case studies of sectoral reform and provides timely insights related to the nature of energy reform within broader political economic dynamics. National and regional structures vary considerably and the dynamics behind energy transitions necessarily reflect the complex structures, power of incumbents and political dynamics of change. This part offers insights on four distinct situations and decision dynamics. Understanding the trade-offs of development, climate and clean energy, habitat protection, and political/economic dynamics from these case studies offers insights for approaching today's and tomorrow's complex decisions.

In particular, Baker examines the contested negotiation of key policies which have been fundamental to the introduction of a renewable energy sector in South Africa and considers how the new renewable energy sector has evolved thus far, raising key challenges and concerns for its future development. Ogunley provides insights on power reform based on Nigeria's complex energy economy. This enquiry provides a comprehensive assessment of the reform by isolating the major challenges facing it, focusing on political economy developments surrounding regulatory, institutional, legislative, and fiscal issues with mainstreaming clean renewable energy being the main theme tying together the analysis. Valenzuela and Studer examine Mexico's energy transitions via a comparison of two recent energy reforms, corresponding to two administrations, and offer insights related to coherence of policy and key policy drivers of energy security and climate mitigation. Their contribution underscores the political economy trade-offs between the need for a strong climate commitment that provides a stable long-term energy transition pathway and the political and economic short-term benefits derived from low electricity tariffs. Bucaram, Fernandez, and Grijalva offer a set of intriguing insights from analysing the situation in Amazon basin in Ecuador and the complex trades offs of protecting precious habitat, development, and the access and provision of energy resources.

19

Post-Apartheid Electricity Policy and the Emergence of South Africa's Renewable Energy Sector

Lucy Baker

19.1 INTRODUCTION

Recent developments in coal-dependent, carbon-intensive South Africa's electricity sector present something of a paradox. On the one hand, the country is now a leading destination for investment in renewable energy. Since the country's Renewable Energy Independent Power Producers' Programme (RE IPPPP) was launched in 2011, nearly 6,327 MW of capacity and 92 projects have been approved, of which almost one-third were connected to the grid by October 2015, constituting approximately 2 per cent of overall capacity (CSIR 2015). RE IPPPP is held up as an international model for the procurement of utility-scale renewable energy (Eberhard, Kolker, and Leigland 2014). In reflection of global trends, since mid-2015 solar photovoltaic (PV) and wind technologies have become cost-competitive with South Africa's new build coal-fired power plants.[1]

On the other hand, large-scale renewable electricity generated by independent power producers (IPPs) is feeding into a grid otherwise controlled by the state-owned monopoly utility, Eskom, which has depended almost entirely on abundant sources of historically low-cost coal. This carbon-intensive electricity sector, a key feature of the country's 'minerals-energy complex' (Fine and Rustomjee 1996) accounts for 45 per cent of national carbon emissions (237 Mt CO_2-equivalent in 2010). While Eskom continues to build more coal-fired power plants, it is also experiencing a financial and supply-side crisis, the culmination of events over decades. The most recent symptoms of this crisis

[1] See also Baker (2016).

have included load shedding in late 2014 and throughout 2015. Described as a 'monster of apartheid',[2] Eskom is now subject to growing indebtedness and received a negative outlook for its credit rating in November 2015. Meanwhile, a highly controversial 9.6 GW nuclear fleet is being pushed by the Presidency and explorations are under way for the extraction of shale gas.

In light of these competing parallel narratives, this chapter asks how, why, and when South Africa's renewable electricity sector has emerged. Focusing on shifts that have taken place in the country's electricity governance and policy-making, from a period of generation surplus in 1980s to the supply-side constraints of the present,[3] the chapter builds on earlier studies of South Africa's electricity policy (Pegels 2010; Baker, Newell, and Phillips 2014) and provides an analysis of complex and constantly evolving developments. This analysis includes an examination of two key developments, the Renewable Energy Independent Power Producers' Procurement Programme (RE IPPPP) and the Integrated Resource Plan (IRP) for electricity, both launched in 2011. In addition to enabling the emergence of a renewable electricity sector, these two mechanisms were carried out under relatively transparent and participatory planning and procurement processes when compared to the secretive nature of decision making in electricity and other sectors during the apartheid era.

The chapter is based on extensive field research on the political economy of South Africa's electricity sector. This includes primary data gathered in approximately 50 semi-structured interviews in 2010 and 2015 conducted with: members of the energy industry (electricity, coal, and renewables); energy-intensive users; government departments; the National Energy Regulator of South Africa (NERSA); the utility Eskom; banks and financial institutions; research institutes; and civil society. The research is also informed by an in-depth content analysis of grey literature, such as national policies and publications by government and industry, and a long-term and systematic consultation of media sources on energy in South Africa and globally, including: *Engineering News*, *Mining Weekly*, and *ESI Africa*.

The structure of this chapter is as follows. Section 19.2 explores the key features of South Africa's minerals-energy complex (MEC) and recent shifts within it, particularly in the electricity sector and the current crisis. Section 19.3 provides a longer-term perspective on the governance and structure of Eskom. Section 19.4 explores the nature of post-apartheid policy-making and significant changes that have taken place since 1980s, particularly as regards the role of the Department of Energy (DoE) and other entities that undertake activities in its name. Section 19.5 examines the negotiation of the IRP for electricity, the country's first process for electricity planning and how

[2] By government member, in interview, May 2010 (anonymous).

[3] Note that 'surplus capacity' is a technical term that fails to account for the social reality that until 1993, only one-third of the population was connected to the grid.

its latest revision has since stalled over the controversial issue of nuclear power. Section 19.6 explores the emergence of the renewable energy independent power producers' procurement programme (RE IPPPP), which is followed in Section 19.7 with an examination of key challenges for the new renewable energy industry. Section 19.8 concludes.

19.2 THE MINERALS-ENERGY COMPLEX AND ELECTRICITY CRISIS

The political economy of South Africa's electricity policy has been characterized by the country's minerals-energy complex (MEC) (Fine and Rustomjee 1996), a system that is both electricity- and carbon-intensive. In descriptive terms this refers to an evolving system of production and consumption based on the country's historical dependence on cheap and abundant coal supplies and cheap labour, to produce cheap electricity for an export-oriented industry based on raw and semi-processed mineral products such as coal, platinum, iron ore, steel, and aluminium. Analytically, the MEC offers a framework that can be used to address the economic legacy of apartheid and the nature of power relations, politics and policy-making in contemporary South Africa (Padayachee 2010: 2).

South Africa's natural resource wealth, however, has not been evenly distributed (Harvey 2015). As one of the most unequal countries in the world, the country faces huge socioeconomic development challenges. These include a history of racial oppression and inequality, high levels of violence, an unemployment rate of 40 per cent, and poor access to economic and social services such as health and education. Approximately 25 per cent of the population (12.3 million people) lack access to electricity (IEA 2011) despite a significant electrification programme after apartheid (Bekker et al. 2008).

As a vertically integrated electricity monopoly, Eskom has been a key player in and beneficiary of the MEC. To date Eskom has controlled the country's high-voltage transmission grid, 90 per cent of coal-fired generation plants, and 60 per cent of distribution, which is consumed by one-third of South Africa's customers.[4] As an illustration of the MEC's carbon and electricity intensity, the country's 31 energy-intensive users consume 44 per cent of the country's electricity. This group includes five major mining companies who also supply 80 per cent of the coal used by Eskom (Eberhard 2011). A number of these energy-intensive users were granted preferential tariffs, or 'special purchasing

[4] Municipal distributors purchase electricity from Eskom Distribution and in turn supply about two-thirds of the country's customers, accounting for 40 per cent of total electricity sales.

agreements' by Eskom during the early 1990s when the utility had substantial overcapacity.

However, in recent decades a number of the MEC's core features have been subject to change. Firstly, this can be identified in the evolving set of relationships and linkages between the country's state-owned companies, corporate capital, and a growing financial sector. As McDonald (2009: 20) surmises, the former MEC model of 'big state negotiating with big capital' has been replaced by a 'fragmented and rescaled state negotiating with more globally dispersed capital, in many different sectors with new technical demands'. Secondly, the contribution that mining makes to the national and international economy has declined dramatically. Until 25 years ago South Africa accounted for 40 per cent of the world's mining industry but now only accounts for 4 per cent (Seccombe 2015). Further shifts in the MEC include an increased contribution by financial and business services to the economy, which now constitute 24 per cent of GDP (Bhorat, Cassim, and Hirsch 2014) and the creation of a black industrial and financial elite in the post-apartheid era (Cargill 2010). Meanwhile, the country's national credit rating is now under threat from long-term weak economic growth rate and large budget deficit.

The country's ongoing electricity crisis represents a further shift in the MEC. Since mid-1980s, South Africa has gone from a period of electricity surplus with some of the lowest electricity prices in the world as a result of over-planning and the construction of excess generation capacity (Dubash 2002), to capacity restraints, a supply side crisis, and imminent deficit. Electricity prices have tripled in real terms since 2005 (IEA 2014: 147). By October 2015, Eskom was facing a funding gap of R280 billion (Creamer 2015a).

The crisis has contributed to lowering growth rates, discouraged private investment, exacerbated the country's large current account deficit, and pushed the cost of electricity beyond the reach of the poorest households which are connected to the grid. Since 2005 Eskom has failed to meet its target for an additional 17000 MW of generation capacity by 2018. To make up the shortfall, during 2014 and 2015 Eskom relied heavily on expensive diesel peaking plants and having exhausted its budget to do this, then sought further tariff increases (Creamer 2016). In addition to a commitment of R23 billion from National Treasury in 2015, the utility also discussed the sale of its assets in order to raise the capital to fill an estimated funding gap of R200 billion (US$17 billion) by 2018 (Reuters 2015).

Reasons for the crisis in South Africa's electricity sector are deep-seated and long-term (Eberhard 2007; Trollip and Marquard 2014; Baker et al. 2015). They include a legacy of mismanagement; a failure by Eskom to enforce contracts with its coal suppliers (Olsen 2007); inadequate maintenance of the utility's older power stations; climate change mitigation commitments pledged in 2009; and a failure by government to approve the construction of new generation capacity in the early 1990s as the following sections discuss.

South Africa's industries have been further affected by rising national electricity prices. The country's coal supply chain has also changed: Eskom is no longer able to rely on ready supplies of cheap national coal resources, due to growing export demands, and the end of long-term coal contracts between tied coal mines and Eskom (Burton and Winkler 2014).

The crisis arguably presents a challenge to the dominance of Eskom as a monopoly utility, which may be further challenged by embryonic processes for the procurement of independently produced power from renewable energy under RE IPPPP, in addition to gas, coal and cogeneration, and small-scale rooftop solar PV (Baker et al. 2015). With this in mind, Section 19.3 now provides a longer-term perspective on Eskom's governance.

19.3 GOVERNING ESKOM

Eskom retained its monopoly status and side-stepped global trends of power sector liberalization in the 1980s and 1990s. This was due firstly to South Africa's political and economic isolation under apartheid, which meant that the country was not beholden to the loan conditionalities of structural adjustment; and secondly to the country's cheap and abundant supplies of indigenous coal and a well-developed transmission network. The 'standard model' of power sector liberalization, as endorsed by the World Bank (Gratwick and Eberhard 2008) was based on the notion that public ownership was hampered by poor technical and financial performance and high investment requirements. The standard model thus advocated a move away from a publicly-owned utility to one based on privatization, competition, and the unbundling of the utility into separate transmission, generation, and distribution companies (Dubash 2002). However, in light of repeated failings, the model is now in demise (Williams and Ghanadan 2006; Yi-chong 2006).

Nonetheless, in the mid-1990s attempts were initiated to unbundle South Africa's electricity sector when the Department for Public Enterprises (DPE) of the new democratic government announced plans to restructure Eskom alongside the country's four largest state-owned enterprises (Eberhard 2005). As a member of the energy-intensive users' group explained (in interview, October 2010), this development 'was informed and motivated by the global privatisation drive of the time and the strong belief that privatisation was better than government', despite fierce resistance to this within the utility. Subsequently the 1998 White Paper on Energy Policy (DME 1998) set out a gradual liberalization of the power sector in line with the standard model, which would see the corporatization and outsourcing of various functions of Eskom. Notably, the White Paper anticipated the creation of a separate transmission utility and system operator, which would be owned by the state in the first instance but with a view to a possible future sale. The White Paper was followed by a cabinet memo in 2001 announcing that 30 per cent of electricity generation, including renewable energy, would come from IPPs, in turn followed by a cabinet ruling that Eskom no longer be allowed to build new electricity generation. Such a move was described by Eberhard (2007: 231) as South Africa's 'self-imposed' structural adjustment programme as part of broader attempts to improve efficiencies in government-owned entities.

The subsequent Eskom Conversion Act of 2001 required a corporate governance structure for the utility, which converted it from a statutory body to a public company and required that it pay tax and dividends for the first time. This would mean the formation of separate transmission and distribution bodies and the creation of different generating companies to create internal competition (Gaunt 2008).

However, key aspects of the 1998 White Paper were not implemented and some are still outstanding. For instance, a separate transmission utility has never been established and the Independent Systems and Market Operator bill that would do this has been continually postponed. Significantly, between 1998 and 2003, no new generation was built. The various reasons attributed to this include: union resistance; resistance from within Eskom (Eberhard 2007); tensions within the Department of Minerals and Energy (DME); the loss of construction and management expertise as white staff members left the country after apartheid; and surplus generation capacity from 1980s, which reduced incentives for any new construction in the post-apartheid era. One energy-intensive user stated (in interview, November 2010), 'the then DME did not put in an appropriate policy, the regulator was not sufficiently empowered to make things happen, the ministry did not fully understand and the DPE failed to step up. There were also drivers within Eskom not to cooperate with the introduction of private generators.'

Furthermore, despite approval for the entry of IPPs under the White Paper, there was no adequate institutional framework, including a lack of clarity over who the buyer of power would be. In addition, no IPP would be able to compete with Eskom tariffs, which at the time were well below cost. Consequently, by 2003 government had begun to rethink its strategy. Faced with falling reserve margins and an imminent electricity crisis, a cabinet memorandum approved that Eskom should be re-allowed to construct more power plants but that 30 per cent of new generation should be built by IPPs. However, in the absence of a regulatory framework, it was not until the introduction of RE IPPPP that this actually took place as discussed in the next paragraph.

According to (Marais 2011: 348), such a move reflects a slowing down in the privatization drive in all sectors in the 2000s due to 'the demise of the Washington consensus globally in the late 1990s'. This, in addition to trade union resistance at the national level. Eberhard (2005: 5309) has described this sequence of events as a move 'from state to market and back again', relating to the constant national tension between the interests of public and private capital, and the broad and diverse political spectrum that falls under the umbrella of the ruling ANC, from the developmental state at one end to market liberalization at the other (Gumede 2007).

Section 19.4 discusses the contested nature of electricity policy and planning in South Africa, including the shifting roles of various different departments and institutions involved, and the relative weakness of the DoE, despite its formal mandate.

19.4 ELECTRICITY POLICY

The excessively secretive nature of apartheid era policy-making 'made rational and public debate on energy policy nigh impossible' (DME 1998: 24). In the post-apartheid era, however, electricity policy-making has been influenced by weak institutional capacity (Newbery and Eberhard 2008) and 'a systemic lack of clarity concerning roles and responsibilities in the electricity sector' (IDASA et al. 2010: 4).

Under apartheid there was no department dedicated solely to energy planning and policy-making and no formal public process for it either. Eskom was responsible for all planning and new build decisions in electricity. Since then responsibility for policy-making on electricity has moved across various different departments. In March 1980 the energy function of the then Department of Environmental Planning and Energy was moved into the newly formed Department of Mineral and Energy Affairs (DMEA) (Fine and Rustomjee 1996: 97). It was not until 2009, following President Jacob Zuma's inauguration that the functions of what had by then become the Department of Minerals and Energy (DME) were separated into two departments: the Department of Mineral Resources (DMR) and the Department of Energy (DoE). Meanwhile, it was not until the Electricity Regulation Act of 2006 that responsibility was formally allocated to the energy minister to approve the construction of new generation capacity and what the source of that capacity should be (see Figure 19.1).

Various different departments are involved in South Africa's electricity governance. The Department of Public Enterprises (DPE) is Eskom's principle shareholder and is responsible for the operability of the entity in accordance with the 2001 Eskom Conversion Act. National Treasury is responsible for Eskom's financial exposure. The National Energy Regulator of South Africa (NERSA), established in 2004, has a mandate to determine electricity tariffs; approve generation, distribution, and transmission licences; and oversee the import, export, and trading of electricity within the Southern Africa Power

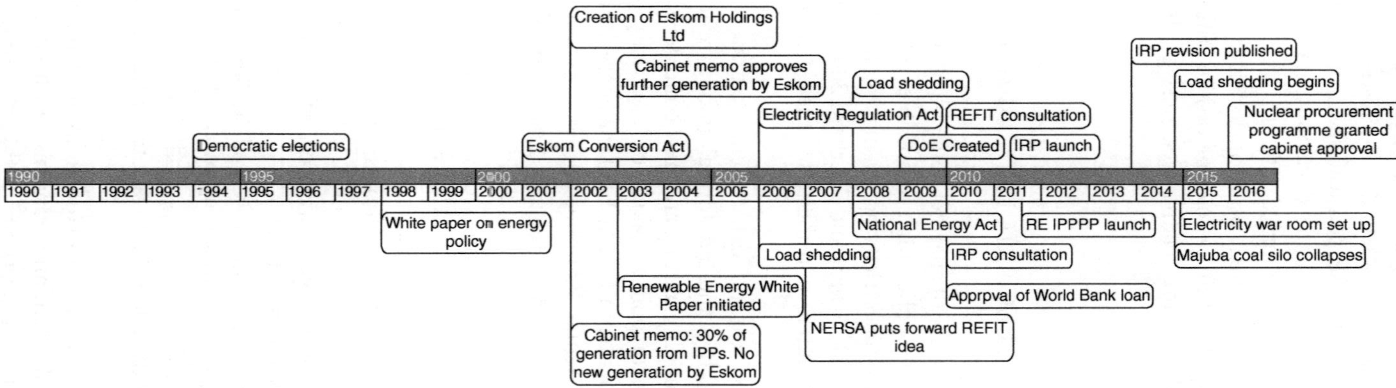

Figure 19.1. Time line of South Africa's electricity policy: 1994–2015.

Source: Author's illustration.

Pool. Both the DoE and NERSA report to the energy minister. Other institutions of national governance that are involved in energy policy include the Department of Environmental Affairs, and metropolitan and municipal governments.

Seven years after it was established, the DoE is still developing capacity and expertise. Its ability to govern electricity has been challenged by lack of knowledge and understanding of new technologies, particularly renewable energy. One of the reasons attributed to this lack of capacity was because energy policy, in addition to many other policy imperatives, such as safety conditions in mining and the growth of mining employment, was marginalized when under the remit of the DME, which was heavily focussed on the distribution of mining rents to the country's recently established group of black capitalists (Baker 2012; Harvey 2015). Therefore, informal and formal influence over many decisions and activities made in the DoE's name is carried out by various departments and institutions under varying degrees of secrecy as the following examples illustrate.

Firstly, and as discussed in Section 19.5, Eskom still carries out electricity planning in name of the DoE. Secondly, the policy capacity for the procurement of privately generated power from renewable energy under RE IPPPP and future IPP programmes is housed within the IPP unit. While the IPP unit acts on behalf of the DoE, the unit was set up by National Treasury's Public–Private Partnership unit together with foreign technical consultants for the specific purpose of managing RE IPPPP. The IPP unit therefore functions outside of formal departmental governmental structures. However, since it was set up in 2011, it has maintained credibility as a high quality, transparent, and secure professional body. Perhaps because it 'did not start out with the level of mistrust of private business that sometimes characterises other government agencies in South Africa' (Eberhard, Kolker, and Leigland 2014: 9).

Thirdly, a 'War Room' was set up in late 2014 by the Presidency in response to the electricity crisis, and in order to ensure greater coordination between all departments that hold a stake in the sector and between which there are evident tensions (Baker et al. 2015). Despite its potential influence, limited information is available with regard to the activities of the War Room and how it makes decisions.

Fourthly, the highly controversial, highly secretive nuclear procurement programme being pushed by the Presidency and a minority within the DoE, was granted cabinet approval at the end of 2015. Should the programme go ahead it is possible that it will be managed by the Energy Security Cabinet Subcommittee (ESCS). While the ESCS reports to cabinet, 'its proceedings and documents are classified under the Minimum Informational Security Standard Act as TOP SECRET [sic]' (Baker et al. 2015: 42). The issue of nuclear power is a key sticking point for the latest revision of the Integrated Resource Plan, now discussed in Section 19.5.

19.5 THE INTEGRATED RESOURCE PLAN FOR ELECTRICITY

South Africa's IRP for electricity is a master plan covering total generation requirements for electricity from 2010 to 2030. The plan anticipates a doubling of national capacity from approximately 41,000 MW–89,532 MW by 2030. Significantly, IRP represents the first time that electricity planning has taken place in South Africa in accordance with the 2006 Electricity Regulation Act, which established the necessary powers for the DoE to conduct an open planning process for electricity. While the IRP's negotiation process was heavily contested, it is still considered a breakthrough in terms of electricity planning. However, as this section explores, this breakthrough is now threatened by the scrapping of the revised IRP released in late 2013, which questioned the need for a national nuclear programme. The scrapping of this revision is seen as a deliberate attempt by government attempt to push through nuclear power (Baker et al. 2015).

Before an electricity generation project can be approved, it must align with the technological allocations set by the IRP in order for NERSA to be able to grant the project a licence (Pienaar and Nakhooda 2010). However, according to the Electricity Regulations on New Generation Capacity updated in 2011, the minister holds the discretion to approve additional generation capacity of a certain technology if s/he considers it appropriate. The IRP allows for approximately 20 per cent (17.8 GW) of installed generation capacity to come from renewable energy (including projects approved under RE IPPPP) which will deliver 9 per cent of supply. The plan claimed to be consistent with a carbon emissions constraint of 275 million tonnes of carbon dioxide annually after 2024 (DoE 2011: 6). Despite this constraint and the positive gains for renewable energy, almost half of the overall generation mix will still be dominated by coal, thereby enabling a low carbon transition to take place in parallel to an already existing high carbon trajectory (Baker 2014).

The IRP 'stakeholder engagement process', which took place throughout 2010 and early 2011 was protracted and controversial. In light of the aforementioned lack of capacity of the DoE, the IRP was put together by Eskom's Systems Operator with inputs from a technical task team that consisted largely of members from government, Eskom, coal companies, and energy-intensive users. This team was heavily criticized for consisting largely of vested interests from the mining and energy spheres and related government institutions, and for excluding the renewable energy industry and civil society. IRP was subject to numerous other criticisms, including over its methodology; a lack of transparency over critical assumptions; technology costs; and the plan's potential impacts on the poor. Notably, the IRP's assumption that national electricity demand would double by 2030, led to one member of the City of

Cape Town to refer to it as 'business as usual on steroids'. Despite this, and while the IRP process was dominated by relatively specialized stakeholders able to engage with the inevitable technical complexity of electricity planning, opening it up to public participation still set an important national precedent. Therefore it was still considered a significant advance on previous electricity planning processes, or lack thereof (Hughes 2010; Mainstream Renewable Power 2010; Nakhooda 2011).

In 2013, a revised IRP was put out for public comment following the requirement that the plan be updated on a biennial basis. Given the decline in South Africa's economic growth since the release of the first IRP in 2010 and declining electricity consumption by the mining and minerals sector, the revised draft made a downward adjustment to the demand forecast by 6,600 MW. The draft also proposed to reduce allocations for new coal (from 6,250 MW to 2,450 MW) and wind (from 9,200 MW to 4,360 MW) while increasing contributions from solar PV (from 8,400 MW to 9,770 MW) and CSP (from 1,200 to 3,000 MW). The wind industry was particularly critical of the assumptions contained within the revised draft, arguing that the technology costs did not reflect the price decreases that had taken place since the start of RE IPPPP (Baker et al. 2015: 24).

However, the draft has yet to be approved by cabinet. The main reason for this is that it challenges the necessity of a 9.6 GW nuclear fleet currently being pushed by the Presidency and members of the DoE (Baker et al. 2015). Significantly, the draft questions the high associated costs of nuclear technology, stating that 'the revised demand projections suggest that no new nuclear baseload capacity is required until after 2025 (and for lower demand not until at earliest 2035)' (DoE 2013: 8). In light of the high costs and implications for technological lock in of a large nuclear fleet, the draft asserts that 'commitments to long range large-scale investment decisions should be avoided' (DoE 2013: 9). Details regarding how much the nuclear programme will cost, how it will be paid for, and who might build own and operate it have been entirely lacking in transparency (Paton 2015c). Current but as yet unconfirmed capital cost estimates are at R1 trillion. National Treasury, which has largely been excluded from the process, has questioned the programme's affordability (Mantshantsha and Marrian 2015).[5] A new draft is anticipated for 2016 that will contain a higher allocation for nuclear energy than the 9.6 GW currently included (Paton 2015b).

The IRP represents a poignant example of the inherently political nature of policy-making in electricity. Despite the 2006 Electricity Regulation Act, discretionary powers held by the DoE may still undermine the plan's apparent

[5] Serious questions over the affordability of any potential programme have been linked to the firing of the finance minister Nene in December, who stated that nuclear power would not be procured if the country could not afford it (Paton 2015a).

neutrality. Furthermore, control over the plan and its implications seem to be as much about a contest over which technologies get prioritized, as a contest over which models should be used to procure and manage them. As one South African project developer clarified [personal communication, November 2014] 'the competitive bid model around RE IPPPP is through private generation, while nuclear is a bi-lateral negotiation led by the head of state'. With this in mind, RE IPPPP as the introduction of the first successful framework to procure renewable energy and privately generated power is now discussed in Section 19.6.

19.6 RENEWABLE ENERGY PROCUREMENT

RE IPPPP is a competitive tender system under which IPPs bid to construct and connect their renewable electricity projects to Eskom's monopoly controlled transmission grid. The programme's initial allocation in August 2011 was for 3,725 MW to be allocated under fewer than five bidding rounds. This was augmented by an additional 3,200 MW of capacity declared by the energy minister in December 2012, a further 6,300 MW in August 2015, and an 'expedited' round to absorb 1,800 MW of projects that failed marginally in previous rounds in November 2015. It is anticipated that a new tender framework for round five and beyond will be introduced in 2016.

Internationally RE IPPPP is held up as example for its high quality regulatory framework, tough qualification criteria, and strong economic development and community ownership requirements, all of which provided a long-awaited positive policy signal to investors and developers. The programme has further been celebrated for the savings it has created for the South African economy. According to the Council for Scientific and Industrial Research (CSIR 2015), solar PV and wind projects collectively generated a R8.3 billion benefit in the first six months of 2015. This is firstly through savings in diesel and coal fuel costs, to a total of R3.6 billion and secondly through savings to the economy by avoiding load shedding or 'unserved energy' (Creamer 2015b).

The initial concept for RE IPPPP began in the form of a renewable energy feed-in tariff (REFIT), for which the idea was put forward in 2006/2007 by individuals within NERSA's Electricity Regulatory Division. This move was supported by the German and Danish bilateral agencies and some individuals within Treasury, the DPE, and the Department of Environmental Affairs. The main aim of REFIT was to create a market mechanism that would 'kick start and stimulate the renewable energy industry in South Africa' in order to meet the target of 10,000 GWh of renewable energy by 2013 as set out in the 2003 Renewable Energy White Paper (NERSA 2008: 4) (see Box 1).

Box 1. **Renewable Energy White Paper**

The 2003 Renewable Energy White Paper was the first national document that referred to the procurement of renewable energy from IPPs. Described by one Eskom employee as 'a visionary blip on the horizon', it was published by the then DME in 2003 with support from the Danish foreign aid agency DANIDA, just missing the World Summit on Sustainable Development hosted by South Africa in 2002. The paper set a minimal target of four per cent of the estimated electricity demand by 2013 (DME 2003), to be achieved through a mixture of biomass, landfill gas, hydro-electricity, and solar water heaters with only one per cent for wind (Edkins et al. 2010). Funded under the World Bank's Renewable Energy Market Transformation Project (REMT) and managed by the Development Bank of South Africa, the paper should have been revised in 2008. Still unpublished by May 2012 it had no influence on the content of the IRP. Reasons for the delay are attributed by one energy analyst to a resistance within the DoE: 'the World Bank paid for the research to be carried out on the renewable energy white paper as part of REMT. This was resisted by the DoE. The irony is that the DoE should be making this policy. Instead it ended up being an activist process funded by the World Bank' (in interview September 2010).

Initially there was opposition to the notion of a feed-in tariff and renewable energy from within the regulator, the DoE and Eskom. However, in pushing for REFIT, NERSA's director succeeded in galvanizing the global renewable energy industry, whose market opportunities in Europe and the US had been undermined by the 2008 global financial crisis. From the outset, REFIT was subject to intense interest from South African renewable energy IPPs, in turn backed by international renewable energy developers and private finance, who were awaiting the policy certainty that would allow them to build their projects and connect to the grid (Baker 2012). During the negotiation process for what is now RE IPPPP there were protracted disagreements over:

- how the tariff levels should be set
- who the buyer of power would be
- mistrust of renewable energy from various factions of government and business
- how political and financial risks should be allocated between government and the private sector.

There was also significant conflict over whether a tender system, which was eventually selected, or a feed-in tariff system should be implemented. Notably, the DME proposed a tender system as the preferred model, while private sector representatives, NERSA, and the Danish Embassy argued in favour of a feed-in tariff system (Renewable Energy Summit 2009). While a feed-in tariff pays generators a fixed price for each unit of renewable energy sold to the grid, which is set at a higher rate than the retail price of electricity generated from conventional resources, under a competitive bidding system potential project developers bid for a renewable energy contract below a certain cap. The latter system is therefore more competitive and is currently the preferred global model (EY 2014).

When REFIT had become so high profile as to be irreversible, in late 2010 the DoE backed by National Treasury's Public Private Partnership Unit and various international technical advisors made changes to the Electricity Regulations on New Generation Capacity, which effectively transferred powers over the procurement process away from NERSA to the DoE and National Treasury. This move also facilitated the shift from a feed-in tariff to a competitive bidding system, which saw the proposed REFIT replaced by RE IPPPP in August 2011 (Baker 2012: 109–13). National Treasury justified the move by stating that the proposed REFIT was illegal given that 'the predetermined tariff would fall foul of South Africa's procurement rules' (Creamer 2011).

Significantly RE IPPPP's negotiation demonstrates a turf battle between NERSA as the initiator of the process and the DoE backed by National Treasury, who altered legislation in order to take control of the programme. While NERSA was undoubtedly 'acting beyond its mandate' (Baker 2012: 105) given that under the 2006 Electricity Regulation Act, it is the DoE's role to make policy and NERSA's to implement it through licensing and regulation, NERSA's lead in pushing for the system was clearly instrumental in its emergence.

19.7 CHALLENGES FOR A NEW INDUSTRY

South Africa's renewable energy sector that has emerged out of RE IPPPP potentially offers a window of opportunity to achieve a clean energy transition. However, as I now discuss, despite RE IPPPP's apparent and rapid success, there are concerns over the way in which it is evolving, particularly as to whether it will be able to bring about social, economic, and environmental benefits beyond the generation of renewable electricity.

The majority of capacity allocated under RE IPPPP is for wind, solar PV and concentrated solar power (CSP). Under RE IPPPP, projects are assessed 70 per cent on price below a tariff certain cap, which decreases with each round and 30 per cent on socioeconomic development requirements. RE IPPPP's economic development requirements are potentially very progressive, and include factors such as participation of historically disadvantaged individuals, job creation, local content, rural development, community ownership, and skills development. A project must meet the economic development requirements before the price submission can be considered. Successful projects sell their power to Eskom under a 20-year power purchase agreement (PPA) which is government-backed and local-currency-denominated. However, partly due to the risk-adverse nature of project finance, many national firms have been prohibited from participating in the market, while ownership of the industry is dominated by large international companies (Baker 2015). Such a scenario raises questions as to who the winners and losers might be in the creation of a new renewable energy sector in a process that has to date favoured international actors over local companies.

The dramatic decrease in the tariffs bid by project developers under RE IPPPP means that many renewable energy projects are now cost-competitive with Eskom's new build coal projects. However, these low tariffs have also been identified as a risk to project 'bankability'. Not least, given that projects are now operating on very tight financial contingencies. While this risk is carried by the private sector rather than Eskom, these costs will inevitably be structured into any successful project and therefore passed on to the country's electricity consumers.

The capacity of Eskom's transmission grid to integrate renewable energy, as an intermittent generation source is a further challenge. While IPPs pay for the connection of their projects, Eskom is required to strengthen the transmission network and upgrade substations in order to do so. However, Eskom has stated that due to funding constraints, it is likely to struggle to connect IPP projects outside of areas of grid strength. It is therefore anticipated by industry that in some locations solar PV and wind will start to compete for access to the same line.

19.8 CONCLUSION

In conclusion, factors that have shaped the nature of policy-making in post-apartheid South Africa's electricity sector include: surplus power capacity from mid-1980s until the early 2000s; the country's economic and political isolation under apartheid; a spectrum of political and economic ideological conflicts within government, Eskom, and the unions with regards to the privatization of the electricity sector; and the loss of skills in the utility. Significantly, in light of the recent introduction of a regulatory framework for electricity planning and in the absence of expertise and capacity within the recently established DoE, activities carried out in the department's name have been undertaken by various other institutions. Electricity policy is therefore embedded within long-standing political and economic forces, which demonstrates the diverse and often conflicting nature of interests within the ruling party.

The contested nature of electricity policy in South Africa relates as much to a struggle over which technology is selected as the model that should procure and manage it. Decision-making over the ideal electricity mix reflects deeper struggles over what gets supported by the state, who gets to build it, and who gets to benefit. While privately generated renewable electricity presents a clear challenge to Eskom's crisis-ridden monopoly utility, coal-fired power is likely to remain the primary energy source for the time being, despite the end of historically cheap prices. Meanwhile, the potential construction of a highly contested 9.6 GW nuclear generation programme appears to represent an attempt by the state to reinforce its power through Eskom, at the same time as precipitating a potential return to an era of secretive decision making characteristic of the apartheid era.

This chapter also contributes to the broader debate over the ideal model of power sector governance. Significantly, my argument does not subscribe to the notion of 'private goods and public bads', which would assume that a liberalized sector is inevitably efficient, competitive, and innovative while a state-owned utility is inherently over-subsidized, corrupt, and resistant to new technologies. Or alternatively that a state-controlled utility always works in the interests of the majority public, while a privatized electricity sector in the interest of a private rentier class. Regardless of the ownership structure, the harder question is whether the sector is transparent, financially accountable, subject to public scrutiny and technically and managerially competent.

On a related point, this chapter raises poignant questions over the significance of transparent and democratic decision making for a clean energy transition. Following a breakthrough as the first public negotiation process for electricity planning, the IRP as a process now appears to have been undermined. This is due to the DoE's stalling of the revised draft in light of the draft's questioning of a potentially unviable nuclear fleet whose costs are unknown. Arguments aside as to whether nuclear power should contribute to a clean energy transition, of greater concern is that it represents a backward step in terms of a transparent process for electricity planning (Baker et al. 2015). This has huge implications for how any other technology, clean or otherwise, and its related procurement model may then be pushed through in future, beyond the view of public scrutiny.

Finally, while this chapter focusses on South Africa's emerging utility scale renewable energy sector, it acknowledges the other processes now emerging for the procurement and development of independent power from other sources, including embedded generation roof top solar PV, coal, co-generation, and gas. This, in addition to a small IPP programme of projects less than 5 MW of which ten winning bidders were announced in early October 2015. The way in which RE IPPPP may continue to develop therefore has significant implications.

REFERENCES

Baker, L. (2012). 'Power Shifts: The Political Economy of Socio-Technical Transitions in South Africa's Electricity Sector'. Unpublished PhD thesis, University of East Anglia, UEA, Norwich.

Baker, L. (2014). 'Renewable Energy in South Africa's MEC: A "Low Carbon" Transition?' *Review of African Political Economy*, 42(144): 245–61.

Baker, L. (2015). 'The Evolving Role of Finance in South Africa's Renewable Energy Industry'. *Geoforum*, 64: 146–56.

Baker, L. (2016). 'Post-Apartheid Electricity Policy and the Emergence of South Africa's Renewable Energy Sector. 2016/15'. UNU-WIDER, Helsinki.

Baker, L., J. Burton, H. Trollip, and C. Godinho (2015). 'The Political Economy of Decarbonisation: Exploring the Dynamics of South Africa's Electricity Sector'. Research Report, Energy Research Centre, ERC, University of Cape Town, Cape Town.

Baker, L., P. Newell, and J. Phillips (2014). 'The Political Economy of Energy Transitions: The Case of South Africa'. *New Political Economy*, 19(6): 791–818.

Baker, L. and H. Wlokas (2015). 'Renewable Energy Procurement in South Africa: A New Frontier?' Research Report, Energy Research Centre, ERC, University of Cape Town, Cape Town.

Bekker, B., A. Eberhard, T. Gaunt, and A. Marquard (2008). 'South Africa's Rapid Electrification Programme: Policy, Institutional, Planning, Financing and Technical Innovations'. *Energy Policy*, 36: 3125–37.

Bhorat, H., A. Cassim, and A. Hirsch (2014). 'Policy Co-Ordination and Growth Traps in a Middle-Income Country Setting'. WIDER Working Paper 2014/155, UNU-WIDER, Helsinki.

Burton, J. and H. Winkler (2014). 'South Africa's Planned Coal Infrastructure Expansion: Drivers, Dynamics and Impacts on Greenhouse Gas Emissions'. Energy Research Centre, Research Report, ERC, University of Cape Town, Cape Town.

Cargill, J. (2010). *Trick or Treat: Rethinking Black Economic Empowerment.* Johannesburg: Jacana Media.

Creamer, T. (2011) 'Renewables Bid Represents "Substantive Progress", But Raises Compliance Burden'. *Engineering News.* Available at: <http://www.engineeringnews.co.za/article/renewables-bid-represents-substantive-progress-but-raises-compliance-burden-2011-08-23> (accessed 8 March 2016).

Creamer, T. (2015a). 'Amid R280bn Funding Gap Eskom Re-Phases Transmission Capex, Delays Grid Code Compliance to 2022'. *Engineering News.* Available at: <http://www.engineeringnews.co.za/article/amid-r280bn-funding-gap-eskom-re-phases-transmission-capex-delays-grid-code-compliance-to-2022-2015-10-16/rep_id:3182> (accessed 17 October 2015).

Creamer, T. (2015b). 'South Africa's Renewables Ramp Up Impressive'. *Engineering News.* Available at: <http://www.engineeringnews.co.za/article/south-africas-renewables-ramp-up-impressive-study-shows-2015-10-13> (accessed 14 October 2015).

Creamer, T. (2016). 'NERSA Seeks to Get to Grips with Eskom's R8 billion Diesel Claim'. *Engineering News.* Available at: <http://www.engineeringnews.co.za/article/nersa-seeks-to-get-to-grips-with-eskoms-r8bn-diesel-claim-2016-02-04/rep_id:3182> (accessed 15 February 2016).

CSIR (Council of Scientific and Industrial Research) (2015). 'Financial Benefits of Renewables in South Africa in 2015'. Available at: <http://www.csir.co.za/media_releases/docs/Financial%20benefits%20of%20Wind%20and%20PV%202015.pdf> (accessed 20 October 2015).
DoE (Department of Energy) (2011). 'Electricity Regulation Act No. 4 of 2006, Electricity Regulations on the Integrated Resource Plan 2010–2030 (Policy-Adjusted IRP)'. Pretoria: Government Gazette.
DoE (2013). 'Integrated Resource Plan for electricity (IRP) 2010–2013, Update Report 2013'. Department of Energy, Pretoria.
DME (Department of Minerals and Energy) (1998). 'White Paper on the Energy Policy of the Republic of South Africa'. Department of Minerals and Energy, Pretoria.
DME (2003). 'White Paper on Renewable Energy'. Department of Minerals and Energy, Pretoria.
Dubash, N. (2002). 'Power Politics: Equity and Environment in Electricity Reform'. World Resources Institute, Washington, D.C.
Eberhard, A. (2005). 'From State to Market and Back Again: South Africa's Power Sector Reforms'. *Economic and Political Weekly*, 40(50): 5309–17. Available at: <http://www.gsb.uct.ac.za/files/FromStatetoMarketandBackAgain.pdf> (accessed 7 March 2016).
Eberhard, A. (2007). 'The Political Economy of Power Sector Reform in South Africa'. In D. Victor and T. C. Heller (eds), *The Political Economy of Power Sector Reform*. Cambridge: Cambridge University Press, pp. 215–53.
Eberhard, A. (2011). 'The Future of South African Coal: Market, Investment, and Policy Challenges'. Program on Energy and Sustainable Development, Stanford University, Working Paper 100, Stanford University, Palo Alto, CA.
Eberhard, A., J. Kolker, and J. Leigland (2014). 'South Africa's Renewable Energy IPP Procurement Program: Success Factors and Lessons'. PPIAF, South Africa.
Edkins, M., A. Marquard, and H. Winkler (2010). 'Assessing the Effectiveness of National Solar and Wind Energy Policies in South Africa'. For the UNEP research project 'Enhancing Information for Renewable Energy Technology Deployment in Brazil, China and South Africa', Energy Research Centre, University of Cape Town, Cape Town.
EY (2014). 'Renewable Energy Country Attractiveness Index (RECAI)'. *EY*, 40.
Fine, B. and Z. Rustomjee (1996). *The Political Economy of South Africa: From Minerals-Energy-Complex to Industrialisation.* London: C. Hurst & Co. Ltd.
Gaunt, C. T. (2008). 'Electricity Distribution Industry Restructuring in South Africa: A Case Study.' *Energy Policy*, 36(9): 3448–59.
Gratwick, K. N. and A. Eberhard (2008). 'Demise of the Standard Model for Power Sector Reform and the Emergence of Hybrid Power Markets'. *Energy Policy*, 36(10): 3948–60.
Gumede, W. M. (2007). *Thabo Mbeki and the Battle for the Soul of the ANC.* London and New York: Zed Books.
Harvey, R. (2015). 'Mineral Rights, Rents and Resources in South Africa's Development Narrative'. SAIIA Occasional Paper 224, November 2015, SAIIA, Johannesburg.
Hughes, A. (2010). 'IRP 2010 Assumptions'. Energy Research Centre, University of Cape Town, Energy Modelling and Analysis Group, Cape Town. Available at:

<http://www.doe-irp.co.za/hearing1/ENERGY_RESEARCH_CENTRE.pdf> (accessed 12 September 2015).

IDASA, World Resources Institute, Prayas Energy Group, A. Andrews, ANSA, DGRU, ERC-UCT, Green Connection, ILRIG, SECCF, SEA, REEEP, and WWF (2010). 'Electricity Governance Initiative of South Africa: The Governance of Power, Shedding a Light on the Electricity Sector in South Africa'. Electricity Governance Initiative, Pretoria. Available at: <http://electricitygovernance.wri.org/files/egi/EGI-SA%20report%20-%20The%20Governance%20of%20Power%20(Feb2010)-cover.pdf> (accessed 7 November 2016).

IEA (International Energy Agency) (2011). 'World Energy Outlook'. Available at: <https://www.iea.org/publications/freepublications/publication/WEO2011_WEB.pdf> (accessed 14 March 2016).

IEA (2014). 'Africa Energy Outlook'. Available at: <https://www.iea.org/publications/freepublications/publication/WEO2014_AfricaEnergyOutlook.pdf> (accessed 7 March 2016).

McDonald, D. (2009). 'Electric Capitalism: Conceptualising Electricity and Capital Accumulation in (South) Africa'. In D. McDonald (ed.), *Electric Capitalism: Re-colonising Africa on the Power Grid*. Cape Town: HSRC Press, pp. 1–49.

Mainstream Renewable Power (2010). 'IRP 2010 Comments and Inputs'. Presented to Department of Energy IRP2010 Hearings, Johannesburg, 2 December 2010. Available at: <http://www.doe-irp.co.za/irpJHB/MAINSTREAM_RENEWABLE_POWER.pdf> (accessed 2 September 2015).

Mantshantsha, S. and N. Marrian (2015). 'Model and Cost of Nuclear Not Yet Decided'. *Business Day*. Available at: <http://www.bdlive.co.za/business/energy/2015/08/31/model-and-cost-of-nuclear-not-yet-decided> (accessed 3 September 2015).

Marais, H. (2011). *South Africa Pushed to the Limit: the Political Economy of Change*. London: Zed Books.

Nakhooda, S. (2011). 'Empowering a Sustainability Transition? Electricity Planning in a Carbon Constrained South Africa'. MSc Thesis, London School of Economics and Political Science, London.

NERSA (National Energy Regulator of South Africa) (2008). 'NERSA Consultation Paper: Renewable Energy Feed-in Tariff'. NERSA, South Africa.

Newbery, D. and A. Eberhard (2008). 'South African Network Infrastructure Review: Electricity, 2007'. National Treasury and DPE, Government of South Africa, Pretoria.

Olsen, S. (2007). 'Confidential Memo to Jacob Maroga from Susan Olsen, Wingfield Consultancy, Warner New Hampshire, USA re: Issues for Today's Meeting Regarding Generation Primary Energy', 19 July, Democratic Alliance, Cape Town.

Padayachee, V. (2010). 'Re-Introducing the Minerals-Energy Complex'. *Transformation: Critical Perspectives on Southern Africa*, 71: 1–2.

Paton, C. (2015a). 'Cabinet Gives Green Light to Nuclear Procurement'. *Business Day*. Available at: <http://www.bdlive.co.za/business/energy/2015/12/14/cabinet-gives-green-light-to-nuclear-procurement> (accessed 5 February 2016).

Paton, C. (2015b). 'Updated IRP May Raise Share of Nuclear Power'. *Business Day*. Available at: <http://www.bdlive.co.za/business/energy/2015/09/02/updated-irp-may-raise-share-of-nuclear-power> (accessed 3 October 2015).

Paton, C. (2015c). 'Key Details of SA's Nuclear Plan Kept under Wraps'. *Business Day*. Available at: <http://www.bdlive.co.za/business/energy/2015/06/02/key-details-of-sas-nuclear-procurement-plan-kept-under-wraps> (accessed 2 February 2016).

Pegels, A. (2010). 'Renewable Energy in South Africa: Potentials, Barriers and Options for Support'. *Energy Policy*, 38(9): 4945–54.

Pienaar, G. and S. Nakhooda (2010). 'The Great Policy Disconnect'. Heinrich Boell Stiftung, Cape Town.

Renewable Energy Summit (2009). 'Conference Report: Towards a Sustainable Renewable Energy Sector in South Africa, 19–20 March 2009', Pretoria.

Reuters (2015). 'NUM Says Would Oppose Eskom Asset Sales'. *Engineering News*, 6 March. Available at: <http://www.engineeringnews.co.za/article/num-says-would-oppose-eskom-asset-sales-2015-03-06/rep_id:3182> (accessed 1 March 2016).

Seccombe, A. (2015). 'South Africa Losing Importance in Global Mining Ranks'. *Business Day*, 16 October. Available at: <http://www.bdlive.co.za/business/mining/2015/10/16/sa-losing-importance-in-global-mining-ranks> (accessed 21 October 2015).

Trollip, H. and A. Marquard (2014). 'Prospects for Renewable Energy in South Africa—Climate Change'. Heinrich Boll Stiftung, South Africa. Available at: <https://za.boell.org/2014/02/03/prospects-renewable-energy-south-africa-climate-change> (accessed 7 March 2016).

Williams, J. H. and R. Ghanadan (2006). 'Electricity Reform in Developing and Transition Countries: A Reappraisal'. *Energy* 31(6/7): 815–44.

Yi-chong, X. (2006). 'The Myth of the Single Solution: Electricity Reforms and the World Bank'. *Energy*, 31(6/7): 802–14.

20

Political Economy of Nigerian Power Sector Reform

Eric Kehinde Ogunleye

20.1 THE CONTEXT

The Nigerian power sector is currently undergoing one of the most ambitious, comprehensive, and bold reforms in the history of Africa. The ultimate aim is to privatize all power assets with a view to ending the country's chronic power shortages and long-standing monopoly of the sector by the state-owned power entity. The reform is based on the 2005 Electric Power Sector Reform Act, 2010 Roadmap for Power Sector Reform, and subsequently the 2013 Roadmap for Power Sector Reform Revision 1, among several other policy documents. While the Act provides a legal backing for the reform, the roadmaps serve as instruments for fast-tracking the proposed fundamental changes to the ownership, control, and regulation of the sector as envisaged in the Act and ensure these are achieved and realized, especially for the ultimate benefit of electricity users. For additional policy initiatives, pronouncements, and documents that have been initiated by the government to guide the power sector reform process, see Ogunleye (2016).

The Act focuses on creating a regulatory agency that would serve as the umpire for ensuring effective enforcement and compliance with the rules of the game in the electricity sector. It emphasizes privatization of a hitherto government-owned electricity company that has monopolized electricity generation, distribution, and transmission for decades. Through this action, the reform seeks to remove obstacles to private sector investment in this critical sector. It also attempts to improve energy independence and diversify energy sources towards cleaner energy such as solar, geothermal, and wind. The objectives of the reform include highly ambitious targets for the power sector. It was projected that power generation would reach over 13,000 megawatts (MW) by 2015 and 40,000 MW by 2020. While the 2015 target has been missed and 2020 is unrealizable, efforts are still being made to scale up and maintain a steady pace and timely implementation of the reform.

The reform was conceived out of obvious challenges facing the country's energy sector with serious negative effects on business. Power outage is a common and daily experience; Nigeria is one of the countries with the lowest electricity consumption per capita in the world at around 100 kilowatt-hour (kWh) per annum. Installed capacity remains low at around 8,000 MW, out of which only 4,000 MW is operable and less than 2,000 MW available to generate energy. The gross inadequacy of energy generation becomes glaring when compared with South Africa's almost 5,000 kWh per capita. Indeed, 60 per cent of the time, there is no access to electricity in Nigeria (Aliyu, Ramli, and Saleh 2013).

The goals of the reform can be broadly categorized into: (1) defining a new national power policy that positions the private sector as driver of the sector on funding, innovation, and leadership; (2) designing an enabling regulatory, policy, and commercial framework for engagements of all stakeholders in the sector; and (3) commercialization of the sector. In more specific terms, the Roadmap for Power Sector Reform outlines specific tasks to include: establishment of a bulk purchaser/trader; strengthening the Nigerian Electricity Regulatory Commission (NERC); provision of Federal Government Credit Enhancement; operationalizing the Nigerian Electricity Liability Management Company (NELMCO); strengthening the training institute, NAPTIN (National Power Training Institute of Nigeria); strengthening technical and managerial capacity of the Transmission Company of Nigeria (TCN); and sale of Nigeria's generating companies and distribution companies to the private sector.

The reform was planned in three phases. Five activities were scheduled for the first phase. These are unbundling and privatization of the long-lasting government-owned monopoly power company, National Electric Power Authority (NEPA) subsequently known as Power Holding Company of Nigeria (PHCN), creation of an independent power sector regulator (NERC) to ensure sanity in the market and carry out other standard regulatory functions, incorporation of the Power Holding Company of Nigeria (PHCN) successor generation and distribution companies, creation of a multi-purpose entity that would have the function of procuring electricity from the independent power producers and newly created generation companies for subsequent sale to the distribution companies, and establishment of the training institute, National Power Training Institute of Nigeria (NAPTIN). The second phase, which has a medium-term horizon, involves developing a cost-reflective electricity tariff to ensure competitive pricing that would attract private sector participation in the sector. The third phase, which is the final long-term phase, focuses on achieving a completely competitive power sector.

It is noteworthy that modest progress has been recorded in the course of implementing the reform. First, the first phase of the privatization process has been successfully concluded with the unbundling of the hitherto state-owned monopoly in the electricity sector—NEPA and later PHCN—giving way to the emergence of 11 distribution companies, six generation companies and a transmission company. Second, the regulatory and institutional framework for the power sector reform has been established and fully functional (see Ogunleye (2016) for a complete list of these institutions). Third, a cost-reflective tariff (multi-year tariff order or MYTO) that would make the sector attractive to both local and foreign investors has reached a very advanced stage of implementation, albeit fraught with some challenges that are discussed later in this chapter. Fourth, there is emergence of several independent power plants with combined capacity to generate additional 2,500 MW of electricity. Fifth, regulatory institutions responsible for implementing the proposed reforms have been established. Sixth, significant efforts are ongoing to improve the gas-to-power infrastructure which is the greatest binding constraint on electricity supply, resulting from pipeline vandalism and gas pricing policy. Lastly, several policies and market rules are being developed to guide the emerging privatized power sector at different stages of the reform.

Despite the progress, challenges remain. The most prominent of these is the political environment surrounding the reform, mainly regulatory and institutional issues. First, it can be argued that the long-lasting monopoly of the sector by government exacerbated by over-regulation was the dominant factor behind the deterioration in the country's energy sector over the years, culminating in its current deplorable state. Second, the reform is generating emerging challenges that require development of an effective post-reform strategy on regulatory, institutional, and human capacity issues. Third, the state of insecurity in the country, especially in the Niger Delta region, induces incessant gas pipeline vandalism with serious implications for the progress of the reform as gas supply to the power stations is being constrained. Fourth, a cost recovery challenge has emerged recently, an outcome of the alleged over-priced power plants and distribution companies during the bidding process by the eventual buyers.[1] Fifth, the MYTOs have also been a thorny issue, with incessant complaints by industrialists while residential consumers engage in sporadic demonstrations and protests over energy bills that they claim do not match supply availability and quality of service. Sixth, weak political will to invest in the sector now further exacerbated by the country's current challenging fiscal revenue position. This is a major concern because transmission remains under government ownership, though managed by an independent private firm. Lastly, there is challenge with energy security due to limited

[1] Some generation and distribution companies have claimed that the power assets purchased during the privatization process were overvalued because the data and information provided to them by the government was wrong and inflated. Examples of such alleged wrong data include the customer base and aggregate technical, commercial, and collection loss level.

investment in clean energy, especially hydro and solar, that would help achieve optimal mix of energy sources.

A few studies have attempted to assess the energy sector reform in Nigeria (see Adenikinju 2003; Oke 2008; Okoro and Chikuni 2007; Sanyaolu 2008; Tallapragada and Adebusuyi 2008; Idris et al. 2013; and Joseph 2014). Some have focused on the investment opportunities and pitfalls associated with the reform (David-West 2014; Onochie, Egware, and Eyakwanor 2015). Others have assessed the impact of the reform (Adoghe, Odighe, and Igbinovia 2009). Yet others have looked at it from consumer perspective (Ochugudu and Onodugo 2013). Generally, most of these studies have focused mainly on the challenges and opportunities offered by the reform without paying particular attention to the political economy of the reform process. Indeed, none provided a detailed, comprehensive, and historical assessment of the reform in within institutional, regulatory, and legislative contexts. This study fills these gaps.

Following this introduction, Section 20.2 outlines the key challenges threatening the reform. An assessment of the reform is the focus of Section 20.3, while Section 20.4 concludes with a way forward.

20.2 KEY CHALLENGES FACING THE POWER SECTOR REFORM

20.2.1 Poor Transmission Network

The transmission segment of the power sector is the weakest link in the Nigerian electricity value chain. Transmission lines are notorious for being very old and weak. Some of the cables are compromised and can no longer withstand the pressures of carrying power lines. Worse still is the fact that the transmission lines are limited in scope and coverage of the country. Currently, the transmission lines are unable to successfully evacuate all the power generated to the point of need as total system collapse is a regular reoccurrence during peak generation period, thus increasing the level of trapped power generated.

What this means is that, even when there is additional capacity to increase power generation, this is constrained by limited transmission capacity that makes evacuating additional power load impossible. One reason for the weak transmission lines is their limited capacity, being originally designed to carry between 3,000 and 3,500 MW. Another factor is lack of regular maintenance of the transmission lines. Security challenges in some parts of the country, and issues associated with obtaining right of way for construction of new transmission lines, constrain expansion.

20.2.2 Inadequate Funding

Funding is one of the most acute challenges facing the Nigerian power sector. Yet, substantial investments are required in the entire electricity supply chain. A rationale for the power sector reform is the need to reduce government funding of the sector through privatization that would leave financing of the sector mainly in the hands of the private sector. While up to 70 per cent of the loans and equity funds that paid for the assets came from the local financial market, the debt obligation on the firms and the banking system is huge. Worsening the situation are the large collection losses being declared by these firms, amounting to a substantial financial haemorrhage. To a great extent, the generation companies are responsible for the loss because they have not done sufficiently well to invest in strengthening the metering system that would discourage power theft and improve the collection rate. Alleged excessive estimated bills that do not reflect power availability and consumption have also discouraged payments by consumers. The poor balance sheets of the banking system resulting from exposure to power sector loans appear to have weakened them and discouraged further extending additional funding required for the reform.

To tackle the financing challenge, the government has taken the lead in proffering solutions through convening investor conferences, investment roadshows, and provision of special funding to stakeholders in the electricity value chain. For instance, on 2 August 2014, the government made available approximately US$1 billion through the Nigerian Electricity Market Stabilization Fund. The purpose of this facility was to make funds available to power generation companies to settle legacy gas debts, address revenue shortfall during the Interim Rule period, and clear debts incurred through infrastructure upgrade, notably metering. As of October 2015, about 30 per cent of this facility has been approved and disbursed. However, this support tilts in favour of the distribution companies to the disadvantage of the generation companies. Earlier, NG₦300 billion was made available through the Power and Airline Intervention Fund. These interventions are temporary stopgap measures for addressing the funding challenge, necessitating need for a more permanent solution that assures sustainable private sector-led financing.

20.2.3 Policy Challenge

The guiding policy for the power sector reform are very numerous and appear uncoordinated, as many of them are not in sync nor do they ‘talk’ to each other (see Ogunleye 2016). They are also characterized by obvious overlaps and conflicts. This is discernible, for example, in the operations of NAPTIN, Standards Organization of Nigeria and National Environmental Standards and Regulations Enforcement Agency in the monitoring of and enforcing compliance with the technical and operating standards in the power sector. Ambiguity is also rampant in some of the policy documents, with generic, broad, and unclear objectives lacking specificity.

20.2.4 Inadequate Gas Supply

Gas as a feedstock for power generation accounts for about 50 per cent of the total cost of producing power in Nigeria. This is significant considering that 80 per cent of total electricity generation in Nigeria comes from gas (Figure 20.1). Around 4.9 billion standard cubic feet (scf) of gas per annum is covered by the agreements between gas suppliers and electric companies, with just about a third of this being delivered. Yet, the country's installed power generation capacity would require gas supply of up to 7 billion scf per annum to adequately power them. Adequate gas supply to power generation is central to achieving the country's objective of electricity availability, affordability, and accessibility. Indeed, distribution companies have named insufficient gas supply as a major challenge constraining their finances because it limits the power supplied to them by the generation companies.

Nigeria ranks among the top ten countries with highest proven gas reserves in the world (Figure 20.2), but rather than put this important resource to good use, a great percentage is wasted through flaring, leading to very low utilization rate (Figure 20.3). Incoherent gas-to-power policy, low profitability of domestic gas supply vis-à-vis exports, and poor gas-to-power infrastructure are all

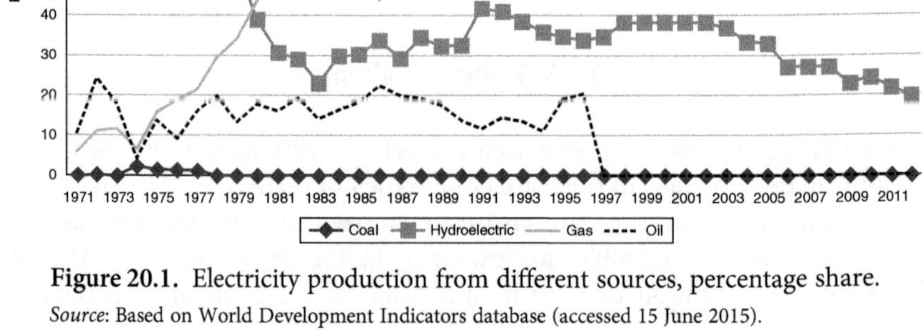

Figure 20.1. Electricity production from different sources, percentage share.
Source: Based on World Development Indicators database (accessed 15 June 2015).

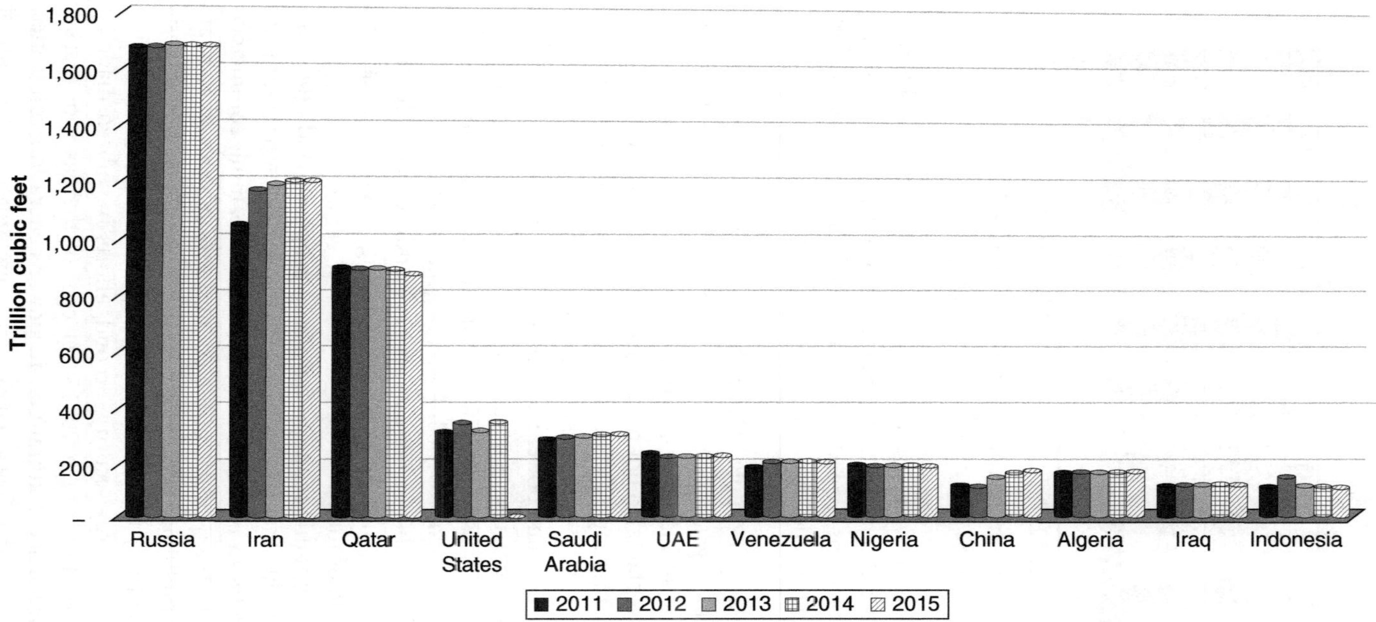

Figure 20.2. Natural gas proven reserves in top ten countries, 2011–15.

Source: Based on IEA database (accessed 22 July 2015).

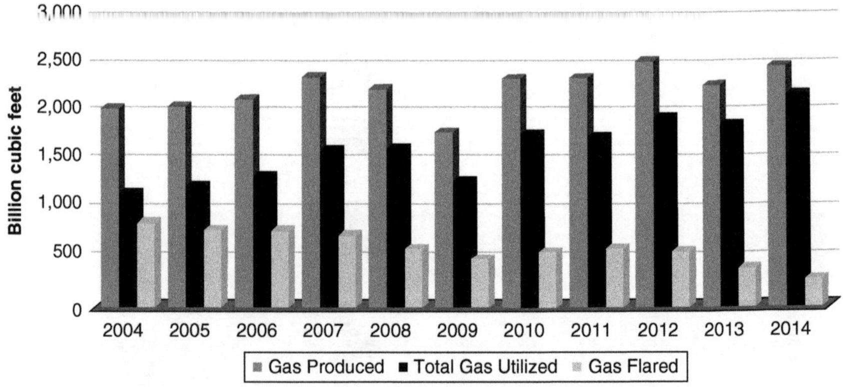

Figure 20.3. Gas production and utilization in Nigeria, 2004–14.
Source: Based on data from PTFP (2015).

challenges affecting its effective utilization to generate power. The problem of gas to power can be broadly classified as gas pipelines vandalism and poor gas project facilities management (Amadi 2015). This boils down to availability, affordability, deliverability, and commerciality of gas supply.

There is a significant gas supply mismatch with some completed power generation plants lacking gas supply while uncompleted projects have gas supply infrastructure available to them. For example, at the time of writing, Geregu and Omotosho plants do not have gas supply infrastructure. Some of the key challenges constraining effective gas supply are political considerations around locating power stations, gas pipeline vandalism, long distances between sources of gas supply and power stations, and weak gas-to-power policy resulting in a misalignment between gas and power policies. The strategic importance of gas to the success of the power sector reform is demonstrated by the relative success of Transcorp Ughelli Power, a generation company in the Niger Delta area that was able to increase its generation capacity from 160 MW to 453 MW within six months of operation owing to its power station's proximity to gas supply locations.

Some political and economic solutions have been proposed, and others are still being sought, to tackle the gas challenge. One is the development of the Gas Master Plan in 2008 to provide a guide for mainstreaming commercial exploitation and management of gas in the country through three key strategies: stimulating the multiplier effect of gas in the domestic economy; positioning Nigeria competitively in high-value export markets; and guaranteeing the country's long-term energy security. Second, efforts have been made by the government to allocate between 10–15 per cent of the gas spot market supplied by Nigeria Liquefied Natural Gas to power generation companies. Third is establishment of the Gas Aggregation Company with the aim of providing a strategic framework for implementing the gas master plan with a view to achieving a wholly competitive and market-driven domestic gas sector. Fourth is removal of an earlier government price cap on domestic gas supply. In August 2014, the government revised the gas-to-power price upward to US $2.50 from US$2/mcf (per thousand cubic feet) and added US$0.80/mcf to cover transportation costs for new capacity. This and future guided policy on pricing are expected to bring the domestic price of gas closer to its international market price, thus encouraging domestic supply instead of exports. Fifth is setting up a special purpose financing facility to settle legacy gas supply debts. Lastly, the power sector regulator, NERC, has made access to gas a compulsory condition for licensing independent power projects.

20.3 ASSESSMENT OF THE POWER SECTOR REFORM

Assessment of the reform is undertaken by analysing the policy documents relevant to the reform vis-à-vis the projected and actual outcomes of the reform. Some of the documents analysed are: the Electric Power Sector Reform Act 2005; the Roadmap for Power Sector Reform; and progress reports by specialized agencies that include NERC, the Presidential Task Force on Power, the Ministry of Power, the National Rural Electrification Agency, Independent Power Plants, the Energy Commission of Nigeria, and so on. Since the study focuses on a contemporary issue, a large number of the information is sourced from real-life observations, speeches made by stakeholders involved in the reform, and presentations in different forums.

20.3.1 Process Infractions in the Management of the Privatization Process

From the outset, several allegations of infractions have surfaced in the power sector reform process. At the point of awarding the management contract of the Transmission Company of Nigeria to Manitoba Hydro International, both the then serving Minister of Power and the Director-General of the agency responsible for the privatization process—the Bureau of Public Enterprises (BPE)—were relieved of their duties as a result of alleged infractions in the bidding process. The minister was found to have a conflict of interest as he had stakes in some of the firms bidding for some of the PHCN successor companies. Geometric Power, in which the minister had a stake, had joined Eastern Electric Nigeria Ltd and O & M Solutions of Pakistan in bidding for the Enugu Distribution Company and Afam Generation Company, respectively. Yet the minister maintained that he had no conflict of interest because he had informed the National Council on Privatization chaired by the Vice President. He asserted that he had transferred his shares to a blind trust and did not participate in the day-to-day operations of the company.

Also, the Director-General of the BPE at the time was believed to have been an accomplice in the non-disclosure of the conflict of interest of the Minister in bidding for the privatized power entities since the BPE was the clearing-house for the bids. In addition, the Senate ad hoc committee on privatization and commercialization had called for removal of the Director-General on the grounds of alleged gross incompetence in the management of the privatization process, and alleged illegal and fraudulent sale of the federal government's 5 per cent residual shares in the Eleme Petrochemicals Company Ltd. In probing the privatization of government enterprises between 1999 and 2011, the committee said that the Director-General of BPE was also implicated in fraudulent disposal of the Aluminium Smelter Company to a Russian firm, Rusal, culminating in the Supreme Court invalidating the sale.

Granted, efforts were made to ensure a transparent and accountable reform process by sticking to the privatization guidelines throughout the privatization process. However, allegations abound of obvious obscurity, opaqueness, and favouritism in the selling process. One key instance of such was the sale of the Kaduna Distribution Company. The preferred bidder, Northwest Power Ltd (NPL), was unilaterally granted deadline extension for the completion of the Share Purchase Agreement rather than reverting to the reserved bidder as provided for in the privatization guidelines. This was an obvious act of favouritism because Section 15 (138) of the Request for Proposals for the Privatization of PHCN Successor Companies stipulated that failure of the preferred bidder to meet its payment obligation before the stipulated deadline would require that the 25 per cent down payment be forfeited while the right of purchase would be given to the reserve bidder. Specifically, Section 15 (140) stated that 'failure to complete the transaction within a mutually agreed timeframe will result in the forfeiture of the down payment as per the terms of the Share Sale Agreement'.

The position was that NPL paid 24 per cent of the total bid on 23 December 2013 and had until 23 June 2014 to settle the 75 per cent balance. When it became obvious that NPL was unable to meet the deadline, the BPE acted against the privatization rules by extending the payment date to 6 August 2014. NPL was unable to meet the 6 August deadline and therefore had to request a further extension to 6 October 2014. Again, this deadline was missed and further extended to 4 December 2014. Yet the BPE finally sold the entity to NPL, despite a court order secured by the reserve bidder to stop the sale. The claim by the BPE was that the National Council on Privatization (NCP) alone had the final say in the sale and had decided to use its discretion in the matter. In fact, in a letter to the Vice-President, who is the chairman of NCP, the BPE warned against handing over the asset to the reserve bidder, claiming that there was no guarantee that the reserve bidder would fare any better than the preferred bidder, and that doing so would set the process back by about 12 months.

20.3.2 Political Interference

This is the most visible and challenging political economy issue facing the power sector reform. In contracting the management of TCN to MHI for US $23 million for three years starting July 2012, one of the key reasons proffered was the need to reduce the bureaucracy associated with public management of the enterprise with a view to improving efficiency of operations devoid of government interference. The firm was mandated to run the entire TCN system and turn it into a technically and financially efficient, sustainable, and stable company within the period of the contract. In addition, MHI was expected to ensure evolution of a TCN that would eventually become a private commercial company by reorganizing the institution so that the Transmission System Provider became a separate entity from the Market Operator and System Operator. In contrast to this thinking, however, management of the TCN appears to be fraught with serious political interference that tends to limit effective performance of the institution and dampen its ability to achieve the stated mandate.

At its inception, the process for awarding MHI the management of TCN was mired in controversies, leading to internal wrangling that bred operational lapses. For example, the chairman of the supervisory board of the institution has changed many times due to dissatisfaction and protests over political interference and alleged 'over-hyped' powers accorded the MHI. Some of the alleged powers include the handover of the day-to-day running of the TCN and operations of its bank account to MHI, as well as allowing it to appoint a Chief Executive Officer) without recourse to the supervisory board. One easy way of interference was in the expectation that MHI would contribute to developing local capacity that would eventually take over management of TCN through local counterparts that are expected to understudy the MHI expatriates. Incessant change in the leadership of the MHI itself that is attributed to the internal wrangling has also been obvious within the short period of the contract.

There were also allegations that the provision for Nigerian counterparts to understudy the MHI expatriates was exploited by the political class to compensate loyalists and political jobbers. One obvious outcome of political interference in TCN activities is that it has made the institution the weakest link in the electricity reform value chain, with its non-performance rubbing off negatively on existing independent power producers, generation companies, distribution companies, and ultimately on attracting much needed investment to the power sector.

Towards the end of the MHI first contract in 2015, the transmission segment of the Nigerian power sector was enmeshed in uncertainty over who was in charge. At the twilight of President Jonathan's administration, the Minister of Power issued several last-minute directives that sometimes appeared contradictory. For example, there was an earlier directive that all actions with respect to the proposed unbundling of TCN be held in abeyance pending the decision of the incoming administration of President Buhari. However, ten days later there was another directive appointing a full management team to a supposedly unbundled TCN (now a System Operator and a Transmission System Provider). Meanwhile, these new entities were yet to be incorporated as legal entities because their operating licences were yet to be issued by NERC and TCN had not been wound up.

Also, proposing a change while the management contract of MHI still subsists and might possibly be extended—it was ultimately extended by one year to July 2016—would only result in confusion and crises in the management of the entity, impinging on its overall performance. It is imperative to ask, however, what the basis was for the contract renewal. There are indications that there was neither careful nor clear assessment and evaluation of the contract terms of MHI vis-à-vis its performance on the agreed milestones that would inform the need to extend the contract or to act otherwise. It appears that the need to avoid a void in this segment of the power sector was the sole factor behind the renewal. About two months to the end of the contract extension, it remains doubtful whether there is any serious plan in place to strengthen the TCN management and it is difficult to ascertain what the future holds for this segment of the power sector. What appears obvious, though, is that a concessioning process has not been activated. Yet, the government would have to decide what to do with TCN. In September 2015, the Federal Government consulted the development partners on what the future of TCN should be. The consensus was that it should be concessioned. But the question remains whether it should be consessioned as an entity or in an unbundled form. The fact that the process of the concession has not commenced is a pointer to government's dilemma on what to do with the TCN.

20.3.3 Corruption

Corruption is one of the key reasons for the collapse of the Nigerian electricity sector and was, therefore, the main rationale adduced for its reform. Yet the reform has been embroiled in corruption allegations from the beginning. First, there were claims that the US$16 billion invested in the National Integrated Power Project (NIPP) under the President Obasanjo administration was largely mismanaged. In fact, the two-year gap in funding NIPP projects under President Yar'adua was due to the intensive legal, political, and financial scrutiny these projects were subjected to with a view to correcting the alleged corruption. Several senior officials of the Rural Electrification Agency and key members of the House of Representatives Committee on Power were arraigned by the Economic and Financial Crimes Commission for alleged embezzlement of NG₦6 billion funds belonging to the Rural Electrification Agency. For a long time during the privatization process that involved reaching a settlement with the erstwhile staff of PHCN, there were allegations that the workers' pension fund of around NG₦88 billion, accruing from the 7.5 per cent deductions from their salaries, was misappropriated.

The current eighth National Assembly has set up a 13-man Senate Committee to probe the power sector reform, based on alleged corruption especially involving unbundling the PHCN. Also being probed is MHI's management of the TCN. It is estimated that approximately US$29.635 billion has been spent on the power sector in the past 16 years with very limited results.[2] While the outcome of this probe is expected, one hopes it does not end up like previous attempted probes of the same issue in 2007 by the House of Representatives and Senate that degenerated into claims and counter-claims of witch-hunting. In the end, in 2010, the chairmen of both committees were accused by the Economic and Financial Crimes Commission of involvement in a NG₦5.2 billion contract scam related to the Rural Electrification Project.

20.3.4 Weak Regulatory, Institutional, and Legal Framework

There was an inherent structural weakness in the institutional framework for the power sector reform. These institutional weaknesses emanate from obvious gaps, overlaps, confusion, and conflicts in the mandates and interactive relationships among these institutions as provided for in their enabling laws. In many cases, the issue of what institution is responsible for playing coordination and leadership role in the power sector is obscure. For example, NAPTIN—a power sector capacity-building institution—is also mandated to oversee effective monitoring of and compliance with the technical and operating standards. Yet, this mandate is also given to the Standards Organization of Nigeria and the National Environmental Standards and Regulations Enforcement Agency, thus revealing a conflict of mandate among these institutions.

The most recent development with respect to possible conflicts in the power sector involves establishment of the Nigerian Electricity Management Services Authority at the twilight of President Goodluck Jonathan's administration. The agency is mandated to carry out inspections of electrical materials throughout the entire electricity supply industry, and to enforce all statutory

[2] It is estimated that Presidents Obasanjo, Yar'adua, and Jonathan spent US$16 billion, US$5.375 billion, and US$8.26 billion, respectively.

technical electrical standards, testing and certification, and regulation of all electrical installations and instruments. In addition, the agency is given the power to ensure that all major electrical materials and equipment used in Nigeria are of the right quality and standards, as well as to specify technical standards for electrical plants, electric lines, and connectivity.

This mandate obviously conflicts with that of the Standards Organisation of Nigeria (SON) that already has the same mandate. Moreover, it also distorts the hitherto regulatory structure and framework of the power sector where NERC was solely responsible for both the economic and technical regulation of the sector. The mandate of the new agency—to promote measures for advancing the skills of those working in the power sector and to issue competency certificates and licences to qualified electrical personnel and contractors—also conflicts with that of NAPTIN. Creating additional regulatory agencies is inconsistent with the philosophy of privatization and negates the provision in section 32 of the EPSRA.

20.3.5 Policy Inconsistencies and Regulatory Uncertainty

The regulatory authorities have manifested policy inconsistencies in the implementation of the power sector reform. One way this has become clear is in the realization of the cost-reflective tariff. For instance, NERC had worked to achieve a viable, stable, and rules-based electricity market where cost-reflective tariff would be a key pillar. This would herald the movement from the Pre-Transition Stage Market to the Transition Stage Market. However, rather than proceed to implement the various contracts, rules, regulations, systems, and procedures involving a move towards a cost-reflective tariff with the advent of the Transition Stage Market on 1 February 2015, NERC rather weakened this pillar by reducing tariffs across board by 24 March 2015—just 11 weeks after introducing those same tariffs. The loss of credibility and reputational damage caused by this policy inconsistency sent a wrong signal to investors and was a big blow to the reform.

20.3.6 Electricity Tariff

Electricity tariff is of concern because pricing should achieve delicate balance between the interests of investors and of consumers. Indeed, for this type of reform to be politically and socially acceptable, there is a need to ensure that the poorer members of society are not made worse off. However, this has been a major issue in recent times. For a long time, the absence of a cost-reflective electricity tariff has been cited as one of the main deterrents to local and foreign investor participation. To correct this anomaly, the power sector reform target is to introduce a cost-reflective tariff through the MYTO. In this framework, consumers are categorized into residential, commercial, industrial, special, and street lighting. The tariff structure is designed to achieve four simultaneous objectives: to ensure fairness to consumers; to allow licensees to recover all reasonable costs within a reasonable period of time; to provide sufficient funding to finance activities of investors; and to allow for reasonable earnings to achieve efficient operations. The review aims to reflect the current market conditions, with cost of supplying electricity to the wide range of consumers, inflation rates, exchange rates, gas supply price, generation capacity, capital expenditure, operations expenditure, and cost recovery and reasonable return on investment for investors as guiding posts.

However, challenges exist with implementation of the cost-reflective tariff. For instance, just one year after the initial roll-out, NERC found two major deficiencies in the new price structure. First, it was realized in 2009 that there had been a substantial shift in the two key variables used to determine the tariff, namely, gas prices and exchange rates. What this implies was that with these changes, the supposedly cost-reflective tariff remained still unattractive to draw investors. Second, the tariff only made provisions for gas as a source of power generation, leaving out other sources such as wind, coal, and solar power. Yet some investors were willing to explore these other sources in their power generation investment bid. This necessitated a review of the tariff by NERC earlier than scheduled. The newest review took effect on 1 February 2016, leading to around 45 per cent hike in tariff. Questions and protests continue to trail this increase that is believed by many to be unfair and unjust, especially since it has not been accompanied by commensurate improved service. One interesting feature in the tariff review is the consideration of fairness and public interest as key factors. The fixed charge that was an intrinsic part of the earlier tariff system has now been removed for vulnerable customers and those engaged in informal activities.

20.4 WAY FORWARD

Gas is the cheapest way to generate electricity per kilowatt unit of power and Nigeria has this resource in abundance. Formulating an effective gas policy that allows gas to play a pivotal role in the power sector reform is imperative to making meaningful progress. One way this could be done is by encouraging public–private partnership and completely private sector-driven gas development initiatives through sufficiently robust incentives. This will further reduce the pressure on government, given its limited financial resources, to undertake this task alone. Another important area would be to encourage private sector investment in gas infrastructure that would help evacuate gas and deliver it to the power stations. Stepping up surveillance on the pipeline infrastructure with a view to increasing the probability of apprehending vandals and ensuring timely punishment that would serve as deterrent to vandalism would contribute to ensuring the security of gas-to-power infrastructure.

Capable, relevant, and sufficient manpower development for the entire power sector value chain is absolutely necessary to push forward the reform agenda and manage post-privatization challenges. In addition to the required system, electrical, hydro, and mechanical engineers and other technical professionals are required to drive the reform and post-reform process. Unlocking the sector through the reform has opened up the sector to a spectrum of power sector value chain activities that necessitate increased human and technical capacity. The reform has created opportunities in the production and maintenance of transformers, transmission lines, meters, cables of acceptable quality, and other facilities in the entire value chain of generation, transmission, and distribution. The government should ensure that NAPTIN plays its role in this respect by providing it with the needed resource support. Private sector players should complement the government's efforts through NAPTIN by undertaking training and capacity development initiatives.

There is a need to push for production, distribution, and dynamic efficiency in the power sector. This can be achieved through a well researched, highly insightful, and incisive evaluation, assessment, and appraisal of the status quo. Probing the power sector reform through politically motivated committees may not achieve this objective. Rather, a holistic technical and economic assessment that looks at the entire value chain is what is required. For instance, it is not clear if any study has examined the equilibrium structure of the entire spectrum of the Nigerian electricity industry with a view to determining the demand, supply, and the existing demand–supply gap in the market. More efforts by both public and private stakeholders should be directed towards research and development in alternative sources of renewable and clean energy. Most studies and investigations have focused on supply-side issues, paying little attention to demand-side solutions. There is need to balance the current focus on supply solutions with demand management interventions that will reduce the pressure on electricity demand.

Political will and policy consistency are necessary to improve on the reform and attract more local and foreign investors. One key area such political will can be demonstrated is in providing leadership by example in the management of the transmission segment of the power sector value chain. This segment is under the control of the government and yet remains the weakest link in the electricity value chain. Rather than engaging in wanton interference and using it for political gain, the focus should be on providing the needed oversight of the management consultant while at the same time giving it the freedom to carry out its assigned responsibilities as contained in the contract signed. The government should ensure sustained investments in the sector.

Concessioning TCN to private operators should be undertaken as a matter of urgency by commissioning a reputable Transaction Adviser to work with the Infrastructure Concessioning and Regulatory Commission to deliver on this. This will address the parallel management of the TCN. Proactive and determined efforts must be geared towards investing in network expansion and completing ongoing transmission projects.

The government should focus on continuously evolving an effective institutional framework that guarantees sustainability of the reform efforts. High priority needs to be given to the regulatory, institutional, and human capacity framework for managing post-privatization challenges. The plethora of policy pronouncements, documents, committees, commissions, and other frameworks for the power sector reform should be reviewed and properly aligned to suit and reflect current realities. Continued development of the power sector through laws and regulations, and enforcement of compliance with these is also necessary. There is also a need for increased investment in clean and renewable power generation to help the country achieve an optimal mix of energy sources, with the emphasis on hydro and solar power sources and ultimately improve the country's green growth.

Evolving a cost-reflective tariff would result in a win-win situation that would help attract the necessary finance from existing and prospective investors while simultaneously making consumers fell contented that they are getting value for money. Unnecessary political interference in energy pricing can at best be counterproductive, as demonstrated by the tariff reduction of 2015 shortly before the presidential elections. NERC's action in reviewing the assumptions for calculating the MYTO 2.1 is in order and should be pursued to a logical end. Consideration for fairness and the public interest in reviewing the tariff is commendable. However, possible loopholes and slippery slopes should be avoided in relation to the issue of aggregate technical, commercial, and collection losses. Distribution Companies (DISCOs) should be compelled to respect their metering obligations to the consumers.

It is necessary to develop special-purpose financial products and vehicles that would help channel funds from surplus sectors to the power sector. The Nigerian pension fund and insurance fund are currently estimated to stand at around US$30 billion and US$2 billion, respectively. These funds are currently under-utilized and will be of limited benefit to the economy if they are not deployed to sectors of real need such as the power sector. Much needs to be done to amend investment rules so that a large part of these resources could be channelled to power generation, especially clean energy and long-term infrastructure projects. Putting the fund into government bonds that are considered to be safe havens would yield limited benefits. The Sovereign Wealth Fund, especially the Infrastructure Fund component, should be effectively utilized in deploying innovative power generation, especially clean and renewable energy.

Tackling the current spate of insecurity would also be of obvious help in moving forward on the power sector reform. The north-eastern part of the country is riddled with insecurity occasioned by an insurgency that makes investment in power generation, transmission, and distribution in the region problematic. Moreover, Niger Delta, the source of the gas needed to fire the power plants, also has its fair share of instability. More recently, there have been sporadic nationwide attacks by Fulani herdsmen. These pose a threat to sustained success of the reform.

Global goodwill and commitment to improve the power sector should be leveraged. In 2013, the US government announced the private sector-focused Power Africa initiative with a commitment to add over 30,000 MW of more efficient electricity generation capacity in six African countries, including Nigeria. The support offered by this initiative is gaining momentum though it has not yet significantly affected choice of power generation technology. In 2014, for example, Power Africa, through USAID's Development Credit Authority, made available US$90 million for on-lending to the newly privatized distribution and generation companies. It also made available US$50 million for a portfolio of solar-battery hybrid power systems distribution in Nigeria. Power Africa is supporting NBET to negotiate and finalize a model Power Purchasing Agreement that ensures effective connection of some wind and solar power projects that are capable of delivering 600 MW to the national grid.

More global and continental development finance, foundations, and concessional private sector financing institutions and South–South partners have indicated interest and even made commitments to help improve the power situation in Nigeria. For instance, during the 2014 US–Africa Leaders Summit, the World Bank announced a US$5 billion commitment in support of the Power Africa initiative. Recently, the African Development Bank has made 'Light and Power Africa' one of its 'High 5s', with a commitment to invest US$12 billion in energy projects between 2015 and 2020. From 2015 until 2020, the European Union (EU) is budgeting for around €2.5 billion in grants and €20 billion investment in sustainable energy provision in sub-Saharan Africa. Additionally, individual EU member countries are expected to provide financial support that will more than double this amount, with France committing US$6.4 billion over the next four years to electrification in Africa. The US and EU have already committed to providing support through synergy by signing a memorandum of understanding in this respect. South–South partners, notably China, are also making inroads into pepping up activities in the power sector. This enthusiasm for investment, and technical support and assistance, should be effectively leveraged to achieve the desired goals of the power sector reform.

REFERENCES

Adenikinju, A. F. (2003). 'Electric Infrastructure Failures in Nigeria: A Survey-based Analysis of the Costs and Adjustment Responses'. *Energy Policy*, 31(14): 1519–30.

Adoghe, A. U., A. Odighe, and S. O. Igbinovia (2009). 'Power Sector Reforms: Effects on Electric Power Supply Reliability and Stability in Nigeria'. *International Journal of Electrical and Power Engineering*, 3(1): 36–42.

Aliyu, A., A. Ramli, and M. Saleh (2013). 'Nigeria Electricity Crisis: Power Generation Capacity Expansion and Environmental Ramifications'. *Energy*, 61(8): 354–67.

Amadi, S. (2015). 'From Megawatts to Actual Delivery: Bottlenecks and Solutions'. Lecture Delivered at the NAPE (Nigerian Association of Petroleum Explorationists) Specialized Change Workshop, Lagos, 14 July.

David-West, A. (2014). 'Nigerian Power Sector: Value Investment Opportunity or Value Trap?' CSL Stockbrokers Power Sector Infrastructure Review, UK.

Idris, A., S. A. Kura, M. A. Ahmed, and Y. Abba (2013). 'An Assessment of the Power Sector Reform in Nigeria'. *International Journal of Advancements in Research and Technology*, 2(2): 1–37.

Joseph, I. O. (2014). 'Issues and Challenges in the Privatized Power Sector in Nigeria'. *Journal of Sustainable Development Studies*, 6(1): 161–74.

Ochugudu, A. I. and V. A. Onodugo (2013). 'Power Sector Reform Deliverables: How Well and How Good to Customers?' *International Journal of Management Technology*, 1(1): 1–14.

Ogunleye, E. K. (2016). 'Political Economy of Nigerian Power Sector Reform'. WIDER Working Paper 2016/9.

Oke, C. A. (2008). 'Resuscitating and Sustaining the Nigerian Power Sector'. Paper presented to the Nigerian Association for Energy Economics (NAEE) at the NNPC Towers, 14 August.

Okoro, O. I. and E. Chikuni (2007). 'Power Sector Reforms in Nigeria: Opportunities and Challenges'. *Journal of Energy in Southern Africa*, 18(3): 52–7.

Onochie, U. P., H. O. Egware, and T. O. Eyakwanor (2015). 'The Nigeria Electric Power Sector (Opportunities and Challenges)'. *Journal of Multidisciplinary Engineering Science and Technology*, 2(4): 494–502.

Sanyaolu, H. A. (2008). 'Electricity Power Sector Reform in Nigeria: Utilising Restructuring and Regulatory Reform as a Means of Achieving a More Efficient and Competitive Sector'. An LLM Dissertation, CEPMLP, University of Dundee.

Tallapragada, P. V. S. N. and B. S. Adebusuyi (2008). 'Nigeria's Power Sector: Opportunities and Challenges'. In P. Collier, C. Pattillo, and C. C. Soludo (eds), *Economic Policy Options for a Prosperous Nigeria*. New York: Palgrave Macmillan, pp. 301–27.

21

Climate Change Policy and Power Sector Reform in Mexico under the Golden Age of Gas

José María Valenzuela and Isabel Studer

21.1 INTRODUCTION: CLIMATE CHANGE, ENERGY TRANSITION, AND ENERGY REFORM

In the last quarter century, Mexico has undergone a series of major institutional and technological changes in the power sector. Since the early 1990s, and as in other restructured markets, Mexico's power sector has gradually evolved into an open-market regime where natural gas has gained a dominant market share (Carreón-Rodríguez, Jiménez, and Rosellón 2006; Fuentes and Bowler 2014). During the same period, climate change emerged as a global priority. However, it was not until the late 2000s that Mexico developed economy-wide mitigation programmes and policies.

Although not having mitigation obligations under the Kyoto Protocol and being responsible for <2 per cent of global carbon emissions, Mexico had assumed voluntary commitments to reduce greenhouse gas (GHG) emissions for 2020 and 2050 even before the 21st Conference of the Parties to the United Nations Framework Convention on Climate Change (COP21—UNFCCC) in Paris. Besides hosting COP16 in Cancún and helping restore the trust in the multilateral climate negotiations, Mexico was the first non-Annex I country to have established a comprehensive climate change law—a far-reaching institutional framework that includes an array of planning instruments—and submitted five national communications under the convention (Fekete, Mersmann, and Vieweg 2013: 78). It was also the first developing country to present its Intended Nationally Determined Contribution (INDC) (Gobierno de la República 2015) with conditional and unconditional pledges (CAT 2015).

Despite its leadership on climate issues, Mexico's ambition to promote clean energy is less certain. Veysey et al. (2015) concluded that achieving Mexico's long-term climate change mitigation goals would require the power sector to become virtually zero emissions by 2050. Yet, observers have noted the lack of certainty about the role that renewable energy will play in the short- and long-term development of the power sector. The 21st Century Power Partnership, under the Clean Energy Ministerial (2015), depicted Mexico's type of process as 'restructuring', committing to deep transformation of the power sector structure, but with a limited or average approach regarding the ambition for clean power transition. Power sector reform is still centred on a large-scale, centralized, fossil-fuel-based generation model (Zinaman et al. 2015).

Mexico's decreasing power sector emission intensity, from 0.608 to 0.499 tCO_2e/MWh between 2003 and 2013 (SEMARNAT et al. n.d.), contributed to a complacency narrative. Emission reductions are still primarily driven by a hustle to power Mexico with natural gas, and marginally by an expansion of wind energy based on ad hoc regulatory benefits rather than a clear, long-term policy for a robust and diversified portfolio of electricity sources.

The energy reform created a series of mechanisms that could be the basis for potential long-term growth of low-carbon technologies. They include an independent system operator, the binding character of clean energy obligations and certificates, and long-term auctions for clean energy. The first of these mechanisms is a key component to the wholesale electricity market, and the other two are fundamental policies to contain the costs associated with increasing clean energy shares. However, the central policy objective of the current administration in Mexico is to lower power production costs, mainly grasping the benefits of low-cost gas in North America. This policy objective discouraged the administration from embracing ambitious climate change mitigation commitments that might limit cost reductions in the short term and thus hurt the main support base of the energy reform coalition—a broad spectrum of industries and businesses eager to access competitive tariffs.

This chapter presents a decade-long outlook on the relation between climate and energy policy in Mexico, putting into perspective the challenges of energy transition policies when fossil fuel prices remain low. This circumstance could be a regular feature in North American energy markets in the near future, if carbon prices are not adopted (IEA 2011). We conclude that a climate policy devised on the basis of the economic co-benefits (including energy security) could be easily undermined if these co-benefits become less clear under a power market driven by low fuel prices, even for a country otherwise considered an international leader in the subject.

21.2 ENERGY TRANSITION AND CLIMATE CHANGE AGENDA AS A POLITICAL NECESSITY (2008–12)

21.2.1 Race between Demand and Energy Reforms

In the last 30 years, Mexico's energy consumption has grown steadily, at a rate of about 2 per cent annually. Fossil fuels represent >92 per cent of the total national energy consumption. Between 2000 and 2010, electricity generation increased by 33 per cent, the population increased by ~66 per cent, and the total energy consumption more than doubled whereas electricity consumption almost tripled. Per capita electricity consumption reached 2 MWh per capita, from 0.9 MWh per capita, as described in detail in Table 21.1. Thirty years ago production almost doubled consumption, a circumstance contributing to a narrative on Mexico's energy abundance position. But by 2013 this position has entirely faded away.

Although the growth in demand comes as no surprise, the pace of Mexico's energy imports merits consideration. Through 2011–13, Mexico faced problems with the reliability of and access to natural gas because of the Mexican oil company Petróleos Mexicanos' (PEMEX) limited ability to increase gas production and expand the natural gas pipeline network (Auditoría Superior de la Federación 2014). At the same time, the fuel mix for power generation shifted away from heavy fuel oil. The share of natural gas in the total electricity output increased from 12 per cent in 1990 to over 50 per cent in 2010.

21.2.2 Climate Change in the Context of Energy Politics

In 1992, the Mexican Congress approved a sweeping energy reform that defined new areas for the participation of private investment in power generation in two

Table 21.1. Key indicators of energy production and demand

	1983	2003	2013
National energy consumption (petajoules)	4383.8	7216.9	9011.8
Population (millions)	71.8	104.7	118.4
Energy consumption per capita (gigajoules/person)	61.1	68.9	76.1
Electricity consumption (gigawatt-hour)	61,183.3	176,718.5	235,158.6
Electricity consumption per capita (kilowatt-hour/person)	852.2	1687.5	1986.2
Energy production (petajoules)	8260.0	10,289.4	9020.2
Relation between production and demand	1.9	1.4	1.0

Source: Based on SENER (2015c).

forms: (i) independent power producers (IPP) with Federal Electricity Commission (Comisión Federal de Electricidad, CFE) as a single buyer; and (ii) generation for self-supply (Hernández 2006). However, from that point on, the political context in Mexico was quite adverse, frustrating efforts to pass a comprehensive energy reform. In 1999, 2002, and 2008, the presidency submitted three energy reform proposals to the Mexican Congress. The first two failed, while the third, proposed by President Felipe Calderón (2006–12), was approved, but only because it was a fairly limited.

The 2008 reform is noteworthy because it was accompanied by two distinctive laws: (i) the Law for Sustainable Use of Energy (Ley para el Aprovechamiento Sustentable de la Energía, LASE) and (ii) the Law for the Use of Renewable Energy and Financing of Energy Transition (Ley para el Aprovechamiento de Energías Renovables y el Financiamiento de la Transición Energética, LAERFTE). The laws shaped a new institution for the promotion of energy efficiency, created a new financial vehicle for the deployment of energy transition initiatives independent from the state-owned enterprises, and firmly provided legal support to many regulatory decisions that could have been contested by future governments from opposing parties, less favourable to private investment in renewable energy.

In an adverse political context and in the face of rapidly growing energy demand and limited energy resources, renewable energy and energy efficiency emerged as key elements of the Calderón administration's strategy to achieve Mexico's energy security and economic development goals. Boosting renewable energy, particularly in the electricity sector where private investment was partially allowed, became a priority goal in the Sectorial Energy Programme for 2007–12 (SENER 2007: 6).

It has been argued that the personal beliefs of President Calderón and a strong sense of global responsibility drove the dynamic climate agenda that unfolded during his presidency (Torres 2014). Mexico is the second largest GHG emitter in Latin America and ranks ten among the world's highest levels of GHG emissions, excluding land-use change and forestry (World Resources Institute 2015). It is estimated that 15 per cent of Mexico's territory, 68 per cent of the population, and 71 per cent of gross domestic product (GDP) 'are highly exposed to climate change risk' (OECD 2013). Embracing the fight against climate change could help improve Mexico's world image, which had been severely affected by the violence linked to Calderón's war against the drug cartels, and could potentially generate concrete benefits, such as financial resources to support the energy transition, thus explaining Mexico's strong promotion of the Green Climate Fund.[1]

[1] As of 2014, Mexico has pledged US$10 million towards the Green Climate Fund (GCF 2015).

Although all of these factors could have played a role in understanding the wide array of new institutions for climate policy-making developed during the Calderón administration, they do not explain the feature that became the backbone of his climate policy. Calderón's power sector transition policy was a child of necessity and a by-product of the political constraints that impeded his government from fully opening Mexico's energy sector to private investment as a means to achieve the twin goals of energy security and economic growth. Climate change mitigation was not a primary goal, but rather, as Fekete, Mersmann, and Vieweg have argued, 'a secondary political goal' that opened up 'new potentials to achieve other primary development goals' (2013: 2). As the former energy minister Georgina Kessel stated in 2009, diversifying primary sources and reducing the uncertainty generated by the volatility of fuel prices reinforce energy security (Kessel 2009).

The new legal mandates of 2008 reform confirmed the Energy Regulatory Commission (Comisión Reguladora de Energía, CRE) powers to push for the rapid development of self-supply contracts, through wheeling fees and energy banking regulation that allowed wind energy to surpass 2000 MW in 2014 (IRENA 2015: 16). Companies such as Grupo Bimbo, Cemex, FEMSA, and Walmart in Mexico leveraged their AAA credit ratings to become off-takers from renewable self-supply generation projects, accruing 5–20 per cent less than the applicable electricity rate (Yale School of Management 2012). Private investment was limited not only by the size of the self-supply market but also by the requirement that surplus production is sold to CFE at a discounted price, the lack of transparency on the administrative proceedings for interconnection to the grid, and CFE's unwillingness to build transmission infrastructure in remote resource-rich areas (WWF, Climate Works, and PwC 2013).

The difference in renewable energy investments from the private sector and the CFE is staggering. By 2014, the CFE installed 597 MW of wind energy, through IPP contacts, whereas self-supply development accounted for 1439 MW (SENER 2015b). This difference underscores the opportunity that renewable energy represented to large consumers willing to invest away from the taxing tariff. These electricity prices incentivized private investment in renewable energy projects for self-consumption, particularly large wind 'demonstration' projects in the southern part of Mexico (Oaxaca), which were co-financed by grants from the Global Environmental Facility, the World Bank, and the Inter-American Development Bank and promoted by the Mexican government as an opportunity to reduce their electricity bills. After all, it was possible that the high tariffs for industrial and commercial users were also based on a form of cross-subsidy.

21.2.3 Climate Change: Much Ado about Nothing

The most significant legacy of the Calderón government was the enactment of the General Climate Change Law (Ley General de Cambio Climático, LGCC) in 2012, which was seen as an avenue to provide continuity to the institutional framework, the programmes and plans, as well as the mitigation commitments adopted since 2006. The LGCC created a comprehensive institutional framework and climate change system by consolidating a number of institutions that had been previously created (e.g., Inter-Ministerial Commission on Climate Change, Council on Climate Change), as well as planning instruments (e.g., National Strategy on Climate Change, PECC) and evaluation and funding instruments, including a climate fund. Beyond actions pertaining to the federal government, the LGCC made it obligatory for sub-national governments to develop their climate change mitigation and adaptation plans.

Other policy instruments refer to the requirement for mandatory emissions reporting and the creation of a public emissions registry for all sectors, as well as the reform of subsidies by 2020. The most important features of the LGCC include the confirmation of pre-existing mitigation targets; that is, the reduction of GHG emissions below a business-as-usual (BAU) scenario by 2020 and by 50 per cent by 2050 from levels in 2000, conditional on international financial support; 35 per cent of electricity from clean energy sources by 2024; as well as the possibility of establishing an emissions market and entering into emission trading agreements with other countries.[2] Similarly, energy-intensive industrial sectors strongly opposed the mitigation targets proposed in the original text of the LGCC. Some industrial associations, especially Mexico's National Steel Chamber (CANACERO 2011), were particularly vocal in their opposition against the strong reductions proposed by the LGCC. They were concerned that ambitious targets would restrict Mexico's ability to continue growing and compromise their competitiveness in the absence of a global agreement that set similar commitments from other countries (CESPEDES 2013).

Although the LGCC received multi-party support in both houses of the Mexican Congress (280 versus 10 in the Deputy Chamber, and 78 versus 0 in the Senate), it took three years to be approved. The legislation underwent changes that made the mitigation commitments stated in the law aspirational and subject to the availability of international finance. The Law also assumed a gradual approach for the adoption of climate change mitigation actions, starting with the least-cost ones. The strong stance of industrial sector would remain a key variable on the limitations of future reforms.

[2] For a more extensive discussion see Valenzuela and Studer (2016).

In spite of the strong climate institutional development, Calderón's climate strategy failed to set a de-carbonization pathway for the Mexican economy. The renewable energy target set at 8 per cent for 2012 was not achieved: non-hydropower renewables for 2012 were only 2.9 per cent of total power capacity (IGS-EGADE 2014).

21.3 ENERGY REFORM AND CLIMATE POLICIES BY LEGAL FIAT (2012–15)

21.3.1 Low Energy Prices: Political Priority of Energy Reform

Once the political constraints faced by President Calderón for comprehensive energy reform were removed, based on the results of the 2012 election and the political manoeuvring that led to a multi-party agreement known as the 'Pact for Mexico' (Pacto por México), a comprehensive energy reform was possible under the incoming government of President Enrique Peña Nieto (2012–18). Simultaneously, the government moved away from the previous administration's strategy to link climate change policy to energy security. By contrast, the reduction of energy prices became the top priority of Peña's energy reform (Gobierno de la República 2013: 3). Climate policies were designed by legal fiat, resulting from implementing the LGCC.

As Alvarez and Valencia (2015) have argued, little over 60 per cent of the Mexican industry's energy consumption consists of gas and electricity, almost in equal share. Most industrial sectors (steel and chemical industries as well as glass, pulp, paper, beer, and plastic production) largely depend on gas for their production processes. The automotive sector, another critical manufacturing industry in the Mexican economy, is more dependent on electricity than gas. Since over 50 per cent of electricity is generated with gas, the Mexican industry has a large stake in maintaining and expanding the availability of low-cost gas.

Seven years before reform, César Hernández, the incumbent Under Minister for Electricity, wrote about subsidies serving the purpose of compensating for the inefficiency of the power industry, and not for gains to consumers: 'The problem is not the subsidy, it is a charade—the problem is the cost of production' (Hernández 2006: 32). As Pollitt (2012) has concluded from a comparative perspective, cost might fall not necessarily because of the market but rather due to other structural reasons. For Mexico, two relevant structural adjustments are evident: (i) the displacement of heavy fuel oil by gas and (ii) the reduction of technical and non-technical losses.[3] In any case, by government

[3] For an extended discussion see Valenzuela and Studer (2016).

estimates, the efficiencies from the operation of the wholesale market could represent electricity service cost reductions between 26 and 48 per cent.[4]

21.3.2 Political Effects of Natural Gas Price Cycles

In 2011, the International Energy Agency (IEA) published optimistic conclusions on the future of gas production in North America. According to IEA (2011), the characteristics of the gas market and the lack of substantial liquefied natural gas (LNG) infrastructure adequate for exports would allow for low prices for the foreseeable future. The following years proved the diagnosis correct overall, with prices rapidly falling by 2012.

The price of natural gas is especially important in restructured markets because of its role in determining the marginal cost of electricity, which influences the prices of bulk energy in the wholesale market. Based on 20 years of comparative data among US states, Borenstein and Bushnell identified that the gap between average cost (under regulated markets) and marginal price (under wholesale markets) generates a specific economic expectation: 'consumer and political sentiment has tilted toward whichever regime (regulation and markets) offered the lower prices at the time' (2015: 1). Hence, they argue that the impetus of reform is stronger when low gas prices are observed because large consumers, in particular, industrial users, observe the benefits competitors can have during those periods.

The consumer and political sentiment explained by Borenstein and Bushnell (2015) could certainly be significant in Mexico on the basis of two considerations: (i) the proximity of competitors across the US border with access to power prices from wholesale markets; and (ii) the fact that the self-supply regime in Mexico allowed for large consumers with access to reliable natural gas supply (mostly in the northeast of Mexico) to access power prices that resembled those of restructured markets, owing to on-site generation using the Henry Hub indexed natural gas.

Clean energy auctions with long-term contracts represent an 'average cost' approach to energy procurement. This model is politically castigated during periods of low gas prices and low marginal costs of generation, but is desirable when natural gas prices increase. Thus, if Borenstein and Bushnell's (2015) argument is right, we should expect opposition to the average cost approach during the period when gas prices are low, at least from large consumers. This scenario is clearly observed in the Mexican case.

[4] The government has been cautious of not releasing the numbers to a wide audience. These were presented as part of the official consultation process to publish the new market rules that will apply to the wholesale electricity market (SENER 2015c).

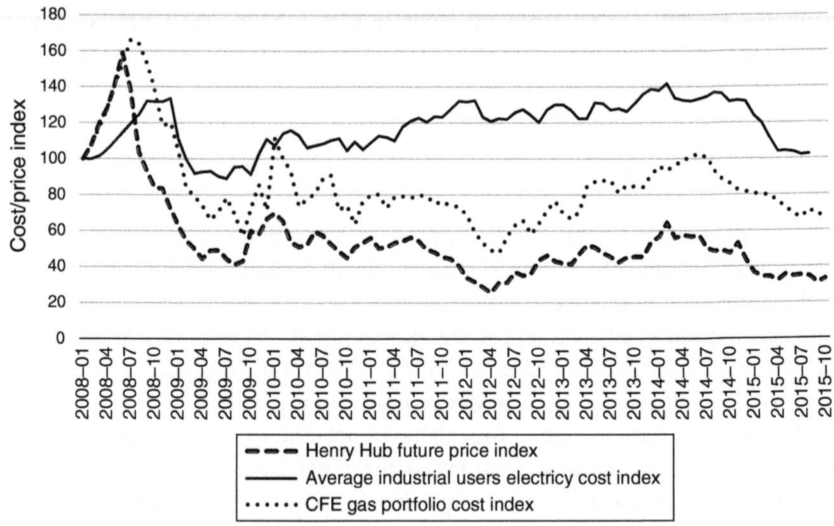

Figure 21.1. Gas prices and industrial tariffs.
Source: CFE (n.d.) and SGM (2015).

As seen in Figure 21.1, since 2008, when the previous reform was drafted and approved, the progression of industrial tariffs and natural gas prices diverged, with natural gas prices increasing to levels that spur debate on the need for a diversified energy matrix. In 2012, the year of the incumbent presidential election, the gap was so significant that gas prices accessible to CFE diminished by >50 per cent, whereas tariffs for industrial users were up by >20 per cent in nominal terms.

In recent years, a gap between the index of gas prices (in Mexican pesos) at the Henry Hub and those accessible to CFE have also grown because of exchange rate and the use of expensive LNG in the generation mix, thus adding to the expectation from power consumers and the incumbent government to benefit from the abundance of natural gas in the region.

21.3.3 Climate Change by Legal Fiat

The 2012 LGCC dictated a battery of policies and programmes aimed at achieving the 2020 and 2050 mitigation goals, which are to be implemented by the government, regardless of political convictions. Most of these policies and programmes, such as the Special Climate Change Programme and the Climate Change Strategy, will be reviewed in a comprehensive manner in 2019 for the first time.

Table 21.2. INDC emissions and pledges (2013–30) ($MtCO_2e$)

	2013	2030 BAU	INDC unconditional	INDC conditional
Oil and gas	87	137	118	Undefined
Power	126	202	139	
Industry	141	202	194	
Domestic and commercial	26	28	23	
Transport	148	229	181	
Waste	31	49	35	
Agro	80	93	86	
Land use, land-use change, and forestry	33	32	−14	
Total	672	972	762	622

Source: Based on Gobierno de la República (2015).

Although Mexico has gained the title of being the first developing country to present its INDC (Gobierno de la República 2015), it is not clear if Mexico's current energy policies and future economic growth will allow for fulfilling these commitments. Mexico proposed an unconditional commitment to mitigate 22 per cent of GHG from a BAU scenario that implies about 45 per cent increase from 2013 emissions. The characterization of emissions in year 2013, 2030 BAU, and the INDC unconditional and conditional pledges are presented in Table 21.2.

The unconditional commitment entails an actual increase of about 15 per cent of economy-wide emission from 2013 to 2030. Within the unconditional pledge, Mexico has published an indicative distribution of emissions and mitigation by sector as part of the INDC. The power sector would then be expected to mitigate 31 per cent from a BAU scenario, resulting in a total emission of 139 $MtCO_2e$ in 2030, rather than the BAU scenario of 202 $MtCO_2e$, compared with 127 $MtCO_2e$ in 2013.

According to Bloomberg New Energy Finance (2015), Mexico's unconditional 2030 target does not rank highly compared with commitments of other countries in terms of absolute emissions reductions. There also remains the question of how different sectors will participate in achieving the conditional goal of 36 per cent reduction of GHG emissions, which would require an additional mitigation of 140 $MtCO_2e$, equivalent to the total power sector emissions in the 2030 INDC. Although the allocation of these additional emissions reductions has not been defined, it is certain that the power sector will have to play a relevant role.

The LGCC also included the possibility of developing a carbon market and linking it to carbon markets in other countries. However, it is unclear that these markets will develop before the end of the incumbent administration in 2018. In October 2015, President Peña Nieto joined the World Bank's initiative to put a price of carbon, stating that Mexico is 'setting a price of carbon that is just' and referring to Mexico's leadership as having already created a carbon tax on the sale of several fossil fuels (Presidencia de la República 2015). The tax rates, which are very low by world standards (less than US$3.50), were set according to the carbon content of propane, gasoline, diesel, and coal relative to gas. Natural gas was exempted from the tax, confirming the federal government's strategy to rely heavily on gas as the 'transition' fuel (Gobierno de la República 2013; Guerra Abud 2014; Pierre-Marc 2014). Also evidencing the alignment of government policies and industrial interests, the originally proposed tax rates were significantly modified to pass through the Mexican Congress, with a 92 per cent reduction for oil coke and 85 per cent for coal (Kossoy et al. 2015). With a negligible environmental impact, the carbon tax in Mexico was introduced mainly for revenue purposes and not as a strategy to invest in clean energy or to affect consumer preferences for fossil fuels.

21.3.4 Walking Half-Way on Clean Energy Deployment

In spite of both Mexico's abundant renewable energy resources and all government efforts on climate policy during the Calderon administration, the share of renewables in electricity generation decreased from 25 to 14 per cent in 1990–2013 (OECD 2015a, b). According to some estimates, it would not be until 2018 that the proportion of renewable energy will return to 1990 levels.

The Law of Electric Industry (Ley de la Industria Eléctrica, LIE), which was enacted in the summer of 2014 with a tight schedule to start trading at the wholesale market in January 2016, introduced a number of mechanisms that would, in principle, facilitate the development of renewable energy. The LIE created an independent system operator, dis-incorporated the dispatching operations unit from CFE, and provided legal and operational capacities to perform traditional system operation activities, while also acting as the clearinghouse for market operations and developing the transmission infrastructure plan. CFE will be unbundled according to guidelines of the Secretariat of Energy (Secretaría de Energía, SENER), and plants will be independently represented in the market to provide for liquidity, with market supervision from SENER and the Energy Regulatory Commission (CRE).

Two instruments—clean energy mandates with tradable certificates and energy auctions—are meant to specifically foster the expansion of clean energy technologies and meet the underlying target of 35 per cent clean energy by 2024. Auctions will be conducted three years in advance of the time for delivering energy, so as to provide sufficient time for the development of projects. They could represent an improvement for large deployment of clean power, as they will be managed by independent entities such as the National Energy Control Centre and CRE, rather than through CFE's IPP bidding processes that were previously used.

The ability of the LIE to promote renewable energy may be constrained by the Peña Nieto administration's strong preference for natural gas as the fuel that could help gain both competitive power prices and reduce emissions. One example of this preference is the ambiguity regarding the definition of clean power in the LIE. The LIE defined clean power with a list of low-carbon technologies that included renewables, large hydropower, nuclear energy, and carbon capture and storage. However, it also included high-efficiency cogeneration, which even if less carbon-intensive per net energy output is still a carbon-emitting technology at a significant rate. With high-efficiency cogeneration expected to grow >7 GW, generation could reach close to 10 per cent of the total power demand by 2024. This would leave low-carbon technologies a potential market share of only 25 per cent out of the 35 per cent of clean energy mandated by law (CESPEDES–PwC 2015: 44). In 2014, large hydropower and renewable energy accounted for almost 17.2 per cent and nuclear energy for 3.2 per cent (SENER 2015b).

The Mexican government's promotion of natural gas as the 'cleanest fuel'[5] contributed to the confusion generated by the inclusion of efficient cogeneration into the definition of clean energy by the LIE. It also nurtured concerns that the government would take advantage of an administrative leeway, included in the LIE, for defining the natural gas combined cycle as clean energy technology. The CANACERO even proposed a definition for clean energy that included technologies with emission levels equivalent to 400 kg of carbon dioxide per megawatt-hour, which would have certainly made new natural gas combined cycle plants qualify as clean energy (CANACERO 2015).

However, the right-wing, largest opposition party, Partido Acción Nacional (National Action Party), presiding over the Special Commission on Renewable Energy and the Commission on Climate Change, presented a proposal for a new Law on Energy Transition (Ley de Transición Energética, LTE) that introduced new obligations for renewable energy and energy efficiency programmes.[6] In addition to limiting the range of potential technologies to be regarded clean by the authority, the LTE also considered intermediate goals for clean energy (25 per cent by 2018 and 30 per cent by 2021) to achieve the 35 per cent by 2024 goal, already stated by the LGCC.

After over a year of its introduction and as COP21 was taking place in Paris, the LTE was finally approved in both chambers of the Mexican Congress. Although the CANACERO aggressively lobbied against its approval,

[5] A common practice by Juan Jose Guerra, environment minister from December 2012 to August 2015, was to focus on natural gas as clean energy during public presentations; this occurred even in fora dedicated to renewable energy (SEMARNAT 2014).

[6] This law replaced the renewable energy and energy efficiency laws (LAERFTE and LASE) that had been approved in 2008 and which were not consistent with the new energy reform. LTE would supersede both laws and integrate them into a coherent legal instrument that would be aligned with the LIE.

an amendment to the LTE stated that no new technology could be accredited as clean energy if emissions surpassed the 100 kg of carbon dioxide per megawatt-hour threshold. At the same time, the Senate approved a flexible mechanism regarding clean energy certificates. During the first four years, those liable will be able to transfer up to 50 per cent of the certificate obligations to the following year at no cost, if either the cost of certificates goes over US$20 or if the number of certificates issued during the given year is <75 per cent (Decreto por el que se expide la Ley de Transición Energética, 2015). If the flexibility mechanism kicks in, the increment of certificate obligations will be postponed and the clean energy goals might not be met.

Surprisingly, markets have proven political fears overstated. The results of the first long-term energy auction concluded in late March 2016 allocated contracts for almost 2000 MW of installed capacity of solar and wind energy. In total, 227 offers from 69 bidders were evaluated, resulting in total winning bids representing 5.4 TWh/years of energy and CEL. The most competitive winning offer on energy and CEL was in the low range of 600 Mexican pesos, or below US$40 per megawatt-hour plus CEL, which is a threshold only seen for natural gas combined cycle. Among the winning projects, seven corresponded to solar photovoltaic and four to wind energy (CENACE 2016). The auction demonstrated a large competitive resource base for renewable energy expansion, and showcased that policy precaution regarding renewable energy expansion goals might be counterproductive to the development of a more competitive power system. The amount of participation in the auction clearly shows that the use of Clean Energy Mandates might be under-utilized at the moment, as these results came weeks after SENER published an increase of CEL obligations by only 0.8 per cent, from 5.0 per cent in 2018 to 5.8 per cent in 2019 (SENER 2015a, SENER 2016).

21.4 THE WAY FORWARD AND THE CHALLENGE FOR POLITICAL COHERENCE

As presented in this chapter, evidence confirms Zinaman et al.'s (2015) conclusion. Mexico's power reform is ambitious about restructuring but modest about the commitment to clean power transition, despite the ratcheting of climate commitments and the favourable trends in renewable energy investment in the country—see Valenzuela and Studer (2016: figure A1).

Mexico's energy reform has indeed created formal means of achieving transition to a low-carbon power system, mainly through clean energy certificates, mandates, and auctions. Nonetheless, the Mexican government strong preference for offering competitive prices risks reducing the pace of the power sector transformation in the future and providing a dubious outlook on a low-carbon transition.

For decades, underinvestment in the oil industry, upstream and downstream, as well as in the electricity sector, created a series of bottlenecks that had direct effects on the availability of natural gas and the price of electricity. Access to abundant and cost-competitive capital and production efficiencies is expected to address these concerns with markets fully opened to private investment in all segments of the power sector, except in distribution and retail. Lower energy costs became the major driver on policy debates—an outcome that could be attainable independently from the power sector reform or long-term diversification strategies—because of historically low gas prices in North America.

One of the fundamental challenges for transitioning to clean power in a liberalized regime is that the cost of ambition will be evident to consumers (Pollitt 2012). This occurs not only because of transparency in operations across industry segments, which were previously embedded into the vertical structure, but also because policy instruments under markets are to be transparent, general, and enduring to provide the right incentives to participants (e.g., clean energy certificates and carbon tax). Ultimately, this is precisely the challenge the current administration faces.

Energy transition under adequate incentives and planning could represent an opportunity to increase energy security and hedge against cost instability that might have economy-wide negative impacts. Yet, under very low prices of gas, the benefits are less clear because of the primacy of delivering on the political promise of low-cost electricity in the very short term. Energy-intensive industries, such as the steel industry, whom are not prepared for the regulatory and economic challenge, have taken this juncture to advocate for further downscaling of Mexico's commitments towards power sector transformation to a low-carbon scenario. These enterprises that benefited from the previous self-supply regime and who have ready access to natural gas imported from the US will need to share the burden of clean energy transition.

In 2008, when a full power sector reform was politically unfeasible and high gas prices further strengthened the argument for diversification through private investment in renewable energy, the Calderón government effectively coupled the emerging climate change agenda with renewable energy investment policies. These included the first renewable energy promotion law and clear long-term non-fossil fuel targets in the power matrix. At the same time, a robust climate change institutional framework was set in place, plausibly securing that mitigation commitments and national goals would become a driving force in the future of energy policy-making and market development. The incumbent government effectively manoeuvred for full reform and benefited from the political windfall profits of lower gas prices in the region.

Although it might seem only natural to continue the hustle to power the country with Texan gas, the government should provide a longer-term vision for the development of low-carbon technologies.

There are three key issues to be addressed. First, power system planning and policy should reflect a clear separation between high-efficiency cogeneration and low-carbon technologies, as this would facilitate linking instruments such as clean energy certificate mandates to mitigation results. Second, risk-mitigating instruments such as auctions should not be compromised by decision-making informed only by low gas prices observed in the near past and future, postponing the expansion of clean energy obligations. Finally, the country should adopt a comprehensive approach for adequate pricing, through a more aggressive use of clean energy obligations and reducing subsidies in the power sector, and for scaling up the carbon taxes economy-wide. These measures can be effectively addressed only if the government is able to assume political commitment to scale up climate ambition, and to fully communicate the economic costs and benefits of rapidly deploying renewable energy in the power sector, at the expense of decelerating the deployment of natural gas-based generation.

The coherence of climate and energy policies, or the lack thereof, will determine the success of Mexico's climate ambition in the 2020s. As discussed, Mexico expects its power sector to reduce a large share of emissions by 2030, whereas specialists expect a zero-emissions power system by the middle of the century. The enactment of the LTE serves to bridge some of the gaps in the energy reform regarding the challenges of efficiency and low-emission technology promotion, allowing for a set of energy sector institutions to work hand-in-hand with climate policy institutions. Developing a solid climate policy framework can have a significant effect in guiding key aspects of power sector restructuring. In Mexico, taking advantage of the existence of a capable independent system operator, clean energy certificates and mandates and long-term auctions could allow for the expansion of renewable energy in a very brief period of time.

The Mexican case offers a valuable example to other economies undergoing or considering power sector restructuring, particularly whenever fossil fuel resources are still more competitive than large-scale deployment of renewable energy. It is possible to observe the political economy trade-offs between the need for a strong climate commitment that provides a stable long-term energy transition pathway and the political and economic short-term benefits derived from low electricity tariffs. It is also possible to observe how these trade-offs are only partially resolved through adequate institutional construction.

Mexico's political elites, interest groups, and society need to recognize and assume that transition could entail additional costs in the short term to some economic actors, but that, in a broader context, these costs could remain below the economic gains of a structural transformation and liberalization of the power sector. More importantly, the lack of coherence between energy policy and climate change policy is unsustainable and will ultimately have an effect on increasing the costs in the power sector owing to late action, with a potential burden to be disproportionately shared by other sectors.

REFERENCES

Alvarez, J. and F. Valencia (2015). 'Made in Mexico: Energy Reform and Manufacturing Growth'. IMF Working Paper WP/15/45, International Monetary Fund (IMF), Washington, DC: Available at: <https://www.imf.org/external/pubs/ft/wp/2015/wp1545.pdf> (accessed 26 March 2016).

Auditoria Superior de la Federación (2014). Importación de Gas Natural Licuado: Auditoría Financiera y de Cumplimiento 13-1-18TOQ-02-0365-DE-174. Informe de Resultados de la Fiscalización Superior de la Cuenta Pública 2013 ['Import of Liquefied Natural Gas: Financial Audit Compliance 13-1-18TOQ-02-0365-DE-174. Report of Supreme Audit Results of Public Accounts 2013']. Comisión Federal de Electricidad, Mexico. Available at: <http://www.asf.gob.mx/Trans/Informes/IR2013i/Documentos/Auditorias/2013_0365_a.pdf> (accessed 26 March 2016).

Bloomberg New Energy Finance (2015). 'How Ambitious Are the Post-2020 Targets? Assessing the INDCS: Comparing Apples with Oranges', p. 15. Bloomberg New Energy Finance, New York. Available at: <http://about.bnef.com/content/uploads/sites/4/2015/10/2015-10-02-How-ambitious-are-the-post-2020-targets-UPDATE-2-Oct.pdf> (accessed 26 March 2016).

Borenstein, S. and J. Bushnell (2015). 'The U.S. Electricity Industry after 20 Years of Restructuring'. NBER Working Paper Series 21113, National Bureau of Economic Research (NBER), Washington, DC.

CANACERO (Cámara Nacional del Acero) (2011). 'Desplegado' ['Communique']. 12 December. CANACERO, Mexico. Available at: <http://www.canacero.org.mx/Es/assets/desplegado_leycambioclimatico_dic2011.jpg> (accessed 4 April 2016).

CANACERO (2015). 'Energy Transition Forum at the Senate', 7 October (PowerPoint slides). CANACERO, Mexico.

Carreón-Rodríguez, V., A. Jiménez, and J. Rosellón (2006). 'The Mexican Electricity Sector: Economic, Legal and Political Issues'. In D. G. Victor and T. C. Heller (eds), *The Political Economy of Power Sector Reform*. New York: Cambridge University Press, pp. 175–214.

CAT (Climate Action Tracker) (2015). 'Mexico'. Available at: <http://climateactiontracker.org/countries/mexico.html> (accessed 23 March 2016).

CENACE (Centro Nacional de Control de Energía) (2016). 'Primera Subasta de Largo Plazo del Mercado Elèctrio Mayorista' [First Long-Term Auction of the Wholesale Electricity Market'], 30 March. CENACE, Mexico. Available at: <http://cenace.gob.mx/Docs/MercadoOperacion/Subastas/32%20Presentaci%C3%B3n%20CENACE%20Conclusi%C3%B3n%201a%20Subasta%20Largo%20Plazo%20v2016%2003%2030.pdf> (accessed 10 April 2016).

CESPEDES (Comisión de Estudios del Sector Privado para el Desarrollo Sustentable) (2013). 'Aportaciones de CESPEDES sobre las oportunidades y retos del Programa Especial de Cambio Climático (PECC), 2013–2018' ['Contributions of CESPEDES to Address the Opportunities and Challenges of the Special Climate Change Programme (PECC), 2013–2018'], November. CESPEDES, Mexico. Available at: <http://www.cespedes.org.mx/wp-content/uploads/2015/03/131121-PECC-CESPEDES-CONSULTA-final-7-0.pdf> (accessed 26 March 2016).

CESPEDES–PwC (2015). 'Estudio de las inversiones necesarias para que México cumpla con sus metas de Energías Limpias' ['Study of the Necessary Investments for Mexico's Clean Energy Goals'], October. CESPEDES, Mexico. Available at: <http://www.estrategiaweb360.com/clientes/cespedes/PwC_CESPEDES_estudio_energias_limpias.pdf> (accessed 26 March 2016).

CFE (Comisión Federal de Electricidad) (n.d.). 'Factores de ajuste mensual a las tarifas de energía eléctrica año 2015' ['Factors of Monthly Adjustments to Electrical Energy Tariffs, 2015']. Available at: <http://app.cfe.gob.mx/Aplicaciones/CCFE/Tarifas/factores/factores_industria.asp?tarifa=ICC2012&anio=2015> (accessed 26 March 2016).

Clean Energy Ministerial (2015). '21st Century Power Partnership: Overview and Goals'. Available at: <http://www.cleanenergyministerial.org/Our-Work/Initiatives/21st-Century-Power> (accessed 23 March 2016).

Fekete, H., F. Mersmann, and M. Vieweg (2013). 'Climate Change Mitigation in Emerging Economies: From Potentials to Actions'. Climate Change Series 19/2013. Umweltbundesamt, Dessau-Roßlau, Germany. Available at: <http://www.umweltbundesamt.de/publikationen/climate-change-mitigation-in-emerging-economies> (accessed 23 March 2016).

Fuentes, R. and J. Bowler (2014). 'In Search of the Mexican Way: How to Kick Start Competition in the Electricity Sector and Achieve Lower Tariffs'. OIES Oxford Energy Comment Paper, May, Oxford Institute for Energy Studies (OIES), Oxford. Available at: <http://www.oxfordenergy.org/2014/05/in-search-of-the-mexican-way-how-to-kick-start-competition-in-the-electricity-sector-and-achieve-lower-tariffs/> (accessed 23 March 2016).

GCF (Green Climate Fund) (2015). 'PledgeTracker: Resource Mobilization'. Available at: <http://news.gcfund.org/pledges/> (accessed 26 March 2016).

Gobierno de la República (2013). 'Reforma Energética' ['Energy Reforms']. Federal Government, Mexico. Available at: <http://cdn.reformaenergetica.gob.mx/explicacion.pdf> (accessed 26 March 2016).

Gobierno de la República (2015). 'Intended Nationally Determined Contribution'. Federal Government, Mexico. Available at: <http://www4.unfccc.int/submissions/INDC/Published%20Documents/Mexico/1/MEXICO%20INDC%2003.30.2015.pdf> (accessed 26 March 2016).

Guerra Abud, J. (2014). 'La Reforma Energética y el Medio Ambiente' ['The Energy and Environment Reform']. Press Conference, 19 August. Federal Government, Mexico. Available at: <http://www.radioformula.com.mx/notas.asp?Idn=434664&idFC=2014#sthash.2XiYqQYy.dpuf> (accessed 26 March 2016).

Hernández, C. (2006). 'La Reforma Cautiva: Inversión, trabajo y empresa en el sector eléctrico mexicano' ['The Captive Reform: Investment, Labour and Entrepreneurship in Mexico's Electricity Sector']. Centro de Investigación para el Desarrollo, AC, Mexico. Available at: <http://reddecompetencia.cidac.org/es/uploads/1/LaReformaCautiva.pdf> (accessed 26 March 2016).

IEA (International Energy Agency) (2011). 'Are We Entering a Golden Age of Gas?' *World Energy Outlook 2011: Special Report.* OECD/ (IEA), Paris. Available at: <http://www.worldenergyoutlook.org/media/weowebsite/2011/WEO2011_GoldenAgeofGasReport.pdf> (accessed 26 March 2016).

IEA (2015). 'Energy Prices and Taxes 2014: Quarterly Statistics'. OECD/ IEA, Paris.

IGS-EGADE (Instituto Global para la Sosteniblidad (IGS)-EGADE) (2014). 'Energías Renovables para la Competitividad en México' ['Renewable Energy Competition in Mexico']. MéxicoIGS-EGADE Business School. Available at: <http://igs.org.mx/sites/default/files/ENERGIAS_RENOVABLES_MAYO23_WEB2.pdf> (accessed 26 March 2016).

IRENA (International Renewable Energy Agency) (2015). 'Renewable Energy Prospects: Mexico'. IRENA, Abu Dhabi. Available at: <http://www.irena.org/DocumentDownloads/Publications/IRENA_REmap_Mexico_report_2015.pdf> (accessed 23 March 2016).

Kessel, G. (2009). 'Speech during the Opening of the Ecological Parks of Mexico Program'. SENER, Juchitán de Zaragoza, Oaxaca, 22 January.

Kossoy, A., G. Peszko, K. Oppermann, N. Prytz, A. Gilbert, N. Klein, L. Lam, and L. Wong (2015). 'Carbon Pricing Watch 2015: An Advanced Brief from the State and Trends of Carbon Pricing Report 2015'. World Bank, Washington, DC. Available at: <http://www-wds.worldbank.org/external/default/WDSContentServer/WDSP/IB/2015/08/26/090224b08309a09a/4_0/Rendered/PDF/Carbon0pricing0e0released0late02015.pdf> (accessed 26 March 2016).

Decreto por el que se expide la Ley de Transición Energética (2015). [Decree on the Energy Transistion Law], *Diario Oficial de la Federación*, 24 December.

OECD (2013). *Environmental Performance Reviews: Mexico 2013.* Paris: OECD Publishing.

OECD (2015a). 'Air and GHG Emissions' (Mexico: Environment Indicator). OECD Data, Paris. Available at: <https://data.oecd.org/mexico.htm> (accessed 26 March 2016).

OECD (2015b). 'Climate Change' (Country Profiles: Mexico). OECD Environmental Data, Paris. Available at: <http://www.oecd.org/site/envind/mexico.htm> (accessed 26 March 2016).

Pierre-Marc, R. (2014). 'México debe de aprovechar el gas natural: Semarnat' ['Mexico Must Take Advantage of Natural Gas: SEMARNAT']. *24 Horas*, 7 January. Available at: <http://www.24-horas.mx/mexico-debe-de-aprovechar-el-gas-natural-semarnat/> (accessed 26 March 2016).

Pollitt, M. G. (2012). 'The Role of Policy in Energy Transition: Lesson from the Energy Liberalization Era'. *Energy Policy*, 50(C): 128–37.

Presidencia de la República (2015). 'México se suma a la iniciativa mundial para reducir las emisiones de bióxido de carbono' ['Mexico Joins Global Effort to Reduce Carbon Dioxide Emissions']. *Blog*, 20 October. Available at: <http://www.gob.mx/presidencia/articulos/mexico-se-suma-a-la-iniciativa-mundial-para-reducir-las-emisiones-de-bioxido-de-carbono> (accessed 16 April 2016).

SEMARNAT (2014). 'Reforma energética aumentará la creación de energías renovables en México' ['Energy Reform Increases the Creation of Renewable Energy in Mexico']. Press Release, 8 September. Available at: <http://saladeprensa.semarnat.gob.mx/index.php/noticias/1833-reforma-energetica-aumentara-la-creacion-de-energias-renovables-en-mexico> (accessed 26 March 2016).

SEMARNAT, CESPEDES, WRI, and WBCSD (n.d.). 'Factor de Emisión Eléctrico 2013' ['Electric Emission Factor 2013']. Secretaría del Medio Ambiente y Rescursos Naturales (SEMARNAT), Comisión de Estudios del Sector Privado para el Desarrollo sustentable (CESPEDES), World Resources Institute (WRI), and World Business Council for Sustainable Development (WBSCD), Mexico. Available at: <http://www.geimexico.org/factor.html> (accessed 26 March 2016).

SENER (Secretaría de Energía) (2007). 'Programa sectorial de energía 2007–2012' ['Sectorial Energy Programme, 2007–2012']. SENER, Federal Government, Mexico. Available at: <dof.gob.mx/nota_to_doc.php?codnota=5030249> (accessed 26 March 2016).

SENER (2015a). 'Aviso por el que se da a conocer el requisito para la adquisición de Certificados de Energías Limpias en 2018' ['Notification on Requirements for Acquiring Energy Certificates in 2018']. *Diario Oficial de la Federación*, 31 March. Available at: <http://www.dof.gob.mx/nota_detalle.php?codigo=5387314&fecha=31/03/2015> (accessed 26 March 2016).

SENER (2015b). 'Programa de Desarrollo del Sistema Eléctrico Nacional (PRODESEN), 2015–2029' ['National Electricity System Development Programme (PRODESEN), 2015–2019']. SENER, Federal Government, Mexico. Available at: <http://www.gob.mx/cms/uploads/attachment/file/54139/PRODESEN_FINAL_INTEGRADO_04_agosto_Indice_OK.pdf> (accessed 26 March 2016).

SENER (2015c). 'Sistema de Información Energética' ['Energy Information System']. SENER, Federal Government, Mexico. Available at: <http://sie.energia.gob.mx/> (accessed 26 March 2016).

SENER (2016). 'Aviso por el que se da a conocer el requisito para la adquisición de Certificados de Energías Limpias en 2019' ['Notification on Requirements for Acquiring Energy Certificates in 2019']. *Diario Oficial de la Federación*, 31 March. Available at: <http://www.dof.gob.mx/nota_detalle.php?codigo=5431515&fecha=31/03/2016> (accessed 4 November 2016).

SGM (Servicio Geológico Mexicano) (2015). 'Seguimiento precio gas natural—Datos' ['Natural Gas Price Data']. SGM, Secretaría de Economía, Mexico. Available at: <http://portalweb.sgm.gob.mx/economia/es/energeticos/precios/701-seguimiento-precio-gas-natural-datos.html> (accessed 26 March 2016).

Torres, B. (2014). 'Mexico and Climate Change: Was the Country a Multilateral Leader?' *Global Governance: A Review of Multilateralism and International Organizations*, 20(1): 147–62.

Veysey, J., C. Octaviano, K. Calvin, S. Herreras Martinez, A. Kitous, J. McFarland, and B. van der Zwaan (2015). 'Pathways to Mexico's Climate Change Mitigation Targets: A Multi-Model Analysis'. *Energy Economics*, 56: 587–99.

Valenzuela, J. María and I. Studer (2016). 'Climate Change Policy and Power Sector Reform in Mexico under the "Golden Age of Gas"'. 2016/33. Helsinki: UNU-WIDER.

World Resources Institute (2015). 'CAIT Climate Data Explorer'. Available at: <http://cait.wri.org/historical/Country%20GHG%20Emissions?indicator[]=Total%20GHG%20Emissions%20Excluding%20Land-Use%20Change%20and%20Forestry&indicator[]=Total%20GHG%20Emissions%20Including%20Land-Use%20Change%20and%20Forestry&year[]=2012&chartType=geo> (accessed 26 March 2016).

WWF, Climate Works, and PwC (2013). 'Plan Integral para el desarrollo de las energías renovables en México, 2013–2018: Propuestas de escenarios y acciones necesarias para su desarrollo' ['Integral Plan for Renewable Energy Development in Mexico, 2013–2018: Proposed Scenarios and Actions Necessary for Implementation']. Iniciativa Mexciana de Energías Renovables, Mexico. Available at: <http://awsassets.panda.org/downloads/130222_plan_integral_para_desarrollo_de_energias_renovables.pdf> (accessed 26 March 2016).

Yale School of Management (2012). 'Walmart de México: Investing in Renewable Energy'. Available at: <http://som.yale.edu/our-approach/teaching-method/case-research-and-development/cases-directory/walmart-de-m%C3%A9xico> (accessed 26 March 2016).

Zinaman, O., M. Miller, A. Adil, D. Arent, J. Cochran, R. Vora, S. Aggarwal, M. Bipath, C. Linvill, A. David, R. Kauffman, M. Futch, E. Villanueva, J. M. Valenzuela, E. Martinot, M. Bazilian, and R. K. Pillai (2015). 'Power Systems of the Future: A 21st Century Power Partnership Thought Leadership '. NREL Technical Report NREL/TP-6A20-62611, February. National Renewable Energy Laboratory (NREL), Golden, CO. Available at: <http://www.nrel.gov/docs/fy15osti/62611.pdf> (accessed 23 March 2016).

22

Sell the Oil Deposits! A Financial Proposal to Keep the Oil Underground in the Yasuni National Park, Ecuador

Santiago J. Bucaram, Mario Andrés Fernández, and Diego Grijalva

22.1 INTRODUCTION

The Yasuni National Park (YNP) is a protected area located in the Amazon region of Ecuador and is recognized as one of the most biodiverse regions in the world (Bass et al. 2010). In 1989 it and much of its adjacent area were designated by UNESCO as a Biosphere Reserve (UNESCO 1989). In recent years, the YNP has received much attention due to the media exposure of the Yasuni-ITT Initiative. This Initiative, announced in 2007 by the Ecuadorian government, proposed a moratorium on oil activities in the Ishpingo-Timbocochа-Tiputini (ITT) blocks within the YNP, in exchange for US$3.6 billion in compensation over a period of ten years. The compensation was supposedly in recognition of the supply of environmental services generated by the YNP from which the entire planet benefited. The Initiative was initially celebrated as an innovative proposal that offered an alternative to global environmental problems as it would promote the transition from the current development model, based on oil extraction, to a new strategy based on equality and sustainability (Acosta, Gudynas, and Vogel 2009; Larrea and Warnars 2009; Rival 2010; Vogel 2010).

The financial mechanism of the ITT Initiative involved contributions to the Yasuni Trust Fund, and may have taken the form of debt-for-conservation swaps, emission permit auctions or conservation projects, and donations from governments, multilateral organizations, non-governmental organizations, private companies, and individuals (United Nations Development Group 2010). In exchange for the contributions, the Ecuadorian government would issue Yasuni Guarantee Certificates (YGCs), which were documents with a face value equal to the contribution in US dollars. The YGCs were intended to represent 'avoided emissions' (measured in metric tons of CO_2e) from keeping oil underground. The avoided emissions were calculated as the ratio between the contribution and the then current price of the European Union Allowances (EUA) from the Leipzig Carbon Market. The maximum total amount of YGCs to issue would be 407 million tons of CO_2e corresponding to the estimated emissions produced from extracting oil from the ITT field. The YGCs did not earn interest and did not have an expiration or maturity date as long as the government maintained its commitment to not exploiting the oil reserves. Any contribution below US$50,000 was considered a donation and no YGC would be issued. By 2013, only US$336 million had been pledged (about 9 per cent of the target compensation) and US$13.3 million actually delivered (0.37 per cent of the target compensation), leading President Correa to terminate the Initiative, arguing that the international community had failed to embrace it.

Some authors, however, argue that the original ITT Initiative is still a coherent and innovative proposal to address climate change (e.g., Vallejo and Friant 2015), and that the failure of the Initiative was not due to poor design but to poor implementation by the policy makers in charge (Pellegrini et al. 2014). Despite the purported advantages of the Initiative, Pellegrini et al. (2014) indicate that its demise depended to a large extent on the inability of policy makers to identify and resolve tensions between the proposal and the institutions that facilitated it; and, the high reliance of the Initiative's implementation on exogenous dynamics (e.g., oil prices) that were not explicitly considered.

Other authors like Haddad (2011) argue that the Initiative took the form of a compensated moratorium, where the compensation was estimated with respect to foregone oil revenues rather than the environmental benefit accrued. Likewise, Harstad (2012a) considers that the Initiative and the oil moratorium could be viewed as the Ecuadorian government holding a hostage (the YNP) and demanding a ransom (the target compensation), so that if the ransom was not received, oil extraction would start. Therefore, the whole Initiative appeared as an arbitrary exercise of power where the stronger party (the Ecuadorian government) demanded a ransom from the rest of the world, who had no alternative but to comply (Williamson 1983).

The objective of this chapter is twofold. First, we conduct a feasibility analysis of the ITT Initiative and show that it was severely flawed from its inception. Subsequently we use the price discounting framework of Bucaram, Fernández, and Grijalva (2016) to assess the revenues that would have hypothetically occurred from the trade of YGCs. We conclude that the ITT Initiative was in fact poorly designed and was bound to fail. Second, we develop a financial approach for a 'New ITT Initiative'. We propose the sale or leasing of the rights of extraction of the oil deposits in the YNP as a feasible strategy to keep the oil underground and, consequently, to protect the YNP's ecosystem services. This financial approach is much more transparent and simplified than the ITT Initiative, and could be easily implemented through existing market mechanisms. Thus, our proposal does not require the creation of financial instruments such as the YGCs, it disregards inaccurate concepts such as 'avoided net emissions', and demonstrates the redundancy of the Initiative with respect to the Kyoto Protocol instruments. We finally emphasize that the Initiative, as originally designed, should be discarded and further efforts should be addressed to the adoption of our new, more useable Initiative.

22.2 FEASIBILITY ANALYSIS OF THE ITT INITIATIVE

This section provides a comprehensive analysis of the flaws of the ITT Initiative as a climate change mitigation instrument and as a financial tool to raise the compensation necessary to motivate the Ecuadorian government to prioritize conservation over oil extraction.

The modelling assumptions are as follows. First, we assume YGCs are financially and operatively equivalent (i.e., fungible) to the Certificates of Emissions Reductions (CERs). The CERs are financial instruments associated with Clean Development Mechanism (CDM) and Reducing Emissions from Deforestation and Forest Degradation (REDD+) projects, which have a payoff structure (i.e., cash flows) similar to the YCGs. That is, CERs are the best proxy or twin security to value the YGCs. This assumption relies on paragraph 26 of the Terms of Reference (TOR) of the Yasuni-ITT Trust Fund of July 2010 (United Nations Development Group 2010), which states, 'the YGCs will also include the metric tons of CO_2 avoided according to the price, at that date, of the European Union Allowances (EUAs) in the Leipzig Carbon Market . . . '. Furthermore, paragraph 27 of the same document asserts that 'if in the future the world carbon market accepts the YGCs as equivalents of Emission Permits, the government will issue YGCs for sale to private and/or public entities in mitigating greenhouse gas emissions . . . '. These statements show a clear and explicit expectation by the Ecuadorian government that the YCGs could become equivalent to the EUAs traded in the European Union Emission Trading Scheme (EU ETS). However, as EUAs are electronic certificates distributed by European governments to firms in the industrial sector, EUAs are by definition not equivalent to those of the YGCs. Hence, we use the CERs for the purposes of our analysis.

Second, we assume donors are rational utility maximizers and, given scarce financial resources, they will choose the best use of their money by evaluating all the alternatives. Donors then use the carbon market as a means to decide the scale of their donations. Third, we assume the Ecuadorian government commits to keeping the oil underground as long as the compensation is incentive-compatible, that is, if the compensation is equal to or greater than US$3.6 billion. Fourth, for the sake of conducting an analysis of the best-case scenario, we ignore the fact that YGCs represent neither a portion of the oil reserves nor avoided carbon emissions as purported but are, in practice, the representation of nothing more than a *bona fide* promise from Ecuador to refrain from exploiting the ITT block. This simplifies the analysis and allows us to approach the problem from a financial perspective and to obtain an upper bound valuation of the YGCs.

22.2.1 Price Discounting

The ITT Initiative did not comply with any of the CDM or REDD+ criteria to be considered a viable mitigation instrument, namely, the permanence, additionality, certainty, and non-leakage criteria of the greenhouse gas (GHG) emissions. The reasons are as follows:

22.2.1.1 Permanence

Oil exports have been Ecuador's main source of revenue for more than 40 years, and oil exploration activities continued to take place in the YNP and its buffer zone even after the implementation of the ITT Initiative (Martin 2011; Arsel and Angel 2012). However, exploration was omitted as a source of potential activities that may affect the YNP. This omission signalled a weak commitment by the Ecuadorian government to the terms of the ITT Trust (Singleton 2000). Moreover, there was always the possibility that Ecuador would need funds to face negative macroeconomic shocks or to boost economic development. Then, the goals of the (current or future) government might switch from seeking international compensation for avoided emissions to prioritizing oil export revenues undermining the long-term credibility of the initiative. Though the Initiative made provisions for the reimbursement of donations if oil extraction occurred, the financial and legal mechanisms that would put this repayment process into operation were not clear. Thus, other concerns arose because of the institutional instability around the design and management of the ITT Initiative (Arsel 2012; Arsel and Angel 2012).

22.2.1.2 Additionality

The ITT Initiative did not provide any baseline or offset estimations to identify truly additional GHG emissions reductions or sequestration. This implied that the claimed equivalence between the YGCs and CERs or EUAs could not hold because credit buyers would pay only for offsets that they could claim as credit under regulatory schemes. That is, credit buyers would not wish to buy YGCs if the offsets might be disallowed because they could not be proven to be truly additional (Kim, McCarl, and Murray 2008; Kim and McCarl 2009). Moreover, the YNP has been a protected area for more than 34 years. Consequently, no additional environmental services are produced by the YNP and, therefore, no further international funding needs to be provided to ensure biodiversity conservation or protection of the indigenous peoples living in voluntary isolation. Although this all depends on whether the legislation is adequately enforced, this is not a matter within the environmental and conservation realm.

22.2.1.3 Leakage

The avoidance of emissions from the combustion of the oil that would be preserved underground was not a sensible aim of the Initiative. Ecuador does not have any control over oil markets, thus its oil is easily substituted in the short term by purchases in other countries or extraction from other fields in Ecuador itself.

22.2.1.4 Uncertainty

Haddad (2011) argues that the Initiative took the form of a compensated moratorium where the compensation was estimated with respect to the value of foregone oil revenue rather than the environmental benefit accrued. In addition to the unavailability of baseline data and data of impacts on ecosystem services from oil extraction, potential credit buyers, donors, and investors could not make decisions regarding compliance with regulatory limits for emissions. Hence, no clear rationale existed of how the compensation requested by the Ecuadorian government was related to offset levels, or whether it was merely an arbitrary calculation. The ITT Initiative was plagued with contradictions and logical flaws. The most important were:

1) the YGCs were not recognized under the Kyoto Protocol mechanisms and no formal procedures were specified to operationalize the equivalence with CERs and EUAs;
2) although the Initiative predicted that the YGCs would be recognized by the US government as a pilot study for carbon offsets (Larrea and Warnars 2009), the transaction costs were too large to make this a reasonable goal.

Bucaram, Fernández, and Grijalva (2016) conduct price discounting simulations assuming that any non-compliance with CDM or REDD+ criteria is partially resolved through discounts to the price of the YGCs if they had been freely traded in the EU-ETS market. They conclude that the longer the time horizon, the lower the price discounts from permanence and uncertainty; but long time horizons have a minimal effect on the heavy discounts related to leakage. In the context of the Initiative, however, it seems fair to emphasize short-run effects. Considering the often volatile political environment in Ecuador, it was not clear whether the government following President Correa would maintain the promise of not extracting oil. Further, the Ecuadorian Constitution allows Correa to remain in office only until 2017 (ten years from the beginning of the Initiative), so donors/investors would probably have looked at this time horizon when evaluating the Initiative. Hence, price discounting is relevant for our evaluation for the following reasons:

1) the unclear and possibly short time horizon of the Initiative, which makes its permanence questionable;
2) the degree of uncertainty concerning the overall design of the ITT Trust and its implementation;
3) the significant leakage in the short run.

22.2.2 Financial Valuation of the ITT Initiative

In this section we evaluate whether the ITT Initiative was an incentive-compatible contract for the Ecuadorian government using as a focal date 13 August 2013 (i.e., when the Initiative was terminated). The target compensation expected by the Ecuadorian government was US$3.6 billion, a figure that, according to the Initiative's designers, was equal to half the foregone oil revenues from the ITT block. Thus, if oil revenues accruing to the government were estimated to be higher than US$3.6 billion, the Ecuadorian government would not have had any incentive to keep the oil underground, resulting in a breach of the ITT commitments. Also, we estimate the revenues that would have been raised from the YGCs trade in the best-case scenario. For this we employ Monte Carlo simulations that incorporate the price discounts calculated by Bucaram, Fernández, and Grijalva (2016).

22.2.2.1 Valuation of Oil Extraction in the ITT Block

The ITT Initiative was drafted under the assumption that the Net Present Value (NPV) of the oil revenues from the exploitation of the ITT block would be around US$7.2 billion, from which the Ecuadorian government would receive 50 per cent. Consequently, if oil revenues exceeded US$3.6 billion, there would no longer be any incentive to comply with the agreements set in the Initiative. The Initiative's designers argued that this compensation value was incentive-compatible, and that the Ecuadorian government would easily be able to comply with it. This conclusion was derived from the following assumptions (Acosta 2007):

1) an expected production of 900 million barrels in total (on average 36 million barrels per year);
2) an average oil price received by Ecuador of US$32 per barrel;
3) a production period of 25 years;
4) production costs of US$12 per barrel;
5) a discount rate of 9 per cent—the official rate of discount of the Ecuadorian government (Ministerio Coordinador de Política Económica 2013); and
6) no initial investment.

However, many of these assumptions were wrong or unrealistic. For instance, West Texas Intermediate (WTI) oil prices were in fact around US $72 per barrel in 2007, and averaged US$80 in 2008 and 2009. Thus, the assumption of an average oil price received by Ecuador of US$32 per barrel was pessimistic and, in fact, actual oil price behaviour during this period changed the government's incentives. High oil prices can be considered as one of the main factors for the termination of the Initiative in 2013.

It is then necessary to determine a more accurate valuation of the revenues of the ITT block through the use of a more realistic and updated set of assumptions. We use as a focal date of our analysis 13 August 2013 (i.e., when the Initiative was terminated). We also apply the official assumptions provided by Petroamazonas (2010), Larrea (2010), and Ministerio Coordinador de Política Económica (2013) as follows:

1) an expected oil production of 846 million barrels;
2) operation and transportation costs of the oil drilling operation of US$15 per barrel;
3) a production period of 23 years;
4) an initial investment of US$5.59 billion; and
5) a participation in the profits of 47 per cent for the Ecuadorian government.

For the average oil price that the government of Ecuador receives for each sold barrel during the life of the project we use three scenarios: US$70 per barrel (pessimistic), US$80 per barrel (most likely), and US$91.70 per barrel (optimistic). We use three different discount rates: 9 per cent, used by the Ecuadorian government (Ministerio Coordinador de Política Económica 2013), 6 per cent, and 12 per cent, used by Larrea (2010). Table 22.1 presents the NPV of oil revenues under these parameters. The results show that there is no case where the ITT Initiative is incentive-compatible (NPV less than US $3.6 billion). In all cases, the compensation would never have been high enough to enable the government to meet its pledge to keep the oil underground indefinitely.

Table 22.1. NPV of oil revenues from exploitation of the ITT block (US$ billions)

Discount rate (%)	Average price of oil (US$ per barrel)		
	70	80	91.7
6	10.93	13.27	16,01
9	8.52	10.38	12.56
12	6.77	8.28	10,04

Source: Authors' calculations.

For the Monte Carlo simulations of the valuation of oil extraction revenues we introduce some additional assumptions as follows:

1) the total amount of oil reserves is uncertain, so we use the range of reserves from Petroamazonas (2010), namely 412 million barrels of proven, 846 million barrels of probable, and 1,531 million barrels of possible reserves;
2) as the bargaining power and contract conditions for oil exploitation are not certain, we assume that participation in the profits for the Ecuadorian government ranges between 40 per cent and 50 per cent;
3) the production and transportation costs range between US$11 and US$19 per barrel;
4) the discount rate ranges between 6 per cent and 12 per cent; and
5) annual average oil prices for the period 2007–12 are derived from the WTI oil price series observed in the New York Mercantile Exchange (NYMEX) market, for 2013–17 from the NYMEX WTI oil forecast, and for 2018 and beyond we assumed that the price of a WTI oil barrel will range between US$70 and US$98 per barrel, for which we apply a discount of 5 per cent to obtain the price received by the Ecuadorian Government.

Simulation results (Figure 22.1) show that the probability of the ITT Initiative being incentive-compatible (i.e., oil revenue less than US$3.6 billion) is less than 0.006 per cent approximately. This confirms our claim that the designers of the ITT Initiative greatly underestimated the potential oil revenues of the ITT block and, as a result, claimed a low level of compensation relative to the true opportunity cost of keeping the oil underground.

22.2.2.2 Valuation of the YGCs Revenue

The other component required to examine the financial feasibility of the ITT Initiative is the valuation of the YGCs. We use as a proxy the market conditions (observed and predicted) for CERs in the EU-ETS market. We explore two scenarios, the first where no price discount occurs and the second where we

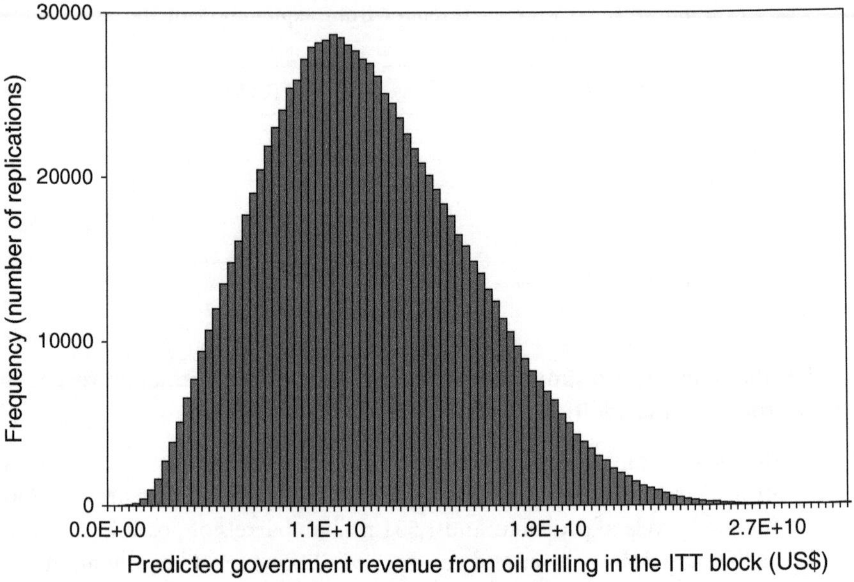

Figure 22.1. Density distribution of the NPV of the expected oil revenue from the ITT block.

Note: The density distribution is the result of one million replications.

Source: Authors' calculations.

apply the discounts calculated by Bucaram, Fernández, and Grijalva (2016). The assumptions used in this section are as follows:

1) the period of YGCs emission or fund collection is 13 years;
2) the maximum number of YGCs issued by the government is 407 million;
3) the sales success rate is 100 per cent (i.e., all the YGCs are sold and traded); and
4) the price of the YGCs is equal to the price of CERs in the EU-ETS market.

For the last assumption we use the historical prices of the instrument from 2007–13 and for the period 2014–20, the forecast price provided by Thomson-Reuters.

Under the no price discount scenario (i.e., best case scenario) and for a discount rate of 6 per cent, the NPV of YGCs revenue is US$616 million. For discount rates of 9 per cent and 12 per cent, the NPVs are US$544 and US$484 million, respectively. That is, even without any price discount, the funds that could have been raised using carbon market mechanisms are much lower than the target compensation of US$3.6 billion.

To simulate the impact of the price discounts on the YGCs valuation, we further assume that the sales success rate varies between 70 per cent and 100 per cent, and that the discount rate ranges between 6 per cent and 12 per cent. Figure 22.2 shows the density distribution of the NPV of the expected YGCs valuation. The left panel shows results when no price discount is applied to the YGCs, while the right panel shows results for price discounts between 60 per cent and 100 per cent. For both cases, the probability of the Ecuadorian government raising US$3.6 billion through the Initiative is zero, regardless of the YGCs price discounting (i.e., using either CER as the twin security of the YGCs).

The most likely amount of funds that could have been raised is US$544 million with no discount on YGCs prices, and US$108 million with price discounts. An interesting result is that the value actually pledged to the Yasuni Trust Fund (US$336 million) was within the range of our valuation of the YGCs (US$108–544 million). In addition, with YGCs price discounting, the probability that the YGCs would be valued at US$336 million is approximately 0 per cent, indicating that international donors discounted the YGCs at a lower rate than Bucaram, Fernández, and Grijalva (2016) considered appropriate (i.e., between 60 per cent and 100 per cent). Specifically, we estimate that donors discounted the price of YGC bonds at 40 per cent approximately. This shows that the willingness of the international community to contribute to the Initiative was mainly for altruistic reasons.

The introduction of additionality to the valuation would render the ITT Initiative completely valueless, reinforcing our findings and conclusion that donors contributed more than could have been expected according to a rational-economic perspective.

22.3 A NEW YASUNI-ITT INITIATIVE

Based on the analysis in Section 22.2, we argue that no practical implementation steps exist that guarantee the success of the original Initiative. Hence, we propose the adoption of a new ITT initiative, as outlined in Section 22.3.1.

22.3.1 The Model

We use the static framework of Harstad (2012b), which entails a coalition of (Kyoto Protocol Annex-I) countries purchasing or leasing the rights to exploit the fossil-fuel deposits in non-coalition countries. That is, the property rights to the oil deposits are transferred to the coalition, which prefers not to extract oil in order to prevent leakage effects when non-coalition countries free ride on the environmental policies of the coalition.

Consider a cost function $C_i(x_i)$ of producing or extracting x_i units of oil, and a benefit function $B_i(y_i)$of consuming y_i units of oil, for all $i$countries.

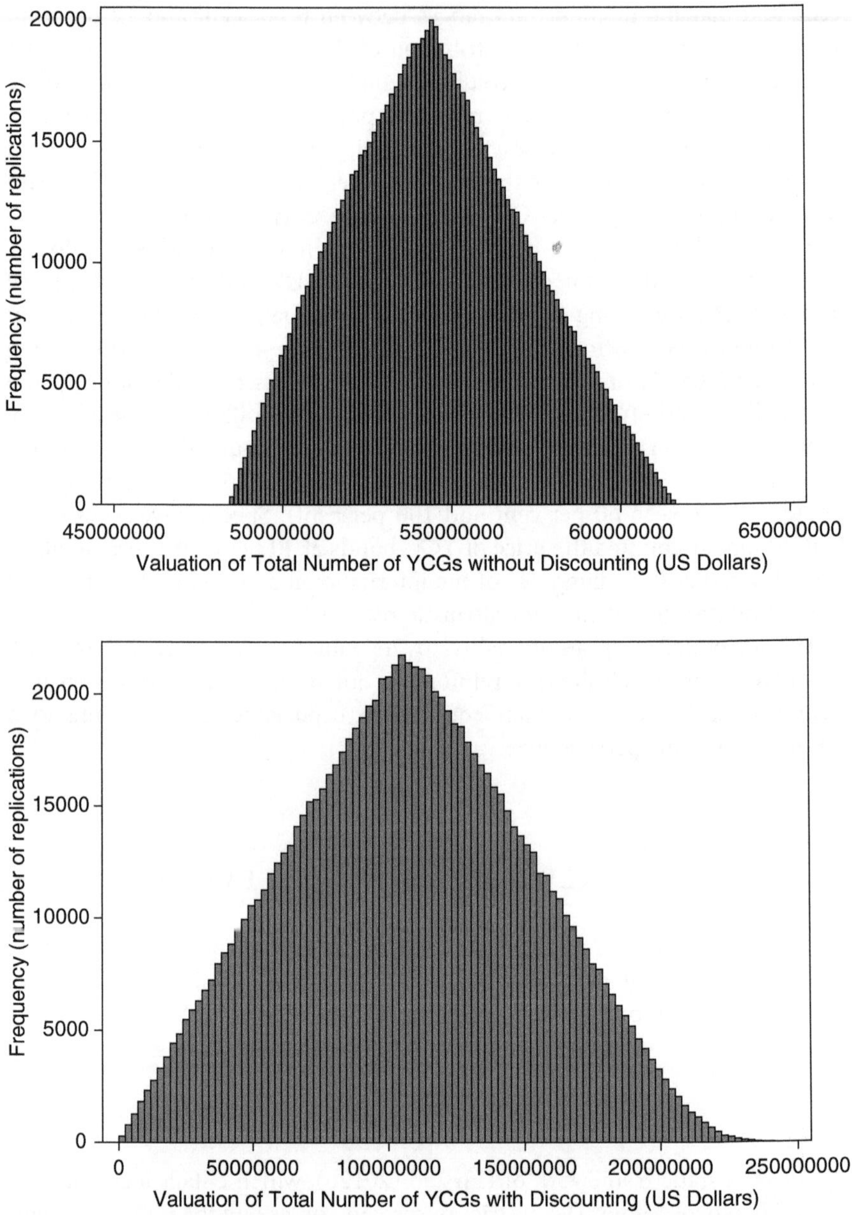

Figure 22.2. Density distribution of the NPV of the expected YGCs revenue.

Note: Each density distribution used one million replications.

Source: Authors' calculations.

$C_i(x_i)$ is an increasing strictly convex function, while $B_i(y_i)$ is an increasing and concave function. $C_i(x_i)$orders a country's deposits according to their extraction costs. This ordering is natural as a country that is extracting x_i units would prefer to first extract the deposits that have the lowest extraction costs. M denotes the coalition countries, which are assumed to act as a single agent, and Nrepresents the set of non-coalition countries. The maximization problems for $i \in N$(e.g., Ecuador) and M are, respectively:

$$\max_{x_i, y_i} B_i(y_i) - C_i(x_i) - p(y_i - x_i), \forall i \in N \tag{1}$$

$$\max_{p, x_M} B_M(y_M) - C_M(x_M) - p(x_M - y_M) - H(x_M + S(p)),$$

where the harm, $H(\cdot)$, experienced by M, is a strictly increasing and convex function. Similar to Harstad (2012b), for simplicity we assume that only M takes the environmental harm into account in the objective function. Equilibrium conditions in the oil market lead the equilibrium price, p, to equal the marginal benefit of consumption and the marginal cost of extraction such that $B_i'(y_i) = p \to y_i = D_i(p) \equiv B_i'^{-1}(p)$, and $p \in C_i'(x_i) \to x_i = S_i(p) \equiv C_i'^{-1}(p) \forall i \in N$. These conditions imply that $y_M = x_M + S(p) - D(p)$ where $S(p) \equiv \sum_N S_i(p)$ and $D(p) \equiv \sum_N D_i(p)$.

Let a deposit ordered between, say, x_i' and x_i'', be characterized by its size or fossil-fuel content, $\Lambda = x_i''$-x_i', and by a marginal extraction cost $c \equiv [C_i(x_i'') - C_i(x_i')]/\Delta$. At this stage, while $C_i'(\cdot)$ describes i's marginal extraction cost, given a set of deposits, c represents the actual extraction cost for a specific but small (marginal) deposit. Different marginal deposits have different c's, and when ordering country i's deposits according to costs, the cost correspondence is given by $C_i'(\cdot)$, whereas $C_i'(x_i)$ is the actual marginal cost when x_i units are extracted. The deposit is owned by $i \in N$, and if the marginal exploitation cost is such that $c < p$, then $i \in N$ would prefer to exploit the deposit.

Coalition $M's$ equilibrium policy means that $B_M'(y_M) - H' \in C_M'(x_M)$, implying that M would prefer to exploit the deposit if and only if $B_M'(y_M) - H' \geq c$. If $c + H' < B'(y_M)$, the deposit will be exploited, whether owned by i or M. If the right to exploit the deposit is transferred from i to M, i saves the extraction cost but loses some profit. Thus, for a given p, the utility of $i \in N$ becomes:

$$U_i = \max_{x_i, y_i} B_i(y_i) - C_i(x_i) - p(y_i - x_i) - (p - c)\Delta, \tag{2}$$

We use the envelope theorem to differentiate (2), such that

$$\frac{dU_i}{d\Delta} = c - p - (y_i - x_i)\frac{dp}{d\Delta}. \tag{3}$$

Similarly, $M's$ utility becomes

$$U_M = \max_{p,\, x_M} B_M(y_M) - C_M(x_M) - p(x_M - y_M) - H(x_M + S(p)) + (p - c)\Delta. \quad (4)$$

We differentiate (4) with respect to Δ, and through the envelope theorem, such that

$$\frac{dU_M}{d\Delta} = p - c. \quad (5)$$

Thus, the transaction of the deposit increases $U_M + U_i$ if $(x_i - y_i)dp/d\Delta > 0$ when $c + H' < B\prime(y_M)$.

If $c \in [B_M'(y_M) - H', p]$, i would exploit the deposit, but M would not. If the deposit is transferred from i to M, $i's$ payoff changes in line with (3). For a given p, the non-coalition's total supply changes from $S(p)$ to $S(p) - \Delta$. Thus, $M's$ utility can be written as:

$$U_M = \max_{p,\, x_M} B_M(y_M) - C_M(x_M) - H(x_M + [S(p) - \Delta]) + p(D(p) - [S(p) - \Delta]). \quad (6)$$

We use the envelope theorem and differentiate (6) to get

$$\frac{dU_M}{d\Delta} = -B_M'(y_M) + H' + p. \quad (7)$$

Hence, the transaction of the deposit increases $U_M + U_i$ if $c - B_M'(y_M) + H' - (y_i - x_i)dp/d\Delta > 0$ when $c + H' \geq B'(y_M)$. For a third country, the transaction between Mand i generates the additional benefit $(x_i - y_i)dp/d\Delta$, $j \in N\backslash i$, where $dp/d\Delta > 0$.

In summary, if $i \in N$ transfers the deposit to M, then

a) $U_M + U_i$ increases if and only if

$$\max\{0,\, c + H' - B_M'(y_M)\} + (x_i - y_i)\frac{\partial p}{\partial \Delta} > 0.$$

b) $\sum_{M \cup N} U_i$ increases if and only if

$$\max\{0,\, c + H' - B_M'(y_M)\} + \sum_N (x_i - y_i)\frac{\partial p}{\partial \Delta} > 0.$$

22.3.2 The YNP in the Market of Deposits

We propose that the YNP (and, consequently, its oil deposits) can be analysed within the framework of Section 22.3.1.

The public debate is divided between the preservation of the YNP and Ecuador's need for the foregone oil revenues. This debate is resolved through the sale or leasing of the rights of extraction to a coalition that may be interested in keeping the oil underground. The revenue raised through this transaction may be taken as a form of compensation for the ecosystem services provided by the YNP. To estimate the price for the transaction of the oil deposit, we rely on Pethig and Eichner (2015). Suppose that the deposit price is p_z. If $p_z = p$, then all non-coalition countries offer for sale all deposits with extraction costs $c \leq p$ because the revenues from selling those deposits are higher than the profit in the absence of deposit trading (i.e., no leasing or sale of oil extraction rights). Thus, the oil price p is an upper bound for the deposit price p_z. If $p_z > p$ the coalition will not accept this price since it can purchase non-coalition countries' profitable deposits at $p_z = p$. Finally, if $p_z = 0$, no country gives away for free any deposit with extraction costs $c < p$, because the extraction of those deposits generates a profit. Hence, $p_z \in [0,p]$, that is, the transaction is linked only to the dynamics of the oil prices. Then, the revenue from the trade of the deposit would need to be negotiated and agreed between the parties on, say, a yearly-basis, for which financial instruments (e.g., futures) may be used to facilitate the transaction. The revenues from the transaction of the extraction rights are thus variable but their calculation is more transparent than the claimed compensation in the original ITT Initiative (i.e., US$3.6 billion).

The cost function $C_i(x_i)$ orders the deposits according to the costs of extraction, and those deposits with high extraction costs are preserved because of their unprofitability. For the YNP we do not know exact extraction costs, but it is reasonable to assume these are not negligible, given the remote location of the deposit. Moreover, given the fragility of the YNP's ecosystem and the risk of significant environmental harm, $c \in \left(B'_M(y_M) - H', p\right)$ still holds as long as the coalition internalizes the harm independently of the oil price.

Finally, we may further assume that the size of the deposit is not large enough to affect oil prices, so that $\partial p / \partial \Delta = 0$. For Ecuador and the coalition to benefit from the transaction of the deposit, $c \in \left(B'_M(y_M) - H', p\right)$. Thus, the coalition will keep the oil underground since the revenues gained by exploiting it are less than the costs to offset/mitigate the environmental harm from its extraction.

Our financial approach incorporates the marginal harm H' and provides a wider set of conditions for which the YNP may be preserved. In this sense, it is a more informative and simplified approach than the original ITT Initiative.

22.3.3 Financial Issues and Implications

The main implications of our theoretical model are as follows:

1) our proposal effectively reduces the counterparty risk (i.e., the risk of default or non-compliance) because of the direct transfer of property

rights (the extraction rights) from the Ecuadorian government to investors or donors (the coalition), and

2) our proposal constructs a realistic appraisal (i.e., mark-to-market) of the ITT-block oil reserves.

The ITT Initiative may be considered a forward contract that is contingent on the price of oil, with indefinite maturity and a strike price of US$36.64 (US $47.24) at a discount rate of 6 per cent (12 per cent) (Figure 22.3). Based on our simulations and the increase in oil prices observed between June 2007 and August 2013, this contract was in-the-money (i.e., it generated value to the Ecuadorian government) only 0.6 per cent (4.4 per cent) of the time. This situation was exacerbated by the non-permanent nature of the commitment to keep oil underground (commonly labelled as Plan B) so that, from the perspective of the investors/donors, there was a high counterparty risk, which made the Initiative unattractive from a financial point of view. Our proposal, however, effectively reduces the counterparty risk because of the direct transfer of the property rights from the Ecuadorian government to the coalition. This transaction aligns the incentives of both parties: the government succeeds in keeping the oil underground for a fair compensation obtained from selling property shares of the ITT block, and the coalition obtains an asset at its fair value with a perspective of revalorization over time. Furthermore, as we proved in Section 22.3.2, for both parties their optimal strategy consists of delaying oil

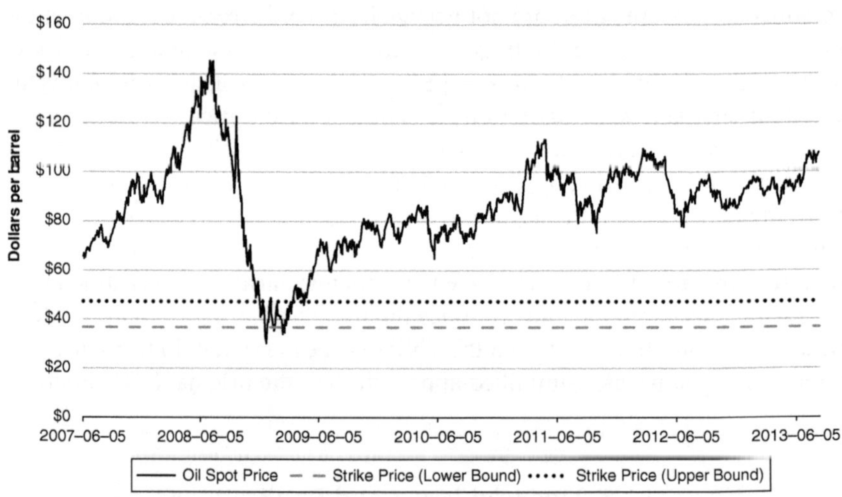

Figure 22.3. WTI crude oil spot price and strike prices of ITT Initiative as forward contract, June 2007–August 2013.

Source: Authors' illustration based on data from Federal Reserve Bank of St. Louis (<https://research.stlouisfed.org/fred2/> (accessed 20 October 2016).

extraction in order to capitalize the value of the ITT block, which was, in fact, the objective of the ITT Initiative in the first place.

In 2010, the ITT Initiative was restructured to make it more attractive to investors/donors. The objective was to reduce the (non-compliance or default) risk by framing the Initiative as a carbon offset instrument to be implemented through an independent Trust Fund. The TOR established that the Initiative should be valued based on the YGCs and that if the Ecuadorian government failed on its promise, the bond holders would be reimbursed at the YGCs' face value. It remains an open question if these provisions actually helped to reduce the counterparty risk, but it is clear that the incentives were never appropriate either for the donors/investors or for the Ecuadorian government.

The main factor that negatively affected the incentives was the structure of the Initiative, and the estimation of its fair value. On one hand, for the Ecuadorian government, the value of the Initiative (particularly its opportunity cost) was linked to the foregone oil revenues from the ITT field. As the price of oil increased, the opportunity cost of the Initiative also increased, making it less likely for the Ecuadorian government to keep its promise. On the other hand, for the donors/investors the value of the Initiative derived from two sources: 1) an intrinsic value related to their altruistic behaviour, and 2) the possibility of trading the YGCs. Figure 22.4 shows that until late 2009 the behaviour of carbon credits followed closely the behaviour of oil prices. From late 2009 onwards, however, the two markets showed a tendency to divorce from one another. In particular, while oil prices showed a tendency to recover following their fall during the financial crisis, the prices of carbon offsets remained flat until 2011 when they experienced a dramatic fall. In other words, from 2011 the value of the ITT Initiative for the Ecuadorian government declined because oil extraction became more valuable, while for the investor/donors the value of the Initiative decreased because hypothetically the YGCs became less valuable in the carbon market. Hence, the strategy of incorporating carbon offsetting did not eliminate an important problem, because by definition the YGCs did not allow mark-to-market of the fair value of the Initiative from both the perspective of the government of Ecuador and that of the investors/donors. Our approach resolves this incompatibility of incentives between the government and the coalition by setting an unambiguous and realistic appraisal of the ITT-block oil reserves: linking its fair value only to oil prices.

Our proposed initiative is also more flexible, as the negotiations for the lease/transfer of the oil fields may involve the terms and time horizon of the contract, so the compensation may be calculated on a period-by-period basis according to the behaviour of oil markets. Furthermore, we gain in simplification as we do not need complex valuations of the ecosystem services of

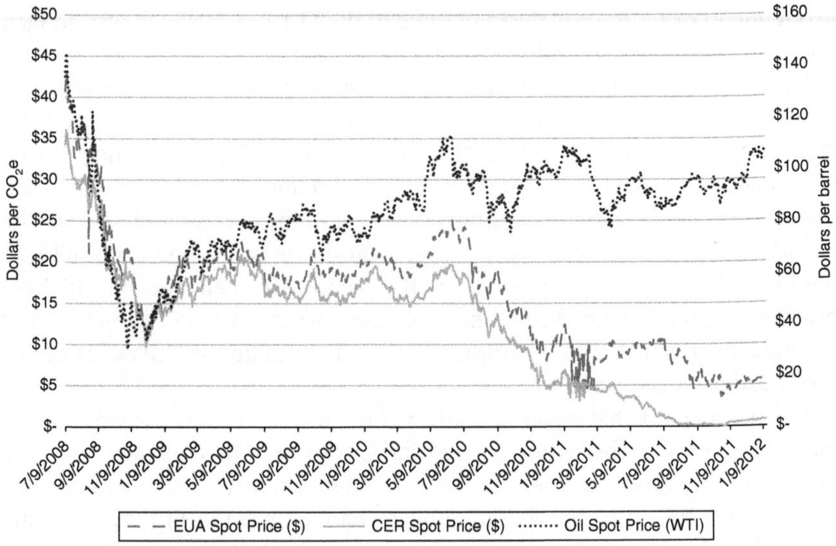

Figure 22.4. Spot prices of WTI crude oil, EUAs, and CERs, August 2008–August 2013.
Source: Authors' illustration based on data from Thomson-Reuters EIKON (http://financial.thomsonreuters.com/en/resources/articles/point-carbon.html last accessed in April 2014).

the YNP in order to set up the transaction of the property rights within this market setting.

22.4 DISCUSSION

The ITT Initiative sought to become an innovative instrument for the mitigation of GHG emissions and climate change. Its main premise was to keep oil underground in exchange for a monetary compensation equivalent to half of the foregone oil revenues, that is US$3.6 billion over a period of ten years. Despite the interest and support it garnered after its announcement, the Initiative was terminated in 2013 because of its inability to collect the requested compensation. Numerous authors still defend the Initiative and argue that it should be reinstated and serve as a model to other countries.

In this chapter we prove that the ITT Initiative was severely flawed since its inception. Not only it was infeasible from a financial perspective, but it also lacked sound theoretical grounds that would make it credible and compatible to the incentives of potential donors or investors. Thus, there was no possibility that the compensation could have been raised. First, the YGCs did not have any financial value because they did not comply with any of the CDM/

REDD+ criteria (i.e., permanence, additionality, non-leakage, and certainty) to be considered equivalent to the CERs and be part of carbon markets. Second, although donors' contributions went further than could have been expected according to a rational-economic perspective, a higher level of altruism would have been required to reach the requested compensation.

We propose a theoretical and practical framework for the development of a 'New ITT Initiative' based on Harstad (2012b). Our main premise is the sale or transfer of the property rights to the ITT fields from the Ecuadorian government to a coalition of countries or parties that would be interested in keeping the oil underground. We develop conditions for the maximization of the utility for both parties. In particular, we prove that the price for this transaction is linked to the behaviour of oil prices.

Several caveats should be mentioned, however. First, the Ecuadorian government claims that oil activities would only affect 0.01 per cent of the Reserve. In this case, the coalition would not be motivated to preserve that share of the Reserve given its negligible size. Second, our approach does not resolve the non-additionality of the ITT Initiative: that is, the YNP has been a protected reserve for decades and no further efforts should be invested in what is supposed to be a commitment from the Ecuadorian government. Third, we acknowledge that our proposal is not free of controversy, particularly for a country and government that might claim that the lease/transfer of the oil fields would be a violation of national sovereignty.

Finally, there are several political and practical implementation considerations. Most importantly, there are two aspects of our proposal that need to be considered: the structure of the coalition and its ability to coordinate, and the guarantee of the rights to oil fields.

Consider first the structure of the coalition. While the proposed transfer of rights reduces the counterparty risk, it comes at the cost of creating a coordination problem on the demand side. From the perspective of the coalition members, it makes sense to acquire rights to the oil fields only if they trust that the other members will also do so and that they will keep the oil underground. If oil starts being extracted by any one of the members, or if they sell their rights to an oil producer, the value of the others' rights will fall. Thus, the counterparty risk is indeed transferred from one actor (the Ecuadorian government) to multiple actors (the members of the coalition). This is important because it can create a hold-up problem: anticipating the inability of other members to commit not to exploit oil in the YNP, each member decides not to acquire rights to the oil fields in the first place. Besides, the non-additionality of oil extraction from the YNP also limits the incentives of the coalition members to buy these rights. We highlight two elements that can help solve the hold-up problem.

First, suppose that before the transfer of property rights takes place there is an opportunity for coalition members to (credibly) signal their commitment to the proposal. This signal could come for instance from their current commitments regarding the Intended National Determined Contributions (INDCs) under the UNFCCC, or from their previous achievements on emission reductions. Using this information, the transfer contract could restrict participation only to those countries that have reduced emissions beyond a particular level in the past and whose promises for further reductions surpass a specific threshold. Alternatively, the contract could provide incentives for the members satisfying these conditions such as deductions from the buying price.

Second, to limit the incentives to breach their commitments, the contracts could establish clauses explicitly disallowing oil extraction by the members of the coalition as well as the sale of rights for that purpose. In the case of violations, the contracts could specify damage payments following a Pigovian structure that would impose a payment on those reneging on their promises equal to the social cost generated (i.e., the fall in the value of other members' rights).

Consider next the guarantee of rights to oil fields. The fact that the fields remain under the geographic domain of the Ecuadorian government implies that a risk of expropriation exists even after the transfer of rights. While it is not easy to eliminate, the experience with oil contracts provides important insights about possible ways to reduce this risk. In the past, oil companies have recurred to international settlement mechanisms to solve disputes with governments in cases of expropriation. An important case is the recent settlement of the Ecuadorian government with Occidental Petroleum Corporation-Oxy. After nine years of litigation, the International Centre for Settlement of Investment Disputes (ICSID) concluded that the actions of the Ecuadorian government vis-à-vis Oxy back in 2006 amounted to expropriation and established a compensation of around US$1 billion that the Ecuadorian government has agreed to pay. While the process was slow and costly, it shows that if the transfer of rights to the oil fields can be framed within the rules of investment protection governing this type of contracts, the ICSID can serve as a third-party guarantor of those rights.

The proposed tools for implementing our proposal have the added advantage of providing a better commitment mechanism for the Ecuadorian government to signal compromise and to establish a reputation for protecting the environment.

22.5 CONCLUSION

In this chapter we presented a new ITT Initiative which is much more transparent and simplified than the failed original ITT Initiative. Our approach is more transparent than the original Initiative because it uses existing financial mechanisms through forward contracts on oil prices. Furthermore, the fair value of our proposal would mark-to-market at oil prices, aligning the incentives of the Ecuadorian government and the coalition members; and, finally, the counterparty risk is reduced because our approach entails the effective transfer of property rights to the coalition.

REFERENCES

Acosta, A. (2007). 'Yasuní, Forjando el Camino Hacia lo Imposible: Dejar el Crudo en el Subsuelo'. Conference, University of Maryland College of Chemical & Life Sciences.

Acosta, A., E. Gudynas, and J. Vogel (2009). 'Dejar El Crudo En Tierra O La Búsqueda Del Paraíso Perdido . Elementos Para Una Propuesta Política Y Económica Para La Iniciativa de No Explotación del Crudo del ITT'. *Revista de la Universidad Bolivariana*, 8: 429–52.

Arsel, M. (2012). 'Between "Marx and Markets"? The State, the "Left Turn" and Nature in Ecuador'. *Tijdschrift voor Economische en Sociale Geografie*, 103(2): 150–63. doi:10.1111/j.1467-9663.2012.00709.x.

Arsel, M. and N. A. Angel (2012). '"Stating" Nature's Role in Ecuadorian Development: Civil Society and the Yasuni-ITT Initiative'. *Journal of Developing Societies*, 28(2): 203–27. doi:10.1177/0169796X12448758.

Bass, M. S., M. Finer, C. N. Jenkins, H. Kreft, D. F. Cisneros-Heredia, S. F. McCracken, N. C. A. Pitman, P. H. English, K. Swing, G. Villa, A. Di Fiore, C. C. Voigt, and T. H. Kunz (2010). 'Global Conservation Significance of Ecuador's Yasuní National Park'. *PLoS ONE*. doi:10.1371/journal.pone.0008767.

Bucaram, S., M. A. Fernández, and D. Grijalva (2016). 'Sell the Oil Deposits!: A Financial Proposal to Keep the Oil Underground in the Yasuni National Park, Ecuador'. 2016/14. UNU-WIDER, Helsinki.

Haddad, R. (2011). 'Un-Conventional Approach: Ecuador's Yasuni-ITT Initiative is in Discord with the UNFCC'. *Sustainable Development Law & Policy*, 12. Available at: <http://digitalcommons.wcl.american.edu/cgi/viewcontent.cgi?article=1509&context=sdlp> (accessed 18 October 2016).

Harstad, B. (2012a). 'Climate Contracts: A Game of Emissions, Investments, Negotiations, and Renegotiations'. *The Review of Economic Studies*, 79(4): 1527–57. doi:10.1093/restud/rds011.

Harstad, B. (2012b). 'Buy Coal! A Case for Supply-Side Environmental Policy on JSTOR'. *Journal of Political Economy*, 120(1): 77–115. Available at: <http://www.jstor.org/stable/10.1086/665405?seq=1#page_scan_tab_contents> (accessed 18 October 2016).

Kim, M.-K., and B. A. McCarl (2009). 'Uncertainty Discounting for Land-Based Carbon Sequestration'. *Journal of Agricultural and Applied Economics*, 41(1): 1–11.

Kim, M. K., B. A. McCarl, and B. C. Murray (2008). 'Permanence Discounting for Land-Based Carbon Sequestration'. *Ecological Economics*, 64(4): 763–9. doi:10.1016/j.ecolecon.2007.04.013.

Larrea, C. (2010). 'Análisis Comparativo de Ingresos Futuros Del Estado Para El Bloque ITT'. Available at: <https://www.sussex.ac.uk/webteam/gateway/file.php?name=analisiscomparadoflujosestadoitt2.pdf&site=449> (accessed 18 October 2016).

Larrea, C. and L. Warnars (2009). 'Ecuador's Yasuni-ITT Initiative: Avoiding Emissions by Keeping Petroleum Underground'. *Energy for Sustainable Development*, 13(3): 219–23. doi:10.1016/j.esd.2009.08.003.

Martin, P. L. (2011). 'Global Governance from the Amazon: Leaving Oil Underground in Yasuní National Park, Ecuador'. *Global Environmental Politics*, 11(4): 22–42. doi:10.1162/GLEP_a_00082.

Ministerio Coordinador de Política Económica (2013). 'Impacto Macroeconómico de la Explotación Petrolera en el 0.1% del Parque Nacional Yasuni'. Ministerio Coordinador de Política Económica, Quito.

Pellegrini, L., M. Arsel, F. Falconí, and R. Muradian (2014). 'The Demise of a New Conservation and Development Policy? Exploring the Tensions of the Yasuní ITT Initiative'. *The Extractive Industries and Society*, 1(2): 284–91. doi:10.1016/j.exis.2014.05.001.

Pethig, R. and T. Eichner (2015). 'Unilateral Mitigation of Climate Damage via Purchase of Fossil Fuel Deposits.' Beiträge zur Jahrestagung des Vereins für Socialpolitik 2015: Ökonomische Entwicklung—Theorie und Politik—Session: Environmental Economics IV, No. D13-V4.

Petroamazonas (2010). 'Prefactibilidad Desarrollo Campos Tiputini-Tambococha-Ishpingo'. Unpublished Manuscript.

Rival, L. (2010). 'Ecuador's Yasuní-ITT Initiative: The Old and New Values of Petroleum'. *Ecological Economics*, 70(2): 358–65. doi:10.1016/j.ecolecon.2010.09.007.

Singleton, S. (2000). 'Co-operation or Capture? The Paradox of Co-management and Community Participation in Natural Resource Management and Environmental Policy-making'. *Environmental Politics*, 9(2): 1–21. doi:10.1080/09644010008414522.

UNESCO (United Nations Educational, Scientific and Cultural Organization) (1989). 'MAB Biosphere Reserves Directory'. Available at: <http://www.unesco.org/mabdb/br/brdir/directory/database.asp> (accessed 18 October 2016).

United Nations Development Group (2010). 'Ecuador Yasuni-ITT Trust Fund: Terms of Reference'. Available at: <http://mptf.undp.org/yasuni> (accessed 18 October 2016).

Vallejo, T. and M. Friant (2015). 'Ecuador's Yasuní-ITT Initiative for Mitigating the Impact of Climate Change'. *Environmental and Planning Law Journal*, 32(3): 278–95.

Vogel, J. H. (2010). *The Economics of the Yasuní Initiative: Climate Change as If Thermodynamics Mattered.* London: Anthem Press.

Williamson, O. E. (1983). 'Credible Commitments: Using Hostages to Support Exchange'. The *American Economic Review*, 73(4): 519–40.

Part VII

Social Inclusion

While clean energy can serve as a robust enabler of economic growth, ensuring socially inclusive development remains a signature political economy challenge for clean energy transitions. With that in mind, this part features chapters that explore the linkages between social development programmes and clean energy aspirations, climate and development implications of energy poverty reduction programmes, and the impacts of energy development actors and interests on local communities and their agendas. Achieving inclusive economic growth in concert with clean energy and climate goals is a complex task; the subsequent case studies intend to illustrate some of the relevant, governing political economy elements to be considered by policy makers in this realm.

Ezeanya and Kennedy explore the various strategies employed by the Government of Rwanda in achieving increased biogas use among the rural poor, particularly through an agricultural direct assistance and poverty reduction programme known as Girinika. This case study, more generally, attempts to shed light on the conditions necessary for successful implementation of clean energy pro-poor reforms in rural communities. McCormick explores hydroelectricity and biomass development in the Brazilian Amazon, and the linkages between the needs and agendas of local communities, the economic interests, and politics at the national level, and international social actors and financial interests. Kruger, Tait, and de Groot characterize the political economy of residential energy choices and energy poverty reduction programmes in developing countries through comparative analysis of South Africa and Indonesia.

23

Integrating Clean Energy Use in National Poverty Reduction Strategies

Opportunities and Challenges in Rwanda's Girinka Programme

Chika Ezeanya and Abel Kennedy

23.1 INTRODUCTION

Poverty reduction programmes set up by national governments and targeted at rural communities have experienced the twin challenges of dearth of adequate energy sources and the effects of climate change on the rural landscape such as forest degradation and increased household air pollution (Barnett 2000). Although the poorest segment of the global population contribute only about 10 per cent to total global emissions, they live in areas that are most vulnerable to climate change. Outside of the general impact of the harmful environmental footprints of the richest 1 per cent globally, which could be as much as 175 times that of the poorest 10 per cent, poor people, especially in rural areas often have to bear the direct consequences of their own natural resource use. Casillas and Kammen (2010: 1) assert that 'mitigating climate change, increasing energy access, and alleviating rural poverty can all be complementary, their overlap defining an energy-poverty-climate nexus'. SSA is a region with many rural communities whose energy use lead to impactful economic, health, and environmental consequences, especially on women and children (Banerjee et al. 2012). Governments across the region must work to reduce dependence on depleting and potentially harmful energy sources while tackling rural poverty.

Integrating clean energy use in poverty alleviation discourse is increasingly gaining global attention, mostly due to the established connection between high-level poverty and its economic, environmental, and health effects on poor communities. The United Nations Development Programme (UNDP) has connected the dots between energy, poverty, and sustainable development. In publications and in policy advice to national governments, UNDP emphasizes the need to focus on generating clean and modern energy that is sustainable, since it 'can be an engine for poverty reduction, social progress, equity, enhanced resilience, economic growth, and environmental sustainability' (UNDP 2015: 1). The International Energy Agency (IEA) notes that in 2014, more than 2.6 billion global citizens were without clean energy for household use and were utilizing fuelwood as well as other forms of resources that result in high levels of household air pollution with its attendant adverse effects; more than 95 per cent of this figure are either in sub-Saharan Africa (SSA) or in developing Asia and 84 per cent are in rural areas (IEA 2014).

The country of Rwanda in East Africa presents a good case in point where poverty levels is synonymous with fast depleting natural energy sources. In 2014, 85 per cent of Rwanda's population relied on wood as main source of energy, a figure that represents a 5 per cent decline from four years earlier when 90 per cent was dependent on fuel burning (FAO 2011; *The New Times* 2015). The high number of Rwandans using fuelwood has led to a recorded national deficit of 8.5 million trees per year, mainly lost to burning wood for household energy use. Forty-five per cent of Rwandans live below the poverty line, representing roughly half of the current population dependent on fuelwood in meeting their energy needs (NISR 2012). By building in the adoption of clean energy use within poverty alleviation programmes, the government of Rwanda has strived to promote clean energy use among rural dwellers, with emphasis on the rural poor.

This chapter focuses on the use of cow dung for domestic biogas among beneficiaries of the Rwandan government's agricultural direct assistance programme, known as Girinka. Girinka entails the gifting of pregnant dairy cows to the poorest of the poor of rural dwellers by the government. Under Girinka, beneficiaries, when they have a minimum of two cows in kraal, are strongly encouraged and assisted—materially and financially—to install domestic biogas plants that use dung and urine to generate energy. The government worked through the National Domestic Biogas Programme (NDBP) and the result has been the migration of a number of rural dwellers who are Girinka beneficiaries away from fuelwood use to using biogas (Bedi, Pellegrini, and Tasciotti 2013). This chapter aims to improve our understanding of the various strategies utilized by the government in achieving increased biogas use among the rural poor, with emphasis on beneficiaries of the Girinka programme. Understanding gained from this study will bring increased enlightenment on the conditions necessary for successful implementation of clean energy promoting pro-poor reforms in rural communities.

Questions guiding this analysis will centrally arise from two key political economy variables—institutions and actors. Along that line, the role of the government of Rwanda and donors as drivers of change in clean energy use among the rural poor will be discussed, and questions that will establish who the main actors are in the promotion of clean energy use among the rural poor in Rwanda will be answered. Other questions along that axis include: What is the current policy environment and what alignments exist with international frameworks? What political and economic factors were at play in the increased use of biogas by rural Rwandans? Further, this chapter will apply the Power Analysis tool of political economy by raising questions around who is included and who is excluded from government of Rwanda's promotion of clean energy use for the rural poor. Finally, social inclusion approach to rural poverty or pro-poor policy-making requires that questions be raised on the perceptions of the rural poor regarding the whole exercise. This is extremely necessary since 'an effective rural development transformation programme that focus on the rural poor requires that a clear understanding of who the rural poor are, where they live, [their aspirations] and the challenges posed by the prevailing poverty levels in their respective habitats is known' (Ewang 2013: 11).

23.2 RWANDA: BACKGROUND INFORMATION

Rwanda is the most densely populated country in SSA and covers a surface area of 26,338 km^2. Rainfall is plentiful, and annual temperature average ranges between 16–20°C, but with a 0.60 ha per household availability, arable land is scarce. According to the National Institute of Statistics of Rwanda (NISR), Rwanda's population density has increased from 321 persons perkm^2 in 2002 to 416 in 2012 (NISR 2012). Overexploitation is the result, accompanied by disastrous environmental consequences. Rwanda's protection areas consist of three national parks: Volcanoes National Park, Nyungwe National Park, and Akagera National Park. Combined, all three have lost more than 50 per cent of their original surface area in the past 40 years (Twagiramungu 2006).

The republic of Rwanda came forcefully and widely into global consciousness with the genocide of 1994 where an estimated 1 million, mostly Tutsi citizens, were killed. At the root cause of the genocide is economic displacement, food, and land insecurity. The end of the genocide marked a transition to a post-conflict nation building stage where there is collaboration between government, private sector, international development partners, and interest groups to advance the interest of the majority poor citizen. Rwanda is classified among the poorest countries in the world; 166th out of 186 (UNDP 2012); the household poverty survey stood at 44.9 per cent in 2010–11 down from 58.9 per cent in 2000–01 (NISR 2012). Extreme poverty in 2012 stood at 24.1 per cent down from the 40 per cent of ten years earlier (IFAD 2012).

Rwanda's 2003 constitution was specifically crafted to reduce poverty and to empower citizens economically (MINECOFIN 2007). Subsequent documents in support of the constitution include the Economic Development and Poverty Reduction Strategy (EDPRS), which came into effect in 2007, and is focused on the reduction of poverty in Rwanda from 64 per cent to 30 per cent, and increasing gross domestic product per capita from US$220 to US$1,240 by 2017 (MINECOFIN 2007). An agrarian country with approximately 90 per cent rural dwellers, poverty reduction strategies in Rwanda, to be effective, must be structured for widespread appeal in rural agrarian settings.

Environmentally, Rwanda's ecosystem is at risk due to high population density and the heavy demand for natural resources. Rwanda's efforts in urban planning have been overtaken by massive post-conflict movements from rural to urban areas, which have placed significant stress on the environment. The nation's industrial sector is also expanding at an appreciable pace, bringing with it additional demands on the already overburdened environment. Deforestation was a serious menace until targeted efforts at reforestation began in 2010 through tree planting and support for cleaner cook stoves (EUEI 2009a). Between 2010 and 2014, Rwanda's sustainable climate management strategy has proved successful; an additional 10 per cent surface area in the country is now occupied by forests and 5 per cent less Rwandans use fuelwood for energy generation (*The New Times* 2015).

23.2.1 Household Energy Use in Rwanda

In 2010, Rwanda's Director of Forestry Field Programmes in the nation's National Forestry Authority warned of a real threat of desertification, with the revelation that in that year, only 533,000 hectares or 20 per cent of the country was covered by forests (FAO 2011). In response, the government embarked on a concerted policy effort aimed at reforestation. Seventy-six per cent of Rwanda's rural land (2,467,000 ha) is used for agricultural purposes, specifically for crops and animal production, while 16 per cent is forested. Currently, 1.6 per cent of Rwanda's rural land mass is designated as 'other land', and is neither used for agricultural purposes nor classified as a forested area, but includes barren, built-up, and wooded land (World Bank 2012: 7). Since firewood and charcoal usage record highest in the rural areas of Rwanda, ways were sought to introduce agricultural rural communities to alternative clean energy sources, with emphasis on biogas that runs on cow dung (FAO 2011).

As far as energy sources are concerned, biomass, including charcoal, dominates other forms of energy in Rwanda. Down from 90–95 per cent in previous years, it is estimated that approximately 85 per cent of Rwandans depend on biomass for use in household cooking. The use of energy sources such as liquid petroleum gas or kerosene is insignificant. Over the past 15 years, energy demand in Rwanda has grown at the same rate as population growth, fast-paced urban expansion, and economic activities (FAO 2011). The increasing cost of petrol, low access to electricity, and the significant growth of industries have combined to place pressure on fuelwood as a reliable and inexpensive source of energy for most households in Rwanda. In the rural and urban areas, all but the wealthiest 5 per cent of households rely on firewood and/or charcoal for household cooking. Wood is sometimes used directly as fuel or converted into charcoal for use in cooking (FAO 2011). About 6 per cent of households use crop residues and peat to cook with, especially in rural areas (FAO 2011).

Annually, total fuelwood consumption in Rwanda is at 2.8 million tonnes, while it is estimated that charcoal, when converted into wood usage, accounts for up to 50 per cent of total wood-fuel consumption in the country (World Bank 2012). In all, Rwanda loses up to 35 per cent of wood used in charcoal-making due to the conversion process (GTZ 2009).

Economically, fuelwood usage is also costly for the average Rwandan household. On average, households spend between 10–15 per cent of their monthly incomes on purchasing fuelwood and charcoal (FAO 2011). In 1991, the average consumption of fuelwood and charcoal in Rwanda was 0.33 kg per person per day (World Bank 1991); by 2000, the figure had risen to 1.93 kg per person per day, with an annual per capita consumption of more than 1m^3 of wood for the entire population (GTZ 2009). The increase is directly proportional to growth in population. In 1991, Rwanda's population stood at 7.1 million while in 2002, the figure was 8.1 (NISR 2016). Food and Agriculture Organization (FAO) estimates that in Rwanda, woodfuel consumption is estimated at 2.7 million t/year, out of which the urban capital accounts for an annual charcoal consumption of 120,000 t, equivalent to 1.2 million m^3 or 850,000 t of wood (FAO 2011). Rwanda's Ministry of Natural Resources projects that between 2009 and 2020, 'the consumption of charcoal and fuelwood in Kigali [Rwanda's capital city alone] will increase from 0.99 to 1.4 million tons' annually (GoR 2013: 18).

23.2.2 Domestic Energy Policy in Rwanda

The need to manage Rwanda's biomass and fuelwood production and consumption adequately has led to the formulation of policies and strategies, and the establishment of specifically mandated government agencies. Rwanda's Energy Policy is based on three major documents: the National Energy Policy, the National Energy Strategy for 2008–12, and the Biomass Energy Strategy (BEST) from 2008 (Bedi, Pellegrini, and Tasciotti 2013). One major reason for policy action on energy in Rwanda has been identified as 'economic growth, which in turn is seen as a prerequisite for tackling poverty' (World Bank 2012: 21). The National Energy Policy was drafted in 2008 as an update of the 2004 Energy Policy statement. With particular concern to this study, the policy among others, aims to 'set the National Energy Policy within Rwanda's long-term development plans and strategies' and to focus on household energy requirement in addition to gender issues arising from such (EUEI 2009b: 1).

The Ministry of Infrastructure (MININFRA) and the Ministry of Environment and Natural Resources (MINIRENA), developed the Biomass Energy Strategy (BEST) in 2008–9. The BEST was designed to address four key areas:

(1) To sustainably increase fuelwood supply including the establishment of new, and management of existing plantations, and the 'professionalization of the charcoal value-chain' (EUEI 2009a).
(2) Improvement of energy use efficiency in Rwanda's households through the extension of all necessary support to manufacturers and importers to produce innovative materials that conserve biomass.
(3) The promotion of alternatives to fuelwood, with particular emphasis on peat, papyrus, and biogas.
(4) Development of institutional capacity within government agencies to be in charge of biomass and energy within the short and medium terms (EUEI 2009a).

Rwanda's National Domestic Biogas Program (NDBP) was established in 2007 and was supported by development partners in Rwanda, specifically Netherlands Development Organization (SNV), and the German Aid Organization known as GIZ. The objective of the project is to improve the provision of biogas to meet the cooking and lighting needs of households that own a minimum of two cows. Rwanda's NDBP aims to install at least 15,000 biogas digesters in rural households, in addition to institutional digesters to be installed in prisons and schools (SNV 2012). Since inception, NDBP has installed domestic biogas in 4,600 households. NDBP is also involved in training of masons, and the project has trained 555 masons in household biogas constructions and maintenance, out of which 195 have registered businesses; 52 masons have been trained in the mechanics of institutional digesters construction and maintenance (REG 2014).

23.3 GIRINKA AND DOMESTIC BIOGAS USE IN RWANDA

The decision of the government of Rwanda to promote biogas use in the country is hinged on the rapid occurrence of deforestation and the availability of cow dung across rural areas of Rwanda. The disappearance of Rwanda's forests, in addition to the change noticed in Rwanda's climate because of greenhouse gas emissions, prompted the government to explore other clean energy alternatives (EUEI 2009b). Biogas was seen as a veritable alternative since it is a renewable fuel cooking solution and is environmentally friendly (EUEI 2009b). Most important in the government's decision to expand biogas access in the country is that unlike at any other times in Rwanda's history, more and more Rwandans in rural areas are becoming owners of cattle. The rise in cattle ownership in Rwanda is a result of a government poverty reduction programme known as Girinka.

Girinka is a government poverty reduction strategy created exclusively for rural communities in Rwanda in 2006. Girinka entails the gifting of a pregnant dairy cow to selected poor and vulnerable households in rural communities by the government. Historically, Rwanda's culture, politics, and economy revolve around cattle. Cattle ownership has played a determining role in the state of Rwanda's economy; indeed, the genocide of 1994 could, in the final analysis, be summarized as a class and economic warfare between the owners of cattle and the non-owners (Mamdani 2001). Selection of beneficiaries for the Girinka programme are left almost entirely in the hands of communities who are given certain criteria from where eligible candidates are presented to the Rwanda Agriculture Board, through the districts. The basic requirements being that a person must be malnourished and destitute, but must be of good standing in community (i.e., not an alcoholic, gambler, or lazy person). The proposed recipient must also have a small patch of land where cattle shed can be built since the programme operates a zero-grazing policy (Ezeanya 2014).

As a modern poverty alleviation strategy, Girinka can be traced back to the ancient practice of richer, cattle owning Rwandans giving cows to their poorer neighbours. Within the Rwandan culture, heads of cattle are usually extended as goodwill gifts during such landmark events as births, deaths, and marriage, while exchange of cattle between previously warring factions often signify a successfully resolved conflict (Rwanda Governance Board 2013). Lifelong friendships are consolidated through the exchange of cows, while no form of recognition for service rendered to the state or a person is ever complete without the gifting of a cow to the recognized individual (Rwanda Governance Board 2013).

Rwanda's Girinka programme is an indigenous knowledge-based poverty alleviation strategy, which was widely accepted by rural Rwandans owing to the cultural significance of cattle. Girinka—which in its loose Kinyarwanda translation means 'may you own a cow'—was established to address the challenges of entrenched malnutrition and extreme poverty, which plagued several parts of rural Rwanda. The idea behind the exercise is that cows will provide milk for starving households, while at the same time the sale from excess milk production could be beneficial in providing a means of income to very low-income households, usually subsistence farmers (Rwanda Agricultural Board 2013). Outside of the direct gift from government to poor farmers, cattle recipients are also mandated to give the second female offspring produced by the original dairy cow to another poor and vulnerable household within the community. Girinka has been supported by several organizations and high net worth individuals in and outside of Rwanda, thereby greatly reducing the burden of sustainability placed on the shoulders of the government.

From the time of its establishment until early 2015, Girinka has bestowed about 200,000 cows upon as many households. This is not inclusive of the mandatory requirement on recipients to pass on the second female offspring to another poor household in the community. The number of households that have received cows because of the Girinka programme is significant in a country of a little less than 12 million.

Among the achievements of the Girinka programme is the reduction in level of malnutrition among Rwandans in rural areas through increased milk production and consumption (Ezeanya 2014). In fact, national production of milk is recorded to have increased by 11.3 per cent during 2009–11, while beef production increased by an estimated 10 per cent (IFAD 2012).

23.3.1 Biogas Use among the Rural Poor in Rwanda

When in 2007 the government of Rwanda established the NDBP, it was with the intention of harnessing the reality of cattle ownership being a growing phenomenon in rural Rwanda. In view of the establishment of the Girinka programme in 2006, there were hopes that more and more villagers would become cattle owners and be able to migrate away from fuelwood to using biogas in meeting household energy needs. A decade after the establishment of Girinka, several Girinka beneficiary households have started using biogas (Kagabo 2014).

Using mostly structured and unstructured interviews, including focus group discussions, the researchers sought to understand the reasons why some Girinka beneficiary households have adopted biogas use while others have not. The lead researcher also conducted interviews among policy makers to determine the efforts of the state, and other supporting institutions, and agencies in promoting the use of biogas across Rwanda.

A total of 130 households were interviewed drawn from three rural provinces of Rwanda: the Northern, Eastern, and Southern provinces. Across the three provinces, 59 Girinka beneficiary households that installed biogas and 71 Girinka beneficiary households that did not install biogas were surveyed. Sample size was selected based on the estimated number of Girinka beneficiaries vis-à-vis those among them who are using biogas within each province, and on the ease of accessibility of the researcher to biogas and non-biogas users. An interview conducted with the director of the Girinka programme, prior to going to the field, was essential in identifying possible respondents, since it was difficult to establish actual numbers of Girinka beneficiaries who are also biogas users. This absence of official records is due to a lack of coordination between the NDBP and the Rwanda Agriculture Board.

Respondents are household heads with mostly only primary education. Average monthly income is about US$15, while the average number of persons in a household is six. Ages of respondents across the provinces ranged from 25 to 72, while average cow ownership stood at two per household. Seventy per cent of those without biogas go to the forest to search for firewood, dry leaves, and splints, while about 30 per cent have to buy firewood or charcoal because of deforestation. A five-foot log of firewood costs about US$7, while a 23 kg bag of charcoal costs approximately US$12.

All respondents attested to how the Girinka programme has assisted in lifting them out of extreme poverty. Many could not afford one square meal and lived off corn porridge, but since receiving a cow from the programme, they have access to nourishing milk, and manure for their farms, leading to increased crop production. Profit from sale of excess milk has often been invested in purchase of health insurance, school fees payment, home improvements, agribusiness investment, starting small businesses, and so on. A few beneficiaries had sick and unproductive cows, but some were able to get replacements while others are still awaiting replacement.

23.3.2 Biogas Use among Girinka Beneficiaries in Rwanda: An Analysis

Girinka beneficiaries who are biogas users noted numerous benefits starting with restoration of neighbourhood forests that were previously over-foraged for firewood. The economic benefit is in a reduction in money spent on buying expensive firewood or charcoal. Although this might not be so in many rural communities across Africa, deforestation has caused some parts of rural Rwanda to become bare of bushes and forests where firewood can be foraged forcing the rural poor in such areas to buy firewood for cooking. A 72-year-old respondent who is not a biogas user notes the extent of deforestation within his vicinity; when asked how he sources firewood for cooking said 'I just pick small splints and leaves. We no longer have a forest where to get firewood.' Biogas also provides manure for the garden saving money on fertilizers and boosting organic agriculture. This is because not all of the dung and urine are converted to gas, and the residue can still be useful in providing nutrition to plants.

Cooking with firewood and charcoal can take its toll on the health of household members. Respondents who use biogas note that their health is better since adopting biogas. This is mostly as a result of the prevention of diseases resulting from smoke such as pulmonary heart disease, respiratory disease, and so on. As a respondent from Eastern province said, 'I was suffering from asthma when I was cooking with firewood, but since I migrated to biogas, I am now OK'. In addition, cleanliness in the kitchen can be maintained more with biogas cookstoves than with firewood or charcoal, leading to improved sanitary conditions. Biogas has also resulted in improved household nutrition as users are able to cook more frequently; firewood and charcoal take a lot of effort and time to light up. Safety wise, household fire incidents that usually occur with the use of firewood and charcoal can be prevented with biogas.

Much of rural Rwanda is yet to be electrified and biogas can provide light at homes without electricity. With biogas, many beneficiaries are able to charge their phones, iron their clothes, and carry out other such household activities that require energy usage. Improved lighting provided by biogas lamps has led to the availability of prolonged study time for children after sunset. With biogas unlike firewood, cooking can be done at any time of the day and any season in the year. With firewood, however, cooking can hardly be done at night since kitchens are located outside of the main building, are usually without electricity or security lock, and are not constructed to offer warmth during cold seasons. Additionally, cooking with firewood during rainy season can be a difficult endeavour once wood becomes wet and difficult to light, and charcoal is very expensive.

Children and women have saved much time that was previously spent on gathering firewood and on the cooking process, since biogas demands little attention during cooking time, therefore users are able to multitask. Cooking with biogas is cheaper and faster than any other method. Money and time saved from buying firewood and charcoal, from gathering firewood, or lighting charcoal can be invested in other more economically productive endeavours. This is more so when account is taken of time spent on gathering firewood, preparing fire hearth, scrubbing soiled pots and pans, and cleaning the kitchen are factored in. Using biogas is much simpler. As a respondent from the Northern Province said, 'even a child can use it because it is convenient'.

Despite the noted benefits of biogas use, many rural households that wish to use biogas are unable to do so. For many of these poor rural households, the major reason for their inability to use biogas is the costs involved in its installation. Further questioning around this revealed discrepancies in government of Rwanda's financial requirements regarding provision of biogas to citizens across rural Rwanda. According to government sources, standard practice is for households to contribute RWF100,000 (US$150) and for government to provide RWF300,000 (US$450). In the Eastern province, several respondents concurred that that they had to pay RWF100,000 to government before biogas plants could be installed. However, others interviewed noted that their contribution was only RWF50,000. The stated amount is in addition to RWF300,000 (US$450) provided by the government as subsidy in the form of material and technical support. However, in parts of the Southern and

Northern provinces, communal biogas plants were provided completely free of charge by government to clusters of families (five each). Some Girinka beneficiaries in the Eastern province who do not use biogas voiced their feelings of marginalization since, according to them, government provided biogas completely free for some, and why not for them as well. Although the general emphasis of the non-biogas users for not installing the plant is as a result of lack of funds, it appears the perceived sense of double standards is partly responsible for preventing some who can afford the co-pay not to have the willingness to contribute towards installation.

Differences in subsidy extended to rural dwellers for biogas installation, according to government sources, is linked to the provision of free biogas installation to clusters of households selected from the poorest of the poor in the rural areas who were previously living in forests and far-flung areas. Providing free collective biogas is part of the incentive for attracting isolated dwellers to come together and live in villages, to make for easier provision of infrastructure. Regarding those who have refused to contribute towards installation of biogas in their households, government sources are of the view that many who can afford it are held back by a sense of entitlement. This is more so since these Girinka beneficiaries received cows free from the government and are very likely expecting to get biogas installed free. However, there is the need for consistency in dealing with beneficiaries in terms of how much subsidy to be paid to government for installation. The fact that some beneficiaries had their biogas plants installed entirely free of charge while others had to pay a certain amount might have discouraged those who wish to adopt biogas from making any financial contribution as they are convinced that by holding out a little longer, government will install biogas in their homes entirely for free. Part of the challenge for government could be the free installation extended to encourage forest dwellers or the plants used as pilots. Such issues should be addressed to prevent perceptions of marginalization and a passive resistance attitude towards co-payment for biogas installation by other equally poor citizens.

Several Girinka beneficiaries who are not biogas users claim that they have made payments to the agency in charge of biogas installation, but as of yet do not have their biogas plants installed. This is in addition to several broken down biogas plants, but for which nothing has been accomplished after months of reporting the fault. Other challenges being faced by users of biogas include difficulties and costs of getting technical support and spare parts in the case of breakdown of biogas plants. Respondents note that they end up going back to cooking with firewood or charcoal until such time when a technician can be found and they can save money needed for spare parts purchase. Additionally, some respondents complained that their plants were not well installed and that it leaks gas. This is a dangerous situation as it can result in household fire incident. In response, government agencies note that the biogas project was stalled as a result of loss of interest on the part of donors who previously funded the programme. At the beginning of the biogas project, donors such as SNV and GIZ actively supported the government in setting up biogas plants in parts of rural areas. However, as most donor funds are short term and project based, the two donors did not continue to renew funding for the project after the term expiration leaving the government to bear the full burden of biogas installation and maintenance. Rwanda's government is dependent on donors for up to 40 per cent of its annual budget (World Bank 2016a); when donors pull out of development projects, progress slows down and such is the case with the biogas installation among the rural poor. The government by itself is not without sufficient funds to continue with the programme, thereby affecting the expansion and sustainability of domestic biogas among Rwanda's rural poor. The result is shortage of materials, equipment, and technicians to install and maintain biogas plants. At fewer than 5,000 installed domestic biogas plants in Rwanda, the government is presently short of the projected estimate of 15,000 installations (SNV 2012; REG 2014).

However, despite the acute shortage of funds to continue with the programme, the government has organized training of technicians with the help of other donors not initially involved in the project. The former situation where only one technician was assigned to a district is being changed. Three technicians are currently being trained per district and two masons per sector. There are plans to train several technicians in each district. In addition, the previous method of a centralized biogas programme implementation is gradually being reversed. Technicians, for instance, are being trained and are working at the district level. When technicians are available, government is convinced that other related issues such as unavailability of biogas plant spare-parts will be taken care of, since the raw materials used in manufacturing biogas plants are 100 per cent sourced from Rwanda. The problem of faulty installations will also be solved by availability of technicians, as there is one-year post-installation warranty on repairs. Donors are also assisting government on a short-term basis to train trainers with trainees being biogas users who are instructed to train others within their communities. Indeed, 5 per cent of respondents in the Eastern province indicated that they travelled to Kayonza district office to be trained on the use of biogas. Others say they were neither adequately trained nor briefed prior to installation of the plants. However, there is no indication that information gained from these trainings was passed on to their neighbours as many stated that there was no formal training before they began to use biogas. Even prior to initiation installation, there is need to enlighten Rwandans on the benefits of biogas use since about 50 per cent of non-biogas using respondents did not understand the need for biogas installation although they had heard that it is beneficial. Efforts ought to be invested into making Rwandans aware of the dangers of firewood use, through widespread awareness campaigns. This will be useful in expanding national biogas reach, through for instance; encouraging citizens who can afford the full price for biogas installation to cooperate with the government in its procurement.

In crafting pro-poor policies, it is important that government finds out what the rural poor really need and what can work best in their environment. It is important for local knowledge to act as the foundation to development policies, this inside-out, rather than outside-in model has been developed into many frameworks and approaches including the participatory development model (Zewde 2010). Another approach known as Participatory Rural Appraisal (PRA) is gaining acceptance within scholarly and development circles, and refers to all methods that 'emphasize local knowledge and enable local people to make their own appraisal, analysis, and plans' (World Bank 2016b: 1). Inability of government and donors to engage fully and actively, rural communities in crafting pro-poor policies can result in several challenges. Under the biogas project such challenges include the fact that the variant of biogas plants installed in rural Rwanda usually come with open ditches to collect dung. Children in villages play around the compound and are not always watched due to the very low crime rate and other cultural factors. There are risks that a child can fall in the ditch and die, and in the Eastern province, respondents noted that a child had fallen in a biogas ditch and died. Government response to this is that the open ditch model is the cheapest and most affordable variety for rural dwellers. There are safer varieties that use covered tanks but that is not within budget. Part of the biogas installation instruction is to locate ditches away from access areas and to cover it with wood planks or other such materials to prevent accidents. However, the possibility of poor rural dwellers to locate biogas open ditches away from access areas is limited due to inadequate living spaces, the country's high population density makes land a coveted commodity across the country.

Other related challenges revolving around adequate consultations with the rural poor include that biogas cannot be used to cook foods that take too long to tenderize such as dry beans (a very affordable staple in rural Rwanda). The daily portion of biogas in the plant does not usually carry the number of hours needed to boil dry beans. Biogas plants need water for cleaning and for mixing with dung in case of insufficient urine. During dry season, there is usually shortage of water in Rwanda and this requires the purchase of water tanks, which could be too expensive for most rural farmers interviewed. Many respondents have large families and have to use large pans to cook. However, such pans cannot be used on the type of biogas cookstoves provided by the government. According to a respondent from the Southern province, 'yes, I still use charcoal and firewood because I have a large family, because we cannot use large pans on the small biogas stoves'. Response of the government official interviewed is that the model being produced at this time can only accommodate smaller cook pans and that is what is available for distribution to all users of biogas.

Further, the government's efforts in providing clusters of households with free biogas as an incentive to move them away from isolated forest living appears well intentioned, but some questions needs to be addressed. It is important for government to note that some of these communal owners of biogas plants are dissatisfied with the situation, and perhaps if they were properly consulted prior to installation might have a different perspective on the issue. Some beneficiaries of communal cow sheds and biogas plants say it is a disadvantage as they do not get enough dung and urine as a community, while some say it is an advantage as it makes sure they do not lack dung and urine. It depends on the availability of cows within each commune interviewed. In addition, there were complaints about co-owners of biogas plants reneging on agreed usage times, and stealthily using and finishing off biogas during periods the plants were supposed to be shut down. These are issues to be taken seriously in further expansion of plans by Rwanda's government. So while it would be beneficial to strengthen biogas 'cooperatives' in order to provide members who individually have few cows the opportunity to pool resources and own biogas, the cooperatives must be structured in such a way that there will be tight control over gas plant usage.

23.4 IMPLICATIONS FOR A TRANSITION TO A CLEAN ENERGY ECONOMY

Exploring Rwanda's experience with integrating clean energy use in national poverty reduction strategies offers several implications for nations with a high number of rural poor. First, the government of Rwanda's dependence on donor support for the success of the programme has brought about a situation where progress is slow due to cutback in funding from donors. Nations intending to integrate clean energy use may plan well ahead using sustainable sources of funding such as domestic revenue mobilization or other more reliable funding options. Donor funding, of course, could provide initial start-up capital including the necessary publicity generation and the funding of some pilot projects, but it is important that medium- to long-term planning on biogas not depend on donors. There is a need to explore further the possibility of public–private partnerships in the installation and maintenance of biogas plants, but this should be done with the full consideration of the economic situation of the rural poor.

Another case to consider seriously is the dependency syndrome, whereby many rural poor interviewed who have yet to pay for biogas installation are convinced that with the passage of time, government will change its mind on the co-pay requirement and extend biogas plants to them free of obligation. Several reasons can be advanced for this belief, including previous experiences with direct assistance from the government and lack of awareness creation on the part of the government.

There is a need for coordination between the government agency in charge of the concerned poverty reduction programme and the government agency in charge of biodiversity. Although many Girinka beneficiaries have biogas installed, there is no structure or platform in place to address specifically or actively promote biogas use among Girinka beneficiaries. Beneficiaries are lumped together among the poor rural dwellers in need of biogas. When request was made for information regarding the number of beneficiaries who use biogas, it was discovered that information was not available within the government office in charge of Girinka. In such settings, however, it will do well for several reasons including promotion of biogas, ensuring compliance, and monitoring progress, for a clearly designated platform of coordination between poverty alleviation strategies that have incorporated clean energy use.

There is a need to make biogas installation grassroots-driven. From 2007 until 2014, the promotion of domestic biogas use was championed mainly by the central government. The provincial and district governments, both of which are closest to the rural poor, were not active in the drive to increase biogas usage among that segment of the population. However, by 2015, this situation was gradually being reversed. There is a need, therefore, for central government, from the onset, to generate and maintain the support of the lowest tier government and if possible, ensure it plays a major role in integrating clean energy use in national poverty alleviation strategies.

23.5 CONCLUSION

Governments around the world in search of strategies to reduce poverty within their borders must of necessity, consider the integration of clean energy in such policy action. Clean cooking is an 'energy issue, a health issue, an environmental issue, a women's empowerment issue, and in many countries—a social and cultural issue' (Calvin and Venkataramanan 2015: 1). The search for policies that represent a nexus between poverty reduction and environmental preservation should collaborate with the masses; it should be fashioned within existing cultural, social, and economic realities of the poor. Rwanda, through the Girinka programme, has tried to leverage the cultural acceptance and prestige attached to cattle rearing to integrate biogas in household energy use of the rural poor, thereby reducing dependence on firewood and charcoal. Through the NDBP, the government is encouraging Girinka beneficiaries to install domestic biogas plants that would use cow dung and urine to generate methane for cooking and lighting. Biogas is much cheaper, easier, convenient, and better for the user and the environment to use than firewood or charcoal. While some beneficiaries have opted to install biogas and report an appreciable level of satisfaction, many have yet to install biogas plants due to several reasons. Mostly financing, and perhaps an additional feeling of entitlement, is at the centre of why many Girinka beneficiaries in rural Rwanda have yet to install biogas. On the side of the government, a dependency on donor funding has crippled the progress and slowed down the drive towards extensive coverage of rural communities with domestic biogas plants. Inconsistencies in government policies regarding funding of biogas plants also appears to be sending the wrong signal to the rural poor, leading to a cautious response to government's call for co-pay prior to installation.

For developing countries wanting to expand the rural poor's access to clean energy use, there is a need to be careful about a dependence on donor funds from the policy-making process down to implementation. The need for strong political will on the part of the government is necessary, in addition to a harmonization of policies regarding provision of biogas plants. Decentralization and appropriate communication of the entire process will be helpful in ensuring grassroots ownership and more widespread coverage.

The 17 UN Sustainable Development Goals approved in September 2015 and in place for the next 15 years has clearly indicated the need for national development agendas to be inclusive of eradication of extreme poverty, promotion of inclusive economic growth, and environmental protection. The three stated goals will be easier to reach by the year 2030 if concerted efforts are made by governments to integrate clean energy use in national poverty reduction strategies.

REFERENCES

Banerjee, M., S. Siddique, A. Dutta, and B. Mukherjee (2012). 'Cooking with Biomass Increases the Risk of Depression in Pre-Menopausal Women'. *Social Science & Medicine*, 75(3): 565–72.

Barnett, A. (2000). 'Energy and the Fight against Poverty'. Livelihood Sector Report, Department for International Development (DFID), London.

Bedi, A. S., L. Pellegrini, and L. Tasciotti (2013). 'Impact Evaluation of Netherlands Supported Programmes in the Area of Energy and Development Cooperation in Rwanda: Impact Evaluation of Rwanda's National Domestic Biogas Programme'. Baseline Report, International Institute of Social Studies, Erasmus University, Rotterdam.

Calvin, K. and R. Venkataramanan (2015). 'Africa: "How?"—Not "How Much?" to Fight Poverty'. Available at: <http://allafrica.com/stories/201510271866.html> (accessed 29 October 2015).

Casillas, C. and D. Kammen (2010). 'The Energy-Poverty-Climate Nexus'. *Science*, 330(6008): 1181–2.

EUEI (European Union Energy Initiative) (2009a). 'Biomass Energy Strategy (BEST) Rwanda'. European Union Energy Initiative, Partnership Dialogue Facility (EUEI PDF), Brussels.

EUEI (2009b). 'Update of the National Energy Policy and the Associated Implementation Plan Rwanda 2008–2012'. European Union Energy Initiative, Eschborn. Available at: <http://www.euei-pdf.org/en/seads/policy-regulation-and-strategy/update-of-the-national-energy-policy-and-the-associated> (accessed 14 November 2016).

Ewang, P. (2013). 'Political Economy of Africa's Rural Transformation'. Inaugural Africa Rural Development Forum (ARDF) 02–04 May. NEPAD, Cotonou.

Ezeanya, C. (2014). 'Indigenous Knowledge, Economic Empowerment and Entrepreneurship in Rwanda: The Girinka Approach'. *Journal of Pan African Studies*, 6(10): 241–63.

FAO (Food and Agriculture Organization) (2011). 'WISDOM Rwanda: Spatial Analysis of Wood-Fuel Production and Consumption in Rwanda: Applying the Wood-Fuel Integrated Supply/Demand Overview Mapping Methodology (WISDOM)'. FAO, Rome.

GoR (2013). 'Rwanda Supply Master Plan for Fuelwood and Charcoal'. Ministry of Natural Resources Rwanda, Kigali.

GTZ (Deutsche Gesellschaft für Technische Zusammenarbeit) (2009). 'Biomass Energy Strategy (BEST)', Volume 1–4. GTZ, Bonn.

IEA (International Energy Agency) (2014). 'World Energy Investment Outlook 2014 . Special Report. IEA, Paris.

IFAD (International Fund for Agricultural Development) (2012). *Enabling Poor Rural People to Overcome Poverty in Rwanda*. IFAD, Rome.

Kagabo, A. (2014). 'Girinka Beneficiaries and Biogas Use'. Interview by Chika Ezeanya on 5 March 2014.

Mamdani, M. (2001). *When Victims Become Killers: Colonialism, Nativism and Genocide in Rwanda*. Princeton, NJ: Princeton University Press.

MINECOFIN (Ministry of Finance and Economic Planning) (2007). 'Economic Development and Poverty Reduction Strategy'. MINECOFIN, Kigali.

NISR (National Institute of Statistics of Rwanda) (2012). '2012 Population and Housing Census'. Kigali.

NISR (National Institute of Statistics of Rwanda) (2016). 'How Many Are We in Rwanda?' NISR, Kigali. Available at: <http://www.statistics.gov.rw/about-us/news/how-many-are-we-rwanda> (accessed 14 November 2016).

Rwanda Agricultural Board (2013). 'Girinka Program: A Success Story', Rwanda Agricultural Board, Kigali.

Rwanda Energy Group (REG) (2014). 'Rwanda Energy Group: Energy Development Corporation Limited'. Biogas Project Brief. Available at: <http://www.reg.rw/images/Documents/Biomass_project_brief.pdf> (accessed 21 September 2015).

Rwanda Governance Board (2013). 'Documentation of Home-Grown Initiatives and Solutions for Policy Makers, Researchers and the Public'. Volume 1, RGB, Kigali.

SNV (2012). 'Domestic Biogas Newsletter'. Issue 7, SNV Netherlands Development Organization, The Hague.

The New Times (2015). 'Local Leaders Allay Fears over Charcoal-Burning', 5 February.

Twagiramungu, F. (2006). 'Environmental Profile of Rwanda'. Government of Rwanda and European Commission, Kigali.

UNDP (United Nations Development Programme) (2012). 'Human Development Index'. UNDP, New York.

UNDP (2015). 'United Nations Development Programme'. Available at: <http://www.undp.org/content/undp/en/home/ourwork/climate-and-disaster-resilience/our-goals.html> (accessed 14 November 2016).

World Bank (1991). 'Rwanda-Commercialization of Improved Charcoal Stoves and Carbonization Techniques: Mid-Term Progress Report'. ESMAP Paper ESM141, World Bank, Washington D.C.

World Bank (2012). 'Establishing a Green Charcoal Value Chain in Rwanda: A Feasibility Study'. World Bank, Washington, D.C.

World Bank (2016a). '2016 Rwanda Country Overview'. World Bank, Washington, D.C. Available at: <http://www.worldbank.org/en/country/rwanda/overview> (accessed 27 January 2016).

World Bank (2016b). 'Participatory Rural Appraisal'. Available at: <http://web.worldbank.org/WBSITE/EXTERNAL/TOPICS/EXTSOCIALDEV/0,,contentMDK:21233809~menuPK:3291499~pagePK:64168445~piPK:64168309~theSitePK:3177395~isCURL:Y~isCURL:Y,00.html> (accessed 27 January 2016).

Zewde, A. (2010). *Sorting Africa's Developmental Puzzle: The Participatory Social Learning Theory as an Alternative Approach.* Lanham, MD: University Press of America.

24

Renewable Energy in the Brazilian Amazon

The Drivers of Political Economy and Climate

Sabrina McCormick

24.1 INTRODUCTION

Brazil has long been a nation leading the usage of renewable energy. Historically, hydroelectricity has been the central energy source (Bermann 2007), and more recently, biomass has developed. While energy resources have long existed outside of the legal Brazilian Amazon, both types of energy development have recently increased within that region (da Silva Soito and Freitas 2011). Yet, expansion of energy development there threatens the rainforest with potentially tremendous ramifications for Amazonian sustainability and climate change (Schaeffer et al. 2013). The expansion of these energy types also affects local populations who are actively attempting to shape energy development. The political economy of energy including local community-level social actors, Brazilian private sector investors, Brazilian government agencies, and international private sector interests intersect to affect this relatively new trajectory of renewable energy. These social actors differentially consider and are differentially affected by climate change in that region, with some taking current risks and future projections into account and others discounting these effects.

While technical factors are important in decisions regarding renewable Amazonian-based energy development, political economic factors are possibly even more influential. Past research has shown that, in the other areas of Brazil and the world, political economic factors are often bigger drivers of energy decision-making than are considerations of economic viability or environmental sustainability (Hochstetler 2011). Current research has only minimally investigated this topic in regards to energy development in the Amazon.

This research investigates how political economy affects the growth of renewable energy in the Brazilian Amazon, and the relationship of this energy development to climate change. In this case, political economy is characterized by: 1) the needs and agenda of local communities, 2) economic interests and politics at the national level, and 3) international social actors and financial interests. This chapter seeks to make unique contributions to the literature on the political economy of renewable energy transitions by including a multi-scalar approach to considering political economic factors, such as considering the perspectives of local and state actors, and by considering the role of climate change, both in terms of how energy development may contribute to it and how climate change may affect renewable energy potential. This is similar to the approach by Tanner and Allouche (2011) who see the intersection of development and climate change across scales, but distinct in focus on political economy. Several bodies of research have previously considered the socioeconomic needs of local Amazonian communities, of the Brazilian state, and the effects of climate change on renewable energy, but few previous analyses have looked at the political economy of renewable energy in the Amazon through a lens that includes them. In so doing, this research seeks to create a comprehensive understanding of how renewable energy in the Amazon may be sustainable from the perspective of diverse social actors and over an extended period of time, such as that in which climate change effects will be felt.

This research focuses on biomass and hydroelectricity since these are the only two forms of renewable energy being developed in the Amazon. Since hydrological planning is expanding more rapidly, this research focuses on large hydroelectric dams in the Brazilian Amazon and on the case of Belo Monte dam, in particular. Belo Monte has recently been constructed after almost thirty years of planning, local contestation, and renegotiated planning.

Understanding the political economic drivers of decision-making in the Brazilian Amazon is critical since the forest is increasingly vulnerable to destruction and related increased poverty. While renewable energy development, and more specifically hydroelectricity, which is the largest area of focus, may increase temporary opportunities for economic development and employment, it is also likely to increase the destruction of rainforest, driving massive in-migration by workers and supporting populations, as well as flooding forest, disrupting livelihoods, and altering ecosystems for decades. By examining the political economy of renewable energy in the Amazon rather than the ecological or economic viability alone, this research seeks to identify how sustainability is being considered in its broadest sense. In other words, this research seeks to answer questions regarding how the current trajectory of renewable energy development in the Amazon can consider long-term impacts it may have on climate change, the effects climate change may have on it, and the effects these developments have on local populations.

This chapter is based on interviews with experts, non-governmental leaders, community members affected by energy development, and governmental representatives working in the area of renewable energy and deforestation in the Brazilian Amazon. Twenty-nine semi-structured interviews were conducted with these actors from summer 2015 through winter 2016. Interviews were recorded or extensive notes were taken. Themes were identified across interviews. Data in this chapter are also drawn from extensive research of government documents and media coverage, when necessary. For example, evidence regarding corruption in the energy sector is largely only available through investigative reporting, and so was drawn from such media venues.

The findings of this research promise the following conclusions relevant to policy makers. First, it will articulate the big picture of how political economy affects energy development in the Amazon. Second, this research will offer a coherent understanding of how energy development in the Amazon accounts for local populations. It, therefore, provides a cohesive picture of how poverty alleviation and sustainability can best be implemented in congruence with their needs. Third, findings clarify the degree to which diverse social actors consider climate change. As such, it will point to specific policies that can better account for climate factors, improving the actualization of real energy potential in the face of climate change. Without such an analysis, it is possible that planners will fail to reach this potential, thus undermining Brazilian economic development and possibly repeating the energy crisis that occurred in 2001.

This research was conducted at a particularly historic moment wherein the country is in transition. While Brazil has been seen as a transitional nation for many decades, the current transition is a relatively new one. This transition is from a model of deforestation that is driven largely by private landholders in the region, to one that accommodates large-scale government energy and infrastructure projects. This moment is one in which a new model of both forest protection and energy development must be made if the Amazon is to be protected. By documenting this moment and exploring the dilemmas therein, this research seeks to demonstrate how unexamined social costs of Amazonian energy generation may undo recent widely-publicized decreases in deforestation, consequently having negative impacts on climate change. I argue that only through addressing these unexamined, and sometimes illegally affected, social costs, can deforestation rates remain relatively consistent or decrease. As such, this analysis offers a warning against the advancement of energy infrastructure in the Amazon without serious consideration of concomitant measures to reduce deforestation and social impacts.

24.2 BACKGROUND

24.2.1 Energy Development in the Amazon

Brazil's energy mix encompasses 82 per cent hydroelectricity, followed by 7 per cent biomass, 5 per cent natural gas, 3 per cent nuclear, 2 per cent petroleum, and 1 per cent coal. While this energy mix is shifting due to urbanization, international trade, and climate conditions, the long-term dependence on hydroelectricity and the growing usage of biomass make these two forms of energy some of the most important to understand.

Brazil has long been dependent on hydroelectricity for the majority of its energy supply. The country has the third largest installed hydroelectric capacity in the world and sees the Amazon as the hydroelectric frontier (da Silva Soito and Freitas 2011). Planning for power generation is therefore generally based on hydroelectric capacity. However, in the past four to five years, hydroelectricity has generated less energy than projected resulting in difficulties in supplying the base load for the economy and also meeting increasing economic growth. As a result, several other forms of energy have advanced in the country, including oil, natural gas, and several other types of renewable energy, particularly solar and wind. Biomass has also grown to 7 per cent of all energy resources resulting in the Amazonian energy mix being focused in these two areas—see McCormick (2016: figure.1). Installations planned in the next eight years will be predominately large dams, small dams, and biomass, in that order of investment. Two areas of petroleum extraction have been developed already. Several ethanol plants will be developed in the north and eastern part of the region. While hydroelectricity is planned across the region, biomass is planned largely for the frontier region in the southern Amazon and one installation in the northern Amazon. The exact number of installations planned is a point of disagreement amongst experts and officials. Figure 1 in McCormick (2016), which is translated from the national energy plan for 2024, or the most recent documentation of energy planning, reflects the planning of five new large dams. However, some other studies have argued that there are up to thirty new dams being planned.

Since the installation of the democratic government, only one large dam complex has been completed in the Amazon. The Rio Madeira Dam complex, constituted by the Santo Antonio and Jirau Dams, are being built in the western Amazon. The dams have an expected 3,150 MW installed capacity that will be consumed in the northern and southern regions of the country. They have had multiple negative localized consequences, such as displacing local indigenous people, threatening the migration of fish populations, and destroying the habitats for multiple species. Most importantly, however, the dam complex reservoirs are meant to be a part of a series of waterways facilitating the export of soy grown in the region (Fearnside 2014). The dam complex began commercial production in 2012, but the second of the two, Santo Antonio, is not yet complete. The second large dam that has been proposed since the 1980s and recently has neared completion is the Belo Monte Dam near Altamira in the western Amazon. Belo Monte will be the third largest dam in the world. It is the most controversial dam project in the history of Brazil and has resulted in numerous violations of human rights laws.

The second largest source of energy in Brazil is biomass. While biomass can be developed in many parts of the country, depending on the type, biomass development in the Amazon has traditionally come from the centre-west region and the north, focused on large-scale conversion of pasture and/or rainforest to crops of carbohydrates, such as sugar cane, soy, and palm. It has grown as an area of investment in recent years with promising future growth (Lora and Andrade 2009). There are several major forms of biomass: sugarcane, soy, switchgrass, and wood. While 61 per cent of Brazil's land is forest, 1 per cent is used for sugarcane, 0.8 per cent is manually forested, and 3.25 per cent is used for soy bean planting (Colodette et al. 2014). Recently, sugar cane has been one of the forms resulting in approximately 30 million tons of sugar and 20 million tons of ethanol annually (Griffin and Scandiffio 2009). Wood production for charcoal, firewood, wood chips, and soy (Walter, Dolzan, and Piacente 2006) has been based around several types of trees, such as eucalyptus and acacia, that can be produced across the country. The development of soy (along with cattle ranching) has been cited as a primary driver of deforestation (Soares-Filho et al. 2006). Other forms of biomass have been found to drive deforestation, although some of these resources can be developed in locations outside of both the legal Amazon and the frontier area, resulting in less deforestation risk.

24.2.2 Energy Development in the Context of Climate Change

Climate change is likely to affect how successful energy development is in the Brazilian region (de Lucena et al. 2009). Droughts in the past decade have been the worst in history, reflecting the effects of climate on forest health (Marengo et al. 2011). The Amazon is an important climate resource as it represents a carbon sink and its destruction could contribute significantly to climate change. The viability of local populations hangs in the balance, as they are affected both by climate change impacts and energy development. Therefore, the Brazilian government and global environmental community currently must identify how to protect the Brazilian Amazon and its local populations while also supporting the expansion of clean energy.

Climate change presents two challenges to energy development in the Amazon. First, energy resources could be taxed or reduced due to a changing climate. Second, the need to reduce greenhouse gas (GHG) emissions could affect energy choices. A central mechanism for the release of GHGs in the Amazon is deforestation. In 2009, Brazil committed to a 36–38 per cent reduction in GHG emissions by 2020, which included a reduction in deforestation by 80 per cent in the legal Amazon. Therefore, energy development should be sensitive to the way it affects deforestation, but may not always be.

Energy development in the Amazon could be plagued by a variety of challenges. Biomass development could be threatened by changes in rainfall that, in turn, affect the growth of such resources. In addition, climate change can reduce the availability of land necessary to grow biofuels and could shift the distribution of insects that affect their growth, in addition to a wide variety of other factors that could negatively affect biomass (de Lucena et al. 2009). This is unfortunate since biomass can be a critical source of energy for the poorest local populations who depend on its burning for cooking and livelihoods. Forest die-back may occur in the frontier region of the Amazon as biomass is developed, reducing the availability of biomass and further driving migration into the forest as the search for viable land to develop continues.

Hydropower resources will likely be taxed by changes in seasonal water availability, drought, dry spells, and changes in air temperature that increase evaporation from reservoirs (de Lucena et al. 2013). Increasingly severe droughts in the Amazon are likely to affect the viability of existing and planned dams, especially if deforestation continues and impacts microclimates driving precipitation patterns.

There are several critical impacts that energy development in the Amazon may have on rainforest resources, possibly demonstrating that they are less sustainable than generally considered for renewable energy resources. Deforestation is not a factor traditionally accounted for in environmental impact assessments (EIAs) that preface the construction of energy installations. However, large infrastructure projects in the Amazon often have unintended and largely unmeasured effects on deforestation. Infrastructure development, such as that for energy and transportation, in the Amazon may present a newly influential factor driving deforestation rates to increase in the Amazon (Southworth et al. 2011; Malingreau, Eva, and De Miranda 2012).

Recent studies of biomass development have demonstrated a clear connection between the planting of new crops to provide lesser-known biomass resources and deforestation (Fearnside et al. 2009). This is a less-known issue in the case of large dams, although a few studies have demonstrated the effects of such installations on deforestation. There are several drivers for this phenomenon. First, large dams require tens of thousands of employees during the construction period. These workers often migrate from other areas to the work site. In addition to individuals who work on the dam itself, populations whose occupations support the workers generally appear in the surrounding area. These include, but are not limited to, physicians, sex workers, farmers, and restaurateurs. After the dam project is complete, its workers and these additional populations often stay in the area, next developing new roads and housing to accommodate their permanence. They, in turn, increase deforestation. In the case of Santo Antonio and Jirau in the western Amazon, unexamined and unintended environmental and social impacts included the expansion of soy production amongst other long-term consequences of development (Fearnside 2014).

In this way, infrastructure and migration patterns into the Amazon are major determinants of deforestation and sustainability. In particular, migration into the Amazon can expand deforestation. Since some cities are already established in the Amazon, some deforestation can be expected as it comes from cities within the Amazon into adjacent areas. This migration is driven by economic need and or familial networks (Randell and VanWey 2014). Many who occupy the Amazon and could move to urban areas, therefore taxing the Amazon less, choose to stay outside of urban areas because their skills are not suited to the urban environment, but rather to rural activities (Macdonald and Winklerprins 2014).

Deforestation rates have been decreasing in the Amazon for the past decade with some slight variation over time. These rates have been widely touted as a success and many believe will remain fairly static now that there is a well-established surveillance system in place. However, infrastructure development there may change this recent trend. There are large-scale plans to develop many forms of infrastructure in the Amazon (Fearnside 2002). Research has demonstrated that the development of such infrastructure will lead to vastly increased deforestation rates due to 'contagious deforestation' wherein one small area of deforestation leads to more deforestation nearby (Laurance et al. 2002). The development of roads is a well-known mechanism for increases in deforestation, yet there is little analysis of how these long-term infrastructure projects may affect deforestation (Barni, Fearnside, and de Alencastro Graça 2015).

Some research has shown that infrastructure development, such as increases in electrification, can decrease deforestation rates because it increases productivity (Assunção et al. 2015). However, cost–benefit analyses of large infrastructure in the Amazon is often insufficient or mischaracterized. For example, the economic value of the Belo Monte dam would change from US$1.62 billion to negative US$3.56 billion if the estimation of water flow were conducted correctly (Amend, Fleck, and Reid 2013).

24.3 DATA AND METHODS

This research is based on a broad-scale literature review, qualitative interviews with officials, local communities, and experts, and ethnographic observations of local communities in the State of Para. The literature review encompassed the fields of climate science, social sciences, energy studies, and sustainability science in order to capture the multiple, intersecting factors affecting the political economy of energy development. Twenty-nine interviews were conducted in person, over the phone, or on Skype from summer 2015 through winter 2016. Interviews were either taped and transcribed, or in-depth notes were taken, depending on the comfort of the interviewee. Interview subjects were selected through a snowball sample approach, starting with experts, officials, and local non-governmental organization representatives who work on related issues. Ethnographic observations were made of local communities and their interactions with state representatives in the State of Para near the site of the Belo Monte dam. This included the observation of a licensing hearing for the dam, local protests, and newly constructed resettlement communities in the City of Altamira.

24.4 POLITICAL ECONOMIC DIMENSIONS OF AMAZONIAN RENEWABLE ENERGY

Energy expansion and related infrastructure development in the Amazon is driven by the political economy of energy in Brazil and internationally. There are several key perspectives that shape the political economy of energy development in the Amazon. They include: the local communities and social movement organizations that represent them, Brazilian government agencies on the local and federal levels, Brazilian financial interests that are sometimes represented by government agencies but also include private interests, and international entities such as lenders and governments. These diverse social actors both conflict and align with one another to shape the political economy and its outcome in the Amazon. Section 24.4.1 outlines the role and effects of each factor in the political economy of renewable energy in the Amazon.

24.4.1 Local Socioeconomic Development: Communities and Social Movement Organizations

The Amazon is an incredibly diverse region composed of over 200 indigenous groups, people of Portuguese descent and of African descent. Energy development is often characterized by conflicts with local communities. This is especially true in Brazil where the long history of dam building has encompassed conflict with communities and social movements in sites across the country (Cummings 2013). The shifting landscape of biomass is also affecting socioeconomic development locally. Both forms of energy development often involve the in-migration of larger economic investors and resources, which have effects on local populations.

Biomass facilities are often characterized by the displacement of small-scale landholders and the concentration of land tenures often involving mechanization

(Tanner and Allouche 2011). Jobs for sugar workers are low-paying and involve difficult working conditions. In order to protect the forest, conservation techniques have been implemented and localized management strategies have been employed (Naughton-Treves, Holland, and Brandon 2005). The Bolsa Floresta programme, a programme in the State of Amazonas, supports the conservation of forests with a targeted approach to engaging forest communities in resource management, such as through payment for ecosystem services. Evaluations of this programme have had mixed results, and few have paid attention explicitly to biomass development in the context of forests.

Conflict with local communities regarding dam development is largely driven by displacement of local populations and community fragmentation (Windsor and McVey 2005). Hydroelectric dams have displaced over one million people in Brazil. They are controversial since many populations have not been resettled, and those who have been resettled by the state often live in conditions of greater poverty than before they were moved (de Araujo 1990). Displacement is very common, which results in impacts on living patterns and kinship systems (Tilt, Braun, and He 2009). Job capacity often actually decreases as mechanization related to dam building takes the place of more labour-intensive industries.

Another human impact of dam construction is increased illness in directly affected and adjacent communities. Often, infant mortality rises, malaria and other illnesses increase at the local level (Lerer and Scudder 1999). Some of these problems are caused by the fragmentation of community and home and decreased access to proper nutrition, among other social determinants. Others are related directly to environmental degradation, such as standing water, causing proliferation of mosquitoes that cause malaria. For example, at Serra da Mesa dam in Goias, schistosomiasis increased dramatically because of dam construction (Thiengo, Santos, and Fernandez 2005). Mercury is also released into the food chain through reservoir leaching.

Since these conflicts have generally occurred in tandem with development of specific projects and most large dams in the country are in regions outside of the Amazon, less conflict has occurred in this region than elsewhere. Local communities are often divided over dam development, with some community members in favour of dam building, anticipating positive effects on the local economy. This is generally true of local political leaders whose governments receive an annual portion of the profits of such energy installations. Local communities affected by flooding or drying of river resources often protest against dam building. This activity is reflected in a long history of the anti-dam movement in Brazil that began in the south of the country in the 1970s and has spread throughout the nation where dams are built (McCormick 2009). The Movement for Dam-Affected People has often led this movement while collaborating with local organizations. In the Amazon, this has included organizations such as the Indigenous Missionary Counsel, the Xingu Lives, and the Socio-Environmental Institute.

In the Amazon, local communities that oppose dam construction and the organizations that represent them are often focused on protecting the rights and livelihoods in the area. These dynamics have occurred in several Amazonian cases, especially in two built since the abandonment of the military government. The first was the Rio Madeira complex, whose impacts were close to the city of Porto Velho. Two companies, FURNAS and Odebrecht, led the development of these dams with support from the newly elected President Inacio Lula da Silva. Viability studies for the Rio Madeira dams began around the same time that the second model of Belo Monte was being developed in 2001 and 2002. When Belo Monte was again defeated in 2003, the Ministry of Mines and Energy proposed a new focus on Rio Madeira rather than Belo Monte. However, populations in the area around Porto Velho, the main town that would be affected, were divided in their support for the dam complex. A number of local organizations protested its construction citing law 10257/01, which guarantees that indigenous communities impacted by development projects will have access to information about it and ability to engage in discussion regarding its planning, but they were defeated.

The Belo Monte dam was first proposed and defeated in 1989 after a massive local protest supported by international organizations and celebrities. In the following 25 years, a variety of models of the Belo Monte dam were proposed, discussed, and rejected. In 2005, it was proposed and it received the first licence. A variety of national and local organizations contested Belo Monte, siting its environmental impacts and the potential it had to illegally displace indigenous communities. MDTX worked in partnership with Forum Carajas, Living Rivers, and the Catholic Church, protesting its construction. The Federal Prosecutor supported these organizations by launching over a dozen lawsuits against the dam. Even after the second licence of the necessary three was granted in 2011, protests continued, including violent occupation of the dam site. As of January 2016, the third operating licence for the dam has been suspended due to the lack of address for impacts on indigenous communities.

Local communities have some influence on the renewable energy portfolio in the Amazon both through vocal opposition to specific projects and through their official inclusion in national programmes for inclusion. However, their increased participation could both bolster their socioeconomic development and improve outcomes of projects through the use of their localized understanding of environment and land use.

24.4.2 Brazilian Governmental and Private Interests

Energy expansion in the Brazilian Amazon is meant to match the growing economy and fill the shortfall of energy generation, in part due to drought conditions reducing the viability of hydroelectricity generation since 2010.

Negative trade balances also drove the further expansion of hydroelectricity and biomass development, especially as a replacement for other oil derivatives in the case of the latter (Griffin and Scandiffio 2009). Renewable energy in the Amazon is also driven by the need for continued economic development in the south to which energy in the Amazon is transmitted, and for electro-intensive industries in the Amazon, such as mineral extraction (Fearnside and Figueiredo 2015).

Biomass has been an important form of energy crops especially in light of addressing climate change and the need to reduce greenhouse gas emissions. Sugar cane has been a central crop in the production of ethanol both for domestic consumption and export to advance the Brazilian agricultural sector. Due to the increasing efficiency of the agricultural sector and especially cattle ranching, sugar cane production has massive potential for expansion in Brazil without interfering with other food resources, which is driving increased national investment (Pimentel and Patzek 2008). Subsidized development of biomass development has been focused on the Amazon and the cerrado (bordering the southwest of the legal Amazon) with a portion of these developments planned as community managed (Walter, Dolzan, and Piacente 2006). National programmes have been developed to support research and development in biomass across the country.

The northern region of Brazil, dominated by Amazonian rainforest, currently has the most hydrological potential and is characterized by the greatest risk (Tundisi et al. 2014). Large dams have clearly caused long-term adverse impacts on biodiversity in the Amazon (Benchimol and Peres 2015). Some legally protected areas present major impediments to dam building. This is particularly true of indigenous lands whose inundation requires approval by the indigenous group living there. Drought conditions in other parts of the country also make this a particularly important region for development. Renewable energy is expanding throughout the country; however, the history of dam building in the country makes this form of renewable energy one of the more appealing approaches to supporting an expanding economy. Existing infrastructure for dam construction and relationships between political representatives and dam builders means that hydroelectricity may be prioritized over other forms.

While the hydrological resources in the Amazon have also played a role in attracting investors, the instability of precipitation patterns and the social conflict surrounding these projects has shifted who is willing to invest. For example, when the state-run public–private partnership for northern Brazil, Norte Energia, won the bid to build the Belo Monte dam, it was constituted by a group of medium-sized production companies including Bertin, Queiroz Galvão, J. Malucelli, Cetenco, Galvão Engenharia, Mendes Junior, and Serveng in collaboration with state-led CHESF (the Hydroelectric Company of San Francisco). Since then, it has evolved into an association between the Brazilian government and pension funds. Today, the central Brazilian bank, the BNDES

(Banco Nacional de Desenvolvimento Econômico e Social), has provided the biggest financial support for the project.

Government leaders have supported the expansion of large dams in the Amazon, including the recent President Lula and the current President Dilma Roussef, who was previously the Minister of Mines and Energy. Both of these presidents campaigned on platforms that included the completion of the Belo Monte dam. Traditionally, billions of dollars have been lent to the Brazilian government by external sources; however, Belo Monte is funded largely by a US$10.8 billion loan from the BNDES (Reuters 2012). Investors like Vale, the second largest iron-ore producer in the world, have supported the project (Leahy 2011).

Recently, the role of corruption in their platforms and political economy of energy generation has gained attention. Corruption in the Brazilian government has influenced the construction of infrastructure and energy projects. The former director of the Brazilian oil company, Petrobras, was arrested and convicted for corruption in dam construction. A consortium of Brazilian companies that won the bid to build the Belo Monte dam complex, such as Odebrecht, and Andrade Gutierrez were discovered to have given around US$6 million in bribes to local officials in order to gain the contracts to build the dam. A former president of Camargo Correa admitted that his company gave millions of dollars to political parties in order to gain 15 per cent of the Belo Monte construction contract (R7 Noticias 2015). These examples have recently revealed the ways in which large private companies are able to influence the planning of dams and energy infrastructure through bribes to public officials.

24.4.3 International Actors and Financial Interests

The political economy of Brazil's energy development is embedded within international markets and financial interests that drive and inhibit particular pathways of energy development. On this global scale, there are three interrelated types of influence: international companies and investors that invest in or gain contracts for the construction of energy installations, international demand for forms of energy, and growing international needs to meet agreements made in the Kyoto Protocol and the Clean Development Mechanism (CDM). International actors that affect renewable energy development in Brazil include investors and financiers. This landscape of these social actors has been shifting. For example, although all private international investors originally interested in Belo Monte dam removed their involvement, some companies were paid for their role in construction. The French construction company, Alstrom, was paid over US$500 million to provide equipment and turbines for Belo Monte. The Chinese play a critical role in Amazonian development in that 82.3 per cent of their foreign direct investment (FDI) was allocated to extractive industries in 2012 (Fearnside and Figueiredo 2015) and 7 per cent of Brazil's FDI was from China between 2005 and 2013. China has invested in renewable energy and related infrastructure development. While China does not have formal decision-making power regarding these measures, other than through the indirect governance framework regarding climate change of the United Nations Framework Convention on Climate Change (UNFCCC), the availability of its resources to advance specific projects in which the country has interests may also catalyse the development of dams, canals, railways, and other transportation infrastructure. China is the largest importer of Brazilian soy, whose export can be made more economically efficient by these developments.

Domestic and global policy frameworks influence decisions made by international investors. Global interest in reducing the effects of climate change have included increased commitment to the Kyoto Protocol and policies therein. As nations around the world attempt to decrease their emissions, demand for ethanol has increased, subsequently increasing its development in Brazil. The United States and India are the biggest importers of ethanol from Brazil. The UNFCCC has provided the international platform for Brazil to commit to a particular approach to the reduction of GHGs, or more specifically, to the commitment of reducing deforestation rather than shifting the energy matrix. The CDM has given Brazil credit for the construction of large dams despite the lack of estimation regarding their impacts (Fearnside 2013). The cases of new large dams already constructed in the Amazon demonstrate that the CDM facilitates the false impression of GHG reductions, when it may actually be increasing them due to methane emissions and other unintended impacts (Fearnside 2015).

24.5 IMPLICATIONS FOR THE CLEAN ENERGY TRANSITION

This research has taken a trans-scalar approach to the political economy of renewable energy development in the Brazilian Amazon to demonstrate how it interrelates with local and global actors across scales and sits within the context of global markets and investors. By using this approach, it is possible to see how the needs of local, statal, and transnational actors are often in competition with one another, and the ways in which unresolved conflicts can decrease the quality of planning. Most centrally, local actors often stake claims that they lose livelihoods, land, and community cohesion due to energy development, often demonstrating the lack of sustainability in energy sources generally called renewable. State and international actors are more driven by market forces that can adversely affect the sustainability of Amazonian resources. While issues of deforestation, sustainability, climate change, and energy are incredibly complex in the Brazilian Amazon, this conceptualization of the political economy begins to highlight how the alignment of interests therein could improve long-term outcomes for both energy development and Amazonian preservation.

This research offers three implications for the expansion and improved implementation of clean energy in the Brazilian Amazon. First, any energy planning for the Amazon should take a more comprehensive approach to assessing environmental and social impacts than current planning methods do. This would include impacts outside of the project itself, such as the surrounding area, and the long-term consequences of each project. This is critical both to the preservation of the rainforest and to the sustainability of livelihoods and culture in the region. Without this, it will be impossible for the advancement of clean energy to take place that considers the needs of local populations who live and work adjacent to such facilities.

Second, renewable energy development in the Amazon must consider the role of climate change, both in terms of how it may affect hydrological resources and energy potential, and how it may affect the needs of local populations. Amazonian energy development has largely ignored both the needs of local populations and climate projections. However, the latter promises to reduce economic viability of renewable energy in the Amazon. Assessing the needs and legal rights of local populations is critical to advancing sustainability in the region. These populations are generally not the recipients of energy generated locally, and may, instead, face negative ramifications of infrastructure development, such as lack of access to water and displacement.

Finally, in the international sphere, Brazil needs to expand its presentation of national impacts on climate change by including the real potential changing deforestation rates that accompany infrastructure development, and the changing energy mix. This presentation should include both the increasing dependence on fossil fuels and, possibly even more critically, present and future methane emissions from hydroelectric dams in the Amazon.

A consideration of these three factors would help support the adoption of renewable energy in Brazil that is truly sustainable for the people of the Amazonian region and for the global community which is indirectly affected by the changes in that region through climate change.

REFERENCES

Amend, M., L. Fleck, and J. Reid (2013). 'Improving Cost-Benefit Analysis in the Assessment of Infrastructure Projects in the Brazilian Amazon'. In M. A. Livermore and R. L. Revesz (eds), *The Globalization of Cost-Benefit Analysis in Environmental Policy*. Oxford: Oxford University Press, pp. 223–34.

Assunção, J., M. Lipscomb, A. M. Mobarak, and D. Szerman (2015). 'Infrastructure Development Can Benefit the Environment: Electrification, Agricultural Productivity and Deforestation in Brazil'. LACEA 2015 Papers. Available at: <http://lacer.lacea.org/bitstream/handle/123456789/53092/lacea2015_infrastructure_development_benefit_environment.pdf?sequence=1> (accessed 1 March 2015).

Barni, P. E., P. M. Fearnside, and P. M. L. de Alencastro Graça (2015). 'Simulating Deforestation and Carbon Loss in Amazonia: Impacts in Brazil's Roraima State from Reconstructing Highway BR-319 (Manaus-Porto Velho)', *Environmental Management*, 55(2): 259–78.

Benchimol, M. and C. A. Peres (2015). 'Widespread Forest Vertebrate Extinctions Induced by a Mega Hydroelectric Dam in Lowland Amazonia', *PloS one*, 10(7): e0129818.

Bermann, C. (2007). 'Impasses and Controversies of Hydroelectricity', *Estudos avançados*, 21(59): 139–53.

Colodette, J. L., C. M. Gomes, F. J. Gomes, and C. P. Cabral (2014). 'The Brazilian Wood Biomass Supply and Utilization Focusing on Eucalypt', *Chemical and Biological Technologies in Agriculture*, 1(1): 1–8.

Cummings, B. J. (2013). *Dam the Rivers, Damn the People: Development and Resistance in Amazonian Brazil*. Oxford: Routledge.

da Silva Soito, J. L. and M. A. V. Freitas (2011). 'Amazon and the Expansion of Hydropower in Brazil: Vulnerability, Impacts and Possibilities for Adaptation to Global Climate Change', *Renewable and Sustainable Energy Reviews*, 15(6): 3165–77.

de Araujo, M. L. C. (1990). 'Na Margem do Lago um Estudo Sobre o Sindicalismo Rural'. Fundacao Joaquim Nabuco, Recife.

de Lucena, A. F. P., R. Schaeffer, A. S. Szklo, R. Soria, and M. Chavez-Rodriguez (2013). 'Energy Security in the Amazon'. Report for the Amazonia Security Agenda Project, COPPE/UFRJ, Rio de Janeiro.

de Lucena, A. F. P., A. S. Szklo, R. Schaeffer, R. R. de Souza, B. S. M. C. Borba, I. V. L. da Costa, A. O. P. Júnior, and S. H. F. da Cunha (2009). 'The Vulnerability of Renewable Energy to Climate Change in Brazil', *Energy Policy*, 37(3): 879–89.

Fearnside, P. and A. Figueiredo (2015). 'China's Influence on Deforestation in Brazilian Amazonia: A Growing Force in the State of Mato Grosso'. BU Global Economic Governance Initiative Discussion Paper 2015-3, Boston University, Boston, MA.

Fearnside, P. M. (2002). 'Avanca Brasil: Environmental and Social Consequences of Brazil's Planned Infrastructure in Amazonia', *Environmental Management*, 30(6): 735–47.

Fearnside, P. M. (2013). 'Credit for Climate Mitigation by Amazonian Dams: Loopholes and Impacts Illustrated by Brazil's Jirau Hydroelectric Project', *Carbon Management*, 4(6): 681–96.

Fearnside, P. M. (2014). 'Impacts of Brazil's Madeira River Dams: Unlearned Lessons for Hydroelectric Development in Amazonia', *Environmental Science & Policy*, 38: 164–72.

Fearnside, P. M. (2015). 'Tropical Hydropower in the Clean Development Mechanism: Brazil's Santo Antônio Dam as an Example of the Need for Change', *Climatic Change*, 131(4): 575–89.

Fearnside, P. M., C. A. Righi, P. M. L. de Alencastro Graça, E. W. Keizer, C. C. Cerri, E. M. Nogueira, and R. I. Barbosa (2009). 'Biomass and Greenhouse-Gas Emissions from Land-Use Change in Brazil's Amazonian "Arc of Deforestation": The States of Mato Grosso and Rondônia', *Forest Ecology and Management*, 258(9): 1968–78.

Griffin, W. M. and M. I. G. Scandiffio (2009). 'Can Brazil Replace 5% of the 2025 Gasoline World Demand with Ethanol?' *Energy*, 34: 655–61.

Hochstetler, K. (2011). 'The Politics of Environmental Licensing: Energy Projects of the Past and Future in Brazil', *Studies in Comparative International Development*, 46 (4): 349–71.

Laurance, W. F., A. K. Albernaz, G. Schroth, P. M. Fearnside, S. Bergen, E. M. Venticinque, and C. Da Costa (2002). 'Predictors of Deforestation in the Brazilian Amazon', *Journal of Biogeography*, 29(5/6): 737–48.

Leahy, J. (2011). 'Vale to Invest $1.4bn in Controversial Brazil Dam'. *Financial Times* (online), 29 April. Available at: <http://www.ft.com/cms/s/0/82c1e266-71f6-11e0-9adf-00144feabdc0.html#axzz439PiMU4X> (accessed 1 March 2015).

Lerer, L. B. and T. Scudder (1999). 'Health Impacts of Large Dams', *Environmental Impact Assessment Review*, 19(2): 113–23.

Lora, E. and R. Andrade (2009). 'Biomass as Energy Source in Brazil', *Renewable and Sustainable Energy Reviews*, 13(4): 777–88.

McCormick, S. (2009). *Mobilizing Science: Movements, Participation and the Remaking of Knowledge*. Philadelphia, PA: Temple University Press.

McCormick, S. (2016). 'Renewable Energy in the Brazilian Amazon the Drivers of Political Economy and Climate'. WIDER Working paper 2016/12, UNU-WIDER, Helsinki. Available at <https://www.wider.unu.edu/publication/renewable-energy-brazilian-amazon> (accessed 18 October 2016).

Macdonald, T. and A. M. Winklerprins (2014). 'Searching for a Better Life: Peri-Urban Migration in Western Para State, Brazil', *Geographical Review*, 104(3): 294–309.

Malingreau, J., H. Eva, and E. De Miranda (2012). 'Brazilian Amazon: A Significant Five-Year Drop in Deforestation Rates but Figures Are on the Rise Again', *Ambio*, 41(3): 309–14.

Marengo, J. A., J. Tomasella, L. M. Alves, W. R. Soares, and D. A. Rodriguez (2011). 'The Drought of 2010 in the Context of Historical Droughts in the Amazon Region', *Geophysical Research Letters*, 38(12): 1–5.

Naughton-Treves, L., M. B. Holland, and K. Brandon (2005). 'The Role of Protected Areas in Conserving Biodiversity and Sustaining Local Livelihoods', *Annual Review of Environment and Resources*, 30: 219–52.

Pimentel, D. and T. W. Patzek (2008). 'Ethanol Production: Energy and Economic Issues Related to US and Brazilian Sugarcane'. In D. Pimentel (ed.), *Biofuels, Solar and Wind as Renewable Energy Systems*. Dordrecht: Springer Netherlands, pp. 357–72.

Randell, H. F. and L. K. VanWey (2014). 'Networks versus Need: Drivers of Urban Out-Migration in the Brazilian Amazon', *Population Research and Policy Review*, 33 (6): 915–36.

Reuters (2012). 'Brazil's BNDES Approves $10.8 bln Loan for Amazon Belo Monte Dam', 26 November. Available at: <http://www.reuters.com/article/bndes-belomonte-idUSL1E8MQ76I20121126> (accessed 1 March 2015).

R7 Noticias (2015). 'Camargo Corrêa paid $20 Million in Bribes in Construction of the Belo Monte Power Plant' (online). Available at: <http://noticias.r7.com/prisma> (accessed 18 October 2016).

Schaeffer, R., A. Szklo, A. F. P. De Lucena, R. Soria, and M. Chavez-Rodriguez (2013). 'The Vulnerable Amazon: The Impact of Climate Change on the Untapped Potential of Hydropower System', *Power and Energy Magazine, IEEE*, 11(3): 22–31.

Soares-Filho, B. S., D. C. Nepstad, L. M. Curran, G. C. Cerqueira, R. A. Garcia, C. A. Ramos, E. Voll, A. McDonald, P. Lefebvre, and P. Schlesinger (2006). 'Modelling Conservation in the Amazon Basin', *Nature*, 440(7083): 520–3.

Southworth, J., M. Marsik, Y. Qiu, S. Perz, G. Cumming, F. Stevens, K. Rocha, A. Duchelle, and G. Barnes (2011). 'Roads as Drivers of Change: Trajectories across the Tri-National Frontier in MAP, the Southwestern Amazon', *Remote Sensing*, 3(5): 1047–66.

Tanner, T. and J. Allouche (2011). 'Towards a New Political Economy of Climate Change and Development', *IDS Bulletin*, 42(3): 1–14.

Thiengo, S. C., S. B. Santos, and M. A. Fernandez (2005). 'Malacofauna límnica da ár auna límnica da ár auna límnica da área de inf ea de inf ea de influência do lag luência do lag luência do lago da usina hidr o da usina hidr o da usina hidrelétrica de Serra da Mesa, a da Mesa, a da Mesa, Goiás, Brasil. I. Estudo qualitativ Estudo qualitativ Estudo qualitativo', *Revista Brasileira de Zoologia*, 22(4): 867–74.

Tilt, B., Y. Braun, and D. He (2009). 'Social Impacts of Large Dam Projects: A Comparison of International Case Studies and Implications for Best Practice', *Journal of Environmental Management*, 90: S249–S257.

Tundisi, J., J. Goldemberg, T. Matsumura-Tundisi, and A. Saraiva (2014). 'How Many More Dams in the Amazon?' *Energy Policy*, 74: 703–8.

Walter, A., P. Dolzan, and E. Piacente (2006). 'Biomass Energy and Bio-Energy Trade: Historic Developments in Brazil and Current Opportunities', IEA Bioenergy Task 40 Country Report for Brazil. Available at: <http://www.bioenergytrade.org/> (accessed 10 February 2016).

Windsor, J. E. and J. McVey (2005). 'Annihilation of Both Place and Sense of Place: The Experience of the Cheslatta T'En Canadian First Nation within the Context of Large-Scale Environmental Projects', *The Geographical Journal*, 171(2): 146–65.

25

The Political Economy of Household Thermal Energy Choices in Developing Countries

Comparing the LPG Sectors in Indonesia and South Africa

Wikus Kruger, Louise Tait, and Jiska de Groot

25.1 INTRODUCTION

The central role of energy in human and economic development is widely recognized (Hall, Lambert, and Balogh 2014). The United Nations has, for example, identified universal access to modern energy as a major policy goal; and sustainable energy has been a widely accepted pillar of the post-2015 development agenda (United Nations Foundation 2015). In the context of developing countries, access to modern energy sources often signifies a shift away from traditional solid fuels (such as wood), which cause significant health and environmental impacts, towards modern energy sources, such as electricity (preferably from renewable sources) and liquid petroleum gas (LPG). In 2014, for example, over 2.6 billion people had no access to clean energy, and used fuel wood and charcoal for thermal energy use (primarily cooking and space heating). The majority of these people live in sub-Saharan Africa (SSA) and Asia (IEA 2014). The resultant household air pollution kills more than 4.3 million people per year (Global Alliance for Clean Cookstoves 2015). Moving away from traditional fuels towards modern cooking and heating fuels is, therefore, a global development priority and a key component of energy transitions in developing countries.

Studies have identified that enabling policies have played an important role in increasing access to and utilization of a key thermal fuel—LPG—in developing countries, including, for example, Ghana and Senegal (Ahiataku-Togobo 2013;

Brew-Hammond, Mensah, and Amponsah 2014). These successes were not solely the result of enabling policies, subsidies, or economic drivers, but part of an interplay between policies, actors, and institutions, their values, and socio-cultural and political factors more broadly, so that it is necessary to understand the prevailing political and economic processes that have affected these outcomes.

Residential LPG usage in South Africa is among the lowest in the developing world (Kojima 2011)—despite the country being one of the leaders in SSA with regard to improving energy access. While South Africa has a relatively limited residential thermal energy policy focus, Indonesia recently implemented one of the largest household energy transition projects to date: the kerosene-to-LPG conversion programme. The majority of Indonesian households switched from using kerosene (the major thermal fuel in most urban poor South African households) to LPG in a relatively short amount of time (Budya and Arofat 2011).[1] It is useful to compare the South African experience with LPG with that of Indonesia, since both countries have a number of significant similarities:

- Indonesia and South Africa are both classified as 'developing nations' or 'middle-income countries', and characterized by good infrastructure (especially transport, telecommunications) and relatively high levels of industrialization.
- Both countries have significant mineral resources that provide(d) an important economic base for their development. These mineral resources continue to play an important part in shaping the energy and economic development trajectories of both countries.
- While they might not be 'hegemonic' powers as such, both are regional 'powerhouses', having the largest economies in their respective areas and the ability to steer regional policy.
- Both Indonesia and South Africa have recently established democratic political systems, coming out of several decades of autocratic regimes that followed the end of colonial rule.
- Both countries have had recent successful energy initiatives. Where South Africa has been remarkably successful in terms of its electrification programme, Indonesia has been particularly successful in terms of LPG provision. Various social grant mechanisms and subsidies are also used in both nations to provide public goods and services to the poor.

[1] At the time this chapter was being written, Hanung Budya was Marketing Director of PT Pertamina (Persero), and Muhammad Yasir Arofat was Senior Analyst Strategy Planning—Marketing Directorate PT Pertamina (Persero).

By exploring and contrasting the Indonesian experience with that of South Africa, the political economic factors that have affected the widespread adoption of certain energy carriers can become more visible.

25.1.1 Political Economy Analysis

The analytical framework employed here is that of political economy analysis, a tool that improves understanding of the drivers of political behaviour, and how this shapes policies and programmes, winners and losers, and the implications for development. This type of analysis, which has a long tradition in the social sciences, has recently gained increased attention in the context of international development, and is promoted by organizations such as the United Kingdom's Department for International Development and the Governance and Social Development Resource Centre. Although there is no single conceptual framework for the analysis, and the concept is sometimes contested, generally political economy concerns 'the social relations, particularly the power relations, that mutually constitute the production, distribution and consumption of resources' (Mosco 2009: 74).

A comprehensive definition of political economy analysis is provided by the OECD-DAC report (Dahl-Østergaard et al. 2005), which describes it as 'the interaction of political and economic processes in a society; the distribution of power and wealth between different groups and individuals; and the processes that create, sustain and transform these relationships over time'. Its main components are: first, politics, understood in terms of contestation and bargaining between interest groups with competing claims over rights and resources; and, second, economic processes that generate wealth, and which influence how political choices are made. In current world affairs, these processes, which are shaped by incentives, relationships, the distribution and contestation of power between different actors (often closely interrelated), constitute a unified set of dynamics that largely influence development outcomes (DfID 2009; Mcloughlin 2014). Crucial to this study, political economy analysis considers historical legacies, social trends, and prior experience (Mcloughlin 2014), and how all of these factors affect or impede the adoption of a particular technology (in our case, the large-scale use of LPG). Political economy analysis allows for the inclusion of these factors, as well as informal institutions and cultural and social practices in the analysis (Williams, Duncan, and Landel-Mills 2007).

In the context of international development, political economy analysis is often employed as a tool to improve development effectiveness by identifying how and where focus should be directed to promote positive change. Here the tool is used to shed light on South Africa's and Indonesia's residential energy transitions and will highlight, through the case studies, that the implementation of energy transitions (in this case LPG) is a highly politicized, locally-driven process. The present analysis, using policy documents, reports, websites, and academic and grey literatures, focuses on the period from when the transition was initiated up to the point where it has become mainstream. Examining this period enables focus on the full adoption of the innovation curve as developed by Rogers (1983). This allows time for economic and political processes to unfold against the backdrop of the country's development path.

The analysis focuses on three broad areas of the political economy:

(1) the *interests and incentives* facing different *groups* in society, and how these generate particular policy outcomes that may encourage or hinder development;
(2) the *role of formal and informal institutions*, as well as social, political, and cultural norms in shaping human interaction and political and economic competition;
(3) the *impact of values and ideas*, including political ideologies, religion, and cultural beliefs on political behaviour and public policy (DfID 2009).

The remainder of this chapter is structured as follows. First, a political economic analysis of Indonesia will be presented, followed by one of South Africa. The case studies will be compared and contrasted and differences and commonalities between the studies highlighted. The chapter concludes with a brief summary of findings and concluding analysis.

25.2 INDONESIA

Indonesia is an archipelago in South East Asia with a population of 255 million people, making it the world's fourth most populous nation. Indonesia's economy is the world's sixteenth largest by nominal gross domestic product (GDP) and eighth largest by GDP at PPP (purchasing power parity)—making it the largest economy in South East Asia. The country emerged from more than three centuries of Dutch colonial rule after the Second World War, and has since gone through various turbulent political, economic, and social transitions. The Indonesian economy is based largely on industry (46.4 per cent of GDP), followed by services (38.6 per cent), and agriculture (14.4 per cent). Economic growth levels have been sitting between 4 and 6 per cent for the past 15 years—weathering the global financial crisis in 2008–9 particularly well. However, poverty remains a persistent problem, with 11 per cent of the population living below the poverty line and unemployment sitting at 5.9 per cent (Central Intelligence Agency 2013).

The country remains a net energy exporter, although imports of oil and oil products have been increasing rapidly. In the 1970s, however, Indonesia was a leading oil exporter, producing more than a million barrels a day, and oil revenues contributed over 70 per cent of GDP. The oil price shocks of the mid-1970s, which saw rapidly rising oil prices, enabled the government to introduce significant fuel subsidies to buoy domestic energy affordability—on the back of increasing state revenues. This enabled the country to maintain the price of petrol at less than US$0.20 per litre up to 2005. Almost half of all energy subsidies went to kerosene, used largely for domestic cooking and lighting by poor households (Diop 2014; Tumiwa et al. 2012; Kumoro and Astriana 2015; Mourougane 2010). Recent years, however, have seen Indonesia importing oil, as domestic production volumes have fallen while, at the same time, domestic consumption has increased. Significant rises in the global oil price placed a great deal of pressure on the Indonesian treasury as it continued its policy of highly subsidized energy provision (Kumoro and Astriana 2015; Tumiwa et al. 2012).

The Indonesian kerosene subsidy is substantial in both absolute and relative terms. By 2006 overall fuel subsidies grew to such a level that they made up more than 10 per cent of state expenditure—despite kerosene consumption actually dropping in this period. Continuing with the subsidy scheme was consequently deemed unsustainable by the Indonesian government (Arofat 2014; ASTAE 2013; Budya and Arofat 2011; Kasih 2012).

The Indonesian government's response was to introduce the kerosene-to-LPG conversion programme in 2007. The programme had distributed more than 55.3 million LPG conversion packages (3 kg filled LPG canister, stove, regulator, pipes) by late 2013, covering most of Indonesia's previously subsidized low-income kerosene users. As a result, LPG use has climbed steadily, while kerosene use has dropped, saving the Indonesian government more than US$1 billion in its first year of implementation (Arofat 2014; ASTAE 2013; Budya and Arofat 2011; Kasih 2012).

The programme required significant investments in distribution and retail infrastructure, and was largely carried out by the state oil company, Pertamina, and coordinated by the Ministry for Minerals and Energy. Pertamina incurred all programme implementation costs, and was subsequently reimbursed by the Indonesian government (Arofat 2014; ASTAE 2013; Budya and Arofat 2011; Kasih 2012).

While this case study is an analysis of the political economy of the LPG conversion programme, to properly appreciate the LPG programme's context it is necessary to understand the nature and need for Indonesia's kerosene subsidies and their impacts on the society and economy. Sections 25.2.1–25.2.3 will examine the political economy factors of the LPG conversion programme, in the context of the overall Indonesian fuel subsidy regime.

25.2.1 Groups, Interests, and Incentives

The dynamics of the Indonesian government's policy choices in respect of residential energy policy has been, in large part, bound up with wider politics. When Sukarno, the country's first post-colonial president, decided to nationalize energy interests, it was reasoned that this was to primarily benefit Indonesian households—the major energy consumers in Indonesia at that time. General Suharto, the successor to Sukarno, continued with this policy. Providing subsidized energy was a way of relieving some of the economic pressures on households, as well as stimulating the Indonesian economy towards rapid industrialization built on the back of cheap energy and energy-based foreign earnings (Ascher 1998; Liddle 1991; Mourougane 2010). Given Indonesia's relative oil resource abundance in the twentieth century, it made sense to use these resources to secure political support and stimulate economic development. At that stage, the most easily provided household fuel was kerosene.

Domestic energy subsidies therefore formed part of a broader strategy to secure and maintain political power. General Suharto learned just how much of a stabilizing role these subsidies played in 1998, when he tried to reduce energy (primarily kerosene) subsidies, and was consequently removed from power by popular protests. While the protests were in part stimulated by the Asian financial crisis of the 1990s, the massive hikes in fuel tariffs played an important role (Mourougane 2010; Seah 2014).

Protests related to kerosene pricing occurred again in 2003 and 2005, although in these cases the opposition was less severe—in part because the country was not facing the same level of economic instability as in 1998. Household energy subsidies have remained politically sensitive in Indonesia, often triggering instability (Braithwaite et al. 2012; Mourougane 2010). The Indonesian government's response to the financial pressure of kerosene subsidies in the first part of this century was therefore the provision of a cheaper and more efficient household fuel alternative—LPG.

Pertamina had to manage various interests and incentives from numerous groups in rolling out this programme. It had to ensure not only that it had the capacity to guarantee the continuous supply of LPG to a geographically challenging populace, but also the overall acceptance of the conversion programme by different interest groups, which had varying objectives:

- The government was ultimately interested in saving money through the reduction of the kerosene subsidy.
- Households and small- and medium-sized enterprises wanted a cheap, reliable, and convenient energy source.

- Retailers—up to that point responsible for selling subsidized kerosene to households—needed to retain their profit margin while switching to a fuel source that required significant new infrastructural investment.
- Various interests groups—non-governmental organizations, political parties, trade associations, and so on—had political and other locally driven agendas that could quite easily be served by undermining the conversion programme or using its failure as political leverage.

The interests and agendas of most of the stakeholder groups were aligned with the success of transitioning from kerosene to LPG. For the government, their support stemmed from the fact that the kerosene subsidy, at that stage, was undermining its fiscal sustainability and endangering the country's economic and political stability. Pertamina was to execute a programme that ultimately served the national oil company's goals too: increasing its production, storage, and distribution infrastructure, while enlarging the consumer base. For households, the switch from kerosene to LPG did not encounter acceptability barriers that often emerge when implementing fuel-switching programmes. LPG is cleaner burning, more convenient, and is seen as a more modern fuel. The provision of free conversion kits (including the LPG cylinder and stove) helped a great deal with regard to acceptability. The most problematic group proved to be the kerosene retailers. Pertamina had to assist retailers in changing business models and practices. Retailers had for years enjoyed the convenience of the kerosene business, but now had to engage in a more complex business with lower profit margins. However, despite their reluctance, of all the groups represented, private sector retailers were also the least powerful, with little collective bargaining power and highly dependent on the state for the provision of subsidized fuel (ASTAE 2013; Budya and Arofat 2011; Kasih 2012).

25.2.2 Institutions

Indonesia's post-colonial political institutions have gone through a number of variations over the years, but have principally conformed to the principles of a unitary presidential system where almost all governing power resides in the centralized government (Macasaquit 2006). The country has, however, moved from authoritarianism and centralized dictatorship to a more decentralized version in the democratic era post-1998 (Ascher 1998; Liddle 1991).

The post-1998 democratic reforms have ushered in significant changes. While a unitary, presidential system is still in place, there has been noteworthy decentralization. The role of the legislature has been restored and strengthened, and it now plays an important role in checking the executive's authority, as well as developing, amending, and approving legislation. The kerosene-to-LPG conversion programme is illustrative of this changing dynamic. It was initiated by presidential decree in 2007, but implementation was delayed by several months due to the Indonesian parliament having several concerns that it wanted to be addressed before the decree would be approved. According to Braithwaite et al. (2012), political alignment on energy issues has become politically loaded. No single party holds an overall majority in parliament, and parliamentary voting history shows that energy reform is being used for political gain. However, stability appears to be even more important for politicians, and the overriding factor that pushed the LPG conversion programme through parliament was the fiscal threat posed by kerosene subsidies. This process contrasts with the introduction of the kerosene subsidy under General Suharto, where there was virtually no chance of parliament delaying or blocking legislation coming from the executive.

Indonesia's political culture has also shaped the form and design of the energy transition programme. As opposed to, for example, using market-based incentives and involving the private sector, it was largely driven, planned, and implemented by central government actors. Private sector involvement was limited to 'last-mile' distribution—serving as the link with customers. According to Karim (2014), there is wide acceptance in most Southeast Asian nations (as well as China) of a (benign) dictatorship as a road to economic development. This seemingly authoritarian political culture, forged in Indonesia's case through more than 30 years of authoritarian rule, appears to be one of the reasons that state-led development planning is not only accepted but also in fact expected by Indonesia's policy makers. It should be noted, however, that the existence of a well-resourced national oil company, the level of investment required, the timeframes required (government consistently pushed for earlier deadlines), and the real political threat of instability were all factors that also influenced the choice of programme design. Nonetheless, the shaping and enabling role of the prevailing political culture needs to be acknowledged as well.

25.2.3 Values and Ideas

Given how deeply rooted energy subsidies are in the Indonesian value system, the pressures on the treasury from kerosene subsidies could not be addressed through merely removing these subsidies. The Indonesian government had to find an alternative means of providing subsidized household fuel that was more affordable for the national budget. LPG was seen as this fuel, largely because it is a more effective fuel that requires a lower level of subsidy for an equivalent amount of energy. As mentioned in Section 25.2, the initial years of the programme seemed to indicate that there were significant savings for the Indonesian state. However, it appears that they are now starting to face the same budgetary pressures with the LPG subsidies as they had with kerosene (Cahyafitri 2015). The longer-term issue of how to decrease energy subsidies therefore remains unsolved—in large part because it is such an integral part of Indonesian society.

25.3 SOUTH AFRICA

South Africa, since emerging from apartheid rule to democracy in 1994, has put service delivery high on the political and economic agenda. The goal of providing universal access to services, including housing, water, electricity, and sanitation, has been a defining feature of the democratic government's rule. It has successfully implemented one of the largest electrification programmes in the world, and almost 6 million households (or 80 per cent of the population) now have access to the national grid—compared to 30 per cent in 1994 (Department of Energy 2012a). This programme has for many years dominated the household energy agenda. Viewed as the solution to all household energy issues, it has crowded out other supply options. But grid electrification is unlikely, on its own, to achieve universal energy access goals. First, the grid is unlikely to reach all households. In remote rural areas, low settlement densities make extension of the grid technically or economically infeasible. In urban areas, informal settlements are not eligible for their own connections due to land-use and planning restrictions related to the illegality of these settlements. Second, poor households are multiple fuel users, employing a range of energy carriers for different end-uses. Pursuing electricity as the only supply solution does not, therefore, match the basic demand profiles of households.

Residential LPG usage in South Africa is among the lowest in the developing world (Kojima 2011). A supply-constrained market and a high retail price have limited LPG's expansion into the low-income market. While household energy policy has focused on electrification, the liquid fuels policy sub-sector has also not incentivized expansion in LPG supply chains. LPG, as a result, plays a minor role in the overall energy mix. Most of what is produced domestically supplies industrial and commercial users.

South Africa has no significant oil or gas reserves. The supply of LPG comes predominantly from local refineries using imported crude oil, but also from coal, through a coal-to-liquids process, and natural gas, in a gas-to-liquids process. South Africa's refineries are old, inefficient, and experience many unplanned shutdowns, causing regular LPG supply shortages in the winter months. Many of the energy sector's resource allocation decisions are rooted in the country's socio-political history. This section aims to explore the key ways in which the legacy of apartheid continues to shape contemporary economic and political decisions in energy policy, first by looking at the legacy of close relationships that exist between the state and the petrochemical sector, and, second, by investigating the politicization of service delivery and how technology supply choices are framed.

25.3.1 Group, Interests, and Incentives

Understanding the framework of existing power relations in the private sector and dominant interests is essential to understanding LPG markets in South Africa. Although a range of supply and demand barriers exist in the low-income market, a key reason for LPG not expanding into the domestic market is that there is insufficient supply to do so, and retail prices are too high. The limited supply appears to be related to maintaining the market power of the dominant oil companies. The refineries, owned by international oil companies, control virtually all of LPG production in South Africa. There is evidence of significant anti-competitive behaviour that has limited market growth and entry by new players, which has spurred the Competition Commission to launch an industry-wide investigation (Ensor 2015). This investigation is not the first of its kind in this broader sector.[2] There is therefore a long-standing and entrenched pattern of market dominance and collusive behaviour in South African oil product markets.

Some of the ways in which barriers to entry are enforced are: preferential supply arrangements with refineries, costs of entry, control over infrastructure, and collusive pricing strategies. There is a high degree of vertical integration in LPG supply chains, from the refineries through to wholesaling, distribution, and retail (Paelo, Robb, and Vilakazi 2014). Refineries tend to sell their product to their own wholesalers. In the absence of adequate importing infrastructure, smaller independent wholesalers find it difficult to source product. Over 80 per cent of the wholesaling market is in the hands of the four major resellers who get their product directly from refineries (Lloyd 2014).

Another barrier to entry in the wholesale market is the cost structure. Wholesaling of LPG is highly capital-intensive, requiring investment in bulk transporter tankers, bulk storage facilities, cylinder-filling plants, cylinders, cylinder delivery vehicles, and installation on customers' premises. There are also anti-competitive long-term selling agreements between wholesalers, distributors, and customers around access to infrastructure—such as pipe networks and storage in a shopping mall—that prevent smaller players from

[2] The commission has also investigated and found cases of collusive conduct by the oil companies in the bitumen, piped natural gas, and diesel markets. Oil companies have been found guilty of sharing information, price fixing, and other coordinated behaviour (Das Nair, Pamela, and Roberts 2015).

entering certain customer markets (Lloyd 2014). The major wholesalers also allegedly under-price cylinders to keep new entrants out of the retail market. Even in the absence of explicit anti-competitive practices, informal institutional norms typical in tight oligopolies—such as relationships, having tacit industry knowledge and inside information that benefit those who have already worked with or for major oil companies—serve to keep out newcomers (Das Nair, Pamela, and Roberts 2015; Paelo, Robb, and Vilakazi 2014). The industry is in general characterized by secrecy and lack of information in the public domain (Lloyd 2014).

The regulatory framework appears to be tied in to a supply limiting market—suggesting regulatory capture. Very specific conditions of market control, price regulation, and limited supply appear to benefit certain market players. Wholesalers appear to enjoy a low-risk market and are potentially drawing significant economic rents from the mark-up between the regulated wholesale and retail prices. Retail prices approximately double the wholesale refinery gate price (Lloyd 2014). For refineries, enabling further growth in the LPG market has no direct benefits. The vertically integrated supply chains suggest they are extracting economic rents through transfer pricing. Since their refineries are old and cannot expand, any increase in supply would have to come from imports, which would be beyond their control. Furthermore, expanding residential LPG usage would only steal market share from their paraffin (kerosene) sales.

Most of the big wholesalers have shown little interest in expanding into the low-income domestic market, and these more peripheral markets have been left to the small players. These retail markets are expensive and difficult to operate in, compared to industrial and commercial markets. Challenges relate to restrictive municipal by-laws around fire safety standards as well as a retail price for LPG, which makes the cost of cooking or heating with it much higher than with electricity or paraffin, both of which are subsidized. Consumers are also not familiar with LPG, and fear explosions. Creating market demand would require running awareness campaigns and advertising, thus raising operating costs. Furthermore, when there are winter supply shortages caused by a refinery shutdown, the residential market is usually the first to experience shortages as distributors try to cushion other sectors. This supply insecurity decreases attractiveness of the energy carrier to low-income households (Mohlakoana and Annecke 2008).

There are, however, current developments which threaten to undermine the status quo and affect the current market power the majors enjoy. The first is the Competition Commission enquiry (Ensor 2015). The second is the construction of new importing and storage infrastructure on South Africa's west coast, with massive investment going into developing a gas terminal at Saldanha Bay. The facility, designed to be open access, will enable smaller players to access infrastructure and help to break the current monopoly hold on product supply. While this import infrastructure is a major step forward for market expansion, it is only part of the infrastructure picture. The need remains for associated investments into LPG cylinders and distributed filling facilities.

25.3.2 Institutions

Existing industry players clearly benefit from current market conditions, but of particular interest is the way the regulatory framework appears to entrench their market power and limit competition. This gives rise to questions around the private sector's influence among decision makers. In many ways this pattern can be traced back to state–business relations under apartheid—which left a legacy of concentration and control in this and many other sectors (Das Nair, Pamela, and Roberts 2015). There is a long history of close relationships between the state and the petrochemical sector in South Africa. During apartheid, petrochemical companies enjoyed guaranteed market returns, tariff protection, and removal of competition in various ways, as the threat of sanctions pushed security of supply in liquid fuels high up the political agenda (Baker, Newell, and Philips 2015; Das Nair, Pamela, and Roberts 2015). An innovative coal-to-liquids process was pioneered by Sasol and, in order to develop this source and protect the profitability of Sasol, competition was removed from the market. Oil companies were forced to cut expansion and buy product from Sasol, but in return they received guaranteed returns in the marketing and retailing sectors. In the post-apartheid dispensation, the state has formally tried to promote liberalization of the sector and a reduction in regulation, for example as laid out in the Energy White Paper (Department of Minerals and Energy 1998) and the Competition Act (Republic of South Africa, 1998). Interestingly, however, the petrochemical sector managed to obtain an exemption from this deregulation.

Regulated markets can lend themselves to regulatory capture. It has potentially been in the interests of oil companies to maintain regulation if they are behaving in cartel-like ways. Competitive markets can be damaging to cartels if they cannot ensure there are barriers to entry. Under more highly regulated regimes, and where dominant actors have sufficient influence, barriers to entry can be enforced through regulation. The South African regulatory framework appears to support the market power of the oil majors in a number of ways, such as through pricing and safety standards. Both wholesale and retail prices are regulated. The regulated wholesale price, known as the maximum refinery gate price (MRGP), has been widely acknowledged to make imports uncompetitive with refinery production. Importing and storage infrastructure is limited in South Africa; as wholesalers can import only small parcels at a time, they incur higher unit costs (Lloyd 2014). A discussion document on reviewing the MRGP released by the Department of Energy in 2012 (Department of Energy 2012b) notes that the MRGP has limited the expansion of import facilities and does not compensate for infrastructural investments. No further mechanism adjustments have been forthcoming since this discussion document, however. Similarly, cylinder safety standards in the industry are set such that only one manufacturer's product complies—thereby keeping others out.

The ways in which various aspects of regulation interact with market power suggest that oil companies may be influencing decision makers. The Department of Energy, the government institution responsible for regulating pricing, is characterized by inaction, which appears to implicitly benefit and maintain the power of certain actors. This is not unique to the liquid fuels energy subsector in South Africa. In an analysis of the electricity sector, Baker (2011) finds that many government departments and private players exert their influence over the energy policy-making process. Industrial stakeholders and heavy industry enjoy privileged levels of access (Baker, Newell, and Philips 2015). Broader structural drivers during the apartheid administration established a certain culture of state–business relations in the energy sector, and close relationships and beneficial arrangements for the petrochemical sector continue to shape their interactions and policy-making today.

Weak institutional capacity also exerts its influence over the policy-making environment (Baker 2011; Newbery and Eberhard 2008). Confusion about roles and responsibilities, and a lack of coordination, often compounds the already weak capacity within the Department of Energy. Different government departments often undertake seemingly contradictory actions that may not be in alignment with other policy objectives. Entities act according to their own goals and, in the absence of a strong guiding policy framework or governing department, different departments may pursue their own strategies. The construction of the new LPG import terminal at Saldanha Bay illustrates this. The tendering, licensing, and construction process has been long drawn out and characterized by multiple disputes. Different government departments with overlapping jurisdictions are pursuing their own agendas according to how they see LPG market expansion developing. The National Energy Regulator of South Africa wants to enable competition in the sector by establishing multiple import terminals and deregulating the market, while the Transnet National Ports Authority wishes to create one large facility to bring down costs. This would be an open access facility but well regulated to reduce players extracting economic rents. Both models have strengths and weaknesses, but currently both players are undermining the actions of the other, and in so doing making for a lack of clarity retarding longer-term market development. Their respective preferred models are likely also related to the particular institutional interests around market influence (e.g., through regulation) or control over infrastructure.

25.3.3 Values and Ideas

For various political reasons, the electrification programme has crowded out other energy access considerations and has dominated the household energy agenda (Matinga, Clancy, and Annegarn 2014; SEA 2014; Tait, Merven and Senatla 2013). Alternative energy carriers for the household market, such as biomass, paraffin, LPG, and off-grid solar, have all received limited policy attention. South Africa has adopted a traditional approach to energy access as a supply-oriented service based on big infrastructure rollouts. Much of the policy focus has centred on the provision of infrastructure rather than on the energy services delivered. As such, issues such as the various institutional and technical barriers to achieving high annual connection rates have dominated. But there has been far less focus on a demand-side appreciation of the services that households derive from electricity and whether these meet households' basic needs. A supply-side orientation tends to embrace standardization and uniformity, to plan technology rollouts that largely aim to be independent of contextual variations (Furlong 2014). The supply-side orientation has arguably blinded decision makers to the ways in which the delivery of an electricity meter may not actually result in users being able to access sufficient safe and affordable energy services to meet all of the household's basic needs. In a technocratic approach to electrical infrastructure planning there is no space to consider an integrated approach to delivering energy services that can consider multiple fuels and technologies, based on the demand context. Thus the fact that many poor households may not cook with electricity and continue to use paraffin gets overlooked in energy policy.

Various political and historic factors have come to shape South Africa's approach to household energy and the favouring of electricity. First, the framing of energy poverty as a policy problem in the early 1990s, and the role key actors played in shaping that agenda; and, second, the values attached to certain technology choices in the post-apartheid socio-political realm. The national electricity utility, Eskom, played a key role in the electrification programme from a very early stage. In the early 1990s, Eskom developed, managed, and financed the 'Electricity for All' programme, aimed at electrifying citizens who had been excluded from service delivery under apartheid. This has largely been seen as a way of demonstrating its commitment to, and alignment with, the values of the new democratic dispensation (Matinga 2010). Eskom at that time also had excess generation capacity and the financial reserves to enable quick rollouts at scale. Their central role at this early stage of the programme meant they became a key actor influencing how energy poverty, as a policy problem, was framed. An engineering focus to infrastructure planning dominated. Even though the programme no longer sits under Eskom's ambit, this legacy remains. The Department of Energy now has an electrification programme, rather than a household energy strategy. Most other household energy initiatives that have been undertaken in South Africa can be traced to the interests of the electricity sector. For example, the only significant programme promoting LPG among poor households, instituted in 2008, came at a time when the country was experiencing blackouts and needed to reduce peak demand. It was discontinued shortly after electricity supply returned to something like normal. From a household welfare perspective, any switching programme should be looking to switch users from paraffin to LPG, rather than electricity to LPG, which has neutral impact on household welfare.

The second consideration of the dominance of electricity has been around how the service delivery agenda of the democratic government has been set. Services such as energy, water, and sanitation have become key features of the government's policy to redress past imbalances. During apartheid, the state systematically denied basic services to non-white citizens, and equality has therefore become a key symbol in current service delivery programmes. Electricity, as the technology enjoyed by the privileged under apartheid, was thus ensured primacy in the government's service delivery agendas, its dominance being as much for its political attributes as its technical ones. Technologies such as electricity represent much more than the direct services they provide, having become symbolic of inclusivity in government policy and of the development aspirations of post-apartheid South Africa (Matinga, Clancy, and Annegarn 2014; Merven and Senatla 2013). Energy options other than an electric grid connection are often perceived to be inferior or not 'modern' (Matinga 2010). When the poor protest about service delivery, their demands relate to electricity and not energy. With the political currency attached to electricity, there is little incentive for politicians to promote alternative energy sources, such as LPG.

The government's electrification programme has also become deeply distorting in household energy markets. The programme consists of a fully subsidized connection as well as a subsidized monthly allocation of free kilowatt hours. Furthermore, electricity tariffs in South Africa were underpriced for many years, leading South Africa to have had one of the lowest electricity prices in the world (Newbery and Eberhard 2008). Even with recent electricity price hikes, the country's electricity tariffs remain among the lowest on the African continent (Fripp 2015). In comparison, LPG's retail price is one of the highest in the world (Kojima 2011) and the upfront costs of purchasing cylinders and appliances are not subsidized. A combination of a bias towards electricity in household energy policy and imperfections in LPG regulations thus make it an uneconomic option for the poor and limit its market penetration. But electricity tariffs are rising substantially as the electricity sector moves towards cost-reflective pricing (Deloitte 2012). The previously stark cost differential between the two energy carriers is therefore fast changing—potentially ushering in a new era of residential energy use patterns in the country.

25.4 CONCLUDING ANALYSIS

This chapter set out to discover the political economy elements and mechanisms influencing residential thermal energy transitions in Indonesia and South Africa, through the lens of the LPG sector. The following key elements emerged from the analysis.

25.4.1 State Power and Control

The relative power of the state, its implementing institutions and the private sector, especially with regard to the energy sector, plays a significant role in residential energy policy. This is, for example, one of the reasons why countries with relatively strong and capable national oil companies—like Pertamina in Indonesia—are able and willing to provide LPG to households at significant scale, and why South Africa, through its national electricity utility company, exhibits a clear, and almost exclusive, preference for household electrification. The more control government is able to exercise over a particular sector or fuel, the more likely it seems that this sector or fuel will be supported for wide-scale adoption. The converse of this argument is, of course, not only that government might not be willing to support sectors where its power is more limited, but also that it might in fact not be able to. The South African LPG sector seems to offer a picture of an industry with enough collective power to not only resist government efforts to entice it into serving low-income, potentially less profitable customers—but to use government regulation to maintain competition-limiting market dominance.

25.4.2 Stability and the Politics of Discontent

Energy provision and, in particular, household thermal energy provision, is distinctly political in both countries, with the potential to cause great instability. Energy cost is recognized as a significant political factor in Indonesia, prompting proactive measures such as the LPG conversion programme. Energy subsidies are also of great financial and political consequence in Indonesia, largely due to the high level of subsidies provided. South Africa's residential energy policy is influenced by the country's socio-political history to such an extent that subsidized electricity provision remains the only acceptable form of energy provision. In both countries extensive political currency is derived from energy subsidies—regardless of the fuel or technology used.

25.4.3 Aligning Interests and Objectives

A key factor influencing the implementation of residential energy policy in both countries is the alignment of policy objectives with the interests and strengths of key actors. This is why, for example, both the Indonesian LPG conversion programme as well as the South African electrification programme were able to achieve such massive rollout in a short time frame. Through actively involving the implementing agency—whether Pertamina or Eskom—in the design of the programme, it was guaranteed that the eventual product would speak to these agencies' strengths and interests. Where this alignment of objectives and interests is not in place—as is the case in South Africa's LPG sector—there is very little chance of successful policy implementation.

REFERENCES

Ahiataku-Togobo, W. (2013). 'National Cooking Energy Strategies—Lessons from Ghana'. Presentation at the WACCA workshop, Ouagadougo, April.

Arofat, M. Y. (2014). 'Indonesia's Kerosene to LPG Conversion Program: The Story and the Way Forward'. Presentation at African LP Gas Summit, Nairobi, Kenya, 2–4 July.

Ascher, W. (1998). 'From Oil to Timber: The Political Economy of Off-Budget Development Financing in Indonesia'. *Indonesia*, 65(April): 37–61.

ASTAE (Asia Sustainable and Alternative Energy Program) (2013). 'Indonesia: Toward Universal Access to Clean Cooking'. East Asia and Pacific Clean Stove Initiative Series. World Bank, Washington, DC.

Baker, L. (2011). 'Governing Electricity in South Africa: Wind, Coal and Power Struggles'. The Governance of Clean Development Working Paper Series, University of East Anglia.

Baker, L., P. Newell, and J. Philips (2015). 'The Political Economy of Energy Transitions: The Case of South Africa'. *New Political Economy*, 19(6): 791–818.

Braithwaite, D., A. Chandra, R. L. Prasetyning Diah, A. Indriyanto, K. Lang, L. Lontoh, N. Siahaan, D. Vis-Dunbar, B. A. Wattimena, U. Widhiantoro, and P. Wooders (2012). 'Indonesia's Fuel Subsidies: Action Plan for Reform'. International Institute for Sustainable Development, Geneva.

Brew-Hammond, A., G. S. Mensah, and O. Amponsah (2014). 'Energy Poverty in Sub-Saharan Africa: Poverty Amidst Abundance'. In A. Halff, B. K. Sovacool, and J. Rozhon (eds), *Energy Poverty: Global Challenges and Local Solutions*. New York: Oxford University Press, pp. 296–315.

Budya, H. and M. Y. Arofat (2011). 'Providing Cleaner Energy Access in Indonesia through the Megaproject of Kerosene Conversion to LPG'. *Energy Policy*, 39(12): 7575–86.

Cahyafitri, R. (2015). 'Rising LPG Subsidies Could Pressure State Budget: LIPI'. *The Jakarta Post*, 6 June. Available at: <http://www.thejakartapost.com/news/2015/06/06/rising-lpg-subsidies-could-pressure-state-budget-lipi.html> (accessed 30 March 2016).

Central Intelligence Agency (2013). 'The World Factbook 2013–14'. Central Intelligence Agency, Washington, DC.

Dahl-Østergaard, T., S. Unsworth, M. Robinson, and R. I. Jensen (2005). 'Lessons Learned on the Use of Power and Drivers of Change Analyses in Development Co-operation'. OECD-DAC, Paris.

Das Nair, R., M. Pamela, and S. Roberts (2015). 'The Inter-Relationships between Regulation and Competition Enforcement in the South African Liquid Fuels Industry'. *Journal of Energy in Southern Africa*, 26(1): 11–19.

Deloitte (2012). 'The Economic Impact of Electricity Price Increases on Various Sectors of the South African Economy: A Consolidated View Based on the Findings of Existing Research'.

Department of Energy (2012a). 'A Survey of Energy-related Behaviour and Perceptions in South Africa—The Residential Sector'. Republic of South Africa, Pretoria.

Department of Energy (2012b). 'Discussion Document on the Review of the Maximum Refinery Gate Price of Liquefied Petroleum Gas'. Republic of South Africa, Pretoria.

Department of Minerals and Energy (1998). 'White Paper on the Energy Policy of the Republic of South Africa'. Republic of South Africa, Pretoria.

DfID (Department for International Development) (2009). 'Political Economy Analysis: How To Note'. DfID practice paper. Available at: <http://www.gsdrc.org/docs/open/PO58.pdf> (accessed 20 October 2016).

Diop, N. (2014). 'Why Is Reducing Energy Subsidies a Prudent, Fair, and Transformative Policy for Indonesia?' Poverty Reduction and Economic Management Network. World Bank, Washington, DC.

Ensor, L. (2015). 'Competition Commission Deepens Probe into LPG Market'. *Business Day*, 28 August. Available at: <http://www.bdlive.co.za/business/energy/2015/08/28/competition-commission-deepens-probe-into-lpg-market> (accessed 7 March 2016).

Fripp, C. (2015). 'South Africa's Electricity Pricing Compared to the Rest of the World'. *Htxt.africa*, 26 June. Available at: <http://www.htxt.co.za/2015/06/26/south-africas-electricity-pricing-compared-to-the-rest-of-the-world/> (accessed 7 March 2016).

Furlong, K. (2014). 'STS beyond the "Modern Infrastructure Ideal": Extending Theory by Engaging with Infrastructure Challenges in the South'. *Technology in Society*, 38: 139–47.

Global Alliance for Clean Cookstoves (2015). 'Five Years of Impact: 2010–2015. Our Story. Our Progress. Our Aspiration'. United Nations Foundation, Washington, DC.

Hall, C., J. Lambert, and S. Balogh (2014). 'EROI of Different Fuels and the Implications for Society'. *Energy Policy*, 64: 141–52.

IEA (2014). 'World Energy Investment Outlook 2014'. Special Report. International Energy Agency, Paris.

Karim, M. F. (2014). 'Indonesian Political Institutions'. Presented at Britzone Community, Ministry of Education of the Republic of Indonesia, Jakarta.

Kasih, T. (2012). 'Providing Cleaner Energy Access for Indonesia: Case Study from Kerosene to LPG Conversion'. Practitioners Workshop on Energy Access for Urban and Peri-Urban Poor. ESMAP—Cities Alliance—World Bank, Washington, DC.

Kojima, M. (2011). 'The Role of Liquefied Petroleum Gas in Reducing Energy Poverty'. World Bank—Oil, Gas and Mining Policy Unit Working Paper. Extractive Industries for Development Series 25, December. World Bank, Washington, DC.

Kumoro, B. and F. Astriana (2015). 'A Polemic against Indonesia's History of Fuel Subsidies'. *Strategic Review*, April–June, 5(2): 60.

Liddle, R. W. (1991). 'The Relative Autonomy of the Third World Politician: Soeharto and Indonesian Economic Development in Comparative Perspective'. *International Studies Quarterly*, 35(4): 403–27.

Lloyd, P. (2014). 'The Status of the LP Gas Industry in South Africa'. In *Proceedings of the Industrial and Commercial Use of Energy Conference*, April 2014, Cape Town.

Macasaquit, M. R. (2006). 'Forming a Government: Parliamentary vs. Presidential System'. *Economic Issue of the Day (EID)*, 6(2): 1–2.

Mcloughlin, C. (2014). *Political Economy Analysis: Topic Guide*, Second Edition. GSDRC, University of Birmingham, Birmingham.

Matinga, M. (2010). '"We Grow Up With It": An Ethnographic Study of the Experiences, Perceptions and Responses to the Health Impacts of Energy Acquisition and Use in Rural South Africa'. PhD thesis, University of Twente.

Matinga, M. N., J. S. Clancy, and H. J. Annegarn (2014). 'Explaining the Non-Implementation of Health-Improving Policies Related to Solid Fuels Use in South Africa'. *Energy Policy* 68: 53–9.

Mohlakoana, N. and W. Annecke (2008). 'Finally Breaking the Barriers: South African Case Study on LPG Use by Low-Income Urban Households'. Istanbul Pre-Conference Workshop on Clean Cooking Fuels, 16–17 June.

Mosco, V. (2009). 'What is Political Economy?' In *The Political Economy of Communication: Rethinking and Renewal.* Thousand Oaks, CA: Sage.

Mourougane, A. (2010). 'Phasing Out Energy Subsidies in Indonesia'. OECD Economics Department Working Papers 808. Paris: OECD.

Newbery, D. and A. Eberhard (2008). 'South African Network Infrastructure Review: Electricity'. National Treasury and the Department of Public Enterprises, Pretoria.

Paelo, A., G. Robb, and T. Vilakazi (2014). 'Study on Barriers to Entry in Liquid Fuel Distribution in South Africa'. Centre for Competition, Regulation and Economic Development, University of Johannesburg, Johannesburg.

Republic of South Africa (1998). 'Competition Act no. 89 of 1998'. Republic of South Africa, Pretoria.

Rogers, E. M. (1983). *Diffusion of Innovations*. Third Edition. New York: Free Press.

SEA (Sustainable Energy Africa) (2014). 'Tackling Urban Energy Poverty in South Africa'. Report for Heinrich Boell Foundation, Cape Town.

Seah, S. H. (2014). 'Can Indonesia's Policy of Reconfiguring its Energy Mix by Increasing Natural Gas Usage Support Its Initiatives to Reform Energy Subsidies?' OIES Paper NG 93, Oxford Institute for Energy Studies, Oxford.

Tait, L., B. Merven, and M. Senatla (2013). 'Investigating the Current and Future Role of Paraffin in South Africa'. Energy Research Centre, University of Cape Town, Cape Town.

Tumiwa, F., L. Lontoh, T. Laan, K. Lang, and D. Vis-Dunbar (2012). 'A Citizen's Guide to Energy Subsidies in Indonesia: A 2012 Update'. International Institute for Sustainable Development, Geneva.

United Nations Foundation (2015). 'What We Do: Achieving Universal Energy Access'. Available at: <http://www.unfoundation.org/what-we-do/issues/energy-and-climate/clean-energy-development.html> (accessed 2 November 2015).

Williams, G., A. Duncan, and P. Landel-Mills (2007). 'Making the New Political Economy Perspective More Operationally Relevant for Development Agencies'. Policy Practice Brief 2. The Policy Practice, London.

Part VIII

Regional Dynamics

This part of the book features chapters that explore how clean energy transitions challenge traditional national boundaries and institutional architectures. As climate and clean energy commitments are made at the national level, a broad observation is that political economy factors may lead to growing territorialization or institutional friction at the sub-national level as regional development and integration dialogues ensue—and so it goes, as international climate commitments are made on behalf of groups of nations. This can occur for a host of reasons, inter alia: conflicts over land and resource use; a desire to localize investment and development benefits; and a desire to avoid the economic negative impacts of fossil fuel infrastructure retirement.

Pham Do and Dinar offer a case study of the Mekong River region, and how the rush to acquire sources of alternative energy and other benefits to meet rapid demand growth have led to circumstances of risk for the broader economic areas, including Cambodia, Laos, Myanmar, Vietnam, and the Yunnan province of China. The struggle over hydropower development and decision-making on water and land across this region offers valuable insights into the complex political economy dynamics at play in regional energy transition efforts. Puppim de Oliveira and Andrade examine the political economy aspects, particularly the influence of Clean Development Mechanisms (CDMs), in clean energy and climate change policies in the Brazilian states of Bahia and Rio Grande do Sul. The case study identifies pitfalls and opportunities for new strategies and mechanisms for boosting clean energy in Brazil, and the efficacy of CDMs and potential future mechanisms for supporting sub-national climate change mitigation efforts. Lepesant reviews tensions and institutional innovations that can arise at local and national levels when implementing regional energy policies—in this case, the European Union's 2009 Renewable Energy Directive. In systematically exploring the conflicting interests and natural tensions, the case study provides insights into how best to encourage adoption and implementation of regional priorities at the local level.

26

The Linkages of Energy, Water, and Land Use in Southeast Asia

Challenges and Opportunities for the Mekong Region

Kim Hang Pham Do and Ariel Dinar

26.1 INTRODUCTION

The Mekong region (MR) is a natural economic area, consisting of Cambodia, Laos, Myanmar, Thailand, Viet Nam, and the Yunnan province of China, bound together by the Mekong River.[1] Originating at an elevation of over 4500 m in the Tibet Qinghai plateau, the Mekong River, with a total catchment area of 795,000 km^2, not only is the major water source in Southeast Asia but also harbours a wealth of natural resources including important stocks of forest, fish, biological biodiversity, transportation, wetlands, and tourism (ADB 2004; Mehtonen, Keskinen, and Varis 2008; MRC 2005). Land and water resources are central to agriculture and rural development, and are intrinsically linked to regional challenges of food insecurity and poverty. For over 20 years, Southeast Asia has been one of the fastest growing regional economies in the world and water resource has assumed a central role in the region's economic growth (ADB 2013). Economy and population, particularly in urban centres, have grown remarkably, leading to increased use of energy and natural resources. The basis for sustainable development is the access to these resources and their sustainable management (FAO 2011).[2]

[1] 'The boundary of the Mekong River region includes the entire Mekong River Basin (MRB) and the coastal area adjacent to the Mekong Delta' (UNEP 2006).

[2] Sustainable energy is one of 17 global goals that make up the 2030 United Nations Development Programme (UNDP) Agenda for Sustainable Development (United Nations, 2015).

In the MR, the linkages between energy, water, land use, and development are numerous and complex, driven by increasing demands for clean energy, cheap electricity, and potential profits. Water is not only lifeblood, it is also an economic resource. The production and consumption of goods and services require energy. Energy development requires water. Food production and land-use planning require water and energy. The fast development of hydropower, mining, infrastructure and urbanization are main factors for change in land use. As the most populous region and the world's manufacturing hub, Southeast Asia and the MR demand water, energy, and land resources in ever-increasing amounts, leading to shortages that are creating serious choke points in the country's development. Pressure on water is at the heart of these resource constraints facing the MR (ADB 2013; Kubiszewski et al. 2012; Matthews and Geheb 2015).

Owing to sustainable development and managing natural resources, especially given the very high economic growth rates in China, the MR is of global importance, in terms of population size, land use, resource base, regime, gross domestic product (GDP) size, GDP per capita, or comparative advantages (Table 26.1). Currently, over 100 larger dams are planned for the main stream of the Mekong River and its tributaries (MRC 2010). Power distribution within the MR is defined by strategic position, making the region a scene for one of the most intensive hydropower developments globally. Dam construction has almost always created conflicts between energy supply and related economic interests and their social and environmental impacts (King, Bird, and Haas 2007; MRC 2010). Life in the MR is facing threats as major dams begin to rise. Deforestation for timber or fuel wood supplies, agricultural expansion, urbanization, and infrastructure creation, all contribute to increased erosion. Consequently, about 21 per cent of the MR area is experiencing problems with

Table 26.1. Selected aggregate indicators for the Mekong region (MR) and China in 2014

	Population (million)	Population growth (%)	Population (per km^2)	Land area (10^3 km^2)	Water area (10^3 km^2)	GDP growth (%)	GNI per capita (USD 2014)[a]
Cambodia	15.18	1.50	84	181.04	4.5	7.0	1020
Laos	6.77	2.00	29	236.80	6.0	7.4	1650
Myanmar	51.42	1.17	76	676.59	23.1	7.7	1270
Thailand	67.09	0.40	131	513.12	2.2	0.7[b]	5370
Viet Nam	91.52	1.05	274	330.95	21.1	6.0	1890
China	1367.82	0.48	143	9569.96	27.1	7.4	7380

Note: GDP, gross domestic product; GNI, gross national income. [a] Data from World Bank (2015). [b] The Thai economy, however, has slowed significantly since the fourth quarter of 2013 because of the delay of a complete election and the limited role of the caretaker government.

Source: Based on ADB (2015) and World Bank (2015).

erosion;[3] only 31 per cent of its original forests have been left intact and only 5 per cent are under regulated protection (UNEP 2006). Together with climate change adaptation and mitigation, the livelihoods of some hundred million rural people across the MR are affected by degradation and depletion of natural resources.

To meet demands of population growth and energy development, the increase in water consumption as alternative energy is intensive. Water needs to be part of energy strategies, and the management of water resources is one of the most urgent development challenges. Hence, it is vital to understand how transboundary rivers can meet water, food, and energy needs of riparian populations while minimizing negative impacts (Bach et al. 2012). This chapter addresses the relationship between energy, water, and land use in the MR.[4] The aim of the chapter is to: (i) provide a comprehensive overview and identify historical indicators of regional conflict and cooperation; and (ii) create a framework to enhance understanding management for sustainable development in the context of the transboundary water resources. It sheds light on the opportunities to mitigate the possible negative impacts of hydropower development for simultaneously protecting the environment and growing the economy by highlighting the role of issue linkages in achieving a regional agreement (Pham Do, Dinar, and McKinney 2012). The results contribute to policy-making as recommendations of how the Mekong River Commission (MRC) should deal with and without the rising power of China and the role of the United Nations as well as the Association of Southeast Asian Nations (ASEAN) in the Greater Mekong Sub-Region (GMS).

The rest of this chapter is organized as follows. Section 26.2 presents an overview of the MR and its water resources. Section 26.3 provides a brief historical development and regional integration. Section 26.4 analyses the role of issue linkages in designing a mechanism of regional and international cooperation. Concluding remarks follow Section 26.5.

26.2 OVERVIEW OF THE MR

With a total land area of 2,334,000 km^2 (Xing 2013: 180), a long history of wars and peace, and the recent challenges in of the potential alteration of complex ecological and social systems (Campbell 2009; Dore and Xiaogang

[3] The 'increased rates of surface water run-off resulting from deforestation and land clearance in upland areas of the MRB are causing increased soil erosion and the consequent entrainment of suspended and bed-load sediments into water courses' (UNEP 2006: 12).

[4] This chapter seeks to update and add to previous studies on transboundary water management (see Pham Do et al. 2012; Pham Do and Dinar 2014).

2004; Kaisti and Kakonen 2012; Pham Do and Dinar 2014), the MR has attracted considerable international attention since 1950.

As seen in Table 26.1, economic disparities are very wide, ranging from four low-income member countries (Cambodia, Laos, Myanmar, and Vietnam) with a low gross national income (GNI) per capita (formerly, gross national product per capita) of US$1020–1890 (at 2014 US$ price) to two middle-income member countries (Thailand and China) with a high GNI per capita of US$5370–7380. Populations range from 6.8 million people in Laos to over 90 million in the combined Yunnan and Guanxi regions. Generally, the average growth of real GDP in the MR has continuously increased in recent years (ADB 2015). However, poverty is a critical issue across the MRB, despite the significant economic growth of China, Laos, Cambodia, and Myanmar. According to UNEP (2008), the proportion of population living below the poverty line exceeds 30 per cent in many different ethnic groups of Laos, Cambodia, and Viet Nam.

26.2.1 Water Resources in the MR

Historically and geographically, the MRB can be divided into two parts: the Upper Mekong Basin (UMB) constituting China and Myanmar (24 per cent of the total catchment area) and the Lower Mekong Basin (LMB) constituting Cambodia, Laos, Thailand, and Viet Nam (76 per cent of the total catchment area). The MRB is home to nearly 75 million people with 90 distinct ethnic groups (Matthews and Geheb 2015). Encompassing a vast range of geographic and climatic zones, the amount of water resources internally renewable annually varies widely by country. Moreover, land and water resources in the region also vary with location and season. For example, although only 16 per cent of the total discharge originates from the UMB, during the critical dry season, China discharges water to most of the Mekong mainstream flow in Laos and Thailand and contributes to almost 45 per cent of the average flow in Cambodia (Goh 2004).

As the longest river in Southeast Asia, the name roughly translating to 'mother of water' in the Lao and Thai languages, the Mekong is the region's largest water resource and provides the largest related resources with power generation potential that support on-going economic development and MRB community livelihoods. Particularly, water use for irrigation is expected to increase in the LMB (FAO 2012). Table 26.2 reports a summary of distributions of water and land resources in the MRB.

Although the MRB is one of the richest areas of biodiversity in the world, the six riparian states have quite different long-term major use patterns of the river. Three primary economic services for the millions who live in the MR are

Table 26.2. Territory of six countries and their contribution to the Mekong's flow

Description	Yunnan (China)	Myanmar	Laos	Thailand	Cambodia	Viet Nam
Area (10^3 km^2)	165	24	202	184	155	65
Area as percentage of MRB	21	3	25	23	20	8
Percentage of total area of country/province	38	4	97	36	86	20
Flow as percentage of MRB	16	2	35	18	18	11
Average flow (m^3/sec)	2410	300	5270	2560	2860	1660

Note: MRB, Mekong River Basin.
Source: Based on MRC (2005).

transportation, renewable freshwater, and electric power (Roland-Holst and Heft-Neal 2012). Water resources are used mainly for hydropower production and irrigation (MRC 2010). Based on the calculated models of four main different sectors of water uses in 2010, Houba, Pham Do, and Zhu (2013) show that (i) water use for irrigation generates the highest aggregate economic value for China and the LMB; and (ii) water use for hydropower generation contributes the second highest economic value for China, whereas fishery is the second highest for the LMB.

26.2.2 Energy Demand and Hydropower Plants in the MR

The MR is endowed with abundant resources that have power generation potential, such as hydropower, natural gas, and coal. However, many countries in the region face difficulties in accessing financial resources and technologies to exploit the energy potential of these resources. Most of the oil and gas for potential development is located in the territorial waters of Cambodia, Myanmar, Thailand, and Viet Nam, whereas coal is in Yunnan province of China (ADB 2013). Owing to the regional diversity in economic development and rise in population, growth in primary energy demand differs by nations (Table 26.3).

As shown in Table 26.3, China has a highest potential for reducing energy demand compared with any other nations in the MR: its annual growth rate reduces from 2.3 per cent in a BAU case to 1.4 per cent in the alternative case. Within the MR, it must be recognized that hydropower is only one of a number of purposes for which water is diverted or stored. Houba, Pham Do, and Zhu (2013) report that hydropower generation in the LMB takes place in the tributaries and produces only 2 per cent of the total economic value of the LMB. This low value reflects the undeveloped hydropower potential. To date, China is the only country to have built dams on the Mekong mainstream.

Table 26.3. Energy demand in business-as-usual and alternative scenarios

	Primary energy demand (Mtoe)		Annual growth rate (%)	Primary energy demand per capita (toe)		Primary energy demand and growth rate with advanced technology	
	2010	2035		2010	2035	2035 (Mtoe)	Growth rate (%)
Cambodia	5.0	8.9	2.3	0.36	0.5	7.4	1.7
Laos	2.8	7.9	4.3	0.45	0.99	7.7	4.0
Myanmar	14.0	30.3	3.1	0.29	0.55	29.2	3.0
Thailand	117.4	204.8	2.2	1.70	2.79	183.5	1.8
Viet Nam	67.7	186.0	4.1	0.77	1.80	167.9	3.7
China	2471.1	4218.1	2.3	1.80	3.05	3418.7	1.4

Note: Mtoe, million tonnes of oil equivalent.
Source: Based on ADB (2013).

According to King, Bird, and Haas (2007), Cambodia has a hydropower potential of 10,000 MW but an installed capacity of only 160 MW. With 90 per cent of land area in the basin, Laos has the greatest potential for hydropower development (13,000 MW) and could become a power hub for Thailand, Viet Nam, and the ASEAN power grid. Whereas Thailand is a major importer of hydropower and a potential hub for the ASEAN power grid, Viet Nam and Myanmar are important both as hydropower developers and as potential importers. The MRC has proposed many plans for developing this potential through dam projects; there are 11 mainstream dam proposals and 30 planned tributary dams to be developed between 2015 and 2030 (Kubiszewski et al. 2012). Moreover, Laos has started constructing the first of 11 planned hydropower projects on the Lower Mekong River mainstream, and hopes to become the 'battery of Southeast Asia', selling electricity to MR neighbours. Yunnan also installed 68 per cent of provincial power generation and is constructing a cascade of hydropower plants on the mainstream of the Mekong with 15,600 MW to be completed by 2025 (King, Bird, and Haas 2007: xi). Construction of dams on the Mekong River may pose immediate and long-term threats to the food security and livelihoods of over 60 million people in the LMB (MRC 2010).

26.2.3 Land-Use Change

Land and water are ecologically linked in a natural system (called a catchment,[5] drainage basin, or watershed). One of the major components that would affect

[5] The catchment surface can be one of three types: water, forest (sparse dry deciduous), or agricultural land (including all non-forest land uses).

future water and energy resources is change in land use. Different drivers cause different changes in land use. The impact that land-use changes can have on water resources is large, but quantifying these impacts presents many challenges. According to Rowcroft (2008), land-use changes are multifaceted and mainly caused by prices of agricultural and forest products. To maintain the MR as an important food production area with irrigation farmland, aquatic agriculture, and pasture, governments in the region have undertaken land reforms and put limits on shifting cultivation for conservation since 1980. However, rapid land-use change has occurred in the region since 2000.

The MRC Strategic Environmental Assessment report identifies expanding irrigation and loss of forest, farmland, and aquatic resource as the main consequences of mainstream dams (ICEM 2010). The loss of land is a direct impact of hydropower projects (MRC 2010) through construction of reservoirs, access roads, and transportations. According to Smajgl and Ward (2013), there are six national developments that would significantly influence land-use change in the MR till 2050: (i) expansion of mainstream dams, (ii) water diversion, (iii) rise in sea level, (iv) adaptation strategies in the Mekong delta, (v) expansion of rubber plantations, and (vi) transnational transport infrastructure (in particular, railway projects linking Kunming to Cambodia).

Having provided an assessment of the impacts of these developments related changes in land use, Xing (2013) shows that during 2000–8 farmland increased in Yunnan (61.3 per cent) and Laos (42.5 per cent) whereas pasture increased in Cambodia (61.3 per cent) but decreased in Yunnan (21.3 per cent). In addition, it is expected that 22,692.42 km^2 of forest will be converted to commercial plantations and mainstream dams may submerge 119.66 km^2 of riverbank gardens. Total area of land-use change is predicted to be 34,068.71 km^2 (see Xing 2013: table 6.10).

26.3 REGIONAL DEVELOPMENT, INTEGRATION, AND CHALLENGES

The MR has a long history of development and dialogue among lower riparian countries. An increasing number of river-based cooperation institutions have emerged in mainland Southeast Asia since the early 1990s. Among these are the MRC, the GMS, and the Mekong Basin Development (MBD) that take place under the overarching framework of the ASEAN.

26.3.1 Regional Development and Integration

The MRC is an intergovernmental organization for regional dialogue and cooperation in the LMB, established in 1995, based on the Mekong Agreement on Cooperation for Sustainable Development of the MR (the so-called 1995 Mekong Agreement). As a successor of the Mekong committee, which was established in 1957, the MRC has the longest history of cooperation in the region. This regional institution works with many different partners, responsible for promoting and developing sustainable management strategies across all sectors including sustaining fisheries, identifying opportunities for agriculture, maintaining the freedom of navigation, flood management, and preserving important ecosystems (see MRC 2016: annex A). However, the MRC has failed to attract China and Myanmar to join.

In 1992, the MR nations launched the GMS Economic Cooperation Program to promote integrative economic links among riparian nations. The GMS comprises Cambodia, Laos, Myanmar, Viet Nam, and the Yunnan province and the Guangxi Zhuang Autonomous Region of China. With support from the ADB and other donors, the GMS programme has become a key for growth and development in mainland Southeast Asia over the past few decades. Particularly, the new GMS Strategic Framework for 2012–22, adopted in December 2011, expands the GMS programme from conventional infrastructure to multi-sector investments designed to allow the implementation of large-scale water infrastructures (such as building commercial relationships in terms of cross-border trade and transportation, energy development, investment, and water resource usage). This was also considered as a peaceful resolution of conflict in Indo-China relationships in the early 1990s: the integration of Cambodia, Laos, Myanmar, and Viet Nam into the ASEAN; the gradual opening of the Yunnan province and China itself to its southern neighbours; and with financial support.

ASEAN was established on 8 August 1967 in Bangkok, Thailand, with the signing of the ASEAN Declaration by five countries: Indonesia, Malaysia, the Philippines, Singapore, and Thailand. It was extended to five other nations: Brunei (1984), Viet Nam (1995), Laos (1997), Myanmar (1997), and Cambodia (1999). To strengthen the interconnections and economic linkages between ASEAN and MR countries, ASEAN set up an ASEAN–MBD Cooperation (AMBDC) on 17 June 1996, consisting of all member states of ASEAN and China. The AMBDC is considered an important framework to enhance and sustain growth of the MR and a policy dialogue for ASEAN and China to foster economic development and cooperation.[6] In addition, the establishment of the ASEAN Economic Community (AEC) in December 2015 is a major milestone for regional economic integration in ASEAN. According to Petri, Plummer, and Zhai (2012), the AEC offers many opportunities to the business community and general public because it permits free movement of goods,

[6] ASEAN's Mekong concept document emphasizes the complementarity of existing development programmes linking them to the Asian Development Bank–Greater Mekong Sub-region and the UNDP–Mekong River Commission (Weatherbee 1997).

services, foreign direct investment, and skilled labour and free flow of capital. However, how Mekong countries decide to pursue future hydropower development is perhaps one of the most challenging strategic decisions they have faced since the signing of the 1995 Mekong Agreement (MRC 2010).

26.3.2 Hydropower Projects in the MRB and Their Impact

Transiting from its headwaters to a delta, with more than 4500 km of flow, the Mekong River provides a major source of energy potential through its basin. Hydropower projects in the MRB were first constructed in Laos and Thailand during the late 1960s and early 1970s. In 1986, China began building the first of a series of dams in the upper mainstream. At present (2015), hydropower opportunities are largely related to the mainstream and its immediate tributaries. The total potential for hydropower generation in the MRB is estimated as ~250,000 MW (ADB 2015). However, only a portion of this potential could be considered economically and environmentally feasible.

Table 26.4 presents an inventory of hydropower development projects in the MRB, revealing 66 hydropower dams that have been commissioned, 37 dams that are under construction, and 93 dams that are planned or proposed. Note that the total number of hydropower projects in China is 46 dams. However, two dams have been cancelled because of environmental problems (WLE Greater Mekong 2015). Before 1990, most hydropower development in the MR was publicly funded. However, since the early 1990s this development has been transferred and led by commercial partnerships between private sector developers and local governments (Hirsch 2011). Hydropower projects in the MR have generally been profitable for both host governments and private sector sponsors. The total monetary profit of hydropower operations in the next 20 years in the region is estimated to be US$15–20 billion, according to Li (2012).

Although dams can help with flood control in the wet season and with increased water supply for irrigation and navigation during the dry season for downstream riparian states, the potential negative consequences for the LMB

Table 26.4. Number of hydropower development projects in the MRB

Country	Commissioned	Under construction	Planned	Total
China	17	10	17	44
Myanmar	0	0	6	6
Laos	20	26	57	103
Thailand	7	0	0	7
Cambodia	0	1	11	12
Viet Nam	22	0	2	24

Source: Based on WLE Greater Mekong (2015).

are multifaceted and likely to materialize in ecological, economic, and negative political outcomes (Biba 2012). Ziv et al. (2012) also show that the completion of 78 dams on tributaries would have catastrophic impacts on fish productivity and biodiversity. According to Kubiszewski et al. (2012), the value of lost capture fisheries, future aquaculture production in the LBM,[7] and the value of lost ecosystem services are estimated to be in the range of US$33 billion to 274 billion.

26.3.3 Challenges and Opportunities in the MR

The rich human and natural resources, as well as the current peaceful political situation in the MR, have attracted many foreign investments and made it one of the world's fast growing regions (UNEP 2008). However, with globalization and population growth, development in the MR has had losses as well as benefits, thereby increasing pressure on land and water use. The series of floods in Laos (during 2011, 2013, 2015) and the current drought in the Mekong delta (one of the world's granaries and Viet Nam's rice bowl) provide dramatic examples of how environmental mismanagement can be costly for the region (for details, see MRC 2012). Particularly, the heavy socioeconomic costs are disproportionately borne by downstream countries, especially Cambodia, Viet Nam, and riverine parts of Thailand (Cronin and Hamlin 2012).

According to Cronin (2012), the cumulative net economic profit was US$33.4 billion over 20 years, and total economic profits for 11 proposed dams ranged from a small positive sum of US$6.6 million to a larger negative (cost) sum of US$274.4 billion. Moreover, under the MBD plan, only Laos has a net profit whereas three other members of the MRC faced a loss ranging from US$50–128.9 billion (Kubiszewski et al. 2012). A particular challenge for the Mekong River is its tremendous diversity of fish species, which contribute to the well-being of more than 60 million people in the LMB. All of the giant fish is highly threatened by pressure on fishing, and will be possibly driven to extinction by the construction of mainstream dams (Matthews and Geheb 2015; Ziv et al. 2012).

The growing human population and increasing energy demand have been accompanied by the spread of land-use change. Managing land resources, therefore, is a major challenge in meeting with demand for agricultural production, urban expansion, and other uses. In terms of electric power trading, exporting countries are China, Myanmar, and Laos (see ADB 2012: tables 1 and A1.1). China and Laos have the most potential for mainstream hydropower generation and are positioned to reap the most benefits from

[7] Further details of future impact on hydropower development in the Mekong region can be found in Kuenzer et al. (2013).

building dams on the Mekong. However, owing to climate change and to meet growing energy needs, the promotion of regional power generation, boosting low carbon source trade, reduction of the dependence on oil and coal imports, and the lowering of greenhouse gas emissions from power generation are not only challenges but also opportunities.

In terms of intra-regional trade dependence and the degree to which China plays a role in that dependence, China has grown faster than the other five GMS countries (Cambodia, Laos, Myanmar, Thailand, and Viet Nam). As the China–ASEAN Free Trade Agreement (CAFTA) came into force in January 2010, it established the third-largest free trade area in the world, following the European Union and the North American Free Trade Area (NAFTA). However, since then China has been facing a challenge in getting the agreement formally implemented because the trade structure between China and ASEAN countries is competitive rather than complementary (Wang 2011).

Although China and Myanmar are not members of the MRC, the ASEAN has played an important role in economic development of the MR and has attracted international attention since 1995 (Hensergerth 2009; Weatherbee 1997). In this regard, the MRC, ASEAN, and GMS play a role in analysing opportunities of issue linkage in reaching a basin-wide agreement and clean energy transaction.

26.4 THE ROLE OF ISSUE LINKAGES IN MANAGING THE MR

As in most transboundary river basins, the relationships between upstream and downstream nations in the MR are politicized and disputed. China views the UMB primarily as a source of hydropower and as a trade route. Laos also considers the MR primarily as a source of hydropower, whereas Thailand seeks cheap energy and water for irrigation and agricultural sectors. More than 90 per cent of electricity in Laos is produced from hydroelectric plants, whereas the main value of the Mekong River for Cambodia is for fishery (Campbell 2009). Cambodia prefers the conservation of the current hydrological region, including the seasonal flooding, which gives rise to its significant fishery industry. Viet Nam, on the other hand, relies on the water to support the Mekong delta's agricultural production.

Owing to the impact of climate change and the rise in sea level, Viet Nam wants to protect its efficient agriculture and aquaculture production in the delta from saltwater intrusion. In addition, it is believed that the highly centralized Chinese government has more grips on its water resources than the fragmented MRC with its less effective management (Houba, Pham Do, and Zhu 2013).

Hence, potential conflicts exist between these demands for water, land use, and energy, which will require trade-offs among water-using sectors.

26.4.1 A Basic Framework

Game theory is the formal study of conflict and cooperation.[8] In the context of the historical development and integration of the MR, Pham Do and Dinar (2014) argue that a negotiation process between upstream (China) and downstream (four LMB countries, represented by MRC) can be considered as a two-stage game. In the first stage, countries (China and LMB) can play at being non-cooperative over independent policy issues (strategies) such as energy (hydropower generation), trading, and the ecosystem (fishery and agriculture) to determine (evaluate) their policy (variables). Final outcomes, as the results of linked issues,[9] are then considered in the second stage for negotiating nations.

Mathematically, let $N = \{1,2\ldots, n\}$ be a set of policy issues. Assume that the upstream (U) and downstream (L) simultaneously make a policy choice or action $a_J=(a_{J1},\ldots a_{Jn}) \in A_J$, where $J = U, L$; and each action (policy) profile $a = (a_U, a_L) \in A = A_U \times A_L$ specifies a policy choice for each player (region) with respect to each $i \in N$. Furthermore, for each issue $i \in N$, each player J has a measurable payoff function w^a_{Ji} on action profile a with the objective function of players being linearly separable in policy issues; that is, $w_J = \Sigma^n_{i=1} w^a_{Ji}$.[10] The corresponding stage game with strategy space $a_J = A_{J1} \times A_{J2} \ldots \times A_{Jn} = \{c,d\}^n$ is denoted by Γ. For example, for policy profile $a = (a_U, a_D)$ and two issues i and k (e.g., water and trading), the two-person games $\Gamma_i(a)$ and $\Gamma_k(a)$ can be described in the following two matrices:

		Lower	
	$\Gamma_i(a)$	C	D
Upper	C	$(w^{cc}_{Ui}, w^{cc}_{Li})$	$(w^{cd}_{Ui}, w^{cd}_{Li})$
	D	$(w^{dc}_{Ui}, w^{dc}_{Li})$	$(w^{dd}_{Ui}, w^{dd}_{Li})$

		Lower	
	$\Gamma_k(a)$	C	D
Upper	C	$(w^{cc}_{Uk}, w^{cc}_{Lk})$	$(w^{cd}_{Uk}, w^{cd}_{Lk})$
	D	$(w^{dc}_{Uk}, w^{dc}_{Lk})$	$(w^{dd}_{Uk}, w^{dd}_{Lk})$

[8] Conflict or non-cooperative strategy refers to a situation in which a binding agreement cannot be achieved; while it is possible in cooperative strategy.

[9] The idea is that linking two (or more) policies (regimes) could allow countries to use surplus enforcement power that may be available in one policy domain to discipline cooperation in other domains.

[10] Such as dam construction plan, trading and energy plan, ecosystem protection, and environmental policy, and so on.

To achieve a basin-wide agreement through linked issues, each player can consider two possible actions: *C* (or *c*) for cooperating or *D* (or *d*) for defection (selfish policy action). For any two independent games, the values of a two-linked game are determined as the sum of two values in these games. Hence, in a linked game, player *J*'s payoff is $w_j = w^a_{Ji} + w^a_{Jk}$. The objective of each player is to maximize the final outcome $w_j = max_a\{w^a_{Ji} + w^a_{Jk}\}$ (for further details, see Pham Do, Dinar, and McKinney 2012). Without loss of generality, it is assumed that both the LMB and China (UMB) are faced with two strategies—cooperation and non-cooperation—in each independent game.

26.4.2 Independent Games in the First Stage

26.4.2.1 A Water Game

China's water resources can be used for industrial and household activities during the wet season and then stored for use in the dry season. China's outflow in the wet season fosters local fish reproduction before it runs to the mainstream of the LMB downstream. During the dry season, water inflow plus the stored water can be used for similar purposes as in the wet season and outflow from the dams can also be used for irrigation. For the tributaries of the LMB, water inflow can be used for similar economic activities as in China; water flows in the LMB are similar to those in the UMB, except for the impact of dams on the flow of tributaries.

Water inflow for the mainstream LMB solely consists of the outflow received from China. In each region (UMB and LMB), the economic values of water use are determined by aggregating four main activities for each season (wet and dry) (for details, see Houba, Pham Do, and Zhu 2013): (i) dam capacity for hydropower generation and mitigation of flood damage, (ii) industrial and household activities, (iii) irrigated agriculture, and (iv) environmental services or damages (i.e., wetland benefits or damage from saltwater intrusion in the estuary during the dry season). From the annual economic net values under cooperation and non-cooperation, a water game can be determined as shown in Figure 26.1 (first stage). In the water game, the dominant strategy is either not to share water (i.e., upstream) or not to pay for the water (i.e., downstream), because sharing or making side payment always costs China/LMB some welfare reduction. The total basin level annual incremental welfare gains are US$2.05 billion (i.e., 24.81–22.76) for moving from *non-cooperation* (22.76) to *cooperation* (24.81) under *weak governance*, and US $0.05 billion (24.81–24.74) under *strong governance*. In general, LMB nations can obtain almost the same joint welfare with strong governance, regardless of China's situation. If LMB nations act individually (*non-cooperation*), however, the total net aggregate economic welfare is lower. From the perspective of

China, the incentives are quite different. China does not perform its best under the Nash equilibrium (2.75, 22.06), although the total economic welfare could be better under the cooperative situation for both parties. Particularly, China can gain more with weak governance in the LMB. This observation could explain why China is interested in signing bilateral agreements rather than multilateral ones, as shown in studies of Naohiro (2012) and Yongqi and Anfei (2013).

26.4.2.2 A Trade Game

Over the last two decades (1990–2010), the trade/GDP ratio is 131 per cent (in 2010) for the region as a whole (Petri, Plummer, and Zhai 2012). ASEAN markets are especially important for Laos and Viet Nam. However, Laos appears to be a 'free rider' in ASEAN and Viet Nam is a loser, although the latter is a potential player for agricultural productions.[11] As trade is an important driver of economic growth, ten members of ASEAN agreed to implement the AEC by the end of 2015, which commits to free movement of goods, services, foreign direct investment, and free flows of capital (ASEAN 2010). Hence, all ASEAN economies are open to trade and investment. With AEC, the strategy of LMB members of ASEAN is either to retain barriers with non-ASEAN partner economies (such as China) or to remove the barriers (i.e., open trade with more partners of the world). According to Petri, Plummer, and Zhai (2012), the region's share pattern is essentially symmetric. About one quarter of overall ASEAN trade is shared within ASEAN as well with the United States, the European Union, China, and Japan, and the rest of the world. China, therefore, is considered as a partner of ASEAN and is involved with AEC only under two arrangements: (i) increased bilateral free trade area with the four LMB states (under CAFTA) where the states are members of AFTA, or (ii) bilateral free trade area with AEC (under AFTA).

To determine the trading game, the welfare gain of the LMB is defined as the aggregated gains obtained from all four LMB nations in ASEAN (for details, see Pham Do and Dinar 2014). From Figure 26.1 (trade game), one can easily see that the LMB has open trade as the dominant strategy, whereas China's dominant strategy is CAFTA. In this game, the Nash equilibrium (*CAFTA, Open*) is not efficient as the total outcome is less than in (*AFTA, Open*). Scrutiny of the MR water game and the MR trade game suggests clearly that playing each game separately will lead to nowhere.

[11] Owing to lack of data in a water game, the values of the trade game is based only on trade results related to the four states of the LMB and China.

First stage: independent games

WATER GAME		LMB	
		Strong governance	*Weak governance*
China	*Cooperation*	(2.75, 22.06)*	(3.76, 21.05)
	Non-Cooperation	(2.73, 22.03)	(2.73, 20.03)

TRADE GAME		LMB	
		Open	*Restrict*
China	*CAFTA*	(–7.8, 15.4)*	(0.4, 2.8)
	AFTA	(–12.2, 52.9)	(–4.6, 12.0)

Second stage: linked games

LINKED GAME		LMB	
		Liberalize (*c*)	Status quo (*d*)
China	Liberalize (*c*)	(–5.05, 37.46)*	(–4.04, 36.45)
	Status quo (*d*)	(–9.25, 74.96)	(–8.24, 73.95)

Figure 26.1. MR two-stage game.
Note: = Nash equilibrium
Source: Authors' illustration based

26.4.3 Linked Game in the Second Stage

As mentioned in Sections 26.4.2.1 and 26.4.2.2, *cooperation* is the dominant strategy in the water issue, whereas *open* is the dominant strategy in the trade issue. Taking into account the two outcomes of the water issue and the two outcomes of the trade issue, a linked game can be presented,[12] as shown in the second stage of Figure 26.1. Note that the values of the linked game are determined from two independent games in the first stage. For example, (−5.05, 37.46)=(2.75−7.8, 22.06+15.4), (−4.04, 36.45)=(3.76−7.8, 21.05+15.4), (−9.25, 74.96)=(2.75−12.0, 22.06+52.9), and (−8.24, 73.95)=(3.76−12.0, 21.05+52.9).

The linked game indicates that the total social welfare will increase when water and trade issues are considered together. Note that the main uses of water are taken into account in the water game, whereas all agricultural products and the electric power trade are considered in the trade game. Hence, with a higher outcome, the LMB could make a side payment to China. The losses and gains are similar for China and the LMB in the linked game. For example, the outcome (−9.25, 74.96) indicates the total payoff of 65.71, which is 74.96−9.25, whereas for the total outcome (−8.24, 73.95) the payoff is 65.71, which is 74.96−9.25. For the others, the outcomes (−5.05, 37.46)* and (−4.04, 36.45) both lead to 32.41. Thus, linkage issue will give more opportunities for countries in the negotiation process.

[12] As we aim to investigate whether or not China will consider joining the MRC in the context of ASEAN, we assume that the LMB states act in one voice in the linked game.

26.5 POLICY IMPLICATIONS AND CONCLUDING REMARKS

Like any developing economy of the world, the MR is influenced in various ways by pressures from global markets, increased demand for natural resource and energy, as well as the impact of climate change and mitigation efforts. The transboundary negative externality nature of the flow of the Mekong River adds an extra dimension of complexity to the debate about equitable sharing of resources in the MR, particularly for hydropower plants as alternative strategies in achieving clean energy transitions. In theory, benefit sharing can improve the livelihood of people impacted by hydropower development. In practice, however, benefit sharing in the MR has shown mixed results.

This chapter demonstrates the advantages of linked issues in bringing together five (or six) countries into a common framework for coordinating and managing the MR. Issue linkages allow balancing the interests of all stakeholders in the MR. For example, the MRC is not a strong and solid organization. However, with international and regional support, LMB nations have the incentive to negotiate with China regarding the trade issue. This opportunity will help the LMB nations (via MRC) to decide on how to strike a balance between hydropower development and the preservation of conditions necessary for sustaining (fish and agricultural production) ecosystems in the future. In addition, further opportunities and investment should be considered and added.

Having examined the ability of facilitating cooperation (by allowing countries to tie together issues in which they have dissimilar interests), this chapter also shows that LMB countries can benefit most from linked issues. Particularly, the MR problem is not a lack of mechanism but of creating an effective body out of a multiplicity. Hence, in managing the MR energy transition, water is just one issue to be taken into account, and is insufficient on its own to establish a viable regime in managing sustainable development. The policy measures, therefore, are to establish a legal framework in which more issues should be considered simultaneously, thus implementing the beneficial procedures of issue linkages.

REFERENCES

ADB (Asian Development Bank) (2004). 'Greater Mekong Subregion: Atlas of the Environment'. ADB, Manila.

ADB (2012). 'Greater Mekong Subregion Power Trade and Interconnection: 2 Decades of Cooperation'. ADB, Manila.

ADB (2013). 'Energy Outlook for Asia and the Pacific'. ADB, Manila.

ADB (2015). 'Renewable Energy Developments and Potential in the Greater Mekong Subregion'. ADB, Manila.

ASEAN (Association of Southeast Asian Nations) (2010). 'ASEAN Regional Guidelines on Competition Policy'. ASEAN Secretariat, Jakarta.

Bach, H., J. Bird, T. J. Clausen, K. M. Jensen, R. B. Lange, R. Taylor, V. Viriyasakultorn, and A. Wolf (2012). 'Transboundary River Basin Management: Addressing Water, Energy and Food Security'. Mekong River Commission, Lao PDR.

Biba, S. (2012). 'China's Continuous Dam-Building on the Mekong River'. *Journal of Contemporary Asia*, 42(4): 603–28.

Campbell, I. (2009). *The Mekong: Biophysical Environment of an International River Basin*. Amsterdam: Elsevier.

Cronin, R. and T. Hamlin (2012). 'Mekong Turning Point: Shared River for a Shared Future'. Henry L. Stimson Center, Washington, D.C.

Cronin, R. P. (2012). 'Laos' Xayaburi Dam Project: Transboundary Game Changer'. GWF Discussion Paper 1216, Global Water Forum, Canberra.

Dore, J. and Y. Xiaogang (2004). 'Yunnan Hydropower Expansion: Update on China's Energy Industry Reforms and the Nu, Lancang and Jinsha Hydropower Dams'. Working Paper. Kunming, Chiang Mai University's Unit for Social and Environmental Research and Green Watershed, PR China. Available at: <http://mpowernetwork.org/Knowledge_Bank/Key_Reports/Research_Reports/Yunnan_Hydropower_Expansion.html> (accessed 16 November 2016).

FAO (Food and Agriculture Organization of the United Nations) (2011). 'The State of the World's Land and Water Resources for Food and Agriculture (SOLAW): Managing Systems at Risk'. FAO/Earthscan, Rome and London. Available at: <http://www.fao.org/docrep/015/i1688e/i1688e00.pdf> (accessed 25 April 2016).

FAO (2012). 'AQUASTAT: FAO's Information System on Water and Agriculture'. FAO, Rome. Available at: <http://www.fao.org/nr/water/aquastat/main/index.stm> (accessed 10 March 2012).

Goh, E. (2004). 'China in the Mekong River Basin: The Regional Security Implications of Resource Development on the Lancang Jiang'. RSIS Working Paper 069/04, Institute of Defence and Strategic Studies, Singapore.

Hensergerth, O. (2009). 'Transboundary River Cooperation and the Region Public Good: The Case of the Mekong River'. *Contemporary Southeast Asia*, 31(2): 326–49.

Hirsch, P. (2011). 'China and the Cascading Geopolitics of Lower Mekong Dams'. *The Asia-Pacific Journal*, 9 (20). Available at: <http://apjjf.org/-Philip-Hirsch/3529/article.pdf> (accessed 16 November 2016).

Houba, H., K. H. Pham Do, and X. Zhu (2013). 'Saving a River: A Joint Management Approach to the Mekong River Basin'. *Environmental and Development Economics*, 18(1): 93–109.

ICEM (International Centre for Environmental Management) (2010). 'Strategic Environmental Assessment (SEA) of Hydropower on the Mekong Mainstream'. ICEM for the Mekong River Commission, Hanoi. Available at: <http://icem.com.au/portfolio-items/strategic-environmental-assessment-of-hydropower-on-the-mekong-mainstream/> (accessed 25 April 2016).

Kaisti, H. and M. Kakonen (2012). 'Actors, Interests and Forces Shaping the Energyscape of the Mekong Region'. *Forum for Development Studies*, 39(2): 147–58.

King, P., J. Bird, and L. Haas (2007). 'The Current Status of Environmental Criteria for Hydropower Development in the Mekong Region: A Literature Compilation'. ADB Technical Report, March 2007, WWF–Living Mekong Programme, Vientiane, Lao PDR. Available at: <http://www.hydrosustainability.org/IHAHydro4Life/media/PDFs/evaluations%20and%20historical%20assessments/king-bird.pdf> (accessed 25 April 2016).

Kubiszewski, I., R. Costanza, P. Paquet, and S. Halimi (2012). 'Hydropower Development in the Lower Mekong Basin: Alternative Approaches to Deal with Uncertainty'. *Regional Environmental Change*, 13(1): 3–15.

Kuenzer, C., I. Campbell, M. Roch, P. Leinenkugel, V. Q. Tuan, and S. Dech (2013). 'Understanding the Impact of Hydropower Developments in the Context of Upstream–Downstream Relations in the Mekong River Basin'. *Sustainability Science*, 8(4): 565–84.

Li, X. (2012). 'Hydropower in the Mekong River Basin'. *Environmental Claims Journal*, 24(1): 51–69.

Matthews, N. and K. Geheb (eds) (2015). *Hydropower Development in the Mekong Region: Political, Socio-economic and Environmental Perspectives.* Earthscan Studies in Water Resource Management Series. London: Taylor and Francis.

Mehtonen, K., M. Keskinen, and O. Varis (2008). 'The Mekong: IWRM and Institutions'. In O. Varis, C. Tortajada, and A. K. Biswas (eds), *Management of Transboundary Rivers and Lake.* Berlin: Springer, pp. 207–26.

MRC (Mekong River Commission) (2005). 'Overview of the Hydrology of the Mekong Basin'. MRC, Vientiane, Lao PDR. Available at: <http://www.mekonginfo.org/assets/midocs/0001968-inland-waters-overview-of-the-hydrology-of-the-mekong-basin.pdf> (accessed 25 April 2016).

MRC (2010). 'Mekong River Commission Strategic Plan 2011–2015'. MRC, Vientiane, Lao PDR. Available at: <http://www.mrcmekong.org/assets/Publications/strategies-workprog/Stratigic-Plan-2011-2015-council-approved25012011-final-.pdf> (accessed 25 April 2016).

MRC (2012). 'The Impact and Management of Floods and Droughts in the Lower Mekong Basin and the Implications of Possible Climate Change'. Working Paper 2011–2015, March, MRC, Flood Management and Mitigation Programme, Lao PDR. Available at: <http://www.mrcmekong.org/assets/Publications/basin-reports/FMMP-working-paper-110820.pdf> (accessed 25 April 2016).

MRC (2016). 'Strategic Plan 2016–2020'. MRC, Lao PDR and Cambodia. Available at: <http://www.mrcmekong.org/assets/Publications/strategies-workprog/MRC-Stratigic-Plan-2016-2020.pdf> (accessed 2 May 2016).

Naohiro, K. (2012). 'China's External Economic Cooperation: Ties to the Mekong Region'. *Nippon.com*, 19 June. Available at: <http://www.nippon.com/en/in-depth/a00803/>(accessed 12 June 2015).

Petri, P., M. Plummer, and F. Zhai (2012). 'ASEAN Economic Community: A General Equilibrium Analysis'. *Asian Economic Journal*, 26(2): 93–118.

Pham Do, K. H. and A. Dinar (2014). 'The Role of Issue Linkage in Managing Non-Cooperating Basins: The Case of the Mekong'. *Natural Resource Modelling*, 27(4): 492–518.

Pham Do, K. H., A. Dinar, and D. McKinney (2012). 'Transboundary Water Management: Can Issue Linkage Help Mitigate Conflicts and Enhance Cooperation'. *International Game Theory Review*, 13(1): 39–59.

Roland-Holst, D. and S. Heft-Neal (2012). 'Food-Water-Energy Nexus: GMS Challenges of Growth for 2020 and Beyond'. In *GMS 2020 Conference Proceedings: 'Session 4. Challenges and Dynamics of Growth in the Next Decade, 2011–2020'*, pp. 484–500. Available at: <http://www.gms-eoc.org/resources/gms-2020-conference-proceedings-draft-> (accessed 30 March 2016).

Rowcroft, P. (2008). 'Frontiers of Change: The Reasons behind Land-Use Change in the Mekong Basin'. *Ambio*, 37(3): 213–18.

Smajgl, A. and J. Ward (eds) (2013). *The Water–Food–Energy Nexus in the Mekong Region: Assessing Development Strategies Considering Cross-Sectoral and Transboundary Impacts.* New York: Springer.

UNEP (United Nations Environment Programme) (2006). 'Mekong River, GIWA Regional Assessment 55', edited by A. Snidvongs and S.-K. Teng. Kalmar, University of Kalmar, on behalf of UNEP, Sweden. Available at: <http://www.unep.org/dewa/giwa/areas/reports/r55/giwa_regional_assessment_55.pdf> (accessed 25 April 2016).

UNEP (2008). 'The Mekong River—Survival for Millions'. Vital Water Graphics 2. GRID–Arendal. Available at: <http://www.grida.no/publications/vg/water2/page/3263.aspx> (accessed 12 June 2015).

United Nations (2015). 'Transforming Our World: The 2030 Agenda for Sustainable Development'. Department of Economic and Social Affairs, United Nations, New York. Available at: <https://sustainabledevelopment.un.org/post2015/transformingourworld> (accessed 25 April 2016).

Wang, L. (2011). 'Is China a Trade Competitor of ASEAN?' *Journal of Contemporary Eastern Asia*, 10(2): 1–23.

Weatherbee, D. (1997). 'Cooperation and Conflict in the Mekong River Basin'. *Studies in Conflict and Terrorism*, 20(2): 167–84.

WLE (Water, Land and Ecosystems) Greater Mekong (2015). 'Dams in the Mekong Basin'. CGIAR Research Programme on WLE, Vientiane, Lao PDR. Available at: <https://wle-mekong.cgiar.org/wp-content/uploads/A4_Mekong_Dams_2015.pdf> (accessed 30 March 2016).

World Bank (2015). 'Data, by Country: Countries and Economies'. World Bank, Washington, D.C. Available at: <http://data.worldbank.org/country> (accessed 25 April 2016).

Xing, L. (2013). 'Land-Use Change in the Mekong Region'. In A. Smajgl and J. Ward (eds), *The Water–Food–Energy Nexus in the Mekong Region: Assessing Development Strategies Considering Cross-Sectoral and Transboundary Impacts.* New York: Springer, pp. 179–90.

Yongqi, H. and G. Anfei (2013). 'China Strengthens Ties with Neighbors'. *China Daily Asia*, 21 June. Available at: <http://www.chinadailyasia.com/business/2013-06/21/content_15076392.html> (accessed 1 April 2016).

Ziv, G., E. Baran, S. Nam, I. Rodríguez-Iturbe, and S. Levin (2012). 'Trading-Off Fish Biodiversity, Food Security, and Hydropower in the Mekong River Basin'. *Proceedings of the National Academy of Sciences of United States of America*, 109(15): 5609–14.

27

The Political Economy of Clean Energy Transitions at Sub-National Level

Understanding the Role of International Climate Regimes in Energy Policy in Two Brazilian States

Jose A. Puppim de Oliveira and Celio Andrade

27.1 INTRODUCTION

There is tremendous potential for investments in clean energy over the next few years to take the energy matrix in a more sustainable direction to tackle climate change and air pollution. The global bill for oil, electricity, and natural gas is around US$5 trillion annually and investments in clean energy reached US$230 billion in 2011 and US$286 billion in 2015 (UNEP, 2016). These annual investments could reach US$500 billion in 2020 (Glemarec and Puppim de Oliveira 2012). However, much larger investments in clean energy are needed, particularly in developing economies, to change the course of unpredictable climatic change (heading towards an average temperature increase of over 2°C) in the near and medium-term future. Even though, in many instances, clean energy is technically and economically viable, the political economy at the national, sub-national, and local levels does not allow the fulfilment of the possibilities for expansion in clean energy production. This could undermine the implementation of the new Paris Agreement on climate change under the United Nations Framework Convention on Climate Change.

The analytical framework of this research focuses on the analysis of the political economy of clean energy transitions, particularly understanding how sub-national and local institutions for tackling climate change are established and evolve, building on previous efforts (Keohane and Ostrom 1995; Bulkeley and Betsill 2003; Puppim de Oliveira 2009, 2011). There is less research on the dynamics of domestic policies and politics, and on the nature of federal systems, in climate change discussions, as most of the research focuses on international dynamics of climate negotiations (Harrison and Sundstrom 2010). The local and domestic political economy can shed light on the factors that determine the effectiveness of international climate and energy regimes locally, and how to align international institutions to national and local institutions and context. The mismatch between the international goals and the national and sub-national institutions already in place leads to ineffective results in climate policy implementation (Puppim de Oliveira 2014). There is often a decoupling between the networks of public and private actors and institutions working on the ground and those coming from the top, particularly at the international level (Pinto and Puppim de Oliveira 2008; Andrade and Puppim de Oliveira 2015). Better connections between the international regimes in climate change and sub-national political economy could identify ways to boost clean energy sources both at demand and supply.

The objective of this chapter is to shed light on discussions about the political economy at the sub-national/local level surrounding the Clean Development Mechanism (CDM) of the Kyoto Protocol.[1] The CDM is a good research object for looking at the political economy of clean energy transitions at sub-national level. Even though it is an international mechanism, its implementation is done at the local level and influenced by sub-national governance. Moreover, CDM could be implemented only in developing countries, where the energy demand will grow most in the next few years. We examine the main political economy obstacles, particularly at the sub national levels, including international–national–sub-national relations, to enabling global climate initiatives, such as the Kyoto Protocol's CDM, to have a larger impact in developing countries. This enables us to recommend possible ways to overcome those obstacles, strengthen the implementation of clean energy policies, and inform the design of the implementation mechanisms for future climate change agreements. We examine the policy process for climate and energy policies in two states in Brazil (Rio Grande do Sul and Bahia), and, in particular, the influence of CDM in the development of those policies.

27.2 METHODOLOGY

The methodology consisted of a qualitative analysis of the political economy of clean energy initiatives in the two states, including the influence of CDM projects mentioned in Section 27.1. The research uses the case study method,

[1] This chapter is also based on long-term research and several publications co-written by the authors, particularly Andrade, Nascimento, and Puppim (2010); Andrade et al. (2010); Silva-Junior et al. (2013a, 2013b); Falleiro et al. (2014, 2015); and Andrade and Puppim de Oliveira (2015).

which is well documented in the literature (Ragin and Becker 1992; Yin 1994). Because quantitative analyses of policy and political regimes have several limitations for the trade-off between accuracy and generalization and need to simplify data (Mitchell 2004), qualitative study was chosen for two main reasons. First, qualitative study is a good approach for analysing complex processes of decision-making, such as for clean energy initiatives, which need an in-depth understanding of the political economy involving different actors and factors not known in advance that have influenced state energy and climate policies. Second, there was not enough information and data to make, for example, a quantitative study statistically significant. A qualitative analysis gives a broad perspective on the political economy of clean energy at the various levels, enabling a comprehensive understanding of the influence of international regimes in clean energy in Brazil at the sub-national level.

The empirical research was conducted in the Brazilian states of Bahia and Rio Grande do Sul. The two states were chosen because they are the leading states in terms of CDM projects in their regions (north-east and south respectively) during the period of analysis. Moreover, the two states have some of the most progressive policies for climate change and renewable energy in Brazil. The authors also had good knowledge of their contexts and access to state information and data.

The analyses focused on examining the political economy at the sub-national level and its links to the national and international political economy. The research aimed to track the main drivers of clean energy initiatives in the two states, and how much they have been influenced by international climate regimes, particularly the CDM. Thus, besides the broad state policies and initiatives, the research analysed the CDM projects in clean energy, especially those with a long history of implementation that have already received carbon credits (Certified Emissions Reductions—CER). The research studied individual CDM projects to understand a typical local dynamic and also investigated, in greater depth, one CDM project in each state involving wind energy (Bahia) and small hydro power plant (Rio Grande do Sul). The chapter examines how the electricity generation and distribution sector is evolving in Brazil and its links with climate change. The study analyses governmental programmes and provides an overview of the past, present, and future developments in wind and hydro energy in the two states and their links with national and international climate regimes, particularly those related to the UNFCCC.

We conducted document analyses using material from the UNFCCC (CDM database and *Project Design Documents*—PDD), Brazilian Ministry of Environment (MMA), the Ministry of Science, Technology and Innovation (MCTI), and the Ministry of Mines and Energy (MME) in Brasilia; and from the state environmental protection and energy agencies in the two states, among other institutions. Field research for conducting semi-structured interviews and further data collection was carried out in the two states. One CDM project in each state was selected to illustrate the CDM impact at the local level. There were also calls for interviews and email exchanges with the different organizations, such as energy companies and consultants involved in clean energy and CDM projects in 2015.

27.3 NATIONAL AND SUB-NATIONAL POLICIES FOR CLEAN ENERGY AND CLIMATE CHANGE

With the rise of scientific evidence about the causes and consequences of climate change, a variety of market and political institutions have been developed to govern those issues globally, nationally, and locally. Despite the low price of carbon, due to recent crises in the carbon market, and criticism of its ability to promote cleaner technologies and sustainable development, there is an important role for carbon markets in national and sub-national climate change regimes (Streck and Lin 2008; Okereke, Wittneben, and Bowen 2012; Michaelowa 2012; CDM Policy Dialogue 2012). The Brazilian government, though not committed to compulsory targets for reducing greenhouse gas (GHG) emissions because the country is not included in Annex B of the Kyoto Protocol, enacted Law 12.187/2009, which established the National Policy on Climate Change, committing to undertake a number of sectoral actions aimed at mitigating and adapting to climate change; one of these is the rise in investments in clean energy.

The different mechanisms for responding to climate change are creating opportunities in some of the 'green' industries, such as renewable and clean energy. Brazil already has an outstanding share of renewable energies, dominated by its naturally endowed huge potential of hydro energy, providing a share of 74.5 per cent of the domestic electricity supply in 2014 with a total hydro power resources and reserves of 110.3 GW (EPE, 2015). The country has created some incentive mechanisms such as the Programme of Incentives for Alternative Electricity Sources (PROINFRA, Programa de Incentivo a Fontes Alternativas de Energia Elétrica) established by Law 10438/2002, which gives subsidies and incentives for clean electricity projects using the special financial mechanisms of the Brazilian National Development Bank (BNDES, Banco Nacional de Desenvolvimento Econômico e Social). However, this 'clean' energy profile could change considerably depending on the growth of electricity demand, the availability of resources for generation, environmental constraints, the distribution networks, and the cost of exploiting resources. Indeed, Brazilian emissions in the energy sector have increased considerably in recent years through the use of non-renewable energy sources (MME 2015). Thus the research examines how the CDM and other markets and policies have stimulated the development of and investments in renewable energies in Brazil. By collecting and comparing different political and economic preconditions, structures, and the overall development in the two states and projects, the research identifies pitfalls and opportunities for new strategies and mechanisms for boosting clean energy in Brazil.

As part of a federal republic, the Brazilian states play a key role in climate and energy initiatives, as they have a certain amount of autonomy in the way they create new policy mechanisms and implement national policies. Several states have advanced their own climate policies, stating the level of emission reductions to be made, with these sometimes being even more stringent than the national targets, and planning to establish different mechanisms such as emission trading (CNI 2011). This was particularly the case around the time of the UNFCCC COP-15 in Copenhagen in December 2009 (Forum Clima 2012). Most of the Brazilian state climate policies are a response to national directive (Law 12.187/2009, which established the National Policy on Climate Change). Only two out of 27 states—São Paulo and Santa Catarina—established their climate policies before the National Policy on Climate Change. However, recently, states have been becoming more active in climate policies. In 2015, in response to targets set nationally, 16 out of 27 Brazilian states set their own voluntary targets for emissions; four states (São Paulo, Rio de Janeiro, Mato Grosso do Sul, and Paraíba) had already defined their compulsory targets for emissions and seven others were in the process of defining their state climate policies (Forum Clima 2015).

One of the main mechanisms in Brazil for motivating GHG emission reductions has been the CDM. There were more than 7,740 projects registered at UNFCCC in December 2016 (UNFCCC 2016). Brazil registered the first CDM project in the UNFCCC and has been one of the leading countries in executing CDM projects with 330 projects at the end of 2014 (MCTI 2014). In Brazil, energy projects comprise the majority of CDM projects (55 per cent), which are spread through various states across the country. Thus, this research seeks to understand the political economy of the sub-national clean energy initiatives in Brazil and the links with international energy and carbon mechanisms such as carbon markets and, particularly, the Clean Development Mechanism (CDM) of the Kyoto Protocol.

27.3.1 The Case of Rio Grande do Sul

Rio Grande do Sul is the southernmost state in Brazil, bordering Uruguay and Argentina. Although it is one of the largest producers of coal in Brazil, it generates around 77 per cent of its electricity from renewable sources (hydro 71 per cent, wind 6 per cent)—more than the national average of 75 per cent (CEE, 2014). The state also enacted the State Climate Change Policy Law (Law 13.594/10) in 2010, which sets out the reduction of GHGs required by

2020, proportional to what is established in the national policy. The Law also aims to promote projects to reduce emissions through CDM (article 5) (Falleiro, Gastaldini, and Andrade 2014).

In 2004, the state governor issued a decree (Decree 43.476) establishing a commission to study CDM and its environmental implications, already pointing to the potential of CDM to foster opportunities in Rio Grande do Sul. The state now has the largest number of CDM projects in renewable energy among Brazilian states (using methodology ACM0002 from the UNFCCC, 'Consolidated baseline methodology for grid-connected electricity generation from renewable sources') in the first period of the Kyoto Protocol (between 2005 and 2012). The state hosts 19 (11 small hydro, 2 hydro, 5 wind, and 1 biomass) of the 85 Brazilian CDM projects in renewable energy (22 per cent of the total) comprising 1.9 million CERs (25 per cent of the total in Brazil from renewable energy) (Falleiro, Gastaldini, and Andrade 2014).

The state government has worked with other stakeholders in mapping and trying to tap the wind energy potential. The state has produced the Wind Energy Atlas to identify opportunities in wind energy and guide public policies. Wind, in particular, has an immense potential in the state as only 634 MW of installed capacity exists compared to the estimated potential of 102.3 GW. This potential is concentrated in the less economically dynamic regions and could be harnessed to generate economic development, especially taking into account the tradition of regional development policies in the state (AGDI and Eletrosul 2014).

One of the CDM projects analysed in this research is the Ferradura small hydropower plant located in Erval Seco in the north-western part of Rio Grande do Sul. The project was managed by BT Geradora de Energia Elétrica SA, a special-purpose company led by a local entrepreneur. Even though the number of jobs generated (two jobs) is small and the technology transfer limited, as it is a well-known technology, the project unexpectedly generated an area for leisure (fishing, swimming, and picnics) in a region with a lack of such facilities.

27.3.2 The Case of Bahia

Bahia is located in the north-east region of Brazil. It generates more than 83 per cent of its electricity from renewable sources (wind and hydropower), more than the national average (75 per cent). As in Rio Grande do Sul, the Bahia state has a huge wind potential of untapped electricity generation. Bahia's Wind Energy Atlas identified only 1.2 GW of installed capacity as compared to the estimated potential of 195 GW (MME 2015). The state only enacted its Climate Change Policy (Law 12050) in 2011, but between 2005 and 2012 it had one of the highest numbers of CDM projects in renewable energy

(ACM 0002 methodology) among all Brazilian states. It hosted 12 projects: 10 wind and 2 hydro by the end of 2012 (Falleiro, Gastaldini, and Andrade 2015).

One of these CDM projects is the Votorantim Hydropower Plant, registered in 2006. It consists of a hydroelectric plant whose dam, reservoir, and flooded area have existed since the 1980s. The project just added sub-stations, turbines, and generators (Fernandez et al. 2014) with no local environmental impact. The hydropower plant is owned by Votorantim Cimentos Ltd and it has an installed electric capacity of 162 MW. Votorantim is one of the largest Brazilian industrial conglomerates in the fields of steel and cement. This CDM project foresees a reduction of 59,485 tons of CO_2e/year (Silva-Junior et al. 2013a). Economically, the project has contributed to the national economy and has generated tax revenues. Socially, this CDM project has generated employment, improved the public infrastructure of adjacent municipalities, and provided professional training for plant operation and maintenance. Direct jobs were created during the construction of the plant and indirect jobs for its maintenance (Fernandez et al. 2014).

During the proposal phase, the company organized public hearings and visits with diverse groups of stakeholders, mainly NGOs, local authorities, and fishermen. Environmentally, besides GHG emissions reduction, the project has contributed to the reforestation of riparian vegetation, the conservation of flora and fauna, the recovery of degraded areas, the management of aquatic ecosystems, the monitoring of water quality, and an environmental education programme in its area of influence (Silva-Junior et al. 2013a).

Some of its carbon credits go to the Votorantim Institute, which is the company's organization for carrying out initiatives in corporate social responsibility, such as income generation projects with the communities around the dam (Fernandez et al. 2014). We also point out the main opportunities and political economy obstacles to the implementation of CDM-like projects to delivering local sustainable development in Section 27.4.

27.4 ANALYSIS OF THE INFLUENCE OF CDM IN THE POLITICAL ECONOMY OF ENERGY TRANSITION

In this section we analyse the impact of the CDM projects on the political economy of clean energy transitions with a focus on the two states, and also the role of the national government and the federal system (the division of responsibilities between national and sub-national entities). We examine how much CDM has been able to influence different processes and stakeholders to facilitate energy transitions domestically, including the adoption of progressive energy and climate policies, development of local technological capabilities in clean energy, and strengthening local planning processes. The research provides some insights into the political economy for managing the transition to a cleaner energy matrix in developing countries, particularly the role of international mechanisms, such as CDM.

How to make climate change goals compatible with the development agenda at the national and sub-national levels in the public and private sectors is key to the implementation of climate-related policies such as clean energy development (Pinto and Puppim de Oliveira 2008; Glemarec and Puppim de Oliveira 2012; Puppim de Oliveira 2013). The research sheds light on the role of CDM in influencing the way different clean energy and climate institutions are being built. In Section 27.4.1, we examine in greater detail the links between CDM and the changes in climate and energy policy and regulation (for example, many states in Brazil are adopting different climate policies and legislation to tackle climate change). We examine its influence on innovation and technological policies in Section 27.4.2 (reducing GHGs from dissemination of innovations in clean energy) and the planning processes involved in Section 27.4.3 (how energy planning decisions are based on local inputs and take into consideration climate change issues and the interests of future generations).

27.4.1 CDM Impacts on Domestic Climate and Energy Policies

International market mechanisms could change the domestic political economy balance towards the adoption of more progressive climate and energy policies in developing countries. Nevertheless, the role of CDM as a global climate and energy governance mechanism seems quite ambiguous. Even though thousands of CDM projects were executed, issuing millions of carbon credits (CERs), the effect on domestic and local policies is unclear. For example, the GHGs in the main CDM host countries (China, India, and Brazil) have continued to grow in the last decade. China, the largest CDM recipient country, has increased its CO_2 emissions during 2000–13 by 192 per cent and its CO_2 per capita emissions in the same period by 169 per cent, and Brazil has increased its emissions by 48 per cent and 29 per cent respectively (EC 2015).[2] Even though countries have enacted laws, targets, and policies on climate change, none of the large developing countries have included mandatory total emission reductions. The additionality component of CDM may even have blocked national and state legislation for controlling GHGs, as this could undermine additionality.

[2] The emissions included are those from 'the fossil fuel use and industrial processes (cement production, carbonate use of limestone and dolomite, non-energy use of fuels and other combustion). Excluded are: short-cycle biomass burning (such as agricultural waste burning) and large-scale biomass burning (such as forest fires)' (EC 2015).

The federal structure of a country can influence the way international regimes influence national and sub-national policies (Harrison and Sundstrom 2010). Even though there is a high degree of decentralization in certain sectors (e.g., education and health), the autonomy of states and municipalities limits the scope of their action in climate change, particularly in clean energy supply and markets. The structure of the Brazilian federation and the utilities allow little space for strong command-and-control regulations and investment decisions at the sub-national level. Many existing state energy companies have had some of their activities privatized, federalized, or discontinued in recent decades. This is the case of the Rio Grande do Sul's State Company of Electric Energy (CEEE—Companhia Estadual de Energia Elétrica), founded in the first half of the twentieth century, which had a large part of its distribution system privatized and power plants (like coal power plants) nationalized in the 1990s (Rotulo Decuadra and Puppim de Oliveira 2008).

Thus, the policies and legislation at sub-national levels in Brazil generally mention, but give little incentive for clean energy and climate mitigation including CDM projects, and there is no obligation to achieve certain targets for clean energy generation either enforced or directly promoted by the states. Most of the incentives come from the national level, such as PROINFRA, which gives subsidized credit to energy projects.[3] Sub-national governments, such as the two states analysed, promote clean energy through studies to map the energy potential and try to attract energy investments from the public and private sectors. Climate change policies are more in line with the national targets, and just a few states (like São Paulo) have been more ambitious. The laws establishing those policies generally mention the intention to encourage or promote CDM, but no specific mechanisms are concretely defined. Rio Grande do Sul and Bahia have been at the forefront in attracting CDM projects, particularly in clean energy, both because of the potential they have (in hydro, wind, and also solar) and the large size of their economies and populations. CDM ends up being mostly business driven (by the private sector) leading to stiff competition among states to attract those projects, as an early decree to create a CDM commission in Rio Grande do Sul in 2004 seemed to target. Thus, the impact of CDM has been mostly to steer state and national policies towards an incentive-based approach to clean energy and to try to attract private investors to tap into the international opportunity.

[3] World Resources Institute website, available at: <http://projects.wri.org/sd-pams-database/brazil/programme-incentives-alternative-electricity-sources-proinfa> (accessed 15 December 2015).

27.4.2 CDM Building Technological Capabilities

CDM can provide an opportunity to change the local political economy to invest in building local technological capabilities in clean energy, as it involves cooperation between developing and developed countries (the latter may provide the incentives or the technical background to build the capabilities). However, the research has not found any direct changes in the political economy to promote capability building in clean energy at the sub-national level. Even though CDM promoted the acquisition of the know-how and equipment in Brazil, the technologies were well-known, and coming from outside the state. For example, in the Votorantim CDM project in Bahia, the exogenous transfer of technology was very limited—only evident in secondary equipment purchased in Germany. Much of the machinery and know-how required for project development were acquired in the domestic market. Thus, technology transfer was mainly endogenous due to Brazil's mastery in hydroelectric technology production (Silva-Junior et al. 2013b). A similar outcome developed in the Ferradura project in Rio Grande do Sul state, where the technology was already dominated by Brazilian companies. The use of well-known technologies comes about because they are tested technologies, posing less financial and technological risks to the investors and promoters of the CDM projects.

The national government is the key actor in driving technology policies in the CDM through the PROINFRA programme, which required that financed projects acquired more than 60 per cent of the equipment and technology domestically. However, PROINFRA did not foster technology development per se as a requirement for funding. On the other hand, the states have very little influence in technology policies, as the climate policies do not specify any concrete technological policies, even though promotion of clean technology is mentioned in the policies, such as the State Climate Change Policy Law in Rio Grande do Sul. The states have some science and technology funds managed by their state research foundations that have tried to promote calls for clean energy, but none of the projects seem to have used those funds.

In conclusion, the research found that the CDM projects in the two states did not contribute significantly to technological development locally. Technology transfer was also limited, but CDM projects have demanded mostly locally produced equipment that indirectly can help technological development (which would be just a spillover effect from the growing domestic demand for clean energy).

27.4.3 Local Planning for Sustainability

International climate policy and market mechanisms could help local planning processes to be more sustainable by, for example, generating incentives or conditions for changing the local political economy to make planning more inclusive and participatory. The evidence collected from different sources (such as interviews, field visits, and secondary information, such as CDM reports in the UNFCCC page) shows that CDM projects in the energy sector generated little participation and local development in the long term. They just followed the common processes already in place in the two states (such as Environmental Impact Assessment) without building more participatory, engaging planning processes. Most of the projects created only a handful of direct jobs in their operations as, once installed, large parts of the operations are automated as in the case of Ferradura in RS (just two jobs), or such as was the case in Votorantim in Bahia, where the facility was already operational. The decisions on the CDM projects have indicated a prevalence of a single bottom-line profile for decision-making (financial returns from carbon market) over the long term, next generations' interests (Falleiro, Gastaldini, and Andrade 2015).

There is also little sign that CDM has created significant opportunities for engagement with local businesses. The direct participation of small organizations, firms, and communities in CDM projects is limited because of the high transaction costs related to all phases of the projects. The complexity of the registration process is beyond the reach of most local organizations. Nevertheless, one of the CDM projects in this research (Ferradura in Rio Grande do Sul) was led by a local entrepreneur, demonstrating the possibility for emergence of institutional arrangements that allow small local organizations to engage in CDM. Thus, a small number of consultants, mostly international or from outside the regions of the projects, dominate the services, generating few opportunities for local firms to be engaged in the projects and develop the expertise in CDM which could boost other CDM projects locally.

27.5 CONCLUSIONS

The development of large-scale clean energy initiatives in developing countries depends heavily on the local political economy, as well as the local technical capacity to carry out technological changes. International political and market mechanisms in energy and climate change regimes could be an opportunity to transform the balance in the domestic and sub-national political economy to promote clean energy initiatives. However, the results of this study showed limited evidence of the lasting impacts of CDM projects on changing the sub-national political economy to boost clean energy initiatives locally, though CDM brought resources to project implementation, mainly through private actors. The CDM projects generated sparse incentives for local climate and energy policies, had little impact on local planning processes, and showed little development of technological capabilities locally.

There are some features in Brazil that may have limited broader impacts of CDM. Countries and states with better technical capacity and more sophisticated industry can more easily adapt to changes coming from international regimes through small incentives or funding (Puppim de Oliveira 2014). The local market structure and its political economy may also influence clean energy efforts. In countries where there is more effective regulation and open competition in the local energy market, economically and technically viable solutions may win quickly without much state intervention. However, in a rigid energy market like the one in Brazil, we have seen, instead, dependence on national government interventions, as the states promote and compete to attract clean energy investments with the few resources they have. The states of Rio Grande do Sul and Bahia are those traditionally with the largest capacity to implement environmental and energy policies in their respective regions (south and north-east), and this has been reflected in the number of CDM projects they were able to attract, particularly in the area of energy.

We have witnessed the national government introducing incentives to reduce risks and make clean energy more attractive (such as PROINFRA), and CDM helped in offering greater incentives for local implementation. Sub-national governments could play a larger role in such transitions if they had more autonomy to push for different sustainable policies, but there is a mismatch between what locals could do and what the CDM projects could fund. In Brazil, urban policies, in areas such as urban transport and land use, are mostly in the hands of local governments, but there are few opportunities for CDM projects because of the difficulties of designing a methodology and transaction costs, as many projects would be small scale. Moreover, in the area of energy generation the capacity of sub-national governments in countries like Brazil is limited. The new international mechanisms in the Paris Agreement, based on the Nationally Determined Contributions, could open more opportunities in those urban sectors, though there is no direct incentive yet and financial incentives would most likely have to be via national governments.

The research also provides a better understanding of the specific features of the political economy of energy transitions in developing countries, where challenges for coping with climate change and energy governance and risks are different from those of more established institutions in more developed countries. Governance aspects, such as law enforcement, inter-sectoral coordination, and participation, tend to be weaker in developing countries affecting how organizations and individuals perceive the problems and act. Even in the two leading states (Bahia and Rio Grande do Sul) there are significant obstacles to advance climate and clean energy policies. For example, there is a lack of coordination capacity between the public and private sectors in Rio Grande do Sul to define concrete targets and mobilize resources for emission reduction (Falleiro, Gastaldini, and Andrade 2014). Thus, priorities in capacity building in management and governance are very distinct between developed and developing countries.

The emergence of political or economic factors at the higher level can change the balance of political and economic forces at the sub-national level, which can facilitate or make difficult the dissemination of clean energy in a particular context/state. Political organizations or networks promoting clean energy can become more powerful or legitimate to pursue their actions or interests. Thus, the research also sheds light on the effectiveness of implementation of international environmental agreements adding the component to the local political economy. International regimes can be important for changing national and sub-national political and market forces to promote the objectives of the agreement. For example, the discussions leading up to Copenhagen in 2009 helped to catalyse state legislations on climate policies and GHG emission reductions in Brazil. Agreements can result in international projects or economic incentives forcing needed changes locally, such as the CDM projects. Nevertheless, in many cases the objectives of the mechanisms and their implementation are not completely allied to the local political economy (Pinto and Puppim de Oliveira 2008). Thus, there is a need for understanding the local political economy and creating flexible mechanisms to adapt global interests to local needs in the short and medium term.

CDM was one of the innovative mechanisms for facilitating the implementation of the Kyoto Protocol. CDM projects would engage developing countries (Non-Annex B) in carbon reduction initiatives while giving flexibility to developed countries (Annex B) to achieve the reduction targets, and at the same time encouraging technology transfer between North and South. However, even though CDM projects brought billions of dollars to projects in developing countries, it looks like they were concentrated in just a few countries and consulting companies, and the large societal goals, such as tackling climate change and promoting sustainable development, were limited.

27.5.1 Moving Ahead: Lessons from the Paris Agreement

The rules for post-Kyoto (the new Paris Agreement of 12/2015) can be an opportunity to develop mechanisms to reduce overall GHG emissions by bringing development opportunities to the poorest and most vulnerable populations of developing countries. Hopefully, Paragraph 2 of Article 6 of the Paris Agreement (December 2015) established a voluntary mechanism in order to: (a) promote the mitigation of GHG while fostering sustainable development; (b) incentivize and facilitate participation in the mitigation of GHG emissions by public and private entities; and (c) contribute to the reduction of emission levels in the host countries, which will benefit from mitigation activities resulting in emission reductions that can also be used by another country to fulfil its nationally determined contribution and deliver overall mitigation in global emissions.

Policy makers need to rethink the way CDM, or its substitute mechanism post-Kyoto in the follow-ups to the Paris Agreement, can bring more development outcomes to society and build technological capabilities in poorer countries, while reducing overall carbon emissions globally. There are several alternatives for incorporating more local development goals in CDM or its follow-ups. For example, local development goals could be an integrated part of the Project Design Documents (PDD) and the final validation of the carbon credits of specific projects, including the participation of the main local stakeholders in the verification and validation processes. Governments and companies participating in carbon financing mechanisms and markets should also have overall carbon reduction goals for access to the credits, and these should not only be based on one isolated project. Moreover, the simplification of the bureaucracy could allow small firms and local organizations to engage in global carbon markets. Finally, the participation of local civil society in the decisions on carbon mechanisms could strengthen planning processes and have a better share of the economic benefits of carbon credits.

On the other hand, greater reforms in the federal systems in some countries like Brazil could encourage states and municipalities to take more responsibilities in energy and climate policies. Providing more autonomy for more stringent command-and-control regulations, incentives for local carbon and energy markets, or even direct participation in energy investments could allow more sub-national policy innovation and action in developing renewable energy generation and markets at the sub-national level. Some reforms in energy distribution and generation could help to boost cleaner energy initiatives with the participation of local actors.

The practice of CDM has been different to its original concept during the Kyoto Protocol process, and we hope this is not repeated with the mechanisms that will follow up the Paris Agreement. The credibility of international mechanisms and agreements is based on their capacity to deliver sustainable development, including reduction of total emissions. Promoters of voluntary market-based carbon initiatives like CDM have focused on making those markets work efficiently to grow in volume and financial returns, and less on total emission reduction and local development, which are the reasons they were created in the first place. National and states policies can play a key role in steering those markets to bring more societal goals, but they require concrete initiatives beyond general laws (as those we have seen in Brazilian states) without grip, or they risk losing good opportunities to boost their own clean energy initiatives.

REFERENCES

AGDI (Agencia Gaucha de Desenvolvimento e Promoção de Investimento) and Eletrosul (2014). 'Atlas Eólico do Rio Grande do Sul'. AGDI, Porto Alegre.

Andrade, C., A. Costa, K. Pasini, L. Farias, F. Góes, A. Ventura, and T. Cairo (2010). 'Cleaner Technology and Sustainable Development in Brazil: Contribution of CDM'. Paper Presented at the 2010 Annual Meeting of the Academy of Management, Montreal.

Andrade, J. C. S., L. F. Nascimento, and J. A. Puppim de Oliveira (2010). 'Has CDM Delivered What It Promised? An Analysis of CDM Projects in Brazil'. Greenhouse Gas Market Report 2010. In International Emissions Trading Association (IETA) (ed.), *Post Copenhagen and Climate Policy: Where Global Emissions Trading Goes from Here*. Geneva: IETA.

Andrade, J. C. S. and J. A. Puppim de Oliveira (2015). 'The Role of the Private Sector in Global Climate and Energy Governance'. *Journal of Business Ethics*, 130(2): 375–87.

Bulkeley, H. and M. M. Betsill (2003). *Cities and Climate Change*. London: Routledge.

CDM Policy Dialogue (2012). 'Climate Change, Carbon Markets and the CDM: A Call to Action'. Executive Summary: Recommendations of the High Level Panel on the CDM Policy Dialogue. Available at: <www.cdmpolicydialogue.org> (accessed 10 February 2015).

CEE—Grupo CEE (2013). Balanço Energético do Rio Grande do Sul 2013: ano base 2012. Porto Alegre: Grupo CEE. Available at: <http://www.ceee.com.br/pportal/ceee/archives/BERS2012/Balanco_Energetico_RS_2013_base_2012.pdf> (accessed 16 December 2016).

CNI (Confederação Nacional da Indústria) (2011). 'Estratégias Corporativas de Baixo Carbono: Gestão de Riscos e Oportunidades'. Report. CNI, Brazil.

EC (European Commission) (2015). 'Edgar-Emission Database for Global Atmospheric Research'. Available at: <http://edgar.jrc.ec.europa.eu> (accessed 24 October 2015).

EPE—Empresa de Pesquisa Energética (2015). Balanço Energético Nacional 2015: Ano base 2014. Rio de Janeiro: EPE. Available at: <https://ben.epe.gov.br/downloads/Relatorio_Final_BEN_2015.pdf> (accessed on 16 December 2016).

Falleiro, A. M., M. C. C. Gastaldini, and J. C. S. Andrade (2014). 'Projetos de energia renovável no âmbito do MDL—Perfil dos projetos localizados no RS'. *Revista de Administração da UFSM*, 7: 100–12.

Falleiro, A. M., M. C. C. Gastaldini, and J. C. S. Andrade (2015). 'Desafios e oportunidades dos projetos de MDL de energia renovável localizados no Brasil'. Annals of the Encontro Nacional de Gestão Empresarial e Meio Ambiente (ENGEMA). ENGEMA, São Paulo.

Fernandez, L., C. de la Sota, J. C. S. Andrade, J. Lumbreras, and J. Mazorra (2014). 'Social Sustainable Development Benefits of Hydroelectricity CDM Projects in Brazil'. *Journal of Sustainable Development and World Ecology*, 1(3): 246–58.

Forum Clima (2012). 'Ação Empresarial sobre Mudanças Climáticas. O Desafio da Harmonização das Políticas Públicas de Mudanças Climáticas'. Available at: <http://www1.ethos.org.br/EthosWeb/arquivo/0-A-d2ePublica%C3%A7%C3%A3o_Forum%20Clima_2012_com%20anexo.pdf> (accessed 24 July 2015).

Forum Clima (2015). 'Observatório das Políticas Públicas de Mudanças Climáticas'. Available at: <http://forumempresarialpeloclima.ethos.org.br/observatorio-de-politicas-publicas-de-mudancas-climaticas/> (accessed 22 December 2015).

Glemarec, Y. and J. A. Puppim de Oliveira (2012). 'The Role of the Visible Hand of Public Institutions in Creating a Sustainable Future'. *Public Administration and Development*, 32(3): 200–14.

Harrison, K. and L. M. Sundstrom (2010). *Global Commons, Domestic Decisions. The Comparative Politics of Climate Change*. Cambridge, MA: MIT Press.

Keohane, R. O. and E. Ostrom (eds) (1995). *Local Commons and Global Interdependence*. London: Sage.

MCTI (Ministério da Ciência, Tecnologia e Inovação) (2014). 'Atividades de projeto de MDL 2014'. MCTI, Brasilia.

Michaelowa, A. (2012). 'Fragmentation of International Climate Policy—Doom or Boon for Carbon Markets?' In UNEP (ed.), *Progressing Towards Post-2012 Carbon Markets: Perspectives Series*. Nairobi: UNEP.

Mitchell, R. B. (2004). 'A Quantitative Approach to Evaluating International Environmental Regimes'. In A. Underdal and O. Young (eds), *Regime Consequences: Methodological Challenges and Research Strategies*. Dordrecht: Kluwer Academic, pp. 121–49.

MME (Ministry of Mines and Energy) (2015). 'Resenha Energetica Brasileira'. MME, Brasilia.

Okereke, C., B. Wittneben, and F. Bowen (2012). 'Climate Change: Challenging Business, Transforming Politics'. *Business & Society*, 51(1): 1–24.

Pinto, R. R. and J. A. Puppim de Oliveira (2008). 'Implementation Challenges in Protecting the Global Environmental Commons: The Case of Climate Change Policies in Brazil'. *Public Administration and Development*, 28(5): 340–50.

Puppim de Oliveira, J. A. (2009). 'The Implementation of Climate Change Related Policies at the Subnational Level: An Analysis of Three Countries', *Habitat International*, 33(3): 253–9.

Puppim de Oliveira, J. A. (2011). 'Why an Air Pollution Achiever Lags on Climate Policy? The Case of Local Policy Implementation in Mie, Japan', *Environment & Planning A* (Pion), 43(8): 1894–1909.

Puppim de Oliveira, J. A. (2013). 'Learning How to Align Climate, Environmental and Development Objectives: Lessons from the Implementation of Climate Co-benefits Initiatives in Urban Asia'. *Journal of Cleaner Production*, 58(1): 7–14.

Puppim de Oliveira, J. A. (2014). 'The Mismatch of Implementation Networks in International Environmental Regimes: Lessons from Different Agreements. Best Practices for Architecture and Agency'. In N. Kanie, P. M. Haas, and S. Andersen (eds), *Improving Global Environmental Governance: Best Practices for Architecture and Agency*, 108–29. New York: Routledge.

Ragin, C. C. and H. S. Becker (eds) (1992). *What Is a Case? Exploring the Foundations of Social Inquiry*. Cambridge and New York: Cambridge University Press.

Rotulo Decuadra, D. E. and J. A. Puppim de Oliveira (2008). 'Learning from the Implementation of International Environmental Agreements between Developing Countries Only: The Case Brazil-Uruguay over the Acid Rain from UTPM Termo Power Plant'. *International Environmental Agreements*, 8(4): 389–408.

Silva-Junior, A. C., J. C. S. Andrade, E. B. S. Leao, and D. D. Wu (2013a). 'Sustainable Development and Cleaner Technology in Brazilian Energy CDM Projects: Consideration of Risks.' *Human and Ecological Risk Assessment*, 19(4): 1–21.

Silva-Junior, A. C., J. C. S. Andrade, L. Napravnik-Filho, A. C. Ventura, A. C. Telesforo (2013b). 'Clean Development Mechanism (CDM) in Brazil: An Instrument for Technology Transfer and Promotion of Cleaner Technologies?' *Journal of Cleaner Production*, 46: 67–73.

Streck, C. and J. Lin (2008). 'Making Markets Work: A Review of CDM Performance and the Need for Reform'. *The European Journal of International Law*, 19(2): 409–42.

UNEP—United Nations Environment Programme, FS-UNEP Collaborating Centre (2016). 'Global Trends in Renewable Energy Investment 2016'. UNEP: Frankfurt.

UNFCCC (United Nations Convention on Climate Change) (2016). Available at: <https://cdm.unfccc.int/Projects/registered.html> (accessed 16 December 2016).

Yin, R. (1994). *Case Study Research. Design and Methods*. Applied Social Research Methods Series, vol. 5. Thousand Oaks, CA: Sage.

28

Implementing EU Renewable Energy Policy at the Subnational Level

Navigating between Conflicting Interests

Gilles Lepesant

28.1 INTRODUCTION

The political economy of energy has undergone significant changes in Europe during the past 20 years. Energy has become a shared competence between the European Union (EU) and member states with the inclusion of a dedicated section of the Lisbon Treaty (signed on 13 December 2007). EU legislation has significantly expanded to promote energy efficiency and renewably generated energy, pursuant to Article 194(2). The 2020 package—a set of binding rules established in 2007 proposed three targets: a 20 per cent reduction in greenhouse gas emissions (compared to the levels in 1990), 20 per cent renewable energy in the EU, and a 20 per cent improvement in energy efficiency. In this context, a directive on renewable energy defined specific targets for each member state in 2009 (European Union 2009).

On the other hand, subnational authorities are also seeking to help shape the path towards a low-carbon economy. Because renewable energy makes local energy solutions easier, bottom-up energy policies have become increasingly common. European subnational actors are not only willing to curb energy consumption, they also wish to be suppliers. As a result, the transition to clean energy increases pressure on decentralization in centralized countries and affects multi-level systems of governance even in federal states. The interaction between the different levels of governance, as well as between stakeholders at the local level, is very often linked to the concept of acceptance, namely to the level of support enjoyed by renewables. The rise of the renewables is indeed the outcome of technological innovation, local initiatives, and legal as well as economic conditions provided at the national and European levels. Hence, the need to better understand how institutional arrangements may impact clean energy transitions at the local level.

Brandenburg (Germany) and Aquitaine (France) offer useful insights into these debates, as both regions are forerunners in renewable energies, but they are also exposed to rising discontent in terms of acceptance. Focusing on these two regions, this chapter calls attention to factors that both inhibit and accelerate the implementation of EU policy for a clean energy transition at the subnational level in a federal state (Germany), and in a unitary one (France). The chapter addresses the specific tensions arising in these two regions and discusses governance issues raised by the surge of renewable energies. We start by presenting theoretical insights related to the three dimensions of acceptance as suggested by Wüstenhagen, Wolsink, and Bürer (2007) and to multi-level governance. We continue with lessons learnt from the comparative assessment of the two regions and conclude by suggesting how addressing the shortcomings of multi-level governance might help defuse tensions.

28.2 METHODOLOGY

Research on global climate change policy tends to focus on international negotiations and national policies. The role of local authorities in the energy planning process has recently become a significant research issue, however, and the need for both centralized and decentralized approaches has been recognized (Comodi et al. 2012). As Schreurs suggests, 'It is at the local and regional level, in urban as well as in rural areas that many policy ideas are first generated and that is where some of the most creative solutions are being tried out' (2008: 346). Local innovations have been the object of specific studies (Kern et al. 2005; Kern and Bulkeley 2009). However, alongside the innovative solutions put forward by subnational actors, tensions have also risen, and the social acceptance of renewable energy infrastructures has become a major concern.

Because the concept of acceptance is not easily definable, Wüstenhagen, Wolsink, and Bürer (2007) suggest differentiating between socio-political acceptance, community acceptance, and market acceptance. Socio-political acceptance is social acceptance on the broadest, most general level. Several indicators demonstrate that public acceptance for renewable energy technologies and policies is high in many countries, especially in Germany. This positive overall picture is, however, misleading. Moving from global to local, from general support for more renewables to effective siting decisions, some conflicts may arise. Community acceptance refers precisely to the specific acceptance of siting decisions and renewable energy projects by local stakeholders. People's actual motives may vary (Bell, Gray, and Haggett 2005; Wolsink 2006) and may not be the same before, during, and after the project is completed. Opposition may increase or decrease with the degree of being directly affected by a specific wind power project (Simon and Wüstenhagen 2006). It can be underpinned by different factors—those related to distributional justice (how are costs and benefits shared?) and procedural justice (is there a fair decision-making process giving all relevant stakeholders an opportunity to participate?) being the most relevant in the case of wind energy in Brandenburg (Gross 2007).

Finally, Wüstenhagen, Wolsink, and Bürer (2007) stress the importance of market acceptance being understood as the process of adoption of an innovation and refer in this respect to literature on diffusion of innovation (Rogers 1995). We would argue that market acceptance can also refer to the growing competition between different uses of wood resources. Forest biomass has long been perceived as non-controversial since the total volumes of wood available in Europe far exceed the demand. However, intensifying the use of forest biomass can affect other forest functions. Conflicts are now common in the governance and management of forests: they can be observed at different levels and with varying dimensions and intensities. Research on forest-related conflicts has thus developed considerably in recent years. Although there is certain convergence in how different authors define conflict in this area (Söderberg and Eckerberg 2013), motives and challenges differ from one case to another. In Aquitaine, we would argue that conflicts refer to 'one group impairing the activities of another' (Glasl (1999) quoted in Mola-Yudego and Gritten (2010)).

This chapter focuses on community acceptance (with respect to wind farms in Brandenburg) and market acceptance (concerning forest biomass in Aquitaine). In our analysis, a mixed approach was adopted consisting of the analysis of secondary data (e.g., project reports, website materials) and around 30 in-depth, semi-structured interviews with key stakeholders (e.g., representatives of forest biomass and wind sectors, civil servants at the local and regional level, business representatives, NGOs) during 2015 both in Aquitaine and in Brandenburg.

28.3 TWO REGIONS WITH HIGH POTENTIAL FOR RENEWABLE ENERGIES AND GROWING TENSIONS

28.3.1 Two Regions with High Potential

Brandenburg is a pioneering region in terms of wind energy (in proportion to the size of its population, it has the greatest number of wind turbines of any German *Land*: there were 3,300 in 2015 and over 100 are being built every year). Indeed, Brandenburg enjoys numerous advantages for the creation of wind farms: there is significant wind power, demographic density is low, and the region's economic fabric has been weakened since reunification (in 2013 GDP/inhabitant was €23,751 in the region, compared to €33,335 in the entire country (Eurostat n.d.)). In 2008, 2010, and 2012, the *Land* was awarded first place in the ranking of the most dynamic regions in Germany for its policy in support of renewable energy (Renewable Energies Agency 2015). The Energy Strategy for 2030 (Ministry of Economy and Energy of Brandenburg 2012) predicts that by that date, final energy consumption will be reduced by 23 per cent with respect to 2007 figures, as 40 per cent of the *Land's* final consumption will be covered by renewable energies and CO_2 emissions will be reduced by 25 million tonnes (a 72 per cent reduction compared to 1990).

The Brandenburg government has enshrined its strategy in a long-standing energy tradition (it pictures itself as an *Energieland*) that dates back to communist times, when lignite mines and thermal power plants located in the Lusatia area provided a significant share of the energy for the former GDR (German Democratic Republic). Open-air lignite mines and power stations still make Brandenburg one of the three principal mining regions in reunified Germany. Thus, the State of Brandenburg has coined the expression 'bridge-technology' (*Brückentechnologie*) to refer to its brown coal industry (von Hirschhausen et al. 2012).

In France, several laws have broadened local responsibilities such as the Brottes law in 2013 (*Loi Brottes*) or the law on energy transition in 2015.[1] The overall political economy of the French energy sector is indeed experiencing structural changes in the context of the opening-up of the energy sector at the European level. The transition to clean energy implies new stakeholders, weakened monopolies, behavioural changes, bottom-up initiatives, and increasingly decentralized management of the whole network. The institutional landscape is thus evolving.

Forest covers 43 per cent of the surface of Aquitaine (Conseil Régional d'Aquitaine, INRA 2012). Although French forests can be characterized by a significant diversity, maritime pines constitute the majority of the Landes Forest

[1] Four major bills have been adopted since 2009: Grenelle I (*Loi de programme relative à la mise en oeuvre du Grenelle de l'environnement*) (2009) stated the principles of the Grenelle process, addressing issues of climate change and environmental policy in France; Grenelle II (*Loi portant engagement national pour l'environnement*) (2010); the law on transition towards a more efficient energy system as well as on water pricing and wind farms (*Loi visant à préparer la transition vers un système énergétique sobre et portant diverses dispositions sur la tarification de l'eau et sur les éoliennes*) (2013); and the law on energy transition for green growth (*Loi sur la transition énergétique pour la croissance verte*) (2015). The law also commits the country to reducing its 1990 levels of greenhouse gas emissions by 40 per cent by 2030 and its 2012 energy consumption by 50 per cent by 2050, and to achieving a 32 per cent consumption of renewables in energy by 2030.

(one of the three wooded areas in the Aquitaine Forest). As Brandenburg, Aquitaine region has set more ambitious targets than the national government's ones: the share of renewable energies must double by 2020 to reach 32 per cent, greenhouse gas emissions must be reduced by 30 per cent, and energy efficiency must improve by 30 per cent.

28.3.2 Two Regions Where Tensions Are Developing in the Context of a Transition to Clean Energy

Since the late 1990s, the question of acceptance has become a major topic in research relating to wind energy in Europe, as evidenced by papers dedicated to Sweden (Carlman 1988), the Netherlands (Wolsink 2000), and France and Germany (Jobert, Laborgneb, and Mimlerb 2007). A number of authors contend that the NIMBY ('not in my backyard') reflex does not fully explain the perception of the proximate population (Musall and Kuik 2011), who are sometimes less hostile to wind farm projects than populations residing further away (Simon and Wüstenhagen 2006). Other researchers argue in favour of a dynamic approach and refer to projects whose acceptance has developed according to a U-shaped pattern between the initial and final stages of the project (Wolsink 2006). Analysing three French and two German projects, Jobert, Laborgneb, and Mimlerb (2007) question the reasons for the success or failure of a project in terms of its acceptance. They argue that two classes of factors prevail: on one hand, the institutional framework (regulations, economic and financial stakes), and, on the other, local conditions (the site's economic and geographical context, the implementation of concertation processes and planning processes).

In Germany, support for renewable energies has remained elevated despite increased energy prices since the *Energiewende* was implemented. Community acceptance may, however, differ from social acceptance. In spite of its role as a pioneer, Brandenburg is the German *Land* with the lowest rates of social acceptance (Figure 28.1). A 2014 survey indicated that 93 per cent of the population of the *Land* believed that it was important to promote renewable energies, but that only a little over 60 per cent of those questioned would accept a plant near their homes (Figure 28.1). No other *Land* had such a low figure. Citizen initiatives have been launched, such as *Rettet Brandenburg*, an umbrella organization that regrouped about 80 opposition movements throughout the *Land* in 2015.

In Aquitaine, conflicting views about competition for forest biomass have emerged since pulp-and-paper industries are determined to protect this resource against producers of biomass-based heat and energy. Forest covers around 30 per cent of metropolitan France, an area that has continuously

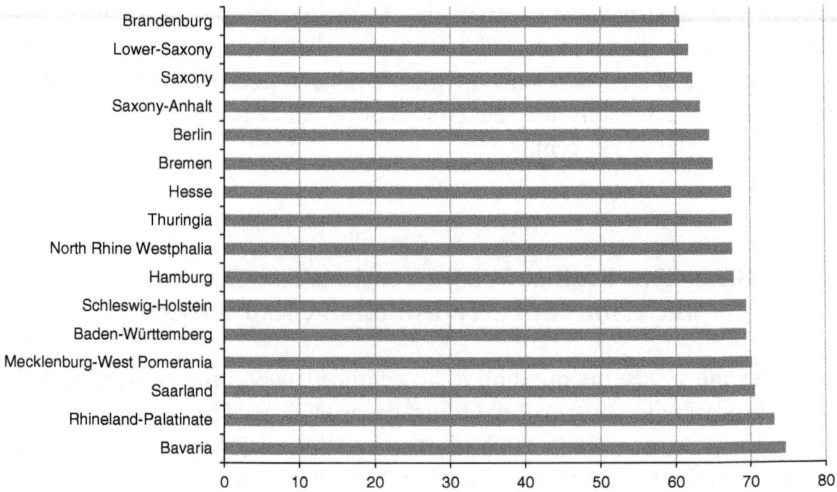

Figure 28.1. Responses to question 'would you agree to a renewable energy facility in the vicinity?' (%), 2013.

Source: Author's illustration based on data from Renewable Energies Agency (2014).

expanded over the past 150 years. Only 60 per cent of the annual natural growth is harvested each year (CEMAGREF 2009).

Although French forestland appears to be statistically under-exploited, the actual quantity of wood is in fact uncertain and only provides a vague indication of the effective available wood resources. Various studies of this issue have been conducted, notably in 2009 (CEMAGREF 2009) and in 2014 (ADEME, IGN, and FCBA 2014). The first study highlights an availability of 28.3 million m^3 IWEW (Industrial Wood and Energy Wood) per year and 8.1 m^3 of wood residue. The second establishes the average yearly availability of wood in the years 2006–20 at 71 million m^3/year, of which 46.1 million m^3/year are readily available, and 14.9 million m^3/year of wood residue, of which 1.6 million m^3/year are readily available. A report by the *École Nationale des Ponts et Chaussées* (2011) has, however, provided a lower estimate of this potential, explaining that an area of fewer than 4 hectares cannot really be considered readily available because management costs render it uneconomical. And finally, the IFN (2010) believes that difficult access to certain plots means that up to 30 per cent of French forests are unavailable to be harvested. The true potential of French forests is therefore unclear, and regional forecasts in Aquitaine, home to one of Europe's largest cultivated forests, indicate that supply is proving inadequate in the face of anticipated demand (see Table 28.1).

In this context, 'market acceptance' appears critical, with competition between wood energy and other uses of forest resources gradually increasing.

Table 28.1. The 'standard' scenario for maritime pine in Aquitaine

	2012–015	2016–17	2018–20	2021–25
Roundwood				
Availability	8,050	5,030	4,730	4,730
Demand	7,600	7,700	7,800	7,950
Softwood lumber	3,650	3,650	3,650	3,650
Industrial roundwood	3,700	3,700	3,700	3,700
Energy wood	250	350	450	600
Sawmill products + forest residue				
Availability	2,540	2,500	2,460	2,440
Demand	2,400	2,800	2,970	3,170
Bark mulch	370	370	370	370
Softwood lumber	1,350	1,350	1,350	1,350
Energy wood	880	1,080	1,250	1,450
Saldo	+390	-2,970	-3,580	-3,950

Source: Author's calculations based on data from ADEME, IGN, and FCBA (2014).

28.3.3 Market Acceptance in Aquitaine

In Aquitaine, the forest products industry is the fourth employer in the region, with 9 per cent of the region's industrial jobs (Dumartin 2009), a far higher share than the renewable energy sector. Wood energy is developing while demand for saw timber is decreasing, and as demand for less noble tree products is increasing among a growing number of the region's economic actors.

Until the 1980s and 1990s, a high level of complementarity between the various uses of forest products prevailed, with high demand for lumber from the furniture and flooring industries and by-products employed by the pulp industry. With increased demand in the wood energy sector, competition has replaced the logic of complementarity, because wood energy and wood for industry both require small wood shavings, trimmings, and residual products.

As a consequence, the packaging and palette sector is concerned by wood price increases and argues that the development of bioenergy may trigger a loss of competitiveness in the long run. According to its representatives, public support for renewable energies has disrupted the market for wood, which has very little elasticity.[2] Indeed, in the forest products sector, strong demand does not necessarily trigger an increase in available supply. Unlike other industrial sectors, outlets do not necessarily structure production, in part because the raw material belongs to a large number of forest owners who are not encouraged to actively manage their forest tracts. The fragmented ownership structure is thus one of several obstacles to mobilization for wood supplies that is

[2] Conversation with the author in Aquitaine, October 2015.

required by the increasing demand. Hence, up-dated forest management policies that tackle supply-side challenges without harming biodiversity and soil protection are requested in parallel with the support provided to clean energies by local and national actors.

28.3.4 In Brandenburg, Unequally Shared Profits Affect Community Acceptance

In Brandenburg, community acceptance is at stake (rather than market acceptance as in Aquitaine), but socioeconomic factors also matter, the key question being how the benefits generated by the surge of wind farms are distributed. Hübner and Pohl (2014) argue that citizen profit-sharing is a far more significant determiner of public perception of wind farms than distance between inhabitants and wind farms. Other studies have also shown that profit-sharing influences acceptance rates, particularly if residents are able to become stakeholders (Brunt and Spooner 1998; Wolsink 2006). Wilfried Bockholt, the Mayor of Niebüll (a town in Brandenburg), confirms that 'participation leads to acceptance' (Renewable Energies Agency 2012), while Zoellner, Schweizer-Ries, and Wemheuer (2008: 4138) contend that economic considerations—'seen as weighing costs against advantages made by each individual'—are important factors in determining acceptance.

In Brandenburg, the landowners of wind farm sites negotiate leases that vary from one case to the next but have a return of between €30,000 and €40,000 per wind turbine per year (Ministry of Economy and Energy of Brandenburg 2014). The plots are sometimes divided among a variety of owners, farmers, and pensioners. In this case, there are numerous beneficiaries in a neighbourhood who stand to enjoy substantial revenue increases over a period of at least 20–25 years. In other instances, beneficiaries are not neighbours but investors that have purchased ex-GDR land after reunification.

Financial implications of wind energy for neighbouring towns can be direct (via professional tax and property tax) or indirect (via the share of income tax and turnover taxes given to the towns). The *Deutscher Städte-und Gemeindebund* (Ministry of Economy and Energy of Brandenburg 2014) conducted a survey, between August and September 2014, of local Brandenburg authorities to determine wind farm profits made by local communities.[3] The major finding was that profits vary considerably but tend to be quite modest. The most significant long-term consequence of wind farms for municipalities might therefore be limited to income taxes on residents with wind farm leases. Property taxes, however, do not generate over €1,000/town and can be

[3] *Deutscher Städte-und Gemeindebund*—German Association of Towns and Municipalities.

considered insignificant. This disparity between the income collected by landowners and that earned by towns (when they do not own the land on which wind farms are installed) illustrates a lack of 'procedural justice' that some authors (Gross 2007) have cited as a factor in determining acceptability.

The Brandenburg example contextualizes the idea that one important attribute of the *Energiewende* is citizens' financial involvement. At the national level, German citizens are highly implicated in the energy transition since they own about 47 per cent of the production capacities for renewable energies (Morris and Pehnt 2012). Citizen involvement has taken the form of investment funds, associations, and cooperatives. However, most of the cooperatives are situated in Bavaria, Bade-Wurttemberg, and Lower-Saxony. They remain rare in Brandenburg and in most ex-GDR States. As a consequence, wind farms are owned by investment funds or developers and the electricity produced by wind farms is essentially injected into the national electricity grid, and not consumed locally.

Moreover, officials can hardly support arguments that energy prices are lower than in states that have invested less than Brandenburg in wind energy. The expensive modernization of the electric grid made necessary by wind farms caused Brandenburg's electric bill to be higher than in some other German states. Grid charges differ, indeed, significantly across Germany. In 2014, the lowest grid charges were in Düsseldorf (4.75 cent €/KwH), the highest in Brandenburg (9.88 cent €/KwH) (RAP 2014). According to the Brandenburg Minister of Economy, Gerber, high electricity costs essentially explain low acceptance rates amongst the population: 'The installation of many wind farms and solar farms in Brandenburg has caused the region to have one of the most elevated electricity costs in Germany. We must renegotiate with the other states about how the costs of reorganizing the network are distributed' (Torsten 2014).

Ultimately, in both Aquitaine and Brandenburg, acceptance is determined significantly by socioeconomic factors. Both regions illustrate the complexity of developing and implementing an energy transition without endangering the economic fabric. This difficulty is not purely economic. It also stems from a lack of appropriate coordination between policies adopted at the local, regional, and national levels.

28.4 ADDRESSING SHORTCOMINGS OF THE MULTI-LEVEL GOVERNANCE SCHEMES

Because governance within the EU involves many levels of interaction between a broad array of officials, some researchers have used the term 'multi-level governance' to describe the EU system since the 1990s. Marks (1993: 392, quoted in Hooghe and Marks (2003)) describes it as a 'system of continuous negotiation among nested governments at several territorial tiers—supranational, national, regional and local'. Section 28.3 explored the socioeconomic factors involved in acceptance, while this section focuses on particular shortcomings in how actions are articulated among actors in each region, as well as between regional and national levels.

28.4.1 Governance Challenges at the Regional Level

Brandenburg State has planned to allocate 2 per cent of its territory for wind farms by 2030, with a projected production level of 10,500 MW. This rate of development will require the construction of about 130 wind turbines per year (BWE 2014). In the face of rising discontent, and to avoid questioning their targets, since 2003, Brandenburg officials have emphasized the notion of acceptance defined as 'a process enabling the development of direct dialogue' (Ministry of Economy and Energy of Brandenburgh 2012).

Imposing an increase in the minimum distance defined by the usual rule (300–500 metres) between wind turbines and housing also represents an option. In response to pressure from Bavaria, the *Bund* granted authority to the states to rule on this issue in 2014, for a period ending on 31 December 2015. The Bavarian government has taken advantage of this opportunity to introduce the 10H rule.[4] Early in 2015, the Brandenburg government announced that it would not adopt a similar provision since it would indeed mean the abandonment of the initiative to install wind turbines on 2 per cent of the region's area.

The support provided by regional authorities to wind energy nevertheless suffers from the parallel support granted to the mining industry (9,000 direct jobs). The coalition contract considers the shutdown of thermal plants by 2040, but the transition period could last longer in view of the reaction provoked by a federal government decision to impose a 'climate levy' on the oldest coal plants. On 27 March 2015, the German Ministry for Economy and Energy proposed to introduce a carbon emissions tax of €18–20 per tonne of CO_2 on old coal power plants in operation for more than 20 years. Such a decision would push out of the market the oldest coal power plants and help to reach the German targets related to greenhouse gas emissions (German Ministry of Economy and Energy 2015). However, the decision sparked strong reaction from coal miners' unions from the industry. Three states with significant mining sectors, amongst them Brandenburg, argued that several

[4] According to this rule, the distance between a wind turbine and the closest accommodation should be factor 10 of the height of the wind turbine. The *Land* of Bavaria nevertheless authorizes some derogations to this rule. See Zaspel-Heisters (2014).

thousands of jobs might be at risk (Ministry of Economy and Energy of Brandenburg 2015). The proposal was withdrawn in June 2015 (Frese 2015). A new and much less controversial one suggested that lignite power plants with a capacity of 2.7 GW (i.e., 13 per cent of the capacity of German lignite power plants) will be put on 'temporary standby'. These power plants will only come into operation in emergency situations and, instead of being taxed as planned by the previous proposal, will receive compensation for remaining on standby for four years, and should then be closed (Ifo Institut 2015). This solution could prove costly—its costs are estimated at €0.5–1 billion annually (Bajczuk 2015)—and should cut emissions to 11 million tonnes only by 2020. Thus, to reach the original objective of the government (reduction in carbon emissions of 22 million tonnes) additional investments, in particular in the area of energy efficiency, have been agreed (Ifo Institut 2015).

Thus, the example of Brandenburg illustrates the fact that a successful energy transition requires more than creating a boom in alternative energies. At the same time, at both the economic and social levels, the gradual marginalization of 'failing' energies must be prepared.

In terms of acceptance, territorial planning also plays an important role. In this respect, Brandenburg can rely on two main documents: the '2030 Energy Strategy', which has little legal constraint, and the land use plan for the entire Berlin-Brandenburg region, ratified in 2009 before being legally declared void and again becoming effective in June 2015.[5] The plan does not specify the eligible areas for wind farms, however, which are defined by the five Brandenburg planning regions (*Regionalen Planungsgemeinschaften*).[6] A planning region has its own legal persona and the decision-making organ is an assembly that meets one to three times a year in which the *Kreise* and towns of more than 10,000 inhabitants are represented.[7] One of the planning region's responsibilities is the adoption of the plan specifying which zones are eligible for wind energy. To this end, the assembly uses a process of elimination that takes various aspects of *Bund* or regional legislation into account—such as protected zones, noise pollution, army radars, and weather forecast services—that enable the progressive limitation of territories that ban wind energy.

Residual areas constitute zones in which wind turbines can be implanted.[8] The territories not included in these zones, in the case of Brandenburg, ban wind farms (other states employ a different approach). As a result, the regional plan is not so much a variation of the state strategy as a document that defines

[5] See <http://gl.berlin-brandenburg.de/> (accessed 2 July 2015).

[6] Havelland-Fläming, Lausitz-Spreewald, Oderland-Spree, Prignitz-Oberhavel, and Uckermark-Barnim.

[7] An intermediary level local authority, between the region (the *Land*) and the municipality (*Gemeinde*).

[8] The planning regions deal exclusively with the issue of wind turbines. Solar energy and biomass requiring less space, their planning is left exclusively to towns.

which zones are authorized for wind turbines under current legislation. Such a plan requires time (at least three years) and can face disagreement between representatives, leading to postponement and eventual cancellation. In the absence of a plan, the regulations of the federal city-planning code according to the 1997 reforms prevail.

Article 35 of the federal urbanism code, modified in 1997, granted a privileged administrative regime to wind farms.[9] Wind farm construction must be authorized for any non-urban zone if it does not prejudice the general interest and has no harmful consequences for environment, protected sites, radars, and so on. In 1998, this provision was amended, with the federal code specifying that an installation can have a negative impact on general interest when a regional or municipal plan for wind turbines on other parts of the same territory has been adopted. Within the framework of a land use plan (*Flächennutzungsplan*), or, in the event of such a plan not existing, of a district plan (*Teilflächennutzungsplan*), a town can therefore specify an area dedicated to wind farms in order to avoid the scattering of installations. In 2002, the Administrative Tribunal nevertheless specified that towns are unable to prevent wind turbines from being installed on their entire territory, because this would violate the principle of the preferred administrative regime given by the federal urbanization code. Ultimately, developers can sue towns determined to restrict the possibility of installing wind turbines (in Brandenburg, this happened to the town of Beelitz in 2015 (Steglich 2015)).

Like their counterparts in Brandenburg, regional authorities in Aquitaine are attempting to redefine their relations with other public actors in terms of energy transition. As in Brandenburg, they are torn between a willingness to develop renewable energies and determination to maintain the industrial workforce (specifically with regard to the forest products industry). They emphasize different ways of managing woodland, considering the fact that the conflict over how such land is used has significantly worsened since two recent storms (Martin in December 1999 and Klaus in January 2009). After the storms, the production potential for maritime pine plummeted from 9.5 million m^3/year prior to Martin to 6 million m^3/year after Klaus. Industrial demand for wood is between 7 and 8 million m^3, however. Finally, it appears that recurrent violent climatic phenomena, which are arguably caused by global warming (Le Treut 2013), are partly responsible for increased tensions concerning forest resources. In this context, regional authorities support new methods of forest management as well as exploiting forest biomass without undermining forest products markets. Since responsibilities are fragmented

[9] Paragraph 35 of the Urbanism Code. The federal Urbanism Code quotes various projects under this regime. The decision to include the implantation of wind turbines was adopted by the German Parliament in 1996.

between different institutions, a highly cooperative approach is needed between stakeholders.

Beyond institutional settings, more fluid forms of governance are developing in the region, whether on the scale of a living area, a planted zone, or even, for example,[10] between urban areas and neighbouring rural areas. The energy transition thus induces on the one hand (slow) progress in terms of decentralization and on the other, informal cooperation at different levels, thus echoing the distinction established by Hooghe and Marks between the Type I and II of multi-level governance (Hooghe and Marks (2003), quoted in Stead (2014)).[11] However, responsibilities are fragmented between numerous local, regional, and national stakeholders, partly because the national authorities have so far been reluctant to devolve more policy competences to the regional level.

28.4.2 The Challenges of Multi-Level Governance: Is the National Government the Leading Actor?

The forest products sector of the economy in France is in second place in terms of external deficit, despite the fact that France's forested territory is the third largest in Europe (Caullet 2013). This illustrates the difficulties faced by public actors as they attempt to structure a diversified, fragmented forest products market. From preparation to post-project phases, the sector suffers from a lack of structure, and supply is limited by problems related to mobilizing resources. The divergence of interests between the various wood products actors, in particular between silviculture professionals and wood industry professionals, impede consensus about the most effective approaches to forest policy. These problems with respect to French forestry also prevail in Aquitaine, where 92 per cent of woodlands are privately owned (AGRESTE Aquitaine 2012).

In addition, the use of resources is constrained by tax policies that encourage long-term woodland ownership through transmission from one generation to the next but that do not encourage active forest management. The basis of this policy goes back to the post-war period, when the government sought to avoid over-exploitation of forest resources at a time when reconstruction generated strong demand for forest products (Caullet 2013). This tax policy currently contributes to weak market fluidity and uneven supply, to the subdivision

[10] As an example, one can mention the 'forest exploitation charts' or the 'rural poles of excellence'.

[11] Hooghe and Marks distinguish between two basic types of multi-level governance, labelled Type I and Type II. Type I governance is designed around territorial communities, while Type II is designed around specific tasks or policy issues.

of forestland through inheritance—France currently has 3.5 million forest owners—and to passive forest management. In short, one of the principal constraints on the development of the wood-based energy market is not the availability of raw materials as it is the tax framework, which, in its current form, does not encourage owners to manage their property actively. However, these two features of the wood market are essentially the national government's responsibility, hence the crucial role of national policies for the success of the regional-level energy transition.

More surprisingly, the national framework is also a key factor in the case of Germany. The legal framework favours wind farms in terms of territorial planning, and when added to the generous feed-in tariffs planned by the EEG,[12] partly explains the flow of investors to Brandenburg. The regulations defined by the *Bund* in terms of purchasing feed-in tariffs and planning rights have contributed significantly to overall increases in wind energy at the regional level. Although there are few cooperatives in Brandenburg, a number of initiatives show that the energy transition has led some local authorities to implement innovative policies. The energy autonomy plan developed in the village of Feldheim, the citizen fund in Brandenburg an der Havel, and the wind farms in Schlalach-Mühlenfließ and Frehne that entail neighbour partners, offer several examples of this promising trend (Becker, Gailing, and Naumann 2012). These experiences have confirmed that citizen ownership of wind energy production capacities dramatically improves community acceptance (Yildiz et al. 2015). The proliferation of local initiatives owes a great deal to a positive financial and legal framework, however. In this respect, the introduction of the new Environmental and Energy State Aid Guidelines (EEAG) in 2014 (European Commission 2014) led to the introduction of new provisions in the national regulative framework (German Ministry of Economy and Energy 2014) that might increase the sensitivity of renewable energy projects to market conditions.

28.5 CONCLUSION

By focusing on two regions, this chapter has attempted to call attention to factors that both inhibit and contribute to the implementation of the EU clean energy transition policy at the subnational level in a federal state (Germany), and in a unitary one (France). The study invites the generalization of some common assertions and the interrogation of others, despite the narrowness of the two cases studied here. It confirms that the infra-government actors in

[12] *Erneuerbare Energien Gesetz* (energy transition law).

Europe are the driving force behind energy transition, it highlights how some have made European and national priorities their own, and that rural areas play a key role in energy transition. Factors proposed by other studies to explain support for or hostility to wind farm projects from local actors can be observed in Brandenburg. The competition with the coal sector, the official rationale of the energy transition—doubts regarding the latter are nurtured by the ambivalent attitude of the Brandenburg *Land* concerning the future of coal mining—and certain attitudes inherited from the communist era and reunification are additional factors that must be considered. The case of Brandenburg also confirms the decisive importance of the spatial planning process.

These case studies nevertheless call for nuanced interpretation concerning certain arguments that have been advanced at international and national levels. For example, the largest cultivated forest in France is in Aquitaine but, despite this abundance, the region continues to suffer from powerful tensions implied by an increased competition over forest resources. The acceptance of renewable energies and strong citizen involvement in the production of renewable energies are two major characteristics of the *Energiewende* at the German level. However, none of them is reflected in the case of Brandenburg, one of the most successful regions in Germany in terms of wind policy. In this specific case, favourable natural and economic conditions added to the very attractive feed-in tariffs agreed at the national level are far more important than acceptance and citizen involvement. Brandenburg can rely on a low population density and on numerous flat and windy areas favourable to wind machines. The adverse economic conditions in rural communities (in comparison with other German *Länder*) have also made it very attractive for investors and developers. The legal framework, namely the national regulations on spatial planning have proved to be crucial too. The vocal support provided by regional authorities and the reluctance to regulate some key issues (such as the distance between wind farms and built neighbourhoods that has been addressed by the State of Bavaria) have been instrumental. However, the case study of Brandenburg demonstrates that low acceptance and low citizen involvement (at least in comparison with other German *Länder* that have seen far more grass- roots initiatives) do not necessarily hamper clean energy transitions.

'Do institutions matter for regional development?' asks Rodriguez-Pose (2013). Institutional arrangements above all are needed across administrative boundaries at the vertical as well as at the horizontal level. We would argue that it is not the institutional system itself that matters, whether unitary or federal, as much as the ability of the various stakeholders to share common visions and to act collectively. Further, the multi-level approach is justified by the fact that the number of actors involved in energy has grown with the implementation of the 3 x 20 package (see Section 28.1). Although decentralized energy production imposes new responsibilities on local actors, the national government is in no way marginalized and remains crucial, particularly in terms of tax policy (for forest lands), spatial planning, and defining of the feed-in tariffs.

These two cases also suggest the need to analyse the implications of growing territorialization of the energy transition for regional actors. Leading this transition calls for reinforced administrative capacities that can develop forest resource by possessing both knowledge of local factors and of the global market (in the case of Aquitaine), or even the ability to build a balanced dialogue with private actors in the wind energy sector (in the case of Brandenburg) in a context of increasingly complex energy laws. More broadly, both of the cases studied here confirm that regional strategies for facing climate change should be coordinated within a vision of territorial development to minimize conflicts over land and resource use (in the case of Aquitaine) or to prepare the conversion of regions negatively affected by the transition (in the case of Brandenburg).

REFERENCES

ADEME (Agence de maîtrise de l'énergie), IGN (Institut géographique national), and FCBA (Forêt Cellulose Bois-construction Ameublement) (2014). 'Etude de disponibilité en bois d'origine forestière en France métropolitaine aux horizons 2020 et 2030'. ADEME, IGN, FCBA, Paris.

AGRESTE (Service de la statistique, de l'évaluation et de la prospective agricole) Aquitaine (2012). 'Enquête CEFIL 2012 sur les propriétaires forestiers privés du massif des Landes de Gascogne'. N 52. AGRESTE Aquitaine, Bordeaux.

Bajczuk, R. (2015). 'The Uncertain Future of the Coal Energy Industry in Germany'. OSW Commentary, Centre for Eastern Studies, 188. Available at: <https://www.osw.waw.pl/sites/default/files/commentary_188.pdf> (accessed 12 January 2016).

Becker, S., L. Gailing, and M. Naumann (2012). 'Neue Energie-Landschaften—Neue Akteurs-Landschaften. Eine Bestandsaufnahme im Land Brandenburg'. Rosa-Luxemburg-Stiftung, Potsdam.

Bell, D., T. Gray, and C. Haggett (2005). 'The "Social Gap" in Wind Farm Siting Decisions: Explanations and Policy Responses'. *Environmental Politics*, 14(4): 460–77.

Brunt, A. and D. Spooner (1998). 'The Development of Wind Power in Denmark and the UK'. *Energy & Environment*, 9(3): 279–96.

BWE (Bundesverband WindEnergie) (2014). 'Die schleppende Regionalplanung in Brandenburg und ihre Konsequenzen für die Energiepolitik, die Branche der Windenergie und die regionale Wertschöpfung', Position Paper, BWE Landesverband Berlin-Brandenburg, Berlin.

Carlman, I. (1988). 'Wind Power in Denmark! Wind Power in Sweden?' *Journal of Wind Engineering and Industrial Aerodynamics*, 27(1–3): 337–45.

Caullet, J.-Y. (2013). 'Bois & Forêts de France: nouveaux défis'. Rapport de Jean-Yves Caullet, Député de l'Yonne, Maire d'Avallon, au Premier Ministre, Monsieur

Jean-Marc Ayrault, Paris, June. Available at: <http://www.developpement-durable.gouv.fr/Le-rapport-de-Jean-Yves-Caullet.html> (accessed 12 March 2015).

CEMAGREF (Centre d'Étude du Machinisme Agricole et du Génie Rural des Eaux et Forêts). IRSTEA since 2011 (Institut national de Recherche en Sciences et Technologies pour l'Environnement et l'Agriculture) (2009). 'Biomasse forestière disponible pour de nouveaux débouchés énergétiques et industriels'. Rapport Final, CEMAGREF, Paris.

Comodi, G., L. Cioccolanti, F. Polonara, and C. Brandoni (2012). 'Local Authorities in the Context of Energy and Climate Policy'. *Energy Policy*, 51(December): 737–48.

Conseil Régional d'Aquitaine, INRA (Institut National de la Recherche Agronomique) (2012). 'Le massif des Landes de Gascogne à l'horizon 2050'. Rapport de l'étude prospective, Conseil Régional d'Aquitaine, INRA, Bordeaux.

Dumartin, S. (2009). 'Les industries du travail du bois, une'. Aquitaine e-publications, January, no.1, INSEE, Bordeaux. Available at: <http://www.insee.fr/fr/themes/document.asp?reg_id=4&ref_id=14253> (accessed 10 July 2014).

École Nationale des Ponts et Chaussées (2011). 'Le développement du bois-énergie: solution durable pour répondre à un défi stratégique?' Groupe d'Analyse des Politiques Publiques. ENPC, Paris.

European Union (2009). 'Directive 2009/28/EC of the European Parliament and of the Council of 23 April 2009 on the Promotion of the Use of Energy from Renewable Sources and Amending and Subsequently Repealing Directives 2001/77/EC and 2003/30/EC'. Available at: <http://eur-lex.europa.eu/legal-content/EN/TXT/PDF/?uri=CELEX:32009L0028&from=EN> (accessed 15 October 2015).

European Commission (2014). 'State Aid for Environmental Protection and Energy'. Available at: <http://ec.europa.eu/competition/sectors/energy/legislation_en.html> (accessed 6 November 2016).

Eurostat (n.d.). Eurostat Your Key to European Statistics (online). Available at: <http://ec.europa.eu/eurostat> (accessed 10 November 2015).

Frese, A. (2015). 'Gabriel will aufräumen', 24 June. Available at: <http://www.tagesspiegel.de/wirtschaft/energiewende-gabriel-will-aufraeumen/11964970.html> (accessed 25 May 2015).

German Ministry of Economy and Energy (2014). 'Die wichtigsten EEG-Neuerungen auf einen Blick'. Available at: <https://www.bmwi-energiewende.de/EWD/Redaktion/Newsletter/2014/20/Meldung/die-wichtigsten-neuerungen-auf-einen-blick.html> (accessed 25 April 2015).

German Ministry of Economy and Energy (2015). 'Der nationale Klimaschutzbeitrag der deutschen Stromerzeugung, Ergebnisse der Task Force "CO_2-Minderung" Berechnungen'. Öko-Institut e.V. & Prognos AG, Berlin.

Glasl, F. (1999). *Confronting Conflict: A First-Aid Kit for Handling Conflict*. Stroud: Hawthorn Press.

Gross, C. (2007). 'Community Perspectives of Wind Energy in Australia: The Application of a Justice and Community Fairness Framework to Increase Social Acceptance'. *Energy Policy*, 35(5): 2727–36.

Hooghe, L. and G. Marks (2003). 'Unraveling the Central State, But How? Types of Multi-Level Governance'. *American Political Science Review*, 97(2): 233–43.

Hübner, G. and J. Pohl (2014). 'Mehr Abstand – mehr Akzeptanz? Ein umweltpsychologischer Studienvergleich', Analyse, Berlin: Fachagentur Windenergie.

IFN (Inventaire Forestier National) (2010). 'La forêt française—les résultats issus des campagnes d'inventaire 2005 à 2009'. IFN, Paris.

Ifo Institut (2015). 'Klimaabgabe für Kohlekraftwerke: Ein richtiger Schritt zur Erreichung des Klimaziels?' *Ifo Schnelldienst*, 14(68): 8–11.

Jobert, A., P. Laborgneb, and S. Mimlerb (2007). 'Local Acceptance of Wind Energy: Factors of Success Identified in French and German Case Studies'. *Energy Policy*, 35(5): 2751–60.

Kern, K. and H. Bulkeley (2009). 'Cities, Europeanization and Multi-Level Governance: Governing Climate Change through Transnational Municipal Networks'. *Journal of Common Market Studies*, 47(2): 309–32.

Kern, K., S. Niederhafner, S. Rechlin, and J. Wagner (2005). 'Kommunaler Klimaschutz in Deutschland—Handlungsoptionen, Entwicklung und Perspektiven'. WZB Discussion Paper, WZB, Berlin.

Le Treut, H. (2013). *Les impacts du changement climatique en Aquitaine, un état des lieux scientifique*. Pessac: Presses Universitaires de Bordeaux.

Marks, G. (1993). 'Structural Policy and Multi-Level Governance in the EC'. In A. Cafruny and G. Rosenthal (eds), *The State of the European Community. Vol. 2, The Maastricht Debates and Beyond*. Boulder, CO: Lynne Rienner.

Ministry of Economy and Energy of Brandenburg (2012). 'Energiestrategie 2030 des Landes Brandenburg'. Ministry of Economy and Energy of Brandenburg, Potsdam.

Ministry of Economy and Energy of Brandenburg (2014). 'Ergebnisse der Umfrage zur Beteiligung der Gemeinden an der Wertschöpfung aus Windkraftanlagen'. Mitteilungen 10/2014: 434, 219. Available at: <http://www.stgb-brandenburg.de/service/mitteilungen/mitteilungen-07-201400/ergebnisse-der-umfrage-zur-beteiligung-der-gemeinden-an-der-wertschoepfung-aus-windkraftanlagen> (accessed 7 November 2016).

Ministry of Economy and Energy of Brandenburg (2015). 'Energieminister Gerber: "Das Eckpunktepapier muss vom Tisch"', 17 April. Available at: <http://www.mwe.brandenburg.de/sixcms/detail.php/bb1.c.395400.de> (accessed 20 April 2015).

Mola-Youdego, B. and D. Gritten (2010). 'Determining Conflict Hotspots According to Academic and Environmental Groups'. *Forest Policy and Economics*, 12(8): 575–80.

Morris, C. and M. Pehnt (2012). 'Energy Transition. The German Energiewende'. Heinrich Böll Stiftung, Berlin.

Musall, F. D. and O. Kuik (2011). 'Local Acceptance of Renewable Energy—A Case Study from Southeast Germany'. *Energy Policy*, 39: 3252–60.

RAP (The Regulatory Assistance Project) (2014). 'Netzentgelte in Deutschland: Herausforderungen und Handlungsoptionen'. Studie im Auftrag von Agora Energiewende. RAP, Berlin.

Renewable Energies Agency (2012). 'Planungsrecht & Erneuerbare Energien'. Renews Spezial Ausgabe, 62. Agentur für Erneuerbare Energien, Berlin.

Renewable Energies Agency (2015). 'Bundesländer mit neuer Energie; Jahresreport Föderal Erneuerbar 2014/15'. Agentur für erneuerbare Energien, Berlin.

Rodríguez-Pose, A. (2013). 'Do Institutions Matter for Regional Development?' *Regional Studies*, 47(7): 1034–47.

Rogers, E. M. (1995). *Diffusion of Innovations*. Fourth Edition. New York: The Free Press.

Schreurs, M. (2008). 'From the Bottom Up: Local and Subnational Climate Change Politics'. *The Journal of Environment Development*, 17(4): 343–55.

Simon, A. and R. Wüstenhagen (2006). 'Factors Influencing the Acceptance of Wind Energy in Switzerland'. Paper presented at the Conference on 'Social Acceptance of Renewable Energy Innovation', Tramelan (Switzerland), February.

Söderberg, C. and K. Eckerberg (2013). 'Rising Policy Conflicts in Europe over Bioenergy and Forestry'. *Forest Policy and Economics*, 33(August): 112–19.

Stead, D. (2014). 'The Rise of Territorial Governance in European Policy'. *European Planning Studies*, 22 (7): 1368–83.

Steglich, J. (2015). 'Windradfirmen klagen gegen Beelitz'. *Märkische Allgemeine*, 20 May. Available at: <http://www.maz-online.de/Lokales/Potsdam-Mittelmark/Windradfirmen-klagen-gegen-Beelitz> (accessed 12 June 2015).

Torsten, G. (2014). 'Kaum noch Jobs durch erneuerbare Energien'. *Märkische Allgemeine*, 26 November. Available at: <http://www.maz-online.de/Brandenburg/Solarkrise-hinterlaesst-Spuren-Kaum-noch-Jobs-durch-erneuerbare-Energien> (accessed 27 November 2014).

von Hirschhausen, C., P.-Y. Oei, C. Gerbaulet, C. Haftendorn, and C. Kemfert (2012). 'Energiestrategie Brandenburg 2030—Erneuerbare forcieren, Braunkohleausstieg fair gestalten'. DIW Wochenbericht 11/2012, 14 March, DIW, Berlin.

Wolsink, M. (2000). 'Wind Power and the Nimby-Myth. Institutional Capacity and the Limited Significance of Public Support'. *Renewable Energy*, 21(1): 49–64.

Wolsink, M. (2006). 'Invalid Theory Impedes Our Understanding: A Critique on the Persistence of the Language of Nimby'. *Transactions of the Institute of British Geographers*, 31(1): 85–91.

Wüstenhagen, R., M. Wolsink, and M. J. Bürer (2007). 'Social Acceptance of Renewable Energy Innovation: An Introduction to the Concept'. *Energy Policy*, 35(5): 2683–91.

Yildiz, Ö., J. Rommel, S. Debor, L. Holstenkamp, F. Mey, J. Müller, J. Radtke, and J. Rognli (2015). 'Renewable Energy Cooperatives as Gatekeepers or Facilitators? Recent Developments in Germany and a Multidisciplinary Research Agenda'. *Energy Research & Social Science*, 6(March): 59–73.

Zaspel-Heisters, B. (2014). 'Länderöffnungsklausel im BauGB—Länderspezifische Mindestabstände für Windenergieanlagen und ihre räumlichen Auswirkungen'. BBSR, Berlin.

Zoellner, J., P. Schweizer-Ries, and C. Wemheuer (2008). 'Public Acceptance of Renewable Energies: Results from Case Studies in Germany'. *Energy Policy*, 36(11): 4136–41.

Part IX

Moving Forward

29

Moving Forward

Douglas Arent, Channing Arndt, Finn Tarp, and Owen Zinaman

29.1 STAGES OF MITIGATION POLICY

In Chapter 1 of this volume, we asserted that the agreement reached at CoP21 in Paris in December 2015 represents a potentially decisive break from the past. In the case where countries vigorously pursue their pledges for emissions reductions codified within their Intended Nationally Determined Contributions (INDCs), then the world will have fully engaged in an era of global mitigation.

This is, in our view, a big step in the right direction. The best available evidence indicates that the emissions reductions implied by the sum of all country level INDCs will significantly reduce the probability of extreme warming outcomes (e.g., an increase in average global temperatures greater than seven degrees Celsius by the end of the twenty-first century) (Webster et al. 2012). This would be a real achievement that should be applauded. At the same time, it is also well understood that the emissions reductions implied by the INDCs pledged at CoP21 are insufficient to maintain the global average temperature rise within the two degrees Celsius target set by the international community. To stabilize the global climate within (or even near) the two-degree threshold, new rounds of mitigation efforts will be required.[1]

At this point, we believe it is possible to identify three stages of mitigation policy/activity. With the passage of CoP21, the globe is now leaving stage 1 and entering stage 2. In stage 1, mitigation activity, while present, was largely ineffective in altering the trajectory of global emissions. There are many reasons for this. An incomplete list, drawn from Chapter 1, would include: a paucity of low emissions technologies that could function at scale, the revealed

[1] These same observations obviously apply to the more ambitious 1.5 degree target put forth at CoP21.

difficulties of implementing 'top-down' approaches to global mitigation at the international level, weak or non-existent developing country participation in proposed mitigation regimes, and a scattered political commitment to mitigation policies with implications for private sector investment decisions.

It is noteworthy that nearly all the collective experience with mitigation policies throughout the world, including but not limited to political economy aspects, is drawn to date from stage 1. This includes the chapters in this book. As highlighted in Chapter 1, this experience is all we have; and we should study it. Importantly, with respect to this book, chapter authors considered their particular issues in light of changing fundamentals, such as the agreement reached at CoP21.

Nevertheless, with rapidly changing fundamentals, we are ushering in a new era, which we are labeling stage 2. Stage 2 is characterized by a relative abundance of attractive low emissions technologies, continued rapid technical advance (including rapid advances in systems knowhow), growing scale of low emissions systems, a functional global agreement to reduce emissions involving pledges from more than 165 countries, and substantial private investment in low emissions energy systems.

We also believe that Stage 2 will be characterized by a high degree of policy experimentation across countries. Countries will pursue their INDCs in their own ways. Doubtless, some approaches will be more effective than others. Especially if the emissions reductions offers reflected in country INDCs are vigorously pursued, stage 2 should over roughly the next five to ten years generate ample information on what works and under what conditions.

It is also clear that stage 2 is not enough for at least three reasons. First, and as already noted, the emissions reductions offered in the country INDCs are insufficient to prevent unacceptably high levels of warming with potentially very damaging environmental and socio-economic impacts. More ambitious emissions reductions will be required in future.

Second, while the architecture of the agreement adopted at CoP21 deserves to be lauded for resulting in a global agreement, it is likely that this architecture will have to evolve in order to achieve the very deep reductions in emissions (or even negative emissions) necessary for stabilization of the global climate. In particular, international trade is governed loosely or not at all under the country-level offer system adopted at CoP21. Under a favourable global policy regime, traditionally high-emitting industries will shift their energy mix to low emissions technologies. The geographical distribution of renewable or other low emitting resources (including nuclear power) may imply a reshaped geographical redistribution of these industries around the globe in the search for low cost sources of energy. Under an unfavourable global policy regime, high emitting industries may concentrate in regions with permissive emissions policies undercutting the global commitment to emissions reductions. Adequate monitoring of total emissions and their sources would greatly assist in avoiding the latter unfavourable outcome. Eventually, a penalty system for non-compliers may be required.

In addition, the policy diversity that appears to be a feature of stage 2 has the disadvantage of adding complexity to international comparisons. It may not be immediately obvious whose emissions reductions policies are more permissive and whose policies are stricter. This opacity is likely to restrict potentially fruitful opportunities for carbon trade or offsets between countries/regions. Restrictions on carbon trade or offsets may also limit climate finance. The financial flows that mirror carbon trade or offsets are widely viewed as a potential source of funds for climate finance that would assist lower income countries to adapt, mitigate, and develop (UNU-WIDER 2014).

Third, emissions reductions goals often remain perceived as aspirations rather than expectations. The debate over stranded fossil fuel assets is indicative in this regard. IPCC (2014) estimates a carbon budget that would limit global temperature rises to 2 degrees above pre-industrial levels, which amounts to between 1/5th and 1/3rd of the world's proven reserves of oil, gas and coal. As noted by Mark Carney, the Governor of the Bank of England, 'if that estimate is even approximately correct, it would render the vast majority of reserves "stranded"—oil, gas and coal that will be literally unburnable without expensive carbon capture technology, which itself alters fossil fuel economics' (Carney 2015).

Despite this observation, considerable debate exists over whether these assets (reserves) are in fact stranded. And, the debate turns largely on mitigation. For example, in an open letter, Royal Dutch Shell plc, a major oil and gas company, states that oil demand will continue to grow with two consequences. First, Shell's oil reserves are not, in fact, stranded (hence, they are valuable); and, second, as a consequence of the emissions from burning these reserves, the two degrees Celsius target is highly likely to be breached by the end of the century (Royal Dutch Shell 2014).

In sum, a further stage of emissions reductions—stage 3—is almost certain to be required. Ideally, stage 3 will chart a feasible path to a stabilized climate and put in place the necessary policy architecture for following that path. Fundamentally, stage 3 marks the passage across a global tipping point where effective climate change mitigation is no longer an aspirational goal but rather becomes an accepted fact on the ground with the associated broad implications for behaviour and decision-making, not least a massive reduction in the resources allocated to prospecting for new fossil fuel reserves.

A clear proximate operational challenge for achieving this tipping point involves effective implementation of country INDCs with attendant information needs. Key research challenges also look further ahead towards achieving this tipping point and entering a third stage of emissions reductions. To this end, Section 29.2 presents four research frontiers.

29.2 RESEARCH FRONTIERS

29.2.1 Frontier I: Enabling Environments for Clean Energy Innovation

Continued technical advance in clean energy generation remains highly desirable. These advances foundationally expand the solution space for confronting climate change. As innovations become ready to enter markets, supporting cost-effective development and deployment remains easily expressed in principle but complex to achieve in practice.

Intentionally or not, energy systems have traditionally been implemented by a handful of institutions—the few serving the many. This was an outgrowth of historical paradigms related to large infrastructure, public services and regulated utilities, and also reflected economies of scale in production of fossil energy resources. Looking forward, the landscape for production and consumption of clean energy may be vastly different. Economic, resilient, distributed resources, combined with smart, individually optimized energy service technology solutions offer an alternative. At the limit, relatively small groups or even families may largely produce and consume their own energy.

At the same time, given the variability in supply associated with most renewable energy generation technologies, there are portfolio effect gains from system inter-linkages. These linkages can occur across energy types and/or across space. The gains from distribution of renewable systems, particularly wind and solar, across large geographic areas augur for grid interconnections between large centres of supply (wind farms, solar farms, and hydropower) and demand (cities and energy-intensive industrial clusters), likely across countries in many settings. Reality will likely evolve to a heterogeneous combination of smart distributed and networked systems—a 'system of systems' both physical and institutional and increasingly interlinked and continuously evolving.

The policy, regulatory, and business model constructs that might allow this suggested future to come about are still nascent, and as with other research areas producing rapid technological advancement, we observe policy and regulation frequently 'playing catch-up' with technology. With the opportunity space for new energy systems growing, institutions in the energy system are increasingly compelled to foster an enabling environment for investment. The enabling environment for emerging technology solutions is highly likely to represent a marked departure from the historical paradigm across a variety of dimensions.

With that in mind, the research agenda should increasingly recognize that innovation in the clean energy space is both broad and interactive, comprising technology, business models, finance, institutional constructs, asset optimization strategies, software/data, and more. While foundational research in individual system components (such as generation, storage, and distribution) remains essential, it is quite possible that the most promising innovations lie in the institutional space. Research into appropriate public policies, regulations, institutions, and financing schemes that efficiently foment and deploy new technologies has high potential.

29.2.2 Frontier II: Big Think: Framing Resiliency Planning

While innovations have been expanding the clean energy solution space, this book has emphasized the need for deliberate action, particularly in the public sphere, in order to initiate, sustain, and eventually complete a clean energy transition. Furthermore, these actions must take place within a framework of ongoing growth and development while simultaneously confronting the warming that is already built into the climate system. Both developed and developing societies must address the question of how to achieve (inherently interlinked) mitigation, adaptation, and economic/social objectives (including environmental objectives not related to climate change). The IPCC has posed this challenge as a quest for 'climate-resilient pathways' to desired economic, environmental, and social outcomes.

Defining these pathways is an integral element of the deliberate action required to achieve an energy transition and thus remains an important research frontier. The goal of constructing these pathways is to establish and update a framework for decision-making that facilitates integrative planning and policy-making with a view to effective policy implementation. Much has been done to identify and investigate the properties of potentially feasible pathways forward. Examples include the pathways developed with respect to global emissions in the Fifth Assessment Report (IPCC 2014); the long-term adaptation strategy process undertaken in South Africa (Department of Environmental Affairs 2013); and the climate action plan for the city of Fort Collins, Colorado (City of Fort Collins 2015). Nevertheless, a great deal remains to be done. As the examples illustrate, pathways are valuable at local, country, regional, and global levels. The approaches to developing and analysing pathways range from broad qualitative goal setting to more specific and detailed quantitative analysis.

29.2.3 Frontier III: From Individual Actions to Global Impacts

In moving towards INDCs, CoP21 represents a shift from historically 'top-down' approaches to mitigating climate change, to an increased focus on participation from nations and regions to sub-national government entities (e.g., cities, counties, provinces), non-governmental assemblages (e.g., social groups, companies/firms) and continuing to the level of families and individuals. Undoubtedly, global climate change efforts will be aided significantly by a society of engaged individuals and groups, armed with credible information and making informed choices. Such a vision—at its core—hinges on strategies to cohesively link individual actions with global climate impacts. These linkages may be between individuals or mediated by the wide array of groupings just mentioned.

We are only now beginning to understand the interlinkages of data and information, informed choice, social networks, formal institutions, and the dynamics of the political economy of change, particularly those related to clean energy and sustainability. What is the appropriate role of institutions to help create informed climate citizens? And, what are the efficacies of various strategies to do so? In what ways should successful initiatives at a local level be extended to other localities or otherwise expanded?

Economists have long argued that a carbon tax represents one means to strengthen perceptions of the climate implications of an individual's choices. It is a relatively clear and comparable signal that could operate via existing pricing systems in nearly all countries in the world. The impacts of prices on behaviour should not be lightly dismissed. At the same time, a carbon tax, particularly a global one, is perhaps more likely to be a product of the tipping point that marks stage 3 mitigation than a catalyst to achieving that tipping point. The relatively recent ubiquity of communication technologies, social networks, and information that increasingly tie together individuals and their associated groupings, both large and small, provides a potentially effective means for converting 'bottom-up' initiative into coherent action.

Understanding the transition pathways, choices and risks along those pathways, and the dynamics of decision-making in this increasingly heterogeneous world of energy choices raises a complex, intriguing portfolio of research areas. These areas span from understanding and informing decisions and the solutions offered from the scale of each human being, to the complex dynamics of social groups, companies, and firms, to sociopolitical constructs such as cities and counties, states, countries, and regions.

29.2.4 Frontier IV: Informing Decision-Making

Informing decisions across the broad landscape of potential climate actors, from individuals to social constructs to governments to international negotiations, will also demand considerable enhancements. While political dynamics continue to strongly influence climate and energy decisions within larger political economy constructs, issues related to energy, electricity, and infrastructure (and their increasingly important roles throughout economies) necessitate credible and objective information to understand the options for and implications of decision-making.

With increasing quantity of data at nearly all scales, questions arise to the efficacy of providing the right data, in combination with the right analyses to inform the decision at hand. In our view, an 'uber model' of billions of agents and millions of social networks and decision bodies would be the wrong direction (and perhaps prove intractable in the timeframes required to mitigate climate change). Focused, analytically rigorous tools with best-in-class data, utilizing state-of-the-art decision science is the most fruitful path forward. Good decisions depend on a solid understanding of today in combination with planning tools that deeply explore various pathways and their implications. Not all planning tools are created equal, and ensuring that well-formulated questions are receiving analytically robust, scale-appropriate, and actionable answers will continue to be a signature challenge of this space.

We now, more than ever before, recognize that energy and the climate are intimately interrelated to our economies, health, education, food and water, and our interdependent relationship with the earth. Modelling tools must continue to develop to represent this complexity (via relevant model hybridizations to characterize sectoral interlinkages) in order to address critical research questions moving forward. Efforts to achieve stated climate goals and further raise climate ambitions will be strongly aided by foundational investments in data, tools, and research to inform our decision-making—advancing the state of knowledge in this space will thus be critical to our future.

29.3 FINAL WORDS

From a climate, clean energy, and political economy viewpoint, our collective knowledge of systems, drivers of change, technology innovation, and decision-making have grown exponentially over the past decade. This book captures some of the leading examples and points towards future efforts.

The political ambition for the world to move forward has been articulated in the Sustainable Development Goals. These include bold ambitions for climate mitigation, codified by leaders across the world. Realizing these goals and becoming true stewards of a sustainable, productive, and vibrant global economy will take continued vision and leadership as well as formidable political will to carry out necessary changes in practice. In pursuing these goals, the pathways taken will reflect varying socio-political, physical, and geographic contexts across societies. Multiple stakeholders within these societies will make diverse pathway choices that will aggregate to collective impacts. Ongoing high-quality research and knowledge-sharing, such that these diverse choices add up to desired outcomes, will remain a key input to addressing the critical, timely issues that we face today and in the future.

REFERENCES

Carney, M. (2015). 'Breaking the Tragedy of the Horizon—Climate Change and Financial Stability.' Speech Given at Lloyd's of London. 29 September. Available at: <http://www.bankofengland.co.uk/publications/Pages/speeches/2015/844.aspx> (accessed 20 October 2016).

City of Fort Collins (2015). 'Fort Collins 2105 Climate Action Plan Framework.' Fort Collins. Available at: <http://www.fcgov.com/environmentalservices/pdf/cap-framework-2015.pdf> (accessed 20 October 2016).

Department of Environmental Affairs (2013). 'Long-Term Adaptation Scenarios Flagship Research Programme (LTAS) for South Africa. Summary for Policy-Makers'. Pretoria, South Africa.

IPCC (Intergovernmental Panel on Climate Change) (2014). 'Climate Change 2014: Synthesis Report. Contribution of Working Groups I, II and III to the Fifth Assessment Report of the Intergovernmental Panel on Climate Change'. IPCC, Geneva.

Royal Dutch Shell (2014). Open Letter in Response to 'Carbon Bubble'. Available at: <http://s02.static-shell.com/content/dam/shell-new/local/corporate/corporate/downloads/pdf/investor/presentations/2014/sri-web-response-climate-change-may14.pdf> (accessed 20 October 2016).

UNU-WIDER (2014). 'Position Paper on Aid, Environment and Climate Change', UNU-WIDER, Helsinki. Available at: <http://recom.wider.unu.edu/article/position-papers-how-does-aid-work> (accessed 20 October 2016).

Webster, M., A. P. Sokolov, J. M. Reilly, C. E. Forest, S. Paltsev, C. A. Schlosser, C. Wang, D. Kicklighter, M. Sarofim, and J. Melillo (2012). 'Analysis of Climate Policy Targets under Uncertainty'. *Climatic Change*, 112(3–4): 569–83.

Index